Reliability in Power Electronics and Electrical Machines:

Industrial Applications and Performance Models

Shahriyar Kaboli
Sharif University of Technology, Iran

Hashem Oraee
Sharif University of Technology, Iran

A volume in the Advances in Computer and Electrical Engineering (ACEE) Book Series

ENGINEERING SCIENCE REFERENCE
An Imprint of IGI Global

Published in the United States of America by
Engineering Science Reference (an imprint of IGI Global)
701 E. Chocolate Avenue
Hershey PA, USA 17033
Tel: 717-533-8845
Fax: 717-533-8661
E-mail: cust@igi-global.com
Web site: http://www.igi-global.com

Library of Congress Cataloging-in-Publication Data

Kaboli, Shahriyar, 1975- author.
Reliability in power electronics and electrical machines : industrial applications and performance models / by Shahriyar Kaboli and Hashem Oraee.
pages cm
Includes bibliographical references and index.
ISBN 978-1-4666-9429-3 (hardcover) -- ISBN 978-1-4666-9430-9 (ebook) 1. Power electronics--Reliability. 2. Electric current converters--Design and construction. 3. Energy conversion. I. Oraee, Hashem, 1957- author. II. Title.
TK7881.15.K33 2016
621.31'7--dc23

2015028560

This book is published in the IGI Global book series Advances in Computer and Electrical Engineering (ACEE) (ISSN: 2327-039X; eISSN: 2327-0403)

British Cataloguing in Publication Data
A Cataloguing in Publication record for this book is available from the British Library.

All work contributed to this book is new, previously-unpublished material. The views expressed in this book are those of the authors, but not necessarily of the publisher.

For electronic access to this publication, please contact: eresources@igi-global.com.

Advances in Computer and Electrical Engineering (ACEE) Book Series

Srikanta Patnaik
SOA University, India

ISSN: 2327-039X
EISSN: 2327-0403

MISSION

The fields of computer engineering and electrical engineering encompass a broad range of interdisciplinary topics allowing for expansive research developments across multiple fields. Research in these areas continues to develop and become increasingly important as computer and electrical systems have become an integral part of everyday life.

The **Advances in Computer and Electrical Engineering (ACEE) Book Series** aims to publish research on diverse topics pertaining to computer engineering and electrical engineering. **ACEE** encourages scholarly discourse on the latest applications, tools, and methodologies being implemented in the field for the design and development of computer and electrical systems.

COVERAGE

- Sensor Technologies
- Qualitative Methods
- Circuit Analysis
- Analog Electronics
- VLSI Design
- Optical Electronics
- Power Electronics
- Computer Architecture
- Algorithms
- Chip Design

IGI Global is currently accepting manuscripts for publication within this series. To submit a proposal for a volume in this series, please contact our Acquisition Editors at Acquisitions@igi-global.com or visit: http://www.igi-global.com/publish/.

The Advances in Computer and Electrical Engineering (ACEE) Book Series (ISSN 2327-039X) is published by IGI Global, 701 E. Chocolate Avenue, Hershey, PA 17033-1240, USA, www.igi-global.com. This series is composed of titles available for purchase individually; each title is edited to be contextually exclusive from any other title within the series. For pricing and ordering information please visit http://www.igi-global.com/book-series/advances-computer-electrical-engineering/73675. Postmaster: Send all address changes to above address.

Titles in this Series

For a list of additional titles in this series, please visit: www.igi-global.com

Sustaining Power Resources through Energy Optimization and Engineering
Pandian Vasant (Universiti Teknologi PETRONAS, Malaysia) and Nikolai Voropai (Energy Systems Institute SB RAS, Russia)
Engineering Science Reference • copyright 2016 • 494pp • H/C (ISBN: 9781466697553) • US $215.00 (our price)

Environmental Impacts on Underground Power Distribution
Osama El-Sayed Gouda (Cairo University, Egypt)
Engineering Science Reference • copyright 2016 • 405pp • H/C (ISBN: 9781466665095) • US $225.00 (our price)

Handbook of Research on Advanced Computational Techniques for Simulation-Based Engineering
Pijush Samui (Centre for Disaster Mitigation and Management, VIT University, India)
Engineering Science Reference • copyright 2016 • 616pp • H/C (ISBN: 9781466694798) • US $310.00 (our price)

Reliability in Power Electronics and Electrical Machines Industrial Applications and Performance Models
Shahriyar Kaboli (Sharif University of Technology, Iran) and Hashem Oraee (Sharif University of Technology, Iran)
Engineering Science Reference • copyright 2016 • 549pp • H/C (ISBN: 9781466694293) • US $255.00 (our price)

Operation, Construction, and Functionality of Direct Current Machines
Muhammad Amin (COMSATS Institute of Information Technology, Wah Cantt, Pakistan) and Mubashir Husain Rehmani (COMSATS Institute of Information Technology, Wah Cantt, Pakistan)
Engineering Science Reference • copyright 2015 • 404pp • H/C (ISBN: 9781466684416) • US $200.00 (our price)

Performance Optimization Techniques in Analog, Mixed-Signal, and Radio-Frequency Circuit Design
Mourad Fakhfakh (University of Sfax, Tunisia) Esteban Tlelo-Cuautle (INAOE, Mexico) and Maria Helena Fino (New University of Lisbon, Portugal)
Engineering Science Reference • copyright 2015 • 464pp • H/C (ISBN: 9781466666276) • US $235.00 (our price)

Agile and Lean Service-Oriented Development Foundations, Theory, and Practice
Xiaofeng Wang (Free University of Bozen/Bolzano, Italy) Nour Ali (Lero- The Irish Software Engineering Research Centre, University of Limerick, Ireland) Isidro Ramos (Valencia University of Technology, Spain) and Richard Vidgen (Hull University Business School, UK)
Information Science Reference • copyright 2013 • 312pp • H/C (ISBN: 9781466625037) • US $195.00 (our price)

Electromagnetic Transients in Transformer and Rotating Machine Windings
Charles Q. Su (Charling Technology, Australia)
Engineering Science Reference • copyright 2013 • 586pp • H/C (ISBN: 9781466619210) • US $195.00 (our price)

IGI GLOBAL
DISSEMINATOR OF KNOWLEDGE
www.igi-global.com

701 E. Chocolate Ave., Hershey, PA 17033
Order online at www.igi-global.com or call 717-533-8845 x100
To place a standing order for titles released in this series, contact: cust@igi-global.com
Mon-Fri 8:00 am - 5:00 pm (est) or fax 24 hours a day 717-533-8661

Table of Contents

Preface

In modern industries, electrical energy conversion systems consist of two main parts: electrical machines and power electronic converters. Electrical machines act in the conversion of electrical energy to mechanical one as a generator and vice versa as a motor. Power electronic converters are used for electrical energy conditioning. It is notable that electrical motors consume about half of the total generated electrical energy in the world. Regarding to the fast and wide usage of electrical energy, it is obvious that these two parts deal with considerable amount of energy. Thus, the uninterrupted operation of these power converters is very important.

Basically, reliability concept is a scale for evaluating the proper operation of systems. Reliability calculation is a method that estimates the effective and useful operative life of the systems. Especially, this scale is very important for the systems which are not practically repairable. In addition, this estimation is an important guideline in design process to design a reliable system. The performance of many industrial processes mainly depends on the quality of electric power converters. Switching power electronic converters and electrical machines are increasingly used for electrical energy conditioning and electromechanical energy conversion, respectively. The existence of high value of energy losses leads to generating hot spots at high temperature in power electronic systems. Temperature rise is one of the most important factors which reduce the operative life. Hence the useful life of such systems with high value of energy loss is decreased. As the effective operative life can not be examined immediately, there are some theoretical and experimental methods for predicting the reliability. In addition, reliability calculations help the designers to estimate the useful life of their designed systems. They can correct their design methodology if the estimated life is smaller than acceptable value. Thus, design for reliability is an important strategy.

On the other hand, methods for improving the reliability such as derating concept can be used in operation process to extend the useful life by proper application of electric power converters. In addition, derating algorithm can be used to continue the operation of an electric power converter under negligible faults.

This book deals with reliability and effective operative life concepts in the field of power electronics and electrical machines. In view of the extensive use of the aforementioned systems in industries, reliable design and an estimation of their effective operative life is considered to be crucial.

The aim of this book is to present a view about reliability in the field of "Electrical Energy Conversion". Based on this view, some of well-known strategies in design of power electronic converters and electrical machines should be reviewed. For example, application of high frequency switch mode power supplies is a common method. But, it may be replaced with a simple linear power supply with poor regulation but with high reliability in a reliable system.

HISTORY OF DEVELOPMENT OF THE BOOK

This book was developed based on teaching the related courses about power electronics and electrical machines in School of Electrical Engineering, Sharif university of Technology. A long term study about these electric power converters shows that a proper power system design and operation procedure is a chain which is led to reliability considerations. The main text of this book is resulted from class notes of related courses. This text core saw enormous changes during developing process of the book during the past 5 years. We tried to present a well-illustrated book to show the practical real examples of each section of the book. These figures were collected during an about 20 years of our activities in this field. Marker arrows were drawn for many figures to emphasize on the related topic of the figure. Developing process of this book was programmed for one year. But it take 5 years of our academic time with two times extension of our contract with IGI Global. It is a disadvantage but we are satisfied because the book in the present form is much more interesting than its initial planned form. Chapters 9 and 11 were not in the first draft and were added during modifications. Contribution of chapter 3 about MIL-HDBK-217 was also added to the final form of the book.

OBJECTIVES OF THE BOOK

This book is presented with the following overall objectives:

- To show the importance of reliability considerations in electric power converters.
- To present the calculation methods of reliability in electric power converters.
- To propose the techniques for improving the reliability in electric power converters

In this publication, methods for reliability calculation in electrical machines and power electronic converters are presented. Furthermore, thermal modeling is explained to determine the hot spot temperature since this temperature is a key factor in estimating the reliability of power electronic converters and electrical machines. In addition, the difference between high reliability and high efficiency systems is described. It is shown that high efficiency is not equivalent to high reliability in complex systems consisting of both the power electronic converters and electrical machines such as adjustable speed drives. Finally, various methods are presented to improve the reliability of the above mentioned systems such as derating method and load sharing method.

In modern industries, there are some new generated problems that affect reliability. Wide usage of adjustable speed drives for speed control of general purpose electrical motors leads to higher loss in these motors because of voltage harmonics fed into the motor. These problems are also considered and discussed in the book.

We should note that this book is not an encyclopedia about reliability. There are many high quality technical references for each chapters of the book. However, none on them deals with complete chain of reliability in the field of electric power converters. We tried to give not only a general system view but also a detailed technical view about complexities in electric power converters.

STRUCTURE OF THE BOOK

All of the materials used in this book are original. All of diagrams were drawn by authors and all of photos were prepared individually. Some of chapters use parts of our previous publications and they supported via proper referring. We use assistance of some companies and organizations via using their publications for presenting in the book. Copyright permission was received from them for all of the items used in the book. Here, we appreciate them for their kindly helps. We also planned to present several examples from other companies but they did not answer to our copyright permission request.

This book consists of 12 chapters which are divided to 5 different parts as shown in the flowchart of the book in Figure 1. Both reliability calculations and reliable design are considered.

Section 1, "Data Preparation," is about fundamental concepts of reliability with the following details:

- Chapter 1, "Electric Power Converters in Industries," presents a brief introduction about importance of electrical energy conversion in the modern industries. The aim of this presentation is showing the dependence of various industrial functions to conversion of electric power. Basic relations of various electrical machines as well as power electronic converters are presented. In each section, some typical industrial examples are presented. This background will be used in the next chapters for reliability calculation and improvement. In fact, this chapter is an introduction about reasons of writing an individual book about reliability of electric power converters. Some examples of reliability importance in various industries are presented.
- Chapter 2, "Fault Mechanism," describes the reason of failure in electric power converters. All of the failure factors which are described in this chapter are catastrophic factors and leads to destructive damage in the systems. Other types of failure without destructive effect on converter like electromagnetic interference will be presented in the next chapters. All of descriptions are based on details of operation of the converters which were presented in the previous chapter. Over temperature, over voltage, Mechanical forces and environmental effects like humidity are the main factors of failure in systems. Origins of these factors are described in this chapter. Over temperature is a special factor among them. Because other failure factors finally act as over temperature in failure process of the converters. Since the over temperature is the main failure factor in electric power converters, loss model of components in electric power converters are presented in details. In addition, the practical technique for measuring the power loss is described. Sample industrial examples of damaged equipments due to these failure factors are shown to give a real sense to reader about failure results.

Section 2, "Reliability Calculation," is about reliability calculations with the following details:

- Chapter 3, "Reliability Prediction," uses probability calculation to predict the failure rate of the converter. The formulation of these calculations are based on the concepts of failure factors which were described in the previous chapter. Some detailed examples are presented to show the power of probability tool for analyzing the behavior of complex systems. This chapter covers the methods for reliability calculation from component to system level. Some standards of reliability are presented. One can use the information from a reliability prediction to guide design decisions throughout the development cycle. MIL-HDBK-217 is described in details as a well-known standard for reliability prediction in component level. Reliability modeling is introduced for calculat-

Figure 1. Flowchart diagram of the book

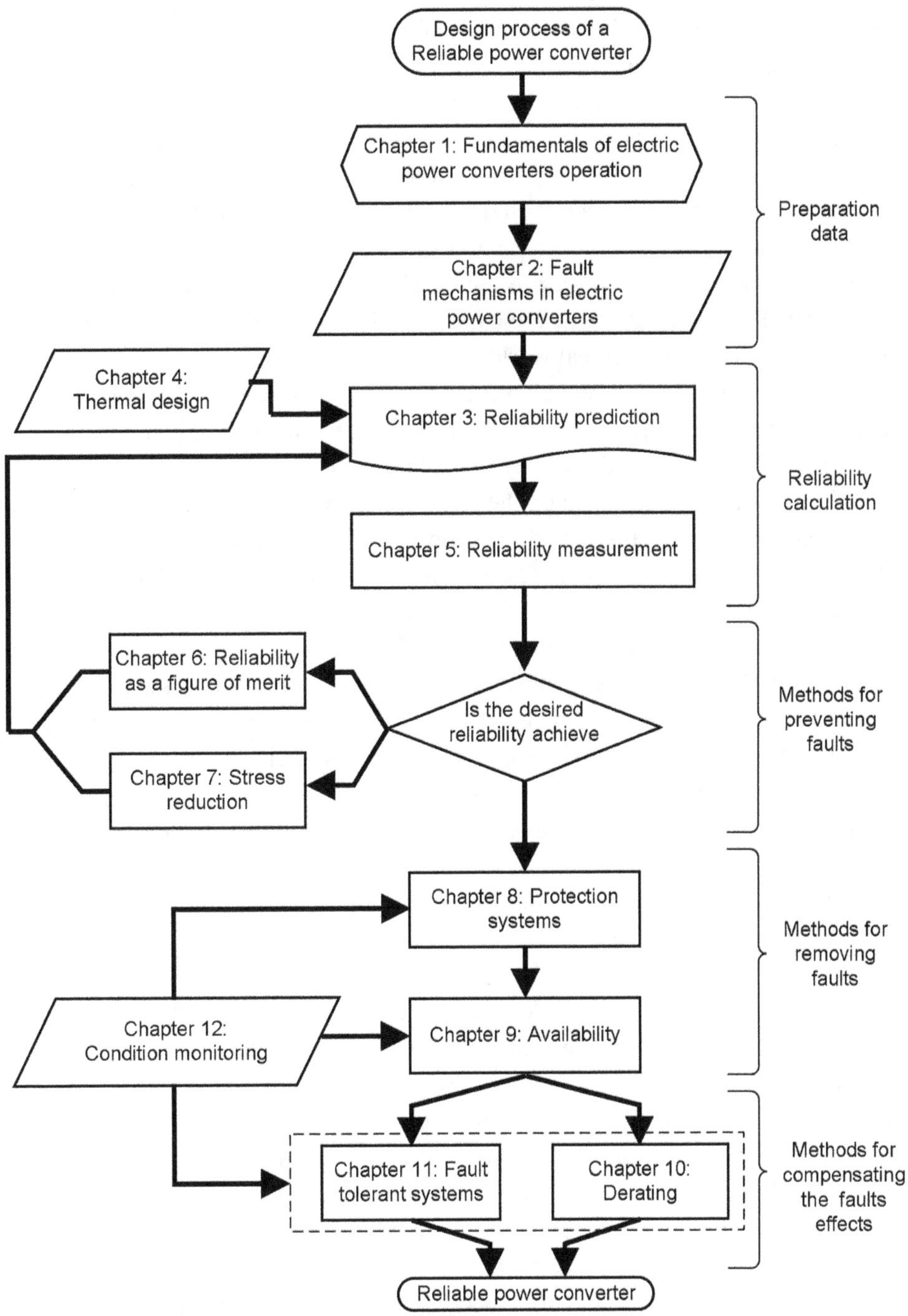

ing the reliability in system level. Difference between system block diagram and reliability model is presented. The reliability models of various static and rotary power converters are expressed. Some sample examples are presented to demonstrate the procedure of calculations for a simple converter with its auxiliary components. This chapter give a quantitative view to reader about evaluation of reliability and its can be used in the next chapters for reliability improvement.

- Chapter 4, “Thermal Analysis,” presents thermal analysis as the most important factors in failure of the converters. Two main approaches for this goal are presented: numerical and lumped mode. Principles of these methods are described with various examples and a comparison is presented. Basic principles of thermal modeling are described and concept of sample node in this model is explained. Methods for thermal management of an electric power converter are described. These methods are in both component and system levels and contain various heat transfer mechanism like conduction and convection. Theoretical methods and practical considerations for heat sink selection and proper mounting of it are presented. Thermal insulation classes and various standards related to thermal management topic are expressed. Industrial samples are presented to show application of theoretical topics in real world.
- Chapter 5, “Reliability Measurement,” presents various methods of tests for this goal. The main approach is accelerated aging test that reduce the time need for failure in a system. In this method, the device is tested under condition beyond its defined nominal specifications. Limits for this harsh condition is determined based on the calculations which are presented in the chapters 3 and 4. If a problem occurs in implementing and operating process of the converter, accelerated aging tests decrease the time to failure. Theoretical concept of accelerated aging tests is described. Standard tests of electric power converters are presented. Equipment and test chambers for standard tests are explained. These tests contain all of four various failure factors which are presented in chapter 2. Sample industrial examples are presented to demonstrate the procedure of the tests. Some of accelerated aging tests may lead to destroy the converter. Difference between destructive and nondestructive tests is presented. Sample devices after accelerated aging tests are shown. Measuring devices for system parameter identification are introduced. Various types of tests are expressed in details for some of the most important tests like electric withstand tests.

Section 3, “Methods for Preventing Faults,” is about reliability improvement in design stage with the following details:

- Chapter 6, “Reliability as a Figure of Merit,” presents reliability as a figure of merit in design of a system and compares it with other indexes. We want to highlight the effect of reliability consideration on the design methodology of a power converter. The most important specification of a power supply or power converter is its robustness. Because any failure in power supply leads to failure of the whole of the system. A power converter may have poor performance but operates very reliable and vice versa. In fact, this is a reliability based design approach to achieve a long useful life. It is shown that in many systems, high efficiency is not a good choice for selection of system operating point. A system can be inefficient but very reliable. Two complex examples are presented to show undesired results of neglecting reliability in design process. Methods for more reliable operation of electric power converters than high performance operation are proposed. A discussion about correct and intelligent optimization of a system parameters and operating set point is presented.
- Chapter 7, “Stress Reduction,” presents guidelines for improvement of reliability. These methods are used in both design and operation process of the converter. The focus of this chapter is on the component stress reduction in design process. Based on background of chapter two, reliability of

a converter increases if it operates in a set point with low stress. It is assumed that the converter is under design process or operates without fault. The methods for reliability improvement in faulty converters are discussed in the next chapters. In this chapter, methods for reducing electric field are described in both system and printed circuit board level. Low temperature operating conditions for an electric power converter are described and tools for this goal are presented. Series connection for voltage sharing and parallel connection for current sharing is explained. Novel control methods of power converters for reducing the complexity and reliable operation are presented. Control of inrush current as a typical transient problem in electric power converters is presented. Methods for preventing the over stress condition on the components in faulty cases are described. Techniques for reducing mechanical and environmental stress are expressed. Mechanical dampers for preventing the high amplitude vibration and insulating colors against humidity are presented. Industrial and real samples are presented to demonstrate application of the proposed methods.

Section 4, "Methods for Removing Faults," is about reliability improvement in operation stage when a fault without damaging effect occurs with the following details:

- Chapter 8, "Protection Systems," assumes that a fault occurs in the converter but there is a short time interval between fault occurrence and catastrophic damaging of the converter. Therefore, the topic of this chapter is the methods for saving the converter in this condition. In this chapter, protection methods for saving the system against damaging faults are presented. Based on background of chapter two, protection systems should be able to bypass the effect of failure factors on electric power converter. Methods for current limiting and voltage clamping as the usual factors of failure in converters are described. Circuit diagram of a snubber is presented and its operation is described based on safe operating area of solid state power switches. Operating diagrams of fuse as emergency circuit breaker are presented. Measurement methods and devices used in protection systems are explained. Experimental samples and standard diagrams are presented to clear the theoretical notes in all cases.
- Chapter 9, "Availability": Protection methods, which are described in the previous chapter, save the converter against non-catastrophic faults. However, this method saves the converter but causes to idle the converter out of the service. Subject of this chapter is about these converters that are not damaged but cannot operate normally. In this chapter, availability of electric power converters as a most important but usually forgotten parameter is described. The concept of availability was originally developed for repairable systems that are required to operate continuously. It is explained that a system may be unavailable while none of its parts damaged. In fact, there is an important difference between reliability and availability. A converter may be very high reliable but very unavailable and vice versa. One of the most important factors for this undesired state is influence of noise. In this chapter, electromagnetic interference and certain methods for reducing its undesired effects on electric power converters are presented. Electric power converters are usually the source of electromagnetic noise due to high operating voltage and/or current. Various techniques for safe operation of sensitive systems that operate close to these converters are described. In the last part of this chapter, alarm management is presented based on availability concept. This method is used to prevent fast shutdown of important systems due to dispensable faults.

Section 5, "Reliability in Operation Process," is about reliability improvement in operation stage with the following details:

- Chapter 10, "Derating," investigates uninterrupted operation of a faulty power conversion system with catastrophic damages in some of its parts. It is shown that a faulty electric power converter can continue to work with degraded specifications. This algorithm named derating for accessibility. This technique can be used for both a faulty system because of its uninterrupted operation and a normal system because of extensive life time. Algorithms for derating of a faulty electric machine and a power supply are described. Derating for increasing the useful life of a motor drive system is presented. A novel method for switching frequency selection in a switching power supply is proposed based on derating concept. Derating is introduced as a technique to compensate additional losses in an electric power converter operating in a harsh environment (for example: a motor drive which is supplied with a non-sinusoidal voltage waveform). Real industrial examples are presented in details for better understanding the derating concept. Some of the presented examples contain novel idea for derating and others are well known in industries.
- Chapter 11, "Fault Tolerant Systems": Fault tolerance is the property that enables a converter to work properly with failure in some of its components. Fault tolerant systems are systems that can be operating after fault occurred with no degraded performance in their basic functional requirements. This is the main difference between fault tolerant systems and derated systems. In this chapter, some of methods for fault tolerance in electric power converters are presented. Fault tolerance is almost the only method for achieving a desired reliability in a converter that operates with non-zero fault probability. There are two main approaches for this aim: re-configuration of the faulty system and using redundant systems. Redundancy is the provision of functional capabilities that would be unnecessary in a fault-free environment. Various types of redundant systems as passive and active redundancy are described and their application in power converter systems is presented. A new approach for a reliable and fault tolerant power supply is proposed and justify with experimental results. Concept of fault tolerance in electrical machines is presented.
- Chapter 12, "Conditions Monitoring": Implementation of all of the previously methods for reliability improvement needs to have an enough information about condition of the converter. This is the topic of the last chapter of this book. Condition monitoring is the process of monitoring a parameter of condition in machinery (vibration, temperature etc.), in order to identify a significant change which is indicative of a developing fault. The use of conditional monitoring allows maintenance to be scheduled, or other actions to be taken to prevent failure and avoid its consequences. In this chapter, commonly used methods for condition monitoring of the converters and electric machines are presented. The aim of this job is producing an alarm in converter before failure factor damage the system. Sensor based and sensor less methods for converter and motor parameter monitoring are described. The data obtained from sensor based methods is real but sensor is a weakness point in a converter. On the other hand, sensorless methods give estimated information but they are very reliable. Temperature as the most important parameter from reliability point of view is a common parameter for monitoring in all of systems. Other parameters like vibration, harmonics and others can be used for monitoring of various faults inside the system. Many typical cases are presented to well demonstrate the techniques.

FEATURES OF THE BOOK

Most of the recent texts on reliability are limited to a particular topics and they are very general without focus on power electronics and electrical machines. These documents do not provide comprehensive coverage of the field. Having a single comprehensive reference for the reliability in power electronics and electrical machines represents a significant advantage for the reader. Indeed, several topics in reliability are routinely encountered in a power electronics and electrical machines design and operation. This book includes the material that after several years of reliability problems has been found both theoretically sound and practically significant.

There are many published books about “Power Electronics” and “Electrical Machines”. However, up to now, there are a few books published specifically in the field of “reliability in power electronics and electrical machines” which may be due to the fact that modern power electronics is young and is only used widely in industries in recent years. However, the importance of publishing such books should not be neglected since there are a number of well-known books published in similar research areas such as “power system”. However, there is a great difference between “power system” and “power electronics”. There are many reliability control tools in power electronics and electrical machines such as “switching frequency” which is fixed in “power systems”. Thus, a dedicated reference is needed to use these tools and estimate the reliability value in power electronics and electrical machines. In addition, it should be mentioned that the other existing books in reliability field usually consider reliability concepts without a focus on lossy systems such as electric power converter or reliability at the device level that they are not applicable for reliability determination at the system level.

Other books in this field deal with only one of the topics in reliability. For example, there are many books in the field of condition monitoring. Springer has book series in the field of reliability. All of these book cover a portion of the chain of reliability in the field of power converters. This book covers the complete chain of failure to reliability in electrical machines and power electronics.

In addition to this main feature, we tried to give some other benefits to the book which are listed in the following.

- This book includes many real industrial examples. There are more than 600 figures and photos in the book. Real examples of faulty electric power converters are presented in details. Real examples of reliability calculations are expressed. This is a different presentation methodology in comparison to similar books. In each chapter, we present an explanation in the beginning of each section and expand our expression in an application example about the subject of the section.
- This book includes some new aspects in this field like chapter 6. In this chapter, traditional methodology of high efficiency in electric power converters is challenged. We show that a converter can be very reliable but with low efficiency.
- We prepared an illustrated presentation of MIL-HDBK-217 which is known as “Bible” of reliability.
- There are many useful references for each chapter to give a fresh state of the art view to reader about topic of the chapter. In addition, a comprehensive list of related documents is prepared for interested readers.
- We study many standards like IEC to give some guidelines to the readers about using standards in reliability study.

- The existing books are divided into two main categories. Some of these books have a general view to reliability like production. Others dedicate to an individual step and not overall view. Both system view and component view are covered in this book.
- In power electronic we have not a book covering reliable design methodology. This book is presentation of design and operation methodology to achieve an electric power converter with high reliability. There are certain steps toward this goal and there are many research books about these steps. However, each of these references focuses on one of these steps. We recognized that there is not proper reference that it covers all of required steps toward a reliable power electronic converter or electrical machine. These books do not give a big picture about the topic.

RELATED READERS

This book can be used by the following groups of readers:

- Electrical engineers: The publication can be used by electrical engineers in operation processes of power electronic converters and electrical machines in industries. For example, derating concept can be used by these engineers to prevent a fault in electrical systems in the near future.
- Designers of power electronic converters and electrical machines: The publication gives useful hints to consider reliability in design processes. Thus, designed systems will be reliable with a long effective operative life.

Since the effect of poor reliability is not seen immediately, reliability is often a forgotten index of quality in electric power converters. Therefore, the challenges in reliability are important especially in the field of electric power converters. Researchers are working with enthusiasm, tenacity, and dedication to develop new methods of analysis and provide new solutions to achieve a reliable converter. In this atmosphere, it is necessary to provide both professionals and students with state-of-the art knowledge on the frontiers in power converter reliability. This book is a good step in that direction

Shahriyar Kaboli
Sharif University of Technology, Iran

Hashem Oraee
Sharif University of Technology, Iran

Acknowledgment

The authors would like to thank all of the people helping the development of this book. We tried to make a complete list of them here. However, the authors present their special acknowledgment to IGI staff and our students, since this book was resulted from their interests and questions to this field. We was supported by many companies and organizations through using their publications under copyright permission. The author thanks the International Electrotechnical Commission (IEC) for permission to reproduce Information from its International Standards. All such extracts are copyright of IEC, Geneva, Switzerland. All rights reserved. Further information on the IEC is available from www.iec.ch. IEC has no responsibility for the placement and context in which the extracts and contents are reproduced by the author, nor is IEC in any way responsible for the other content or accuracy therein. The copyright permission of IEC was obtained by efforts of Dr. Mehdi Mortazavi, IEC National Committee of Iran. We also like to thank International Rectifier Corporation, EPCOS, ON Semiconductor, Moorecorp Ltd, Fluke Corporation, Power guru, Seraj Company, Energy conversion laboratory of Sharif University of Technology, Vibration laboratory of Sharif University of Technology, Grouc Company and Iran Ministry of Energy for providing us with permission of using their product datasheets. Some parts of the book were developed with the help of Mohammad Kazem Jannati, Amin Khakparvar, Ramin Parvari, Esmaeel Athari joo, Mostafa Zarghani, Morteza Aghaei, Mohsen Mortazavi, Mohammad Reza Reiahi, Saeed Haghbin, Dr. Mohammad Reza Zolghadri, Mohammad Reza Sadriyeh, Abbas Mohammadi, Hasan Azad, Hamidreza Teymouri, Dr. Hossein Mokhtari, Dr. Reza Kaboli, Dr. Shirin Kaboli, Mehrnaz Khiabani, Vahid Javadian, Dr. Ali Mehrizi, Dr. Shaahin Filizadeh, and Amir Hossein Azadi.

Section 1
Data Preparation

Chapter 1
Electric Power Converters

ABSTRACT

In this book, we discuss reliability in electrical energy converters. The first step is introducing these devices and recognizing their main functions as well as their importance. Electrical energy conversion systems consist of two main parts: Electrical machines and Power electronic converters. Electrical machines are used for converting electrical energy to mechanical one in the generator state and vice-versa in the motor state. To emphasize the importance of these devices, it may be noted that electrical motors consume about half of the total generated electrical energy in the world. On the other hand, power electronic converters are essential equipments which are used for electrical energy conditioning. These equipments have observed considerable growth in modern industries in recent years. Because energy conditioning allows us to use energy with higher efficiency and better performance, in this chapter, importance of electric power converters in modern industries is presented. The aim of this presentation is showing the dependence of various industrial functions to conversion of electric power. Basic relations of various electrical machines as well as power electronic converters are presented. In each section, some typical industrial examples are presented. This background will be used in the next chapters for reliability calculation and improvement. In fact, this chapter is an introduction on reasons of writing an individual book about reliability of electric power converters.

INTRODUCTION: IMPORTANCE OF RELIABLE POWER CONVERTERS

This book is about reliability in the field of power electronic converters and electrical machines which are named “electric power converters”. Why reliability? and why in electric power converters? The main goal of this chapter is clarifying the importance of the title of this book. So we start the first chapter of the book with some essential questions:

- What is the importance of reliability in the field of electric power converters?
- Why must they work reliable?
- Which parts of the world are affected if electric power converters are unreliable?

DOI: 10.4018/978-1-4666-9429-3.ch001

In this chapter, we answer to these questions by presenting enormous applications of electric power converters in modern industries. In this chapter, terms of "power converter" or "converter" mean "electric power converter" for summary. Power electronic converters and electrical machines are two main parts of electric power conversion field and consume considerable amount of energy. Regarding to the fast and wide usage of electrical energy, it is obvious that these two parts have a key role in normal operation of industries. Thus, the uninterrupted operation of these power converters is very important. Many problems in electric equipments are due to neglecting reliability considerations in design process of their power converter (Song, & Wang 2013). As a senior researcher, we saw many electric systems with unsafe operation because of lake of reliability. Unsafe operation means:

- Damaging in equipment without pre-alarm
- Consecutive shutdowns of equipment
- Interference with other devices

In this book, we talk about reliability in electrical energy converters. Therefore, the first step is introducing these devices and recognizing their main functions as well as their importance. Figure 1 shows the state of this chapter in the book.

Figure 1. State of chapter 1 in flowchart of the book

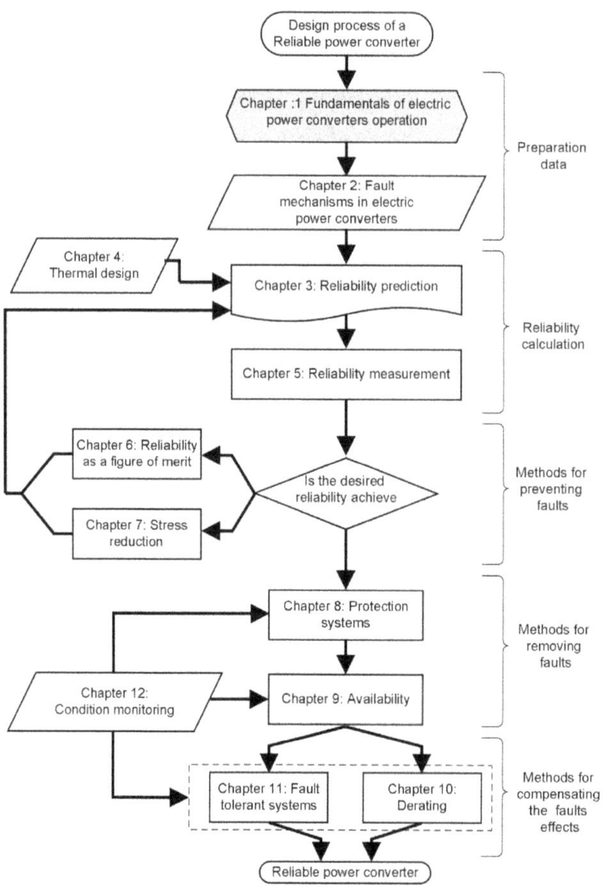

VARIOUS TYPES OF RELIABLE POWER CONVERTER

Risk analysis and reliability studies are important tools in designing electric power converters. There are various types for reliability importance in power converters (Yang, Bryant, Mawby, Xiang, Ran, & Tavner, 2011). There are:

a. Power converters with long time repair process
b. Needs to high level safety
c. Non-stop power converters
d. Mass production of a power converter
e. Power converters with impossible repair process
f. Enormous power converters with difficult access

Each of the above mentioned requirements is a view of reliability in a converter. Figure 2 shows some applications of electric power converters with different reliability considerations. Figure 2(a) shows a power line with transformers and other equipments in a rural zone. Repair and maintenance of this power line is very difficult. In addition, the time interval for repair is long and this leads to long time blackout. All of components of this line should be very reliable to prevent blackout in load side. Figure 2(b) shows another type of reliability importance in power converter. The power converters used in an airplane should be reliable because of safety considerations. Figure 2(c) shows a plant turbogenerator as a non-stop power converter. This generator should be very reliable to work without interrupt for a long time. Reliability is important from cost point of view in a power converter which is produced with mass production scale. Figure 2(d) shows a lamp with its ballast power circuit. It is produced in large scale. Any problem in this products leads to considerable cost. Therefore, it is very reliable. Figure 2(e) shows a satellite as a system with power converter without possibility of repair. Figure 2(f) shows a wind farm. Regarding to enormous number of these wind generators, they should be reliable with long life (Tohidi, Oraee, Zolghadri, Shiyi, & Tavner, 2013).

Power Converters as the Most Reliable Part in a System

Here, there is a key question: Why should the power converter sections be very reliable? and not other parts in the above mentioned applications? Why do we focus on power converter sections in the systems? The answer is that all parts of the equipments should be reliable. However, power converter is a special section. First, it provides the electrical energy for other parts to work. Any failure in power converter leads to interrupt in whole of the system (Motor Reliability Working Group, 1985). Another reason relates to the position of power converters in the equipments: All of equipment energy passes through the power converter. Therefore, the generated heat in power converters is generally high. As we will describe in the next chapter, heat is the most important factor in failure process.

Electric Power Converters in Industries

Electric power converters deal with electrical energy and electrical energy is a commonly used type of energy in industries (Boglietti, El-Refaie, Drubel, Omekanda, Bianchi, Agamloh, Popescu, Di Ger-

Figure 2. Electric power converter applications with high reliability requirement, (a): Electric power network in rural region as a difficult, (b): An airplane needs high reliability power converters from safety viewpoint, (c): Power plant as a nonstop system, (d): Ballast circuit of a lamp as a part with mandatory long life time in wide scale usage, (e): OMID satellite as a system with impossible maintenance, (f): Wind farm as a system with enormous and distributed parts

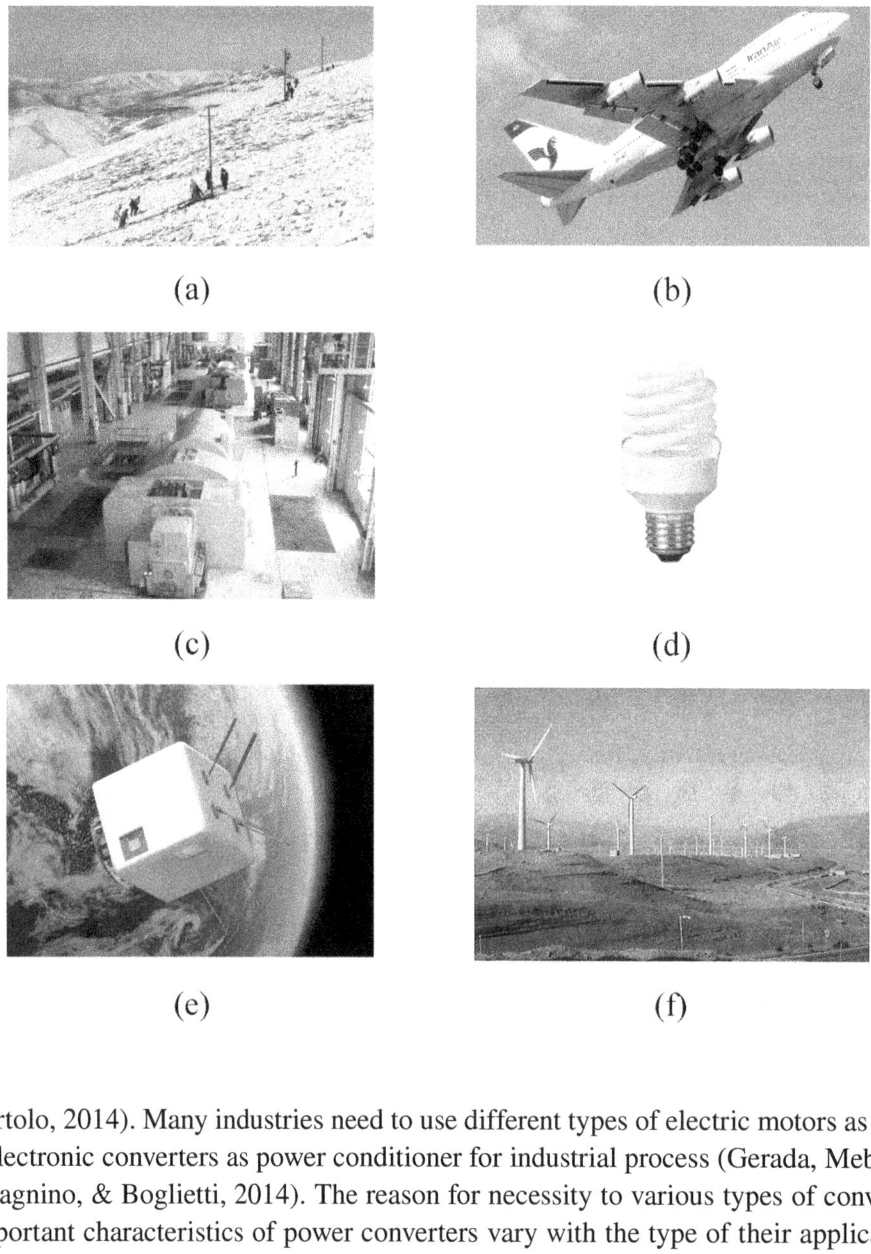

lando, & Bartolo, 2014). Many industries need to use different types of electric motors as prime mover and power electronic converters as power conditioner for industrial process (Gerada, Mebarki, Brown, Gerada, Cavagnino, & Boglietti, 2014). The reason for necessity to various types of converters is that the most important characteristics of power converters vary with the type of their application and the type of task they are expected to perform. There are some important industries that the electric power converters have a key role in their operation. Some of them are listed in the following. These are the most important industries in any country and they consume a great amount of electricity.

a. **Chemical industries:** In a chemical factory, there are many applications for electric motors as a power converter. Electric motors are used in pumps for transferring the liquid and gas. Fans and blowers are other user of electric motors in a chemical process for ventilation applications or in chillers and coolers. Power electronic converters are also used in a chemical factory as a power conditioner for example for controlling the furnace temperature.
b. **Food process:** Electric motors are used for driving the conveyers in a food factory. Other applications are in fans and compressors. In addition to similar applications of power electronic converters for power conditioning in a chemical factory and in a food factory, some types of power electronic converters have special applications in food industries. These are the converters with special output voltage waveform which are used for sterilizing the foods. For example, a high power pulsed voltage is used in a commercial accelerator for sterilizing the vegetables and fruits without using any disinfectant.
c. **Pulp and Paper industries:** In a paper factory, there are many motor driven conveyers for transferring wood. Power electronic converters are also used for controlling the process temperature and speed control of electric motors.
d. **Metal forming industries:** High power presses and rolling systems are driven with electric motors. Precise speed control of these motors is a key factor in quality of the process and this task is performed with power electronic converters. Power electronic converters are also used for temperature control.
e. **Petroleum process:** Power Electronics has been present in recent years in a wide number of applications within the oil and gas industry. Power electronic converters like variable speed drives and soft starters are suitable for extraction plants and chemical industries based on petroleum derivatives. Nowadays, under seas variable speed drives have an important role in petroleum extraction which are built based on power electronic converters.
f. **Electronics:** Advanced electronic industries are established based on robotic process. A robot is driven and controlled with electric motors and power electronic converters, respectively.
g. **Transportation:** Nowadays, environmental pollution of the petroleum-based transportation vehicles has led to interest in electric transportation. An electric vehicle or electric train use electric motors as mover and power electronic converters as controlling device of these motors.

Statistics for Usage of Electric Power Converter

Here, we present brief statistics about application of electric power converters in the world. These statistics are mainly about electric motor usage in industries. Electric motors consume electrical energy and convert it into mechanical energy. Therefore, they are consumer of energy with specified statistic and their power consumption is saved by counters. In the opposite, power electronic converters do not consume electric energy. They operate as electric power conditioner. Electric power converters are used in two different levels in modern industries and homes: low power converters are widely used in general applications like homes and high power converters are used in smaller scale in particular applications like heavy industries.

Electric motors use about half of the manufacturing delivered electricity and 8% of the total fuel consumption. Figure 3 shows a statistic about machine drive electricity use in industries.

Figure 3. Percent of machine drive electricity use in industries

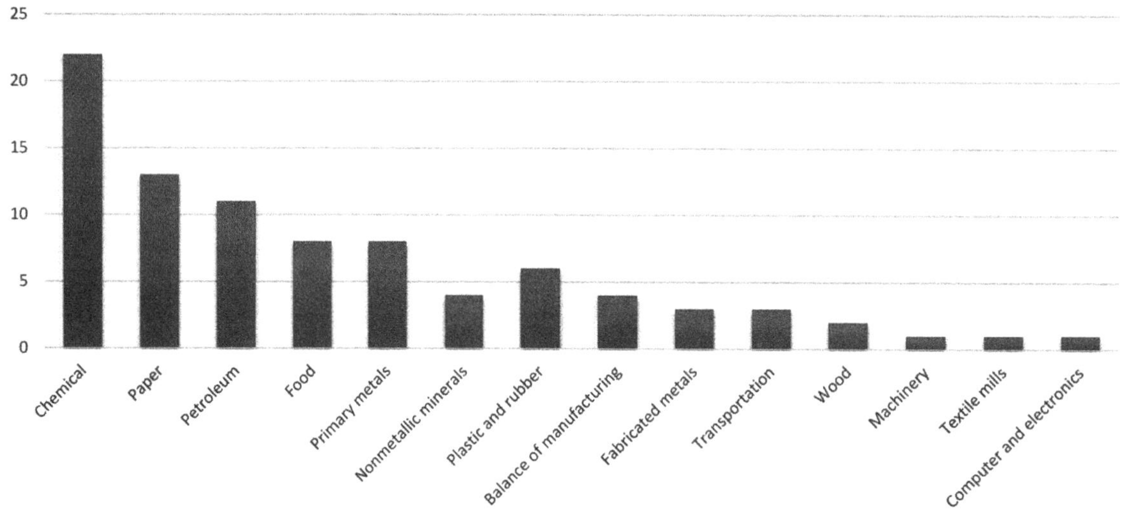

Power electronic converters are also used in modern industries for power conditioning in very wide scale. The power electronic converters are not the end of chain of electricity usage (in the opposite of motors). However, a look at their wide scale applications in industries, show their importance. In the following, we describe the operation principles of the most important types of electric converters.

Power Electronic Converters

Power electronic converters use solid-state devices for conditioning of electric power. The important role of these devices is that their power loss is about zero in switching operation state. Therefore, power electronic converters usually have high efficiency. A brief introduction on basic operation of these switches is described in the following. We use this description in the next chapter when we investigate the reasons of failure in power electronic converters.

Solid State Power Switches

Solid state devices are important tools in the design of power electronics converters. Power electronic devices may be used as switches, or as variable resistor. In switching operation state, an ideal switch is either open in off state with zero current or closed in on state with zero voltage. As the voltage or current of the switches is zero in these two states, they have no power dissipation. The real semiconductor switches approximately show this ideal property and so most power electronic applications rely on switching devices on and off, which makes systems very efficient. The losses that a power electronic device generates should be as low as possible because of importance of efficiency. However, in this book, we try to give a new view to the reader from reliability view point. Based on this new view, efficiency is not the most important concern of designer. A power converter can be inefficient but very reliable. We describe more about this meaning in the next chapters.

There are various types of solid state switches. Diode is a device which is turn on and turn off regarding to the polarity of its current and voltage. Power devices such as thyristors have the ability of control of the start of conduction. Therefore, they are named semi controlled devices. Devices such as BJT and MOSFET provide full switching control. They are named fully-controlled devices.

Another difference between solid state switches is their difference from switching time view point. Devices are different in switching speed. Some diodes and thyristors are operated for low switching frequency. Other devices such as MOSFETS and BJTs can operate at higher switching speed.

Diode

Diode is an uni-polar and uncontrolled switching device. It means that diode is turn on and turn off according to its voltage and current in the circuit and it is not under control of user. The switching role of an ideal diode is presented as the following.

$$\begin{cases} i_D = 0 & for \quad v_D \leq 0 \\ v_D = 0 & for \quad i_D \geq 0 \end{cases} \tag{1}$$

Where i and v are the diode current and voltage, respectively.

For a real diode, the current-voltage characteristic is presented as the following.

$$i_D \approx I_s\left(e^{v_D/nv_T} - 1\right) \text{and } V_T = \frac{k \cdot T}{q} \tag{2}$$

Where T is the temperature. This equation shows the dependence of switches characteristics to temperature. We use this dependence in the next chapters.

The main function of diode in the converters is rectification. Symbol of a diode and its typical structure as well as a real diode characteristic are shown in Figures 4 and 5, respectively.

Thyristor

Thysristor is a semi-controlled switching device. This switch is turn on when a gate current pulse is applied while the anode-cathode voltage is positive. In on state, thyristor operates as a unidirectional switch like a diode. The device can not be turn off via gate voltage and it is turn off when its current falls to zero. Other similar devices in the family of thyristor are GTO and Triac. Symbol of a thyristor and its typical structure as well as a real thyristor forward characteristic are shown in Figures 6 and 7, respectively.

Bipolar Junction Transistor

The Bipolar Junction Transistor (BJT) is a fully-controlled switching device. It can be turn on and turn off according to user command. The BJT cannot be used at high power because they are slower and have more resistive losses when compared to advanced type devices. Symbol of a BJT and its typical structure as well as a real BJT characteristic are shown in Figures 8 and 9, respectively. In a BJT, the collector current is determined based on Base-Emitter voltage as shown in the following.

Figure 4. Typical semiconductor layers in a (a) diode; and its (b) symbol

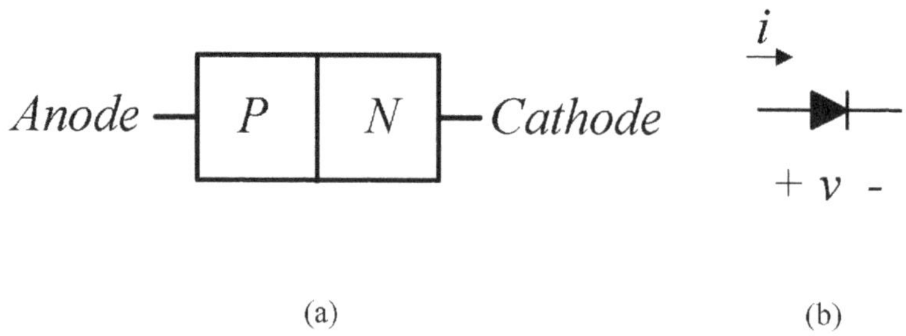

Figure 5. Real forward V-I characteristic of a power diode, Source: ON Semiconductor corp. (with permission)

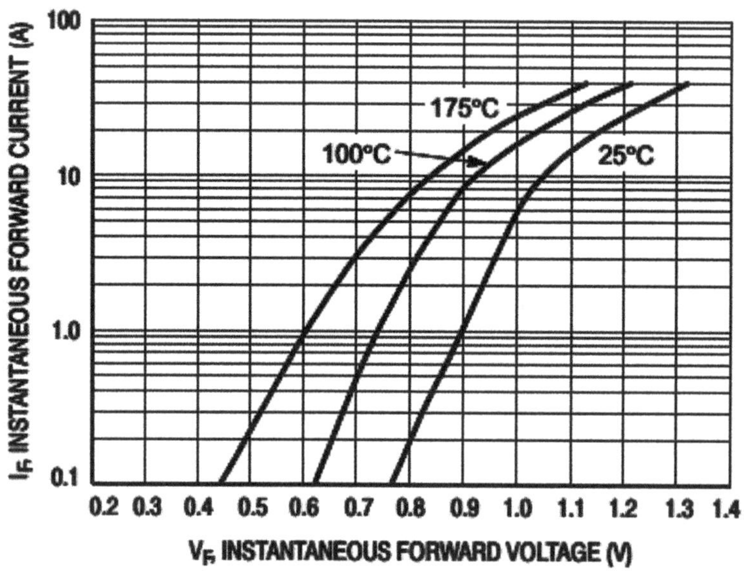

$$i_C \approx I_s e^{v_{BE}/n.V_T}\left(1+\frac{v_{CE}}{V_A}\right) \text{and } V_T = \frac{k \cdot T}{q} \tag{3}$$

Where T is the temperature. It can be seen that the characteristics of other switches also have a dependence to temperature.

MOS

The main advantages of the power MOSFET is higher switching speed of MOSFET in comparison to BJT as well as simpler gate drive. Gate voltage creates a conduction path from drain to source. Symbol of a MOSFET and its typical structure as well as a real MOSFET characteristic are shown in Figures 10 and 11, respectively.

Figure 6. Typical semiconductor layers in a thyristor (a) and its symbol (b)

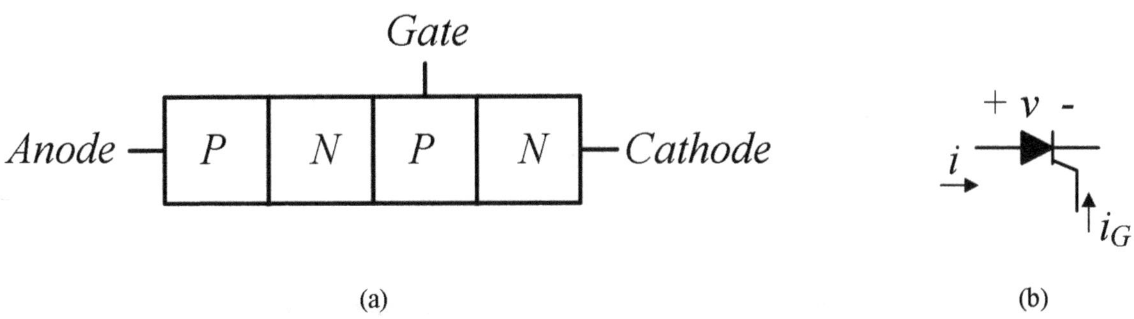

Figure 7. Real forward V-I characteristic of a thyristor, Source: ON Semiconductor corp. (with permission)

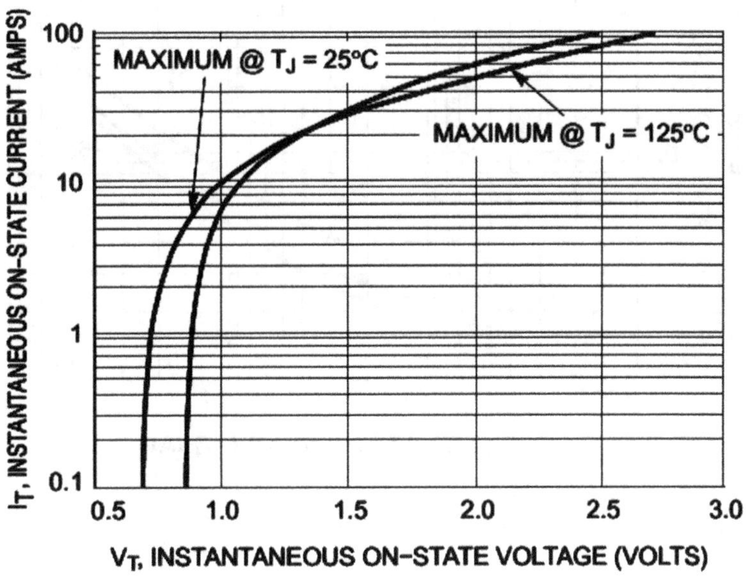

IGBT

This device has the best characteristics of MOSFETs and BJTs. IGBT has high gate impedance like MOSFET devices, thus simple gate driver requirements. On the other hand, this device has low on state voltage drop like BJT, thus low power loss across the switch in on state. Symbol of an IGBT and its typical structure as well as its real V-I characteristic are shown in Figures 12 and 13, respectively.

MAIN TYPES OF POWER ELECTRONIC CONVERTER

There are two general types of power converter: linear and switching. In switched mode power electronic converters, solid state devices operate as a switch. In linear type, solid state element operates as a variable resistance.

Figure 8. Typical semiconductor layers in a BJT (a) and its symbol (b)

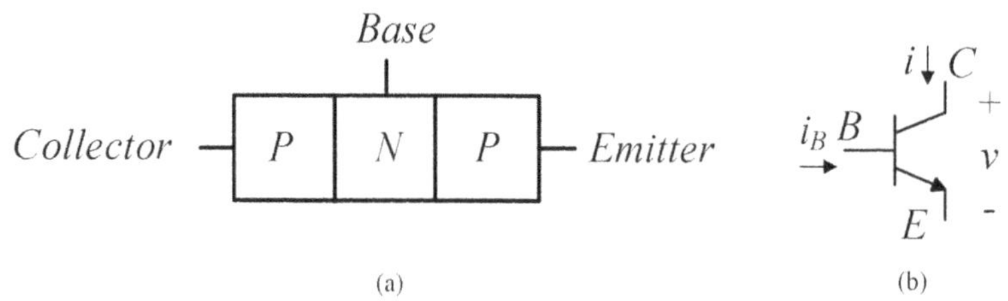

Figure 9. Real forward V-I characteristic of a BJT Source: ON Semiconductor corp. (with permission)

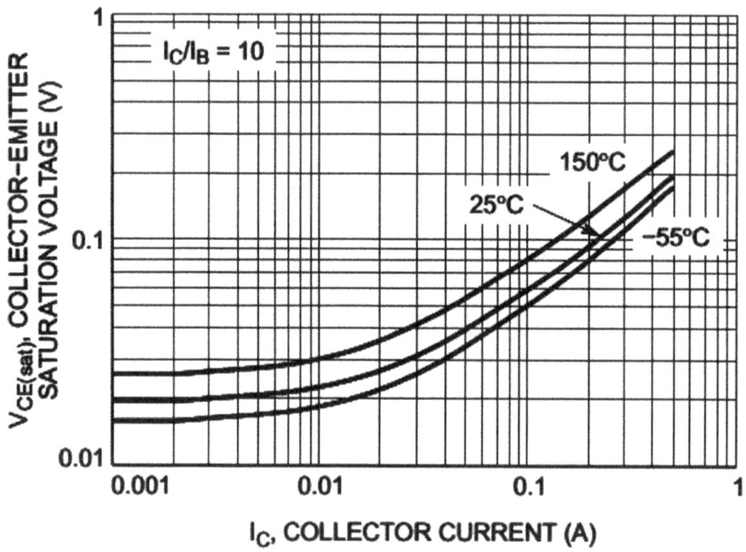

Linear Converters

The voltage provided by many energy sources will vary with changes in load impedance. When an unregulated power supply is the energy source, the amplitude of its output voltage will also vary with changing input voltage. To solve this problem, some power supplies use a linear voltage regulator to maintain the output voltage at a regulated value. The function of a linear power converter is to convert a varying voltage to a regulated voltage.

Advantages

The main advantage of linear power converters is that they can dramatically reduce the magnitude of ripple and noise on the output voltage.

Disadvantages

Huge power loss is the main drawback of linear converters. This power loss increases considerably when there is a great difference between input and output voltage of the converter. Based on the presented efficiency relation, lower power losses leads to higher efficiency.

Figure 10. Typical semiconductor layers in a MOSFET, Source: ON Semiconductor corp. (with permission) (a) and its symbol (b)

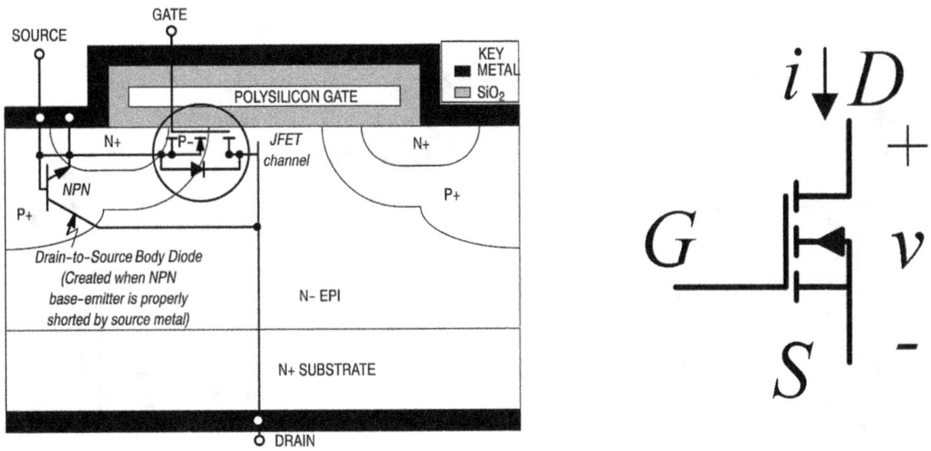

(a) (b)

Figure 11. Real forward V-I characteristic of a MOSFET, Source: ON Semiconductor corp. (with permission)

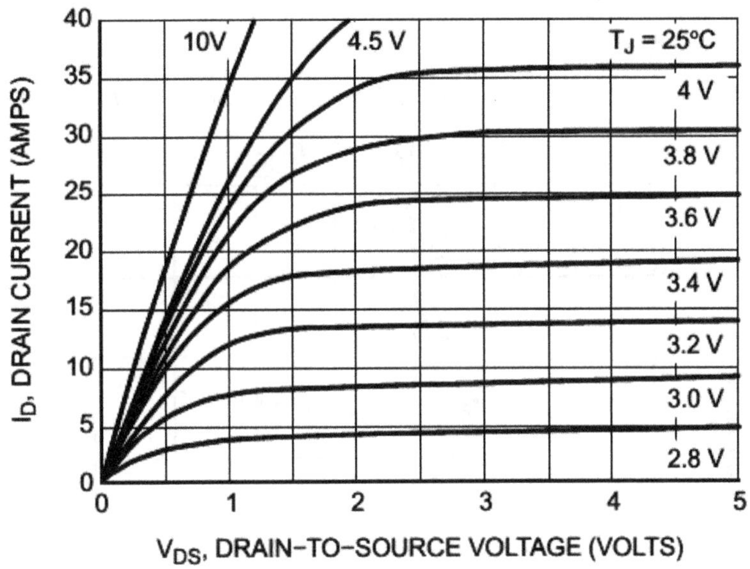

$$\eta = \frac{P_o}{P_o + P_{loss}} \qquad (4)$$

Where P_o is output power and P_{loss} is the power losses.

Figure 12. Typical semiconductor layers in an IGBT, Source: ON Semiconductor corp. (with permission) (a) and its symbol (b)

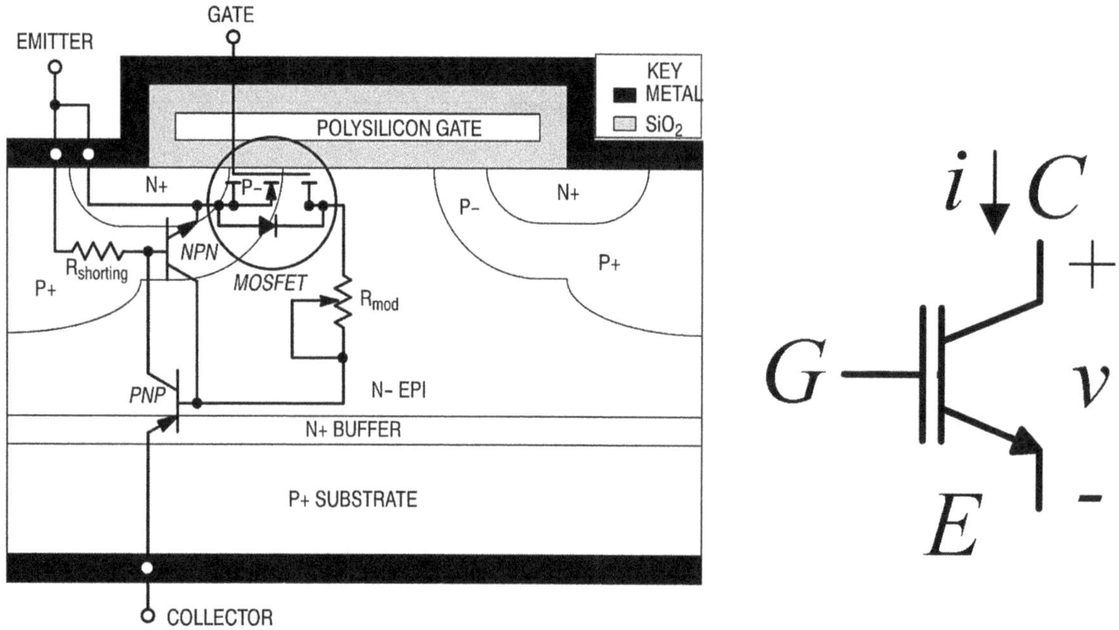

(a) (b)

Figure 13. Real forward V-I characteristic of an IGBT, Source: ON Semiconductor corp. (with permission)

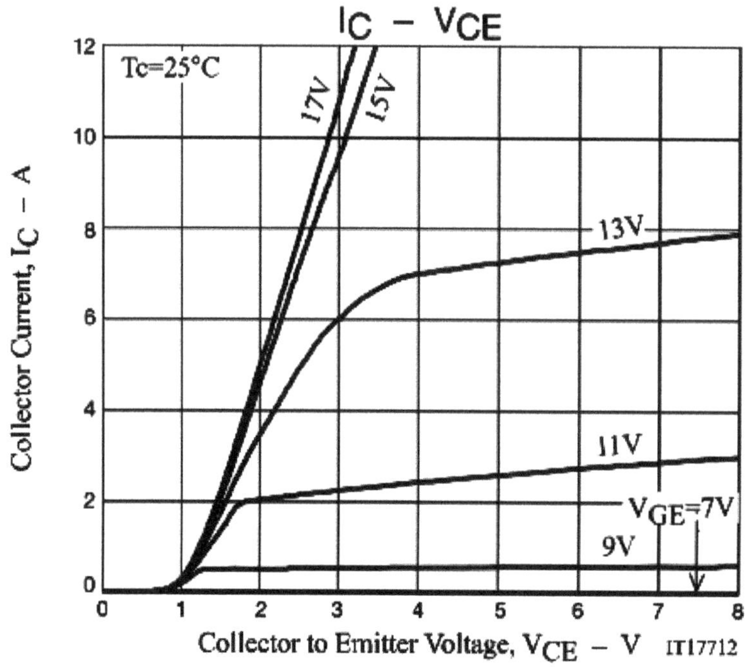

Switching Converters

In a switched-mode power converter, all of the solid state devices operate as switch. Switching occurs at a very high frequency and enabling the use of transformers and filter capacitors that are much smaller, lighter, and less expensive than those found in linear power supplies operating at low frequency. The ripple voltage relation of a rectifier is presented in the following. It can be seen that increasing the frequency leads to decreasing the capacitance value of the converter.

$$V_{r_{PP}} \approx \frac{V_P}{R \cdot C \cdot f_o} \tag{5}$$

Where f_o is the frequency.

Advantages

In switching power converters, the solid state elements operate in switching state. Therefore, switching power converters have a high efficiency. This is the main advantage of these types of electric power converters. The voltage and current relation of an ideal switch is presented in the following. It can be seen that the power losses is zero in switching operation.

$$\begin{cases} i_D = 0 & Switch \quad OFF \\ v_D = 0 & Switch \quad ON \end{cases} \tag{6}$$

Where i and v are the current and voltage of the switch, respectively.

Disadvantages

The current and voltage waveforms of these types of converters are chopped. Therefore, high output ripple and electromagnetic interference are the main problems of switching power converters.

Regarding to the critical energy demand, operation with high efficiency is the main requirement of electric power converters. Thus, switching power converters are used much more than linear types. In the following, we described this type of electric power converters in details. There are four different groups in this field:

- AC to DC converters which are usually named as rectifiers
- DC to AC converters which are usually named as inverters
- AC to AC converters which are usually named as AC voltage controllers
- DC to DC converters which are usually named as choppers

AC to DC Converters

The main application of AC to DC converters is rectification. A rectifier converts alternating current to direct current. These converters are usually used for producing a constant DC voltage from AC network. They are classic rectifiers and modern types.

Rectifier circuits may be single-phase or multi-phase. Most low power rectifiers for domestic equipment are single-phase, but three-phase rectification is very important for industrial applications. Figure 14 shows single phase and three phase classic rectifiers.

Due to time to time increase in demand of rectifiers with specific characteristics in voltage regulation and harmonic elimination, new AC/DC power conversion topologies have been proposed (Kolar, & Friedli, 2013). Among these, Vienna Rectifier, because of ability in improving power quality and resistive main behavior, have received wide interest, in appliances such as feeder of communicational power supply module. Combination of a boost DC/DC converter series with a three phase rectifier provides a new topology called Vienna-type rectifier which is shown in Figure 15. Technical and economical advantages of this type of rectifier can briefly be noted as follow.

Figure 14. Single and three phase classic rectifiers

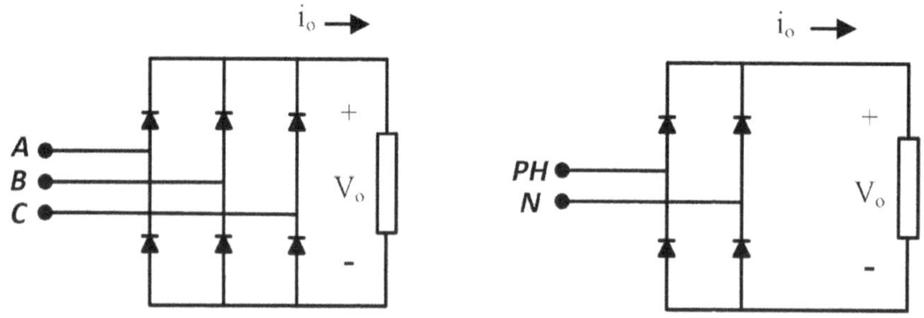

Figure 15. Schematic structure of the Vienna Rectifier

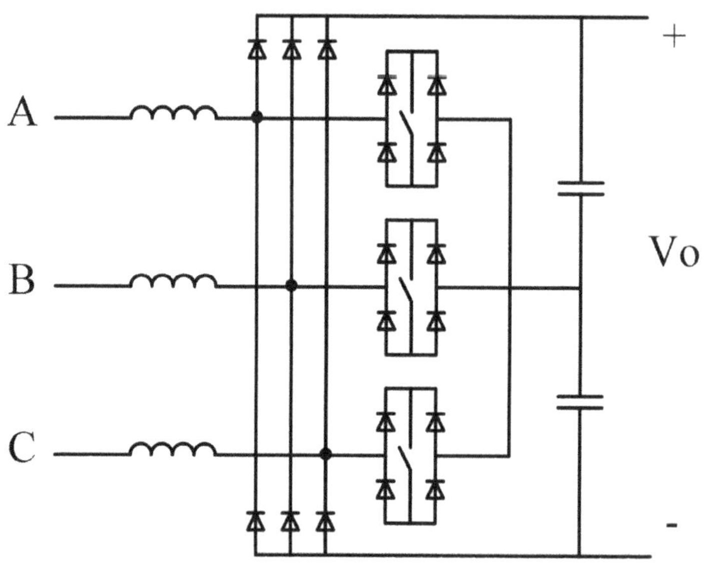

- Low harmonic injection to the main
- Controlled output voltage
- High efficiency
- High reliability
- Low voltage stress on power semiconductors

Application Example: High Power Electrolysis

Electrolysis is commercially highly important as a stage in the separation of elements from natural sources using an electrolytic cell. Electrolysis is a method of using a direct electric current to drive a chemical reaction for separation of elements. The process need high level of DC current and it can not be supplied by the sources with limited life time such as battery. Thus, high current rectifiers are used to produce the required DC voltage for the process via rectification of AC network voltage.

Application Example: Dynamic Braking

Electric motors change to generator during braking process. The energy generated may be returned to the supply or dissipated by some resistors. Figure 16 shows two types of elevator motor drives using classic and regenerative rectifiers.

Figure 16. Conventional (a) and regenerative (b) drive systems

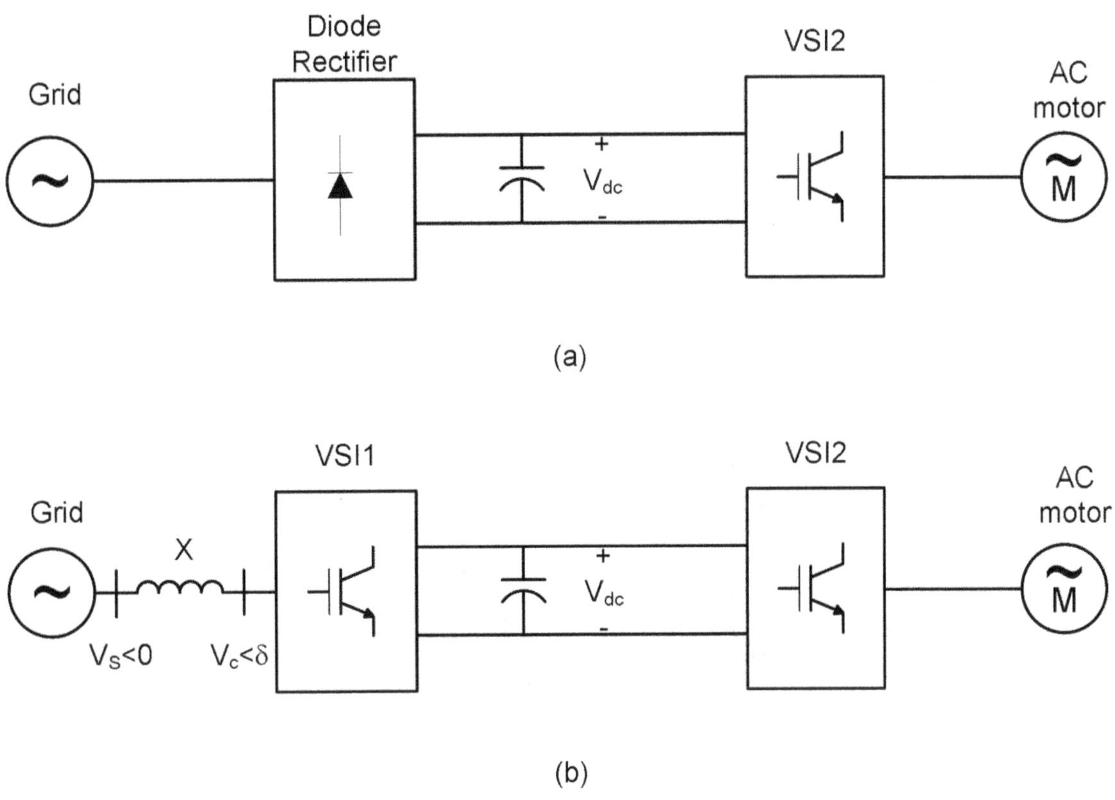

DC to AC Converters

DC to AC converters produce an AC output waveform from a DC source. Some applications of these types of power electronic converters include adjustable speed drives, uninterruptible power supplies, active filters, Flexible AC transmission systems, voltage compensators, and photovoltaic generators.

Topologies for these converters can be divided into two categories: voltage source inverters and current source inverters. Voltage Source Inverters use a constant-voltage source (Abusara, & Sharkh, 2013). Similarly, in current source inverters the controlled AC output is a current waveform (Nag, & Mishra, 2014). The DC to AC power conversion is commonly carried out by fully controllable semiconductor power switches like IGBT and MOSFET. There are some modulation techniques for controlling the output voltage of the inverter (Hamzeh, Karimi, Asadi, & Oraee, 2014). Voltage source inverters have practical uses in both single-phase and three-phase applications. Single and three phase type of an inverter and its typical waveforms are shown in Figures 17, 18, and 19, respectively.

Figure 17. A single phase inverter and relative signal and output voltage waveforms

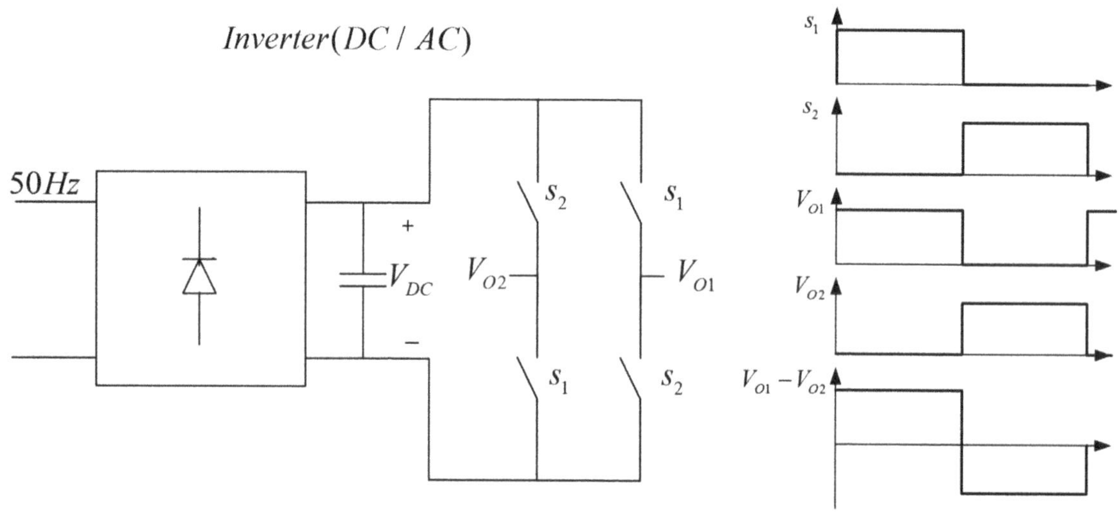

Figure 18. A three phase inverter

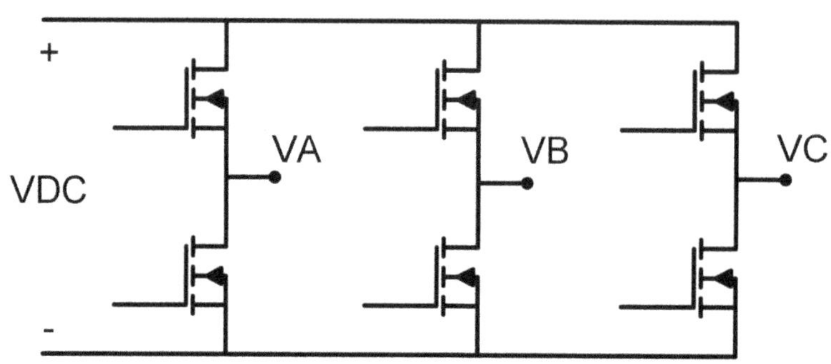

Figure 19. Output voltages of a three phase inverter

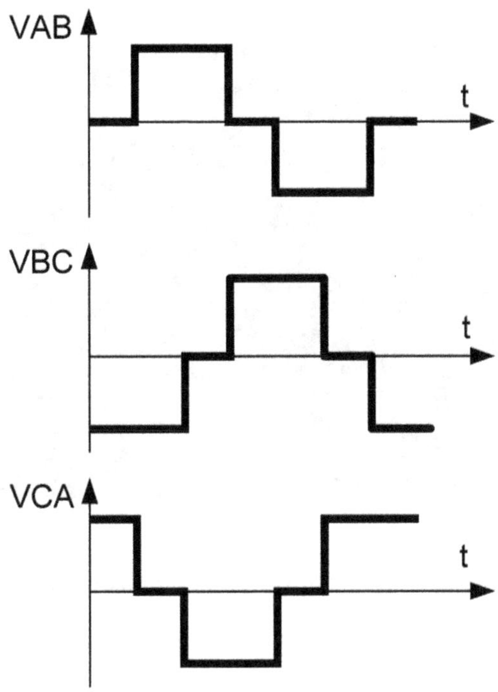

Application Example: Energy Saving

One of the important applications of power electronic converters is their application as a unit of energy saver. Figure 20 shows a driver for an elevator with regenerative performance. While the elevator is going down, the elevator stored energy is delivered to the AC network via an inverter. This inverter operates as a modern fully controlled rectifier during going up. Figure 21 shows operation of this driver during elevator going up and going down.

Application Example: Motor Drive

Combination of power electronic converters in motor applications leads to a high performance level in electric motor applications (Collins, 1992). Figure 22 shows a three phases inverter used as a motor drive. A variable speed drive is an equipment that regulates the speed and torque of an electric motor (Jang, 2013). Many industrial processes must operate at different speeds for different products. In starting a motor, a drive initially applies a low frequency and voltage, thus avoiding high inrush current associated with direct on line starting. However, motor cooling deteriorates and can result in overheating as speed decreases such that prolonged low speed motor operation with significant torque is not usually possible without separately-motorized fan ventilation. We talk more about this problem in the next chapter. Figure 23 shows a typical drive which is used for an induction motor.

Figure 20. An elevator driver with regenerative performance, inverter stack (up), control board (down), series inductors (right)

Figure 21. Input voltage and current of a drive system in motoring (a) and braking (b) operation mode

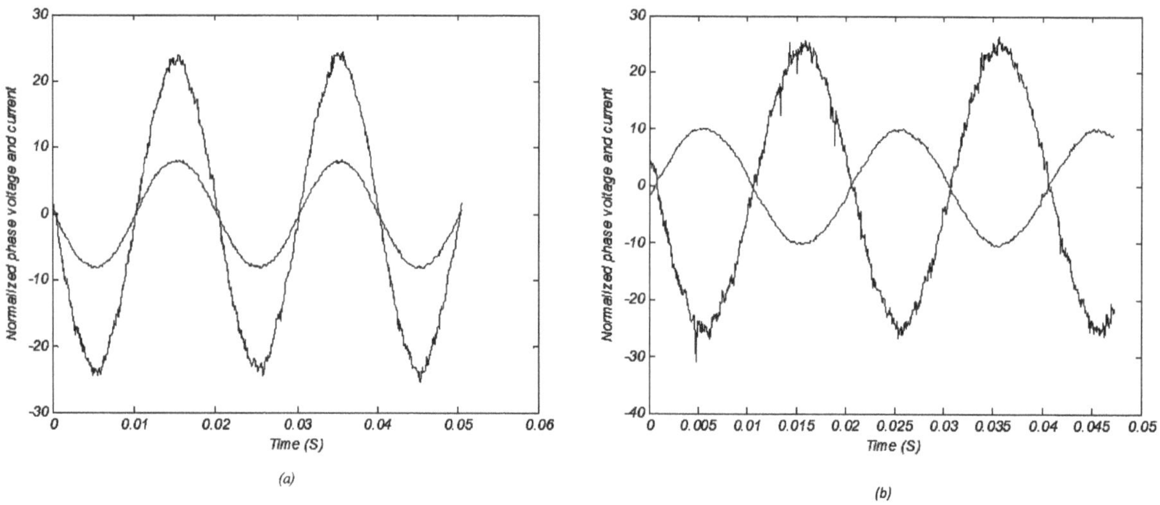

Figure 22. Block diagram of a three phase inverter used as a motor drive

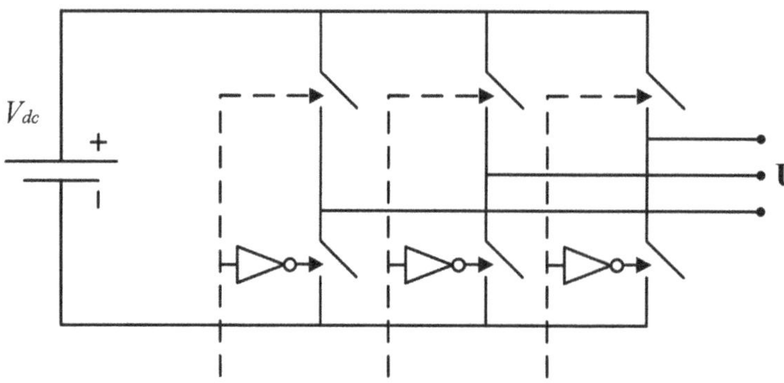

Figure 23. An ASD system for an induction motor: three phase inverter (up), DSP controller (down), induction motor (right)

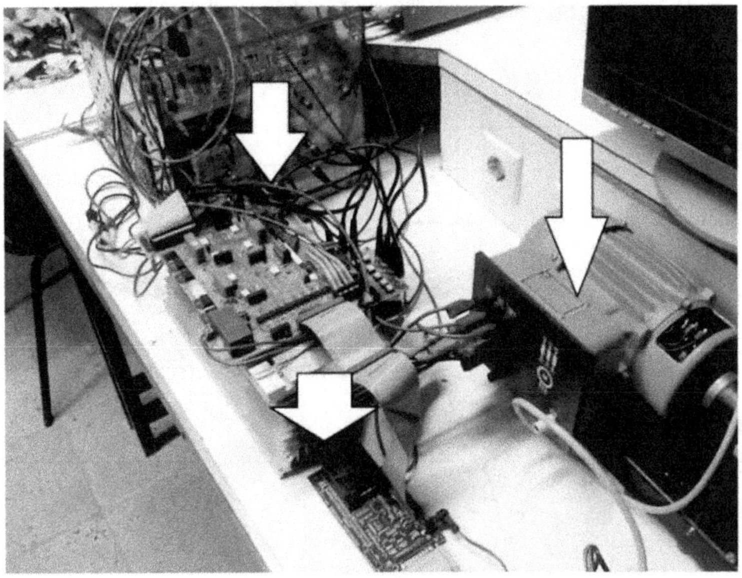

Application Example: Power Active Filter

There are many ways in which electric power can be of poor quality and many more causes of such poor quality power. For example, a fault on the network may cause a dip that will affect some customers; the higher the level of the fault, the greater the number affected. A problem on one customer's site may cause a transient that affects all other customers on the same subsystem. Problems, such as harmonics, arise within the customer's own installation and may propagate onto the network and affect other customers. Inverters are a key tool in power quality improvement. Figure 24 shows a typical circuit diagram of an active filter which is used for low frequency harmonic compensation and power factor correction of a rectifier. Figure 25 shows its waveforms for operation in capacitive and inductive region. Figure 26 shows application of an active filter for ripple cancellation of a DC power supply.

Application Example: High Efficiency Power Amplifiers

A class-D amplifier or switching amplifier is an electronic amplifier in which the amplifying devices (transistors, usually MOSFETs) operate as electronic switches, instead of as linear gain devices as in other amplifiers (Kaboli,, Moayedi,, & Oraee, 2008). The analog signal to be amplified is converted to a series of pulses by pulse width modulation applied to the amplifier. After amplification, the output pulse train can be converted to an analog signal by passing through a passive low pass filter consisting of inductors and capacitors. The major advantage of a class-D amplifier is that it can be more efficient than analog amplifiers, with less power dissipated as heat in the active devices. Figure 27 shows circuit diagram of a power amplifier using an inverter.

Figure 24. Schematic diagram of an active filter used for a power factor correction of a rectified load

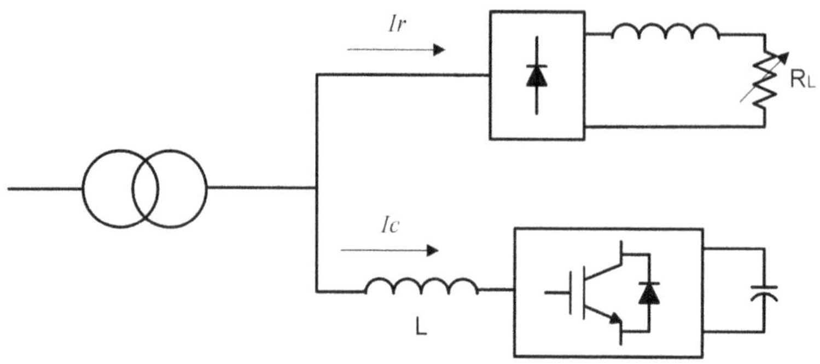

Figure 25. Input voltage and current of an active compensated nonlinear load: operation with lead power factor (a), operation with lag power factor (b)

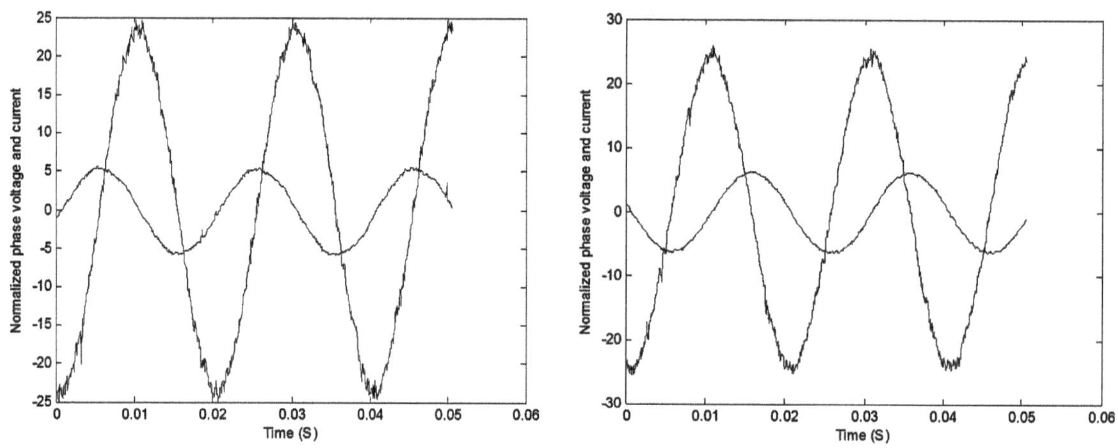

(a) (b)

Figure 26. Operation of an active ripple compensator in a DC power supply: before employment of active filter (a), after applying active compensator (b)

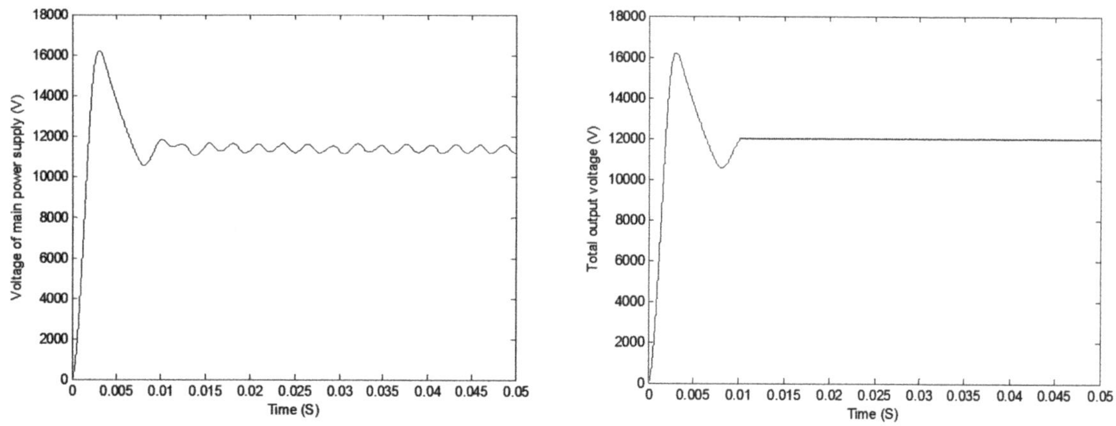

(a) (b)

Figure 27. Schematic diagram of a Class-D amplifier

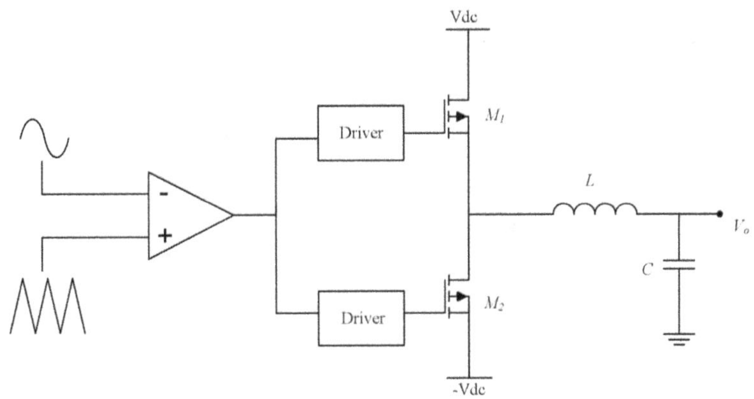

Application Example: Static AC Power Supply

A power inverter can be entirely electronic or may be a combination of mechanical effects (such as a rotary apparatus) and electronic circuitry. Static inverters do not use moving parts like bearing in the conversion process. Therefore, they benefit long time interval between maintenances. Figure 28 shows schematic diagram of a static inverter used for railway applications.

Application Example: Static DC Power Supply

As the inverter AC output voltage is high frequency, the volume of this DC power supply is very small. So, it is used for portable applications where the volume and weight of the system is a limiting factor. Figure 29 shows a diagram of a static DC power supply with an inverter core.

AC to AC Converters

Direct converting AC to AC electric power allows control of the voltage, frequency, and phase of the waveform applied to a load from a supplied AC system. This single power conversion benefits high efficiency especially for extra high power conversion like power transmission between two countries (Cipriano, Jacobina, da Silva, & Rocha, 2012).

There are various different types of AC to AC converters like cycloconverters and matrix converters. Figure 30 shows simplified circuit diagram of a single and three phase AC voltage controller.

Application Example: Motor Soft Starter

A motor soft starter is used in series with AC electric motors to reduce current surge of the motor during startup. This reduces the mechanical stress on the motor and shaft, as well as the electrodynamic stresses on the attached power cables and electrical distribution network, extending the lifespan of the system.

Figure 28. Block (a) and Schematic (b) diagrams of a typical AC power supply used in portable railway applications

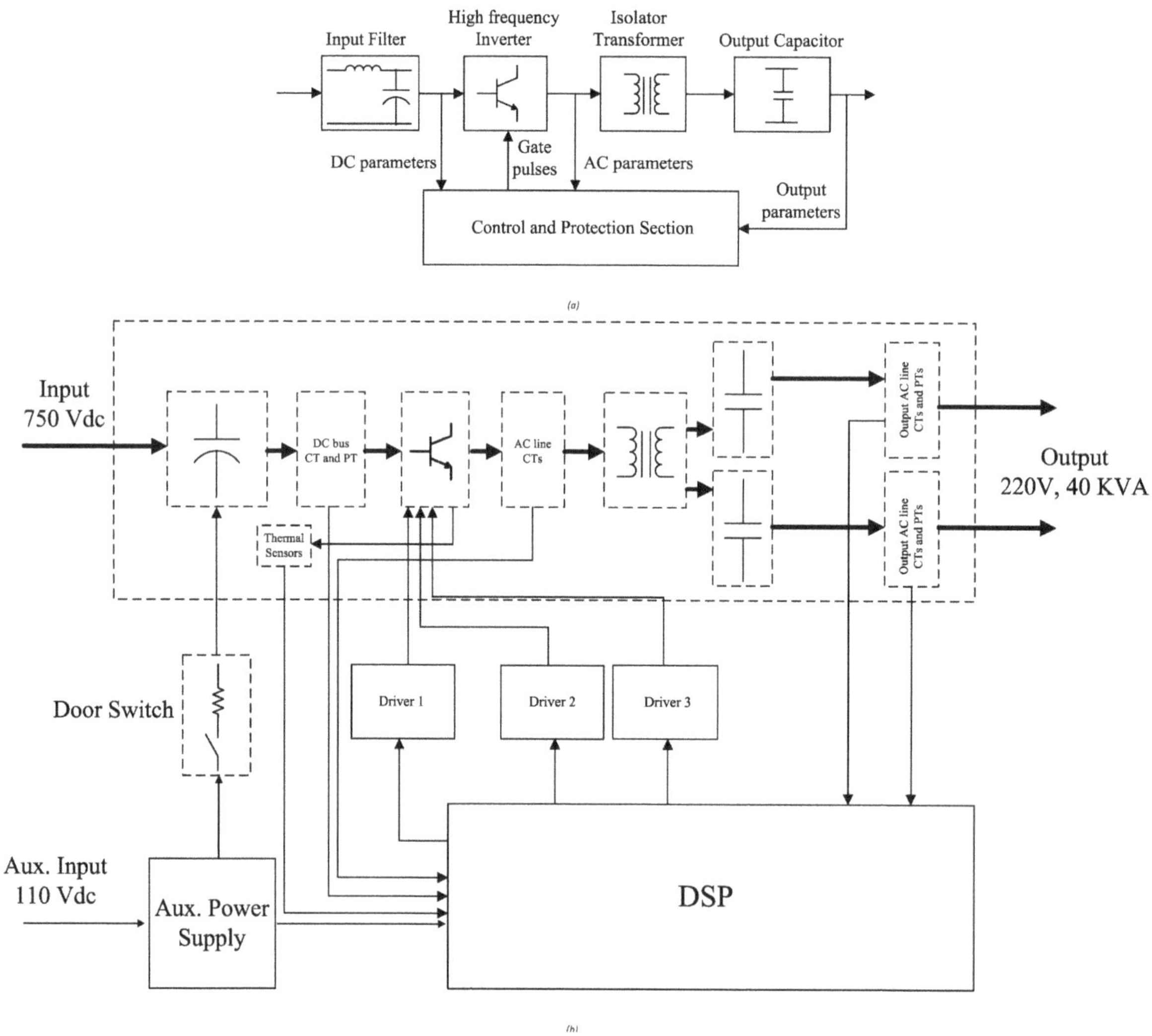

Figure 29. Schematic diagrams of a static DC power supply with an inverter core

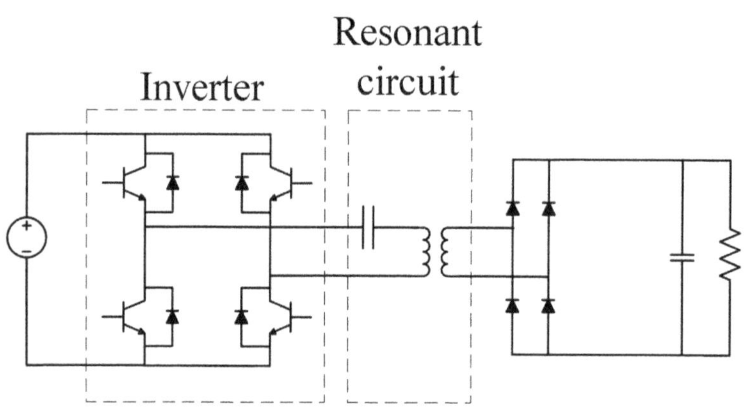

Figure 30. Single phase and three phase AC controller

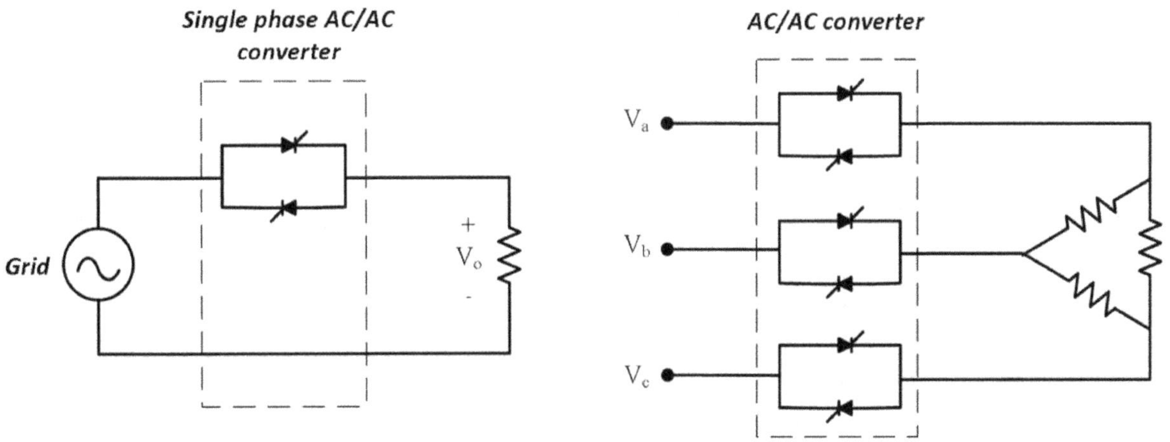

Application Example: High Power Transmission

An AC to AC converter acts as a single stage converter. Therefore, it has a high efficiency in comparison to other multi stage methods.

Application Example: Wind

An AC/AC converter for direct conversion of low-frequency AC from the wind generators to high-frequency AC network is widely used.

DC to DC Converters

DC to DC converters are important in portable electronic devices. Most DC to DC converters regulate the output voltage.

DC to DC converters can be categorized as the following:

- Isolated or on isolated
- Single or multi output

Application Example: Power Distribution Unit

Many power electronic converters are used as adaptor of voltage levels between two or more subsystem. As an example, suppose the power distribution of a satellite. The power subsystem consists of solar panels to convert solar energy into electrical power and batteries that store power and supply the satellite when it passes into the shadow. The satellite has other essential parts: the telemetry subsystem monitors the on-board equipment operations, transmits equipment operation data to the earth control station, and receives the earth control station's commands to perform equipment operation adjustments. The thermal control subsystem helps protect electronic equipment from extreme temperatures due to intense sunlight or the lack of sun exposure on different sides of the satellite's body. The attitude and orbit control sub-

Figure 31. Block diagram of a satellite power supply

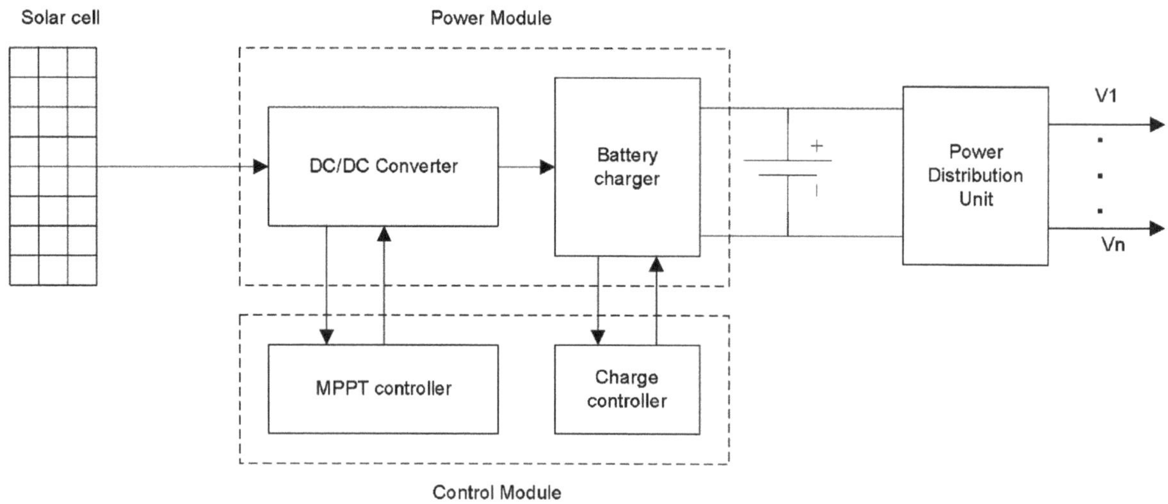

system consists of sensors to measure vehicle orientation; control laws embedded in the flight software; and actuators to apply the torques and forces needed to re-orient the vehicle to a desired attitude, keep the satellite in the correct orbital position and keep antennas positioning in the right directions. Power distribution unit receives input DC voltage from solar panels or batteries and converts it into some DC voltages with different amplitude for the above mentioned subsystems. Figure 31 shows a block diagram for a power distribution unit.

Application Example: Battery Charger in Solar Energy Conversion

Battery charger based on DC to DC power conversion is used to charge various types of rechargeable batteries such as Li-Ion and Ni-Cd by extracting power from power network. Using a PIC microcontroller provides the ability to implement maximum power control of solar panel. The charger is able to provide a regulated output voltage while the battery is full-charged and/or disconnected from the system, therefore the charger output can be used as a voltage bus for the system while battery is failed or not used. Figures 32 and 33 show the typical and real V-I characteristic of a PV panel, respectively.

Application Example: Renewable Energy Converters

Figure 34 shows a solar panel using a battery charger for delivering the power to AC network. Wide application of solar panels is impossible without this converter.

Application Example: Power Supply

Power conditioning and voltage regulating are one of the other applications of power electronic converter. This application is named as power supply. Figure 35 shows a DC power supply.

Figure 32. V-I characteristic of a typical solar panel (a), measuring setup (b)

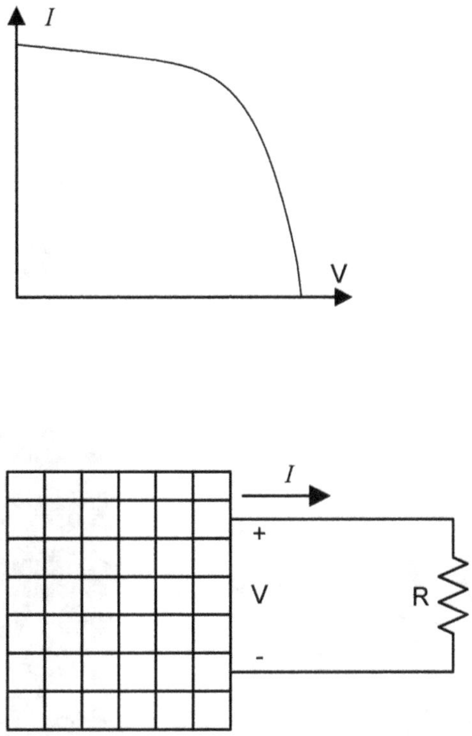

A power supply may be implemented as a discrete, stand-alone device or as an integral device that is hardwired to its load. Examples of the latter case include the low voltage DC power supplies that are part of desktop computers and consumer electronics devices.

A switched-mode supply of the same rating as a line-frequency supply will be smaller, is usually more efficient, but would be more complex. Figure 36 shows an integrated step down DC to DC converter. Figures 37 and 38 show circuit diagrams of a classical voltage mode and novel current mode DC to DC converters.

ELECTRICAL MACHINES

Electric machine converts electrical energy to mechanical energy and vice versa. Transformer is also an electric machine but it is used for changing the level of voltage and isolation without mechanical movement. An electric motor is an electric machine that converts electrical energy into mechanical energy. In normal motoring mode, most electric motors operate through the interaction between an electric motor's magnetic field and winding currents to generate force within the motor.

Electric motors found in applications as diverse as industrial fans, blowers and pumps, machine tools, household appliances, power tools, and disk drives, electric motors can be powered by direct current sources, such as from batteries, motor vehicles or rectifiers, or by alternating current sources, such as from the power grid, inverters or generators. There are 4 types of electrical machines which are used in almost all of industries.

Figure 33. A 200W solar panel, Source: Aria Solar Co. (with permission)

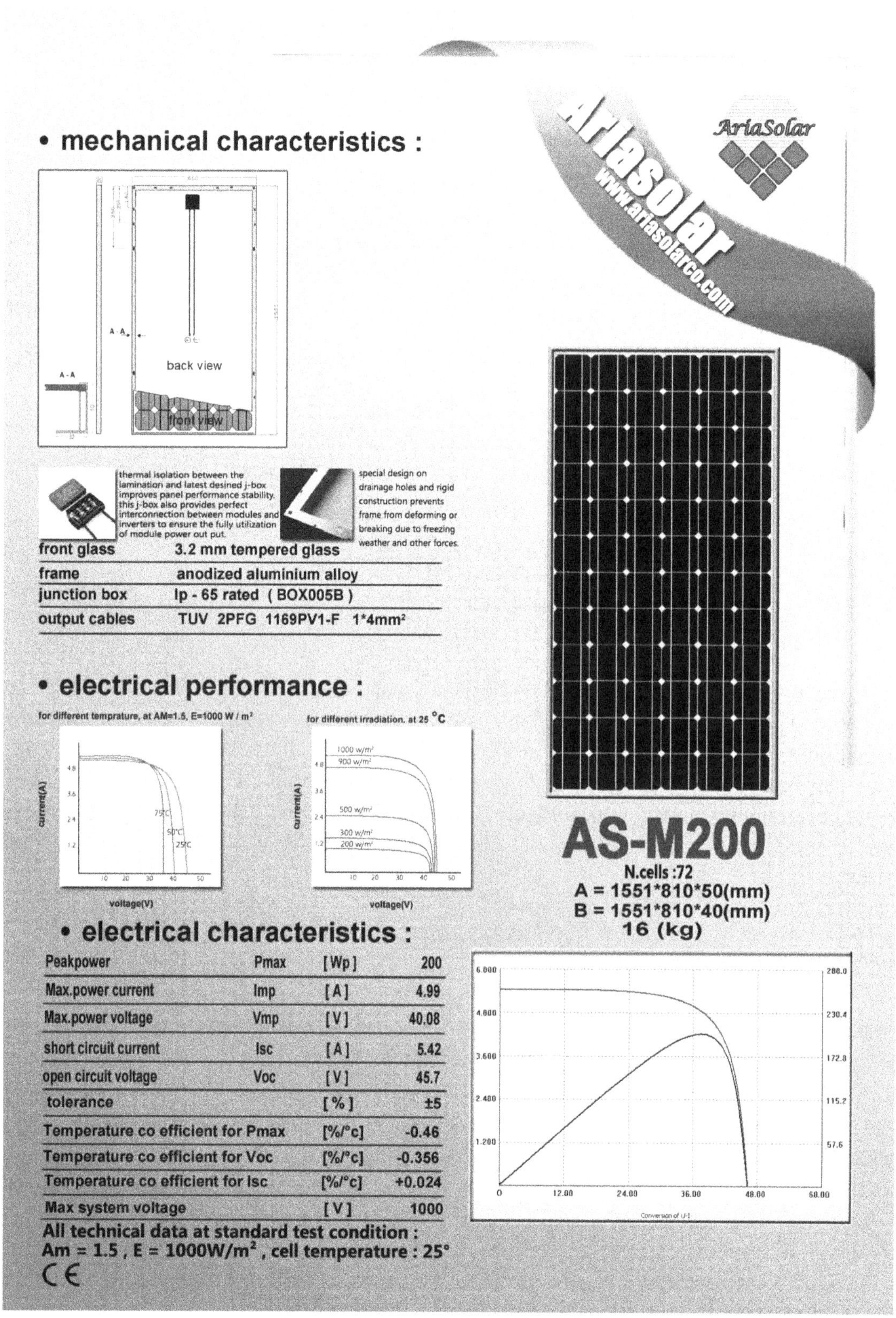

Figure 34. A power system with solar panel as power source: Solar panel (up), Batteries (left), Charger (right)

Figure 35. A DC Power supply

Figure 36. An integrated DC power supply

Figure 37. Circuit diagram of a step down DC to DC converter

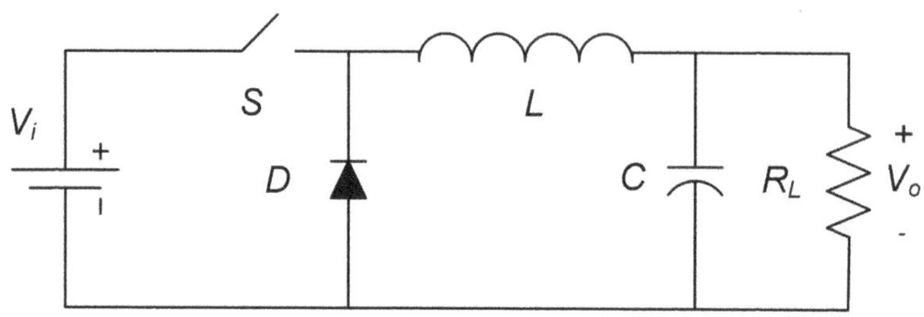

Figure 38. Schematic diagram of a current mode DC to DC converter

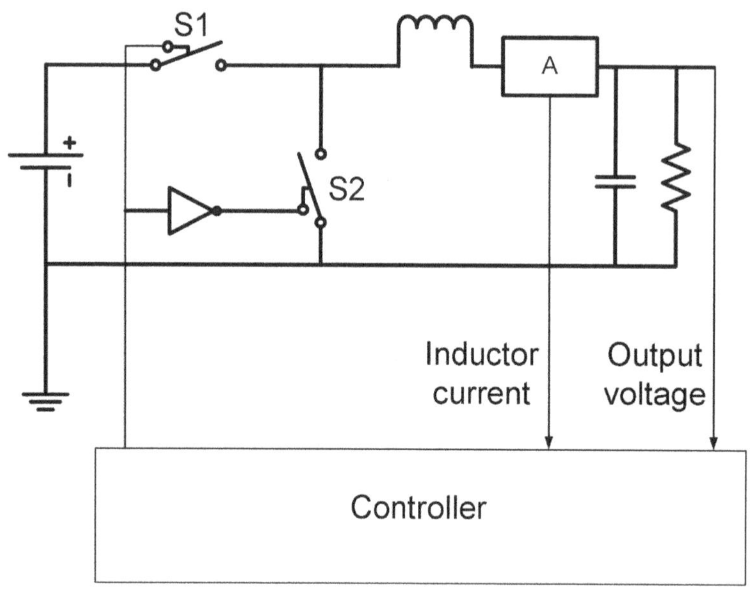

- DC machines
- AC synchronous machines
- AC asynchronous machines
- Transformers

DC Machines

A commutated DC motor has a set of rotating windings wound on an armature mounted on a rotating shaft. The shaft also carries the commutator, a long-lasting rotary electrical switch that periodically reverses the flow of current in the rotor windings as the shaft rotates. The brushes connect an external source of electric power to the rotating armature. Many of the limitations of the classic commutator DC motor are due to the need for brushes to press against the commutator. We talk in details about this problem in the next chapter. This creates friction. Sparks are created by the brushes making and breaking circuits through the rotor coils as the brushes cross the insulating gaps between commutator sections. This sparking limits the maximum speed of the machine, as too-rapid sparking will overheat, erode, or even melt the commutator. There are some types of brushed DC motor:

- DC shunt-wound motor
- DC series-wound motor
- DC compound motor
- PM DC motor
- Separately excited

Figures 39, 40, 41, and 42 show basic parts of a DC machine.

Equivalent Circuit and Characteristics

The equivalent circuit for a DC motor consists of an inductor, a resistor and a voltage source in series. These represent the coil inductance, coil resistance, and back EMF respectively. The back EMF voltage source is dependant on the speed of the motor. Figures 43, 44, 45, 46, and 47 shows equivalent circuit of various types of DC motors.

Figure 39. Basic operation of a DC motor

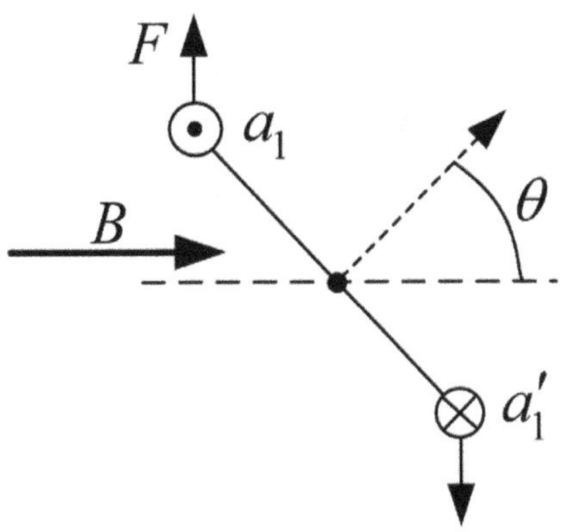

Figure 40. One turn winding

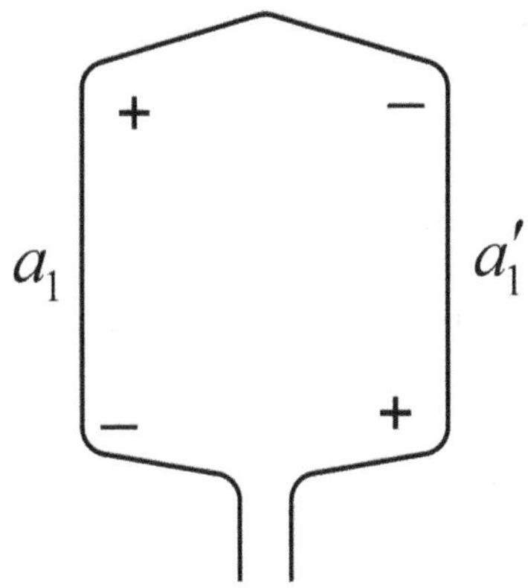

Figure 41. Simplified structure of rotor of a DC motor

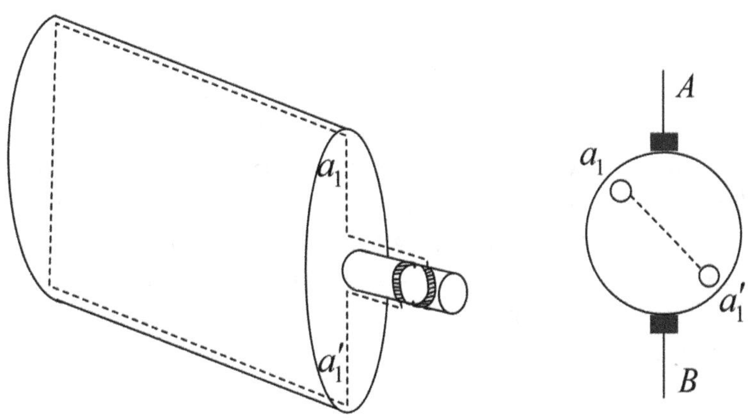

Figure 42. A DC motor with auxiliary windings

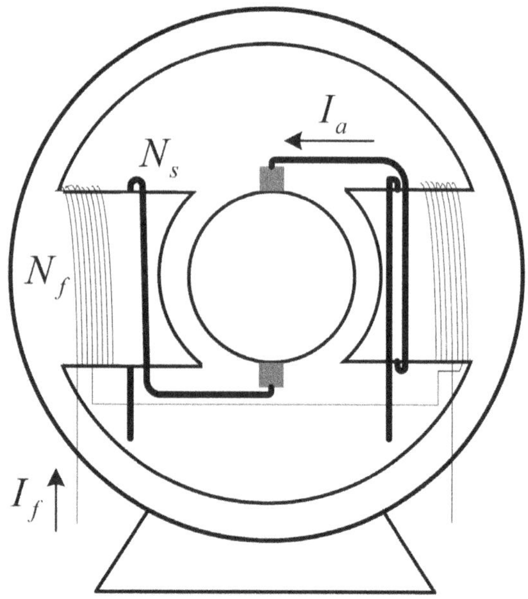

Figure 43. Equivalent circuit of field winding

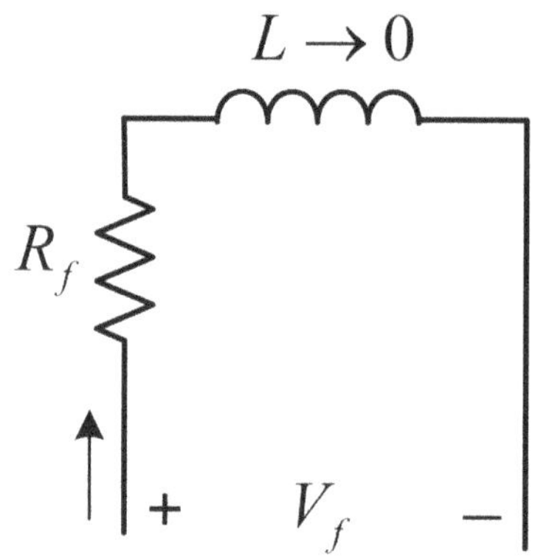

Figure 44. Equivalent circuit of rotor of DC motor

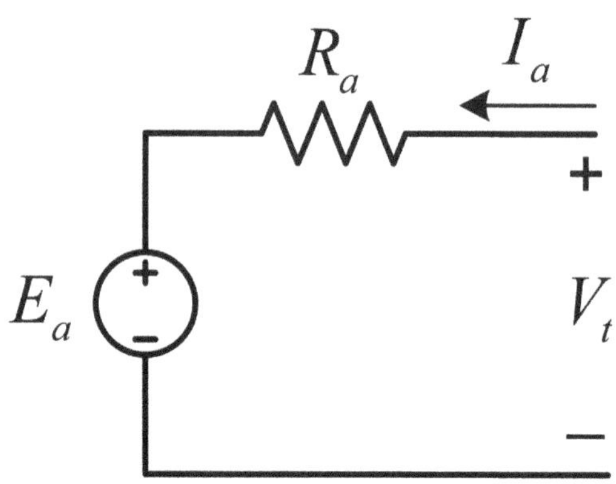

Figure 45. Detailed equivalent circuit of rotor of DC motor

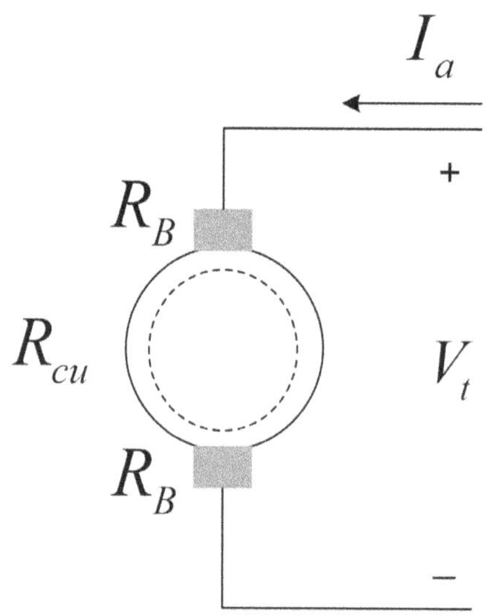

Features

Historically, DC motors were the first type widely used, since they could be powered from existing direct-current lighting power distribution systems. A DC motor's speed can be controlled over a wide range, using either a variable supply voltage or by changing the current in its field windings. DC motors are fully stable over whole range of speed. Speed is a direct and almost linear function of armature voltage. Thus, speed control of DC motors is much simpler than other motors. Therefore, DC motors were almost exclusive choice of robotic applications for many years. Figures 48 and 49 shows output characteristic of various types of DC motors. In a shunt or separately excited DC motor, the basic relation between rotor variables is expressed as the following.

Figure 46. Completed equivalent circuit of separately excited of DC motor

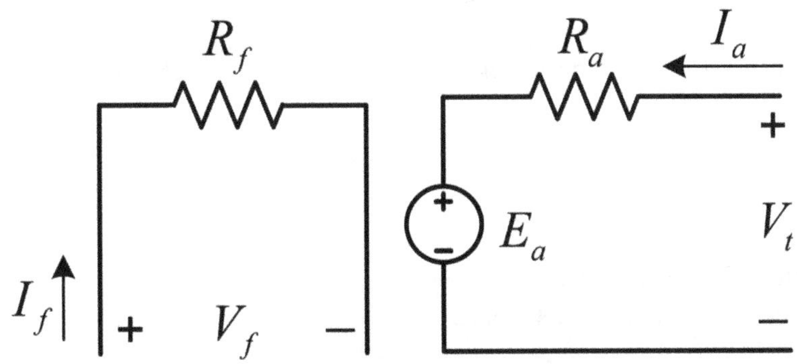

Figure 47. Completed equivalent circuit of series DC motor

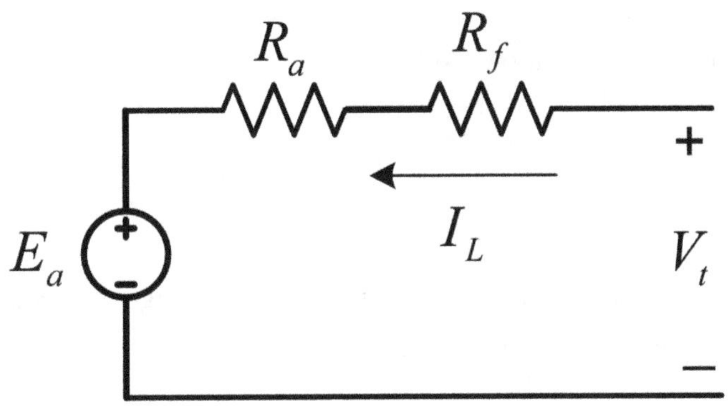

Figure 48. T-ω characteristic of a series DC motor

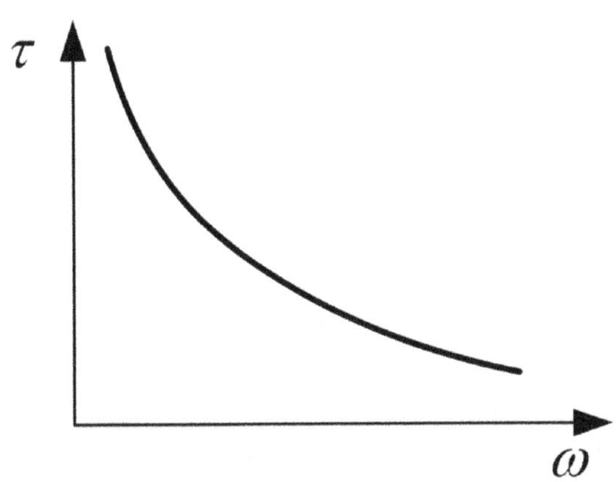

Figure 49. V-I characteristic for various types of DC generator

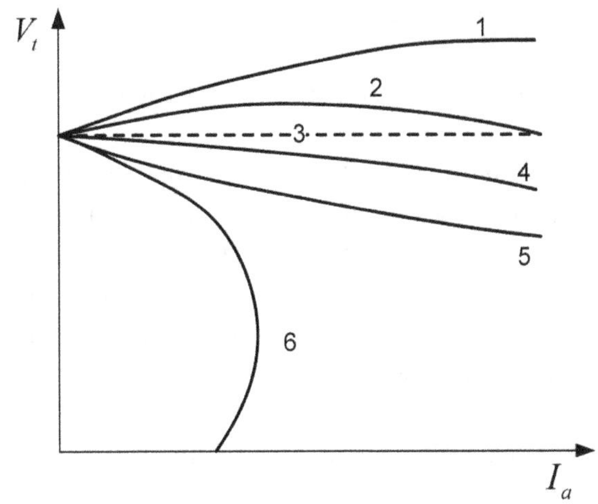

$$E = V - R_a \times i \tag{7}$$

Where R_a is stator resistance, E is motor back electromagnetic force (EMF) and V is the motor terminal voltage.

$$E = K\varphi\,\omega \tag{8}$$

Where ϕ is the motor flux and ω is the motor speed.

The electromagnetic torque of the DC motor is:

$$T_e = K\varphi\, i \tag{9}$$

Where T_e is the electromagnetic torque.

Solving the above mentioned equation results:

$$T_e = \frac{K\varphi V}{R_a} - \frac{K^2\varphi^2\,\omega}{R_a} \tag{10}$$

This equation shows the stability of the DC motor for whole range of speed. Torque-speed characteristic of a series DC motor is also expressed as a stable characteristic. This type of DC motor is also fully stable.

Application Example: Traction

Traction motors are used in electrically powered rail vehicles Figure 50 shows application of a series DC motor in traction.

Application Example: Servomotor

A servomotor is a rotary actuator that allows for precise control of angular position, velocity and acceleration. It consists of a suitable motor coupled to a sensor for position feedback. It also requires a relatively sophisticated controller, often a dedicated module designed specifically for use with servomotors. Servomotors are used in applications such as robotics, CNC machinery or automated manufacturing. At the simplest version, brushed permanent magnet DC motors are used, owing to their simplicity and low cost as well as stability over whole range of speed. Figure 51 and 52 show a typical servo motor and its voltage and current waveforms during speed control. Figure 53 shows application of a servo motor for sun tracking in a PV system. Figures 54 and 55 show typical block diagrams of servo systems.

Figure 50. A series DC motor used in railway traction system

Figure 51. A DC servo system (left) and its driver (right)

AC Asynchronous Machines

An induction or asynchronous motor is an AC electric motor in which the electric current in the rotor needed to produce torque is obtained by electromagnetic induction from the magnetic field of the stator winding (Levi, Bojoi, Profumo, Toliyat, & Williamson, 2007). An induction motor therefore does not require mechanical commutation, separate-excitation or self-excitation for all or part of the energy

Figure 52. Typical current (up) and voltage (down) waveforms of a DC servo motor used in position control

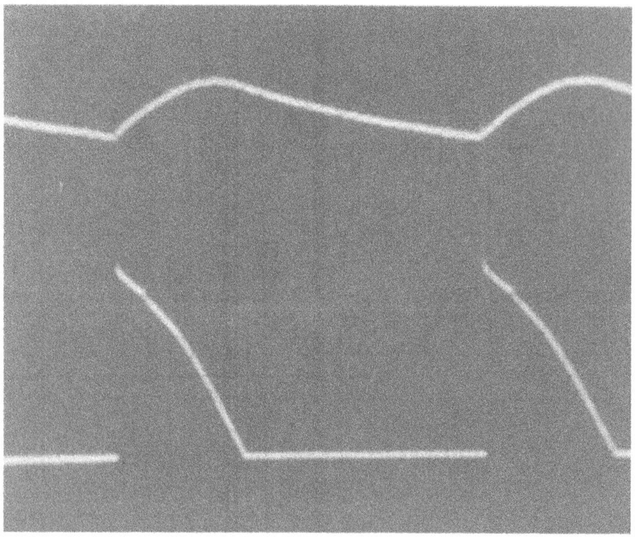

Figure 53. A servo motor used in sun tracking system of a solar panel

Figure 54. A typical block diagram of a servo system

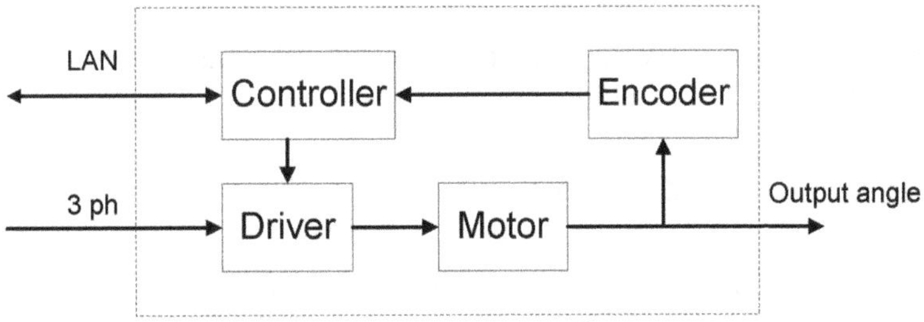

Figure 55. Block diagram of a two-axis servo system

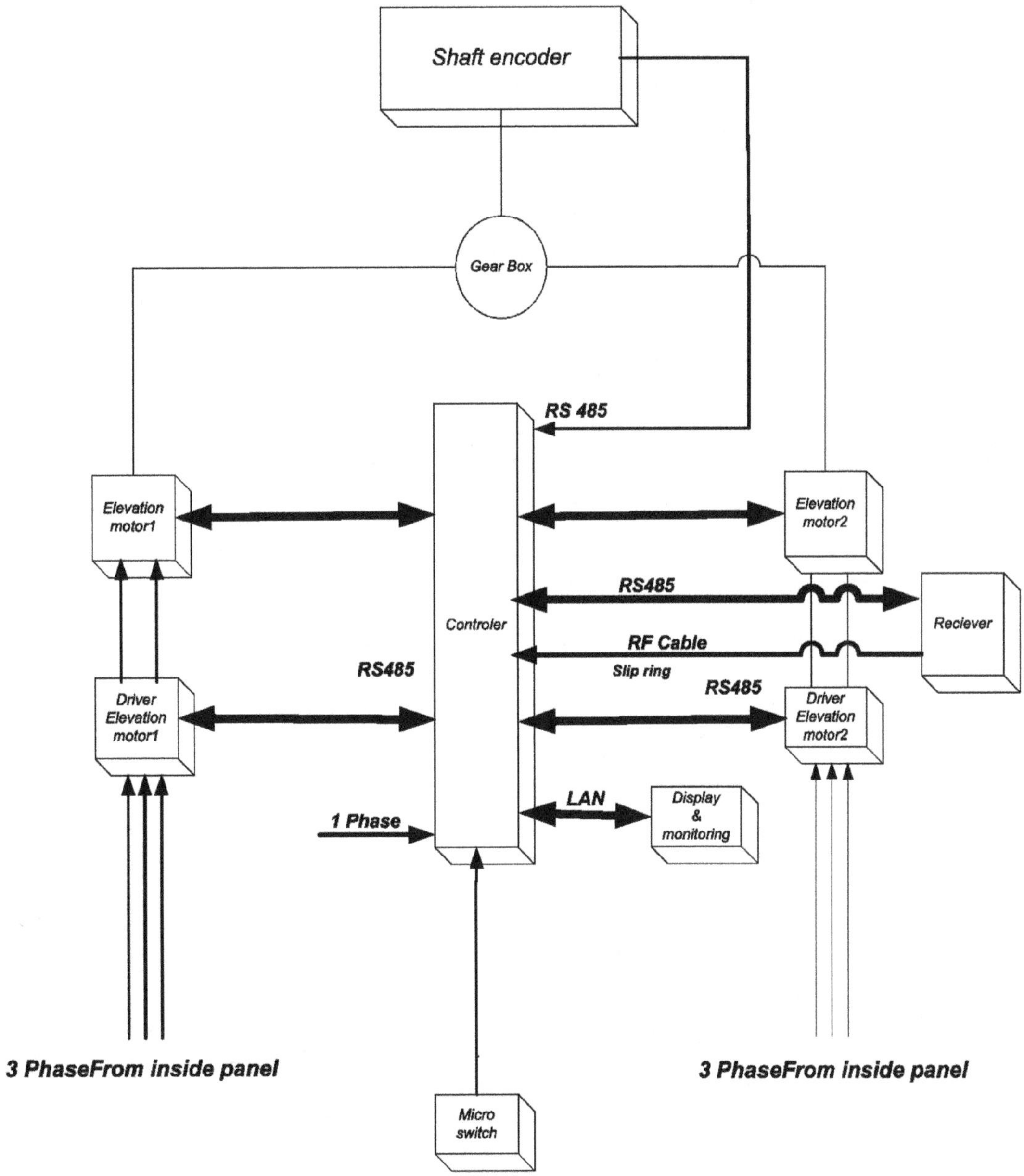

transferred from stator to rotor, as in DC and large synchronous motors (Barati, Shiyi, Abdi, Oraee, & McMahon, 2011).

Induction motors may be further divided into squirrel cage and wound rotor. The first type has a heavy winding made up of solid bars, usually aluminum, joined by rings at the ends of the rotor.

Equivalent Circuit and Characteristics

In an induction motor, electrical power is transferred to the rotor by electromagnetic induction, much like transformer action (Barati, & Oraee, 2010). An induction motor resembles a rotating transformer, because the stator (Gorginpour, Jandaghi, Oraee, & Abdi, 2014) is essentially the primary side of the transformer and the rotor (rotating part) is the secondary side. Figures 56, 57, 58, 59, 60, 61, 62, 63, 64, 65, and 66 show basic parts of an induction machine as well as its circuit diagram and output characteristic.

Figure 56. Simplified winding diagram of a three phase induction motor

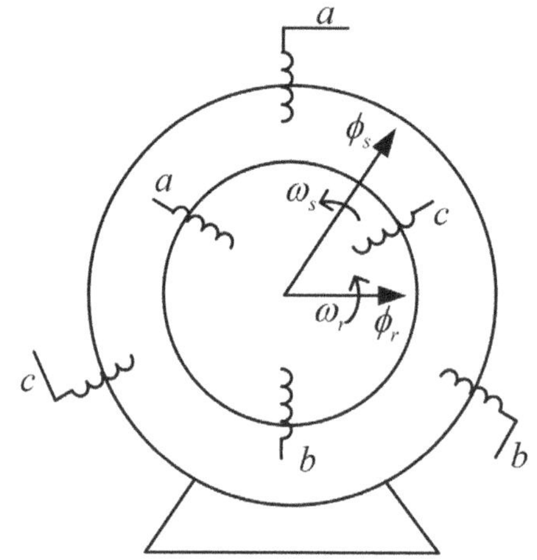

Figure 57. Four poles (left) and two poles (right) stator of an induction motor

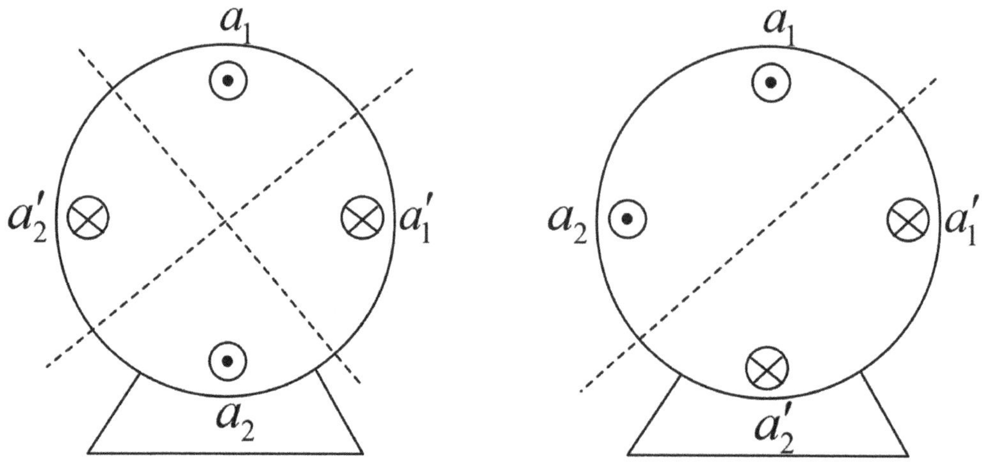

Figure 58. Equivalent circuit of an induction motor

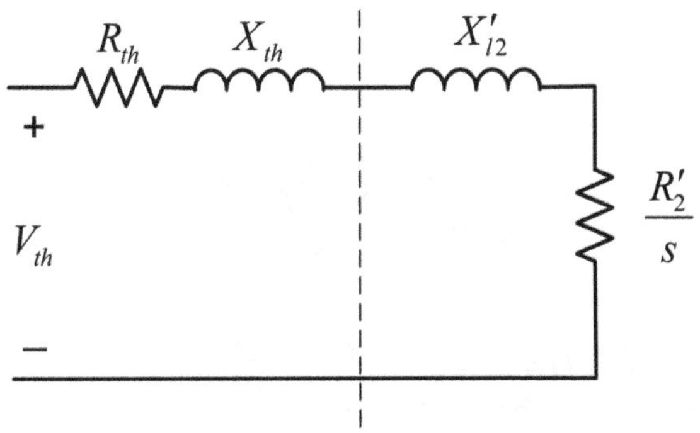

Figure 59. Typical T-ω characteristic of an induction motor

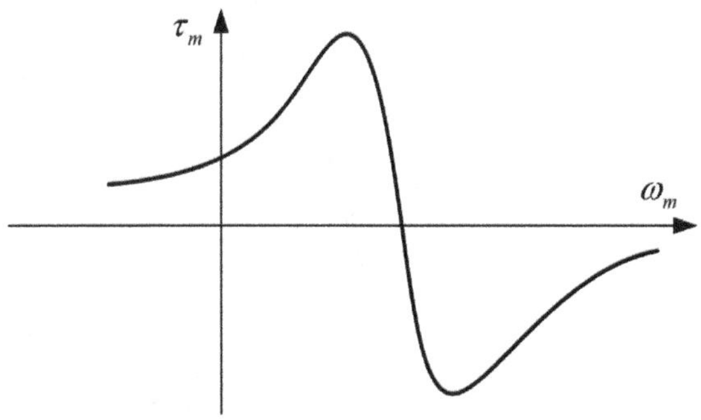

Figure 60. Various types of rotor conductor in induction motor

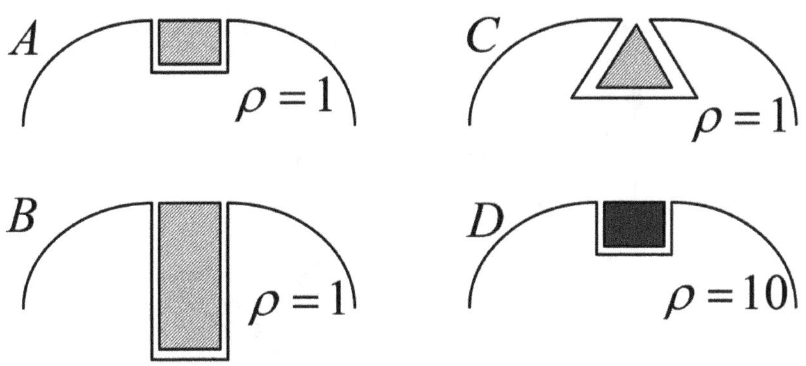

Figure 61. Torque-speed characteristic of four commonly used classes of induction motor

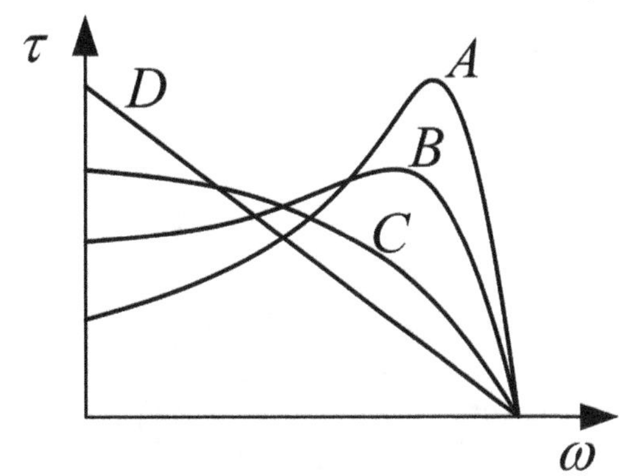

Figure 62. Induction motor speed controller by variation in rotor resistance

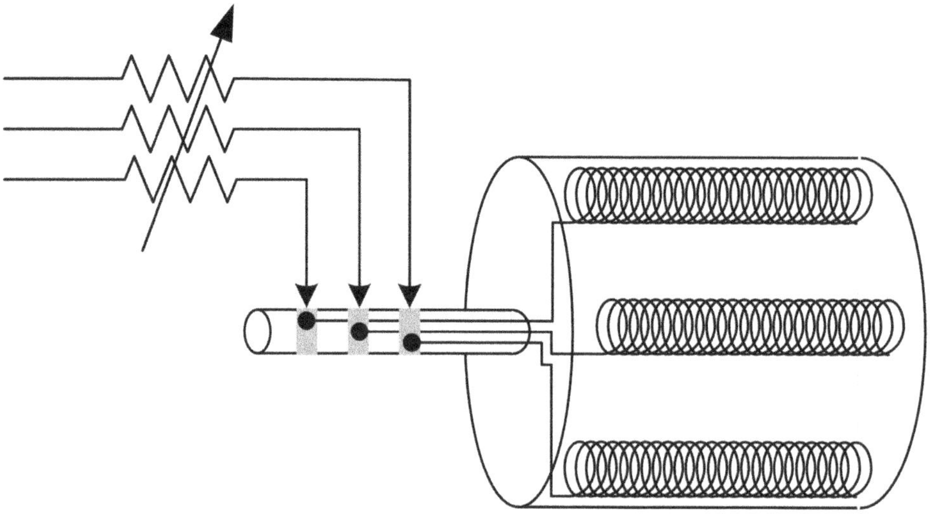

Figure 63. Effect of different rotor resistances on T-ω characteristic of induction motor

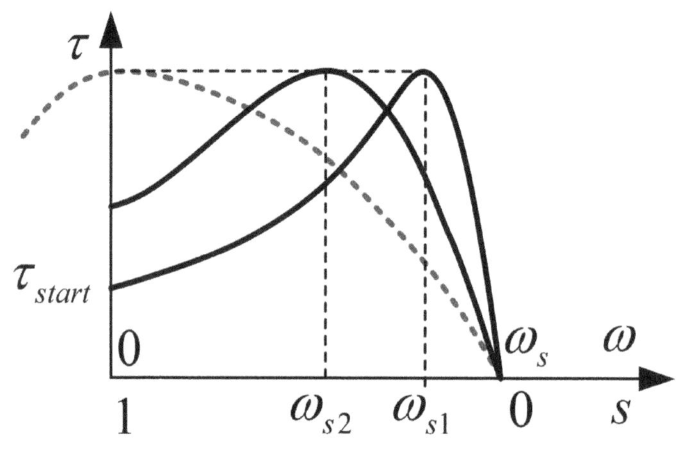

Figure 64. Effect of stator voltage variation on T-ω characteristic of induction motor

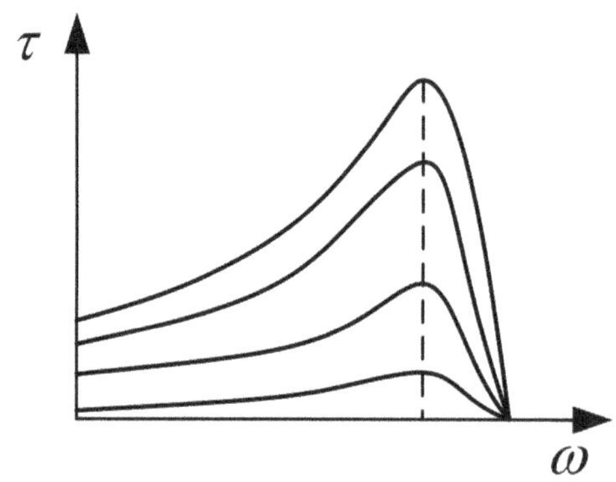

Figure 65. Speed control of induction motor by variable frequency

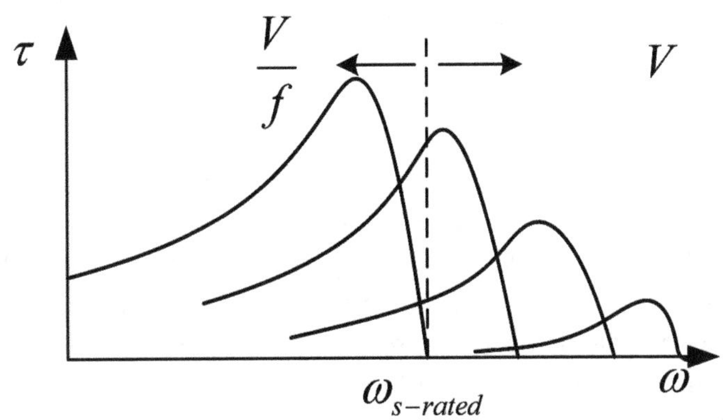

Figure 66. Real characteristic of a 250W induction motor

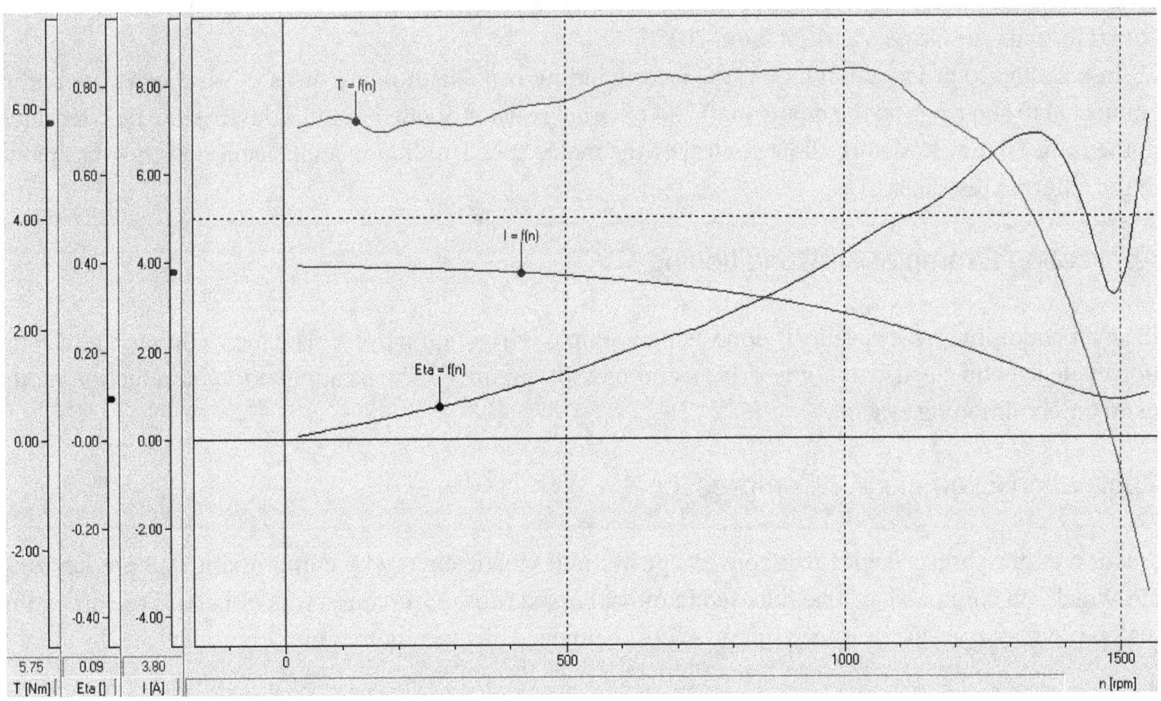

Torque-speed characteristic of an induction motor can be expressed as the following.

$$T_e \approx \frac{3}{\omega_s}\frac{R_2'}{s}\frac{V^2}{(R_1 + R_2'/s)^2 + (x_1 + x'_2)^2} \text{ and } s = \frac{\omega_s - \omega}{\omega_s} \tag{11}$$

Where s is the motor slip and ω_s id the synchronous speed.

Figure 59 shows the illustrative expression of this equation. It can be seen that the motor is stable in a small range of speed. This is a problem and we will see that this causes instability in a synchronous motor. But, non-zero starting torque of an induction motor saves it against instability (Abdi, McMahon, Malliband, Shiyi, Mathekga, Tavner, Abdi, Oraee, Teng & Tatlow, 2013).

Features

Polyphase induction motors are widely used in industry. They are rugged, reliable and economical (Oraee, Abdi, Abdi, McMahon, & Tavner, 2015). Single-phase induction motors are used extensively for smaller loads, such as household appliances like fans (Wang, Zhong, Yang, & Mu, 2010). Electronic drives offer especially important energy savings opportunities for existing and prospective induction motors in variable-torque centrifugal fan, pump and compressor load applications. The shape of the rotor bars determines the speed-torque characteristics. At low speeds, the current induced in the squirrel cage is nearly at line frequency and tends to be in the outer parts of the rotor cage. As the motor accelerates, the slip frequency becomes lower, and more current is in the interior of the winding. By shaping the bars to change the resistance of the winding portions in the interior and outer parts of the cage, effectively a variable resistance is inserted in the rotor circuit. However, the majority of such motors have uniform bars (Gorginpour, Oraee, & McMahon, 2013).

In a wound rotor induction motor, the rotor winding is made of many turns of insulated wire and is connected to slip rings on the motor shaft. An external resistor or other control devices can be connected in the rotor circuit. Resistors allow control of the motor speed, although significant power is dissipated in the external resistance.

Application Example: Air Conditioning

In an air conditioner, the cooling is done using a simple refrigeration cycle. The most common refrigeration cycle uses an electric motor to drive a compressor. Figure 67 shows application of induction motor in an air conditioning system.

Application Example: Air Compressor

An air compressor is a device that converts power into kinetic energy by compressing and pressurizing air, which, on command, can be released in quick bursts. Most air compressors either are reciprocating piston type, rotary vane or rotary screw. All of them need electric motors for operating.

Application Example: Metalworking

In metalworking, rolling is a metal forming process in which metal stock is passed through one or more pairs of rolls to reduce the thickness and to make the thickness uniform. Figure 68 shows a simplified diagram of a metal forming system.

Application Example: Conveyor System

A conveyor system is a common piece of mechanical handling equipment that moves materials from one location to another. Conveyor systems are commonly used in many industries, including the automotive, agricultural, computer, electronic, food processing, aerospace, pharmaceutical, chemical, bottling and canning, print finishing and packaging.

Figure 67. Induction motors used in an air conditioning cooling tower

Figure 68. A schematic diagram of a metal forming system

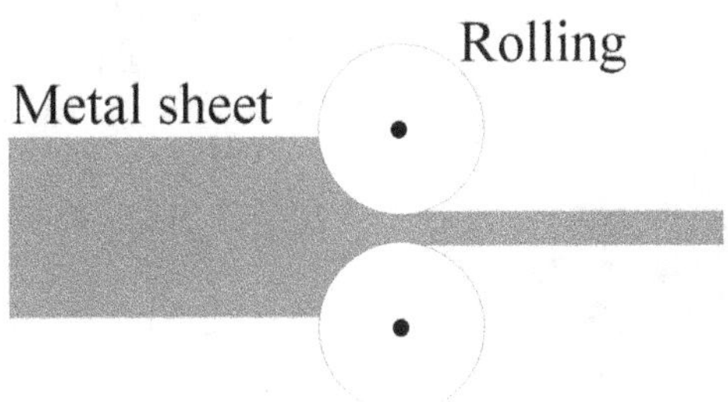

Application Example: Pump

Pumps are used throughout society for a variety of purposes. Early applications include the use of the windmill or watermill to pump water. Figure 69 shows a typical pump and its motor.

Application Example: Mechanical Fan

Most fans are powered by electric motors, but other sources of power may be used, including hydraulic motors and internal combustion engines. Fans produce air flows with high volume and low pressure, as opposed to compressors which produce high pressures at a comparatively low volume. The motor is either hidden in the fan's center hub or extends behind it. For big industrial fans, three-phase asynchronous motors are commonly used, placed near the fan and driving it through a belt and pulleys. Smaller fans are often powered by shaded pole AC motors or brushed or brushless DC motors (Sorrentino, & Fernández, 2011). Cooling fans for computer equipment always use brushless DC motors, which generate much less electromagnetic interference than other types. Figure 70 shows application of a single phase induction motor for cooling fan.

Figure 69. Application of an induction motor in a water pump

Figure 70. Application of a single phase induction motor for cooling fan

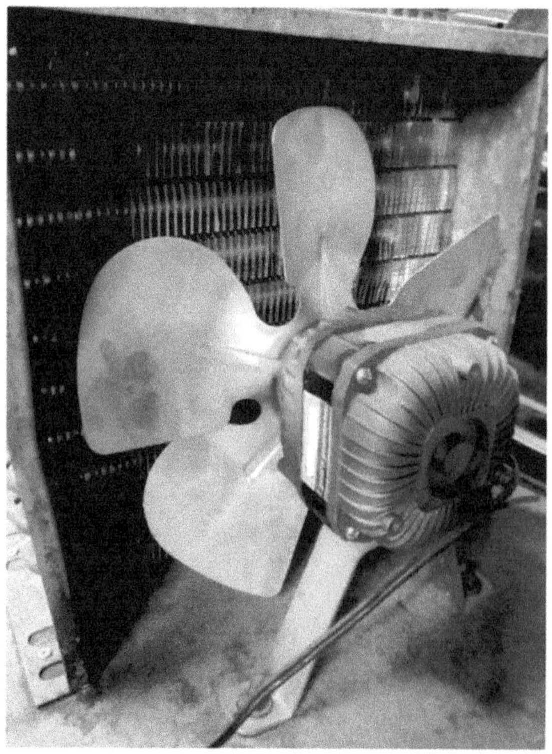

Figure 71. Induction generators used in a wind farm, Source: IRAN SANA (with permission)

Application Example: Wind Energy

The result of over a millennium of windmill development and modern engineering, today's wind turbines are manufactured in a wide range (Tohidi, Zolghadri, Oraee, & Oraee, 2012). Arrays of large turbines, known as wind farms, are becoming an increasingly important source of renewable energy and are used by many countries as part of a strategy to reduce their reliance on fossil fuels. Figure 71 shows a wind farm in Neyshaboor, Iran.

AC Synchronous Machines

A synchronous electric motor is an AC motor distinguished by a rotor spinning with coils passing magnets at the same rate as the AC and resulting magnetic field which drives it. Slip rings and brushes are used to conduct current to the rotor.

Brushless DC motors, stepper and switched reluctance motors are the other various types of this machine. Switched Reluctance Generators (Nasirian,, Kaboli,, Davoudi, & Moayedi, 2013) have emerged as suitable alternatives for permanent magnet and induction machines in renewable energy-related industries, and can be used in wind turbine generators, electric and hybrid electric vehicles, and aeronautics (Bilgin, Emadi, & Krishnamurthy, 2013). Figures 72, 73, 74, 75, and 76 show basic parts of a 6/4 switched reluctance motor as well as its driver circuit (Nasirian,, Kaboli, & Davoudi, 2013).

Equivalent Circuit and Characteristics

The simple equivalent circuit corresponding to steady-state operation of the round rotor synchronous machine and its associated phasor diagram are shown in Figure 77 (Rahman, & Hiti, 2005).

One of the most important relations of this machine is its power-angle relation:

$$P = 3\frac{EV}{X}\sin\delta \tag{12}$$

Figure 72. Cross section of a typical three-phase 6/4 SRG

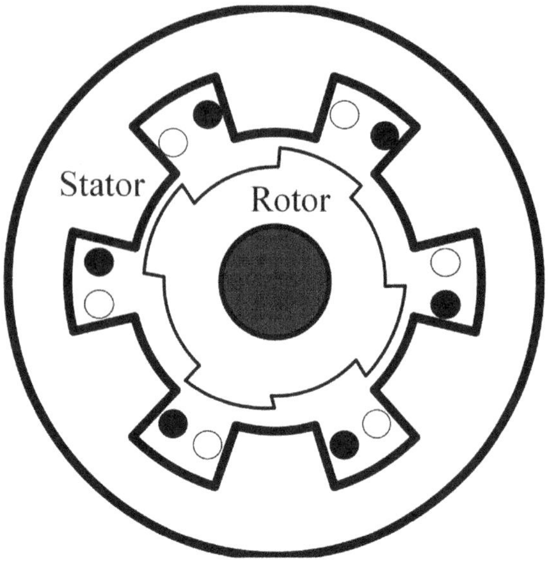

Figure 73. Circuit diagram of a driver of each windings of SRM

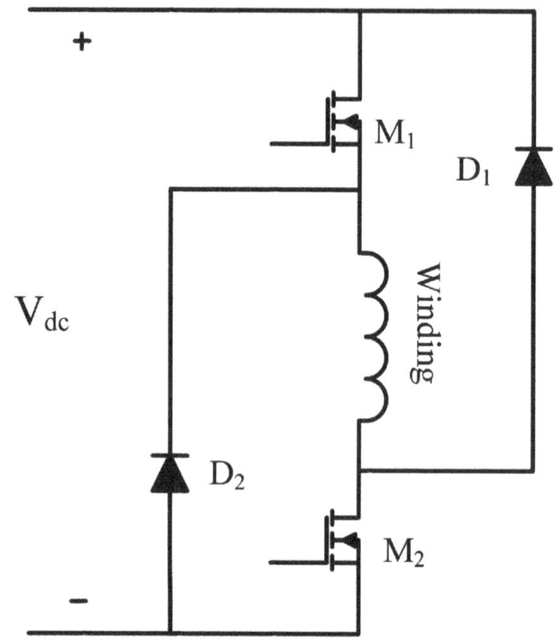

Figure 74. Circuit diagram of a completed driver of SRM

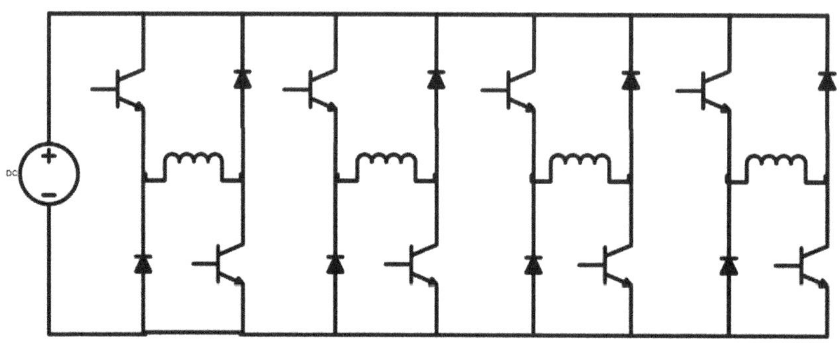

$$Q = 3\frac{E(V_2 \cos\delta - E)}{X} \tag{13}$$

Where P and Q are real and reactive power, respectively

If the load exceeds beyond this limitation, the machine becomes instable and this is an important problem of this type of electrical machine.

Features

In the fractional horsepower range, most synchronous motors are used where precise constant speed is required. In high-horsepower industrial sizes, the synchronous motor provides two important functions.

Figure 75. Various parts of a servo system based on SRM

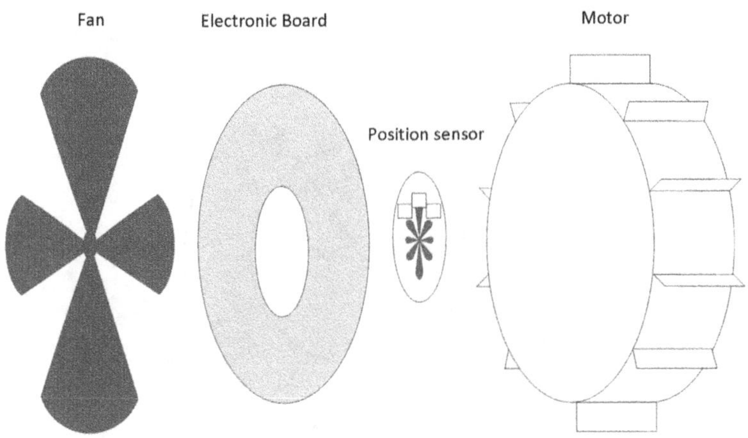

Figure 76. A real SRM driver

First, it is a highly efficient means of converting AC energy to work. Second, it can operate at leading or unity power factor and thereby provide power-factor correction (Pellegrino, Vagati, Boazzo, & Guglielmi, 2012). Some of other features of these machines are listed in the following.

- The best ratio of power to volume
- Excellent control capability among AC motors
- Stable voltage frequency in generator mode
- Generation of capacitive reactive power

Figures 78, 79, and 80 show the characteristics of a synchronous machine.

Figure 77. Basic structure and equivalent circuit diagram of a synchronous machine

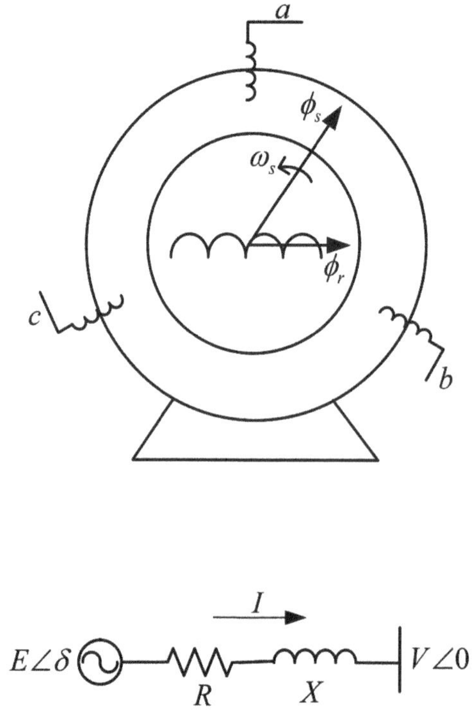

Figure 78. Inductive and capacitive regions of a synchronous motor

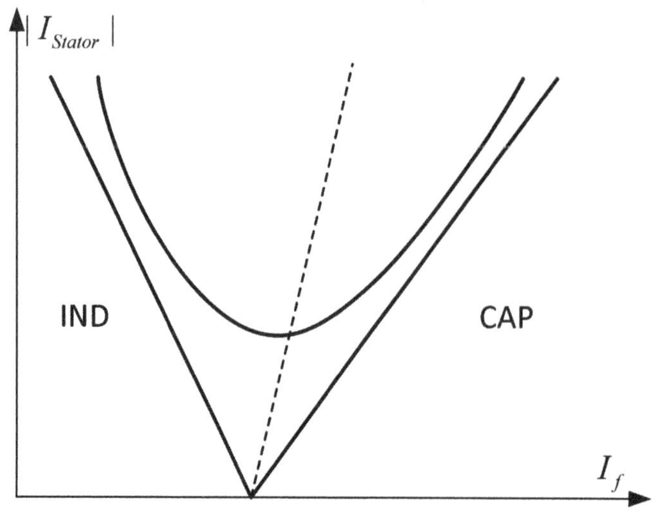

Figure 79. A real V curve of a synchronous motor

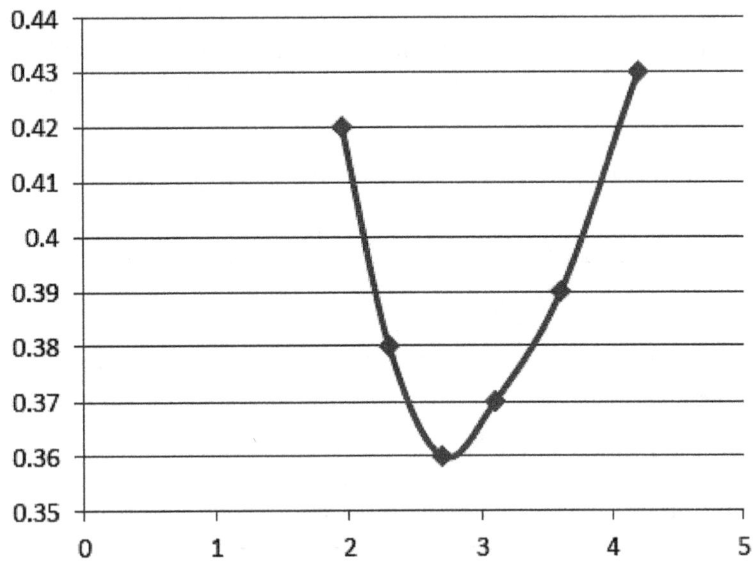

Figure 80. Effect of output power on real V curve of a synchronous motor

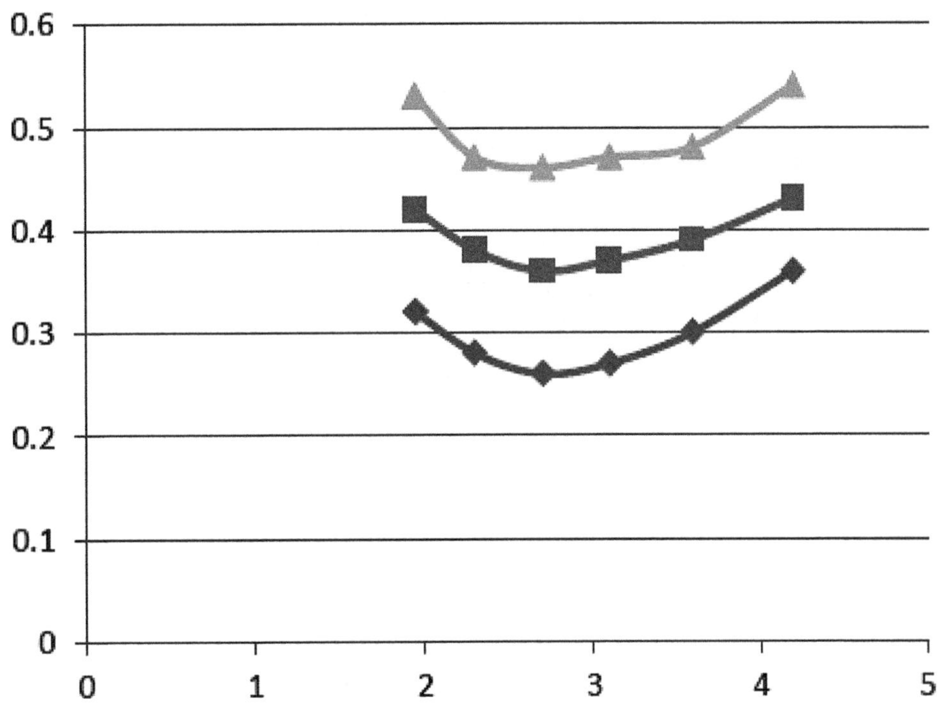

Application Example: Power Stations

In electricity generation, synchronous generators provide nearly all of the power for electric power grids. Figures 81, 82, and 83 show output characteristic of a synchronous generator for various loads.

Figure 81. A real V-I characteristic of a synchronous generator with resistive load

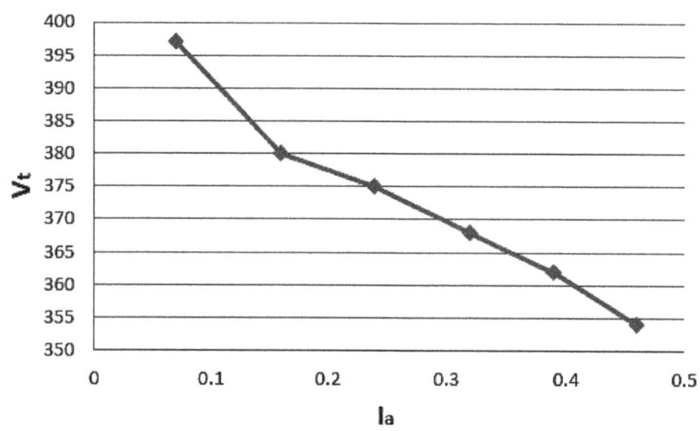

Figure 82. A real V-I characteristic of a synchronous generator with inductive load

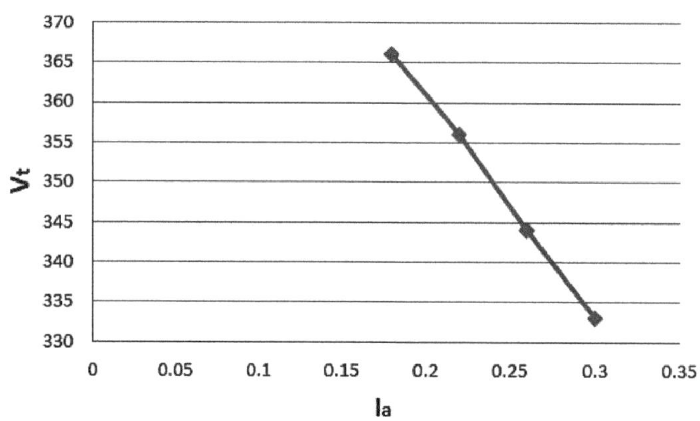

Figure 83. A real V-I characteristic of a synchronous generator with capacitive load

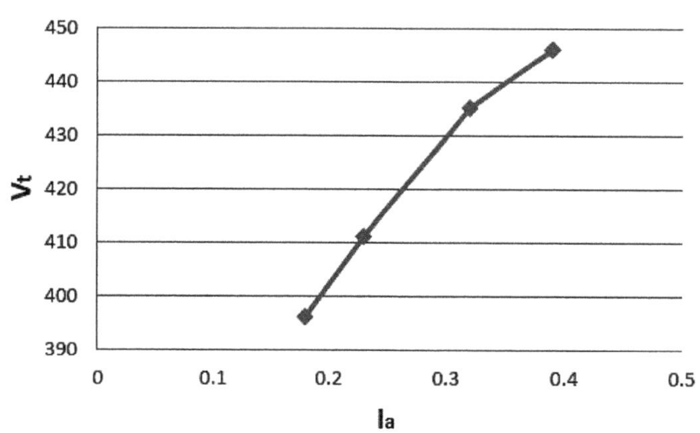

Application Example: Electric Vehicle

Plug-in cars have several benefits compared to conventional internal combustion engine vehicles. They have lower operating and maintenance costs, and produce little or no local air pollution. They reduce dependence on petroleum and may reduce greenhouse gas emissions from the onboard source of power, depending on the fuel and technology used for electricity generation to charge the batteries.

Application Example: Rotary Power Converters

Figures 84 and 85 show a rotary power converters consists of a coupled motor-generator with railway applications. This converter is replaced with static power electronic converters today.

Figure 84. A rotary inverter consists of a mechanically coupled motor and generator

Figure 85. A rotary power converter used a UPS: DC motor supplied with batteries (right), three phase AC generator used as AC power supply (left)

Transformers

A transformer is an electrical device that transfers energy between two circuits through electromagnetic induction. A transformer may be used as a safe and efficient voltage converter to change the AC voltage at its input to a higher or lower voltage at its output. In a transformer, the windings are wound around a common core to provide tight electromagnetic coupling between the windings as shown in Figures 86 and 87. The core material is often a laminated iron core (Jassal, Polinder, & Ferreira, 2012).

Figure 86. A single phase transformer, Source: Seraj Co. (with permission)

Figure 87. Primary and secondary windings of a transformer magnetically coupled with core

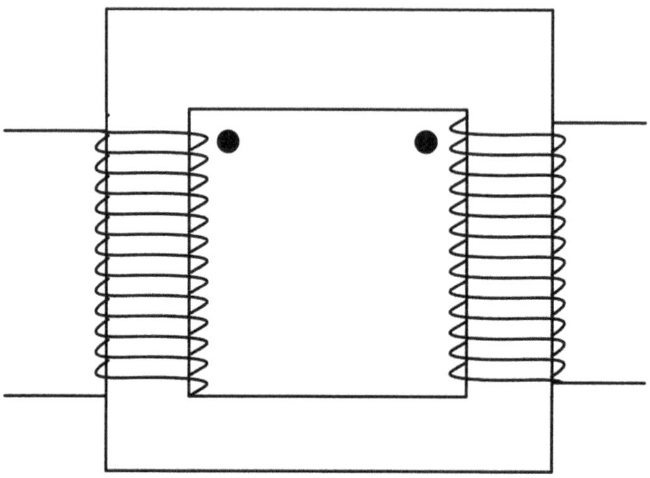

Equivalent Circuit

Winding losses and leakage reactances are represented by the series loop impedances of the model. Core loss and reactance is represented by the shunt leg impedances of the model. Figures 88, 89, 90, and 91 show its equivalent circuit.

Figure 88. Opposite polarities of primary and secondary winding current

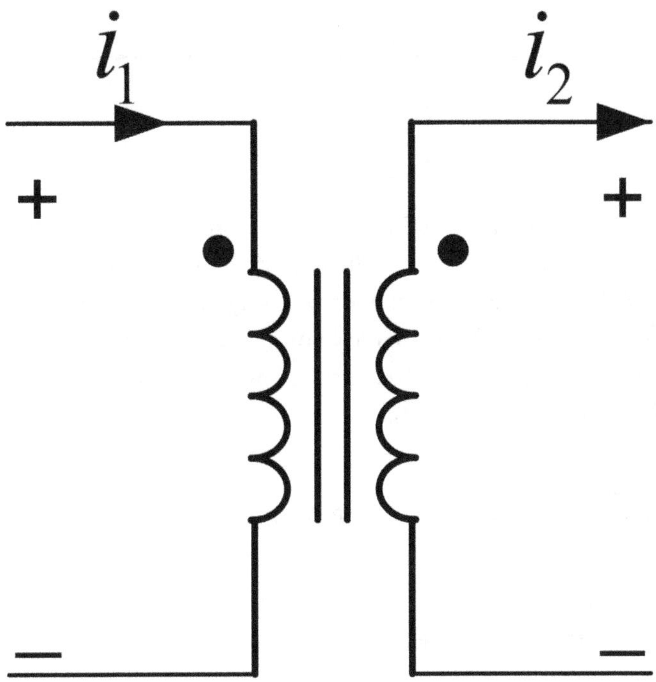

Figure 89. Typical B-H characteristic of an Iron sheet

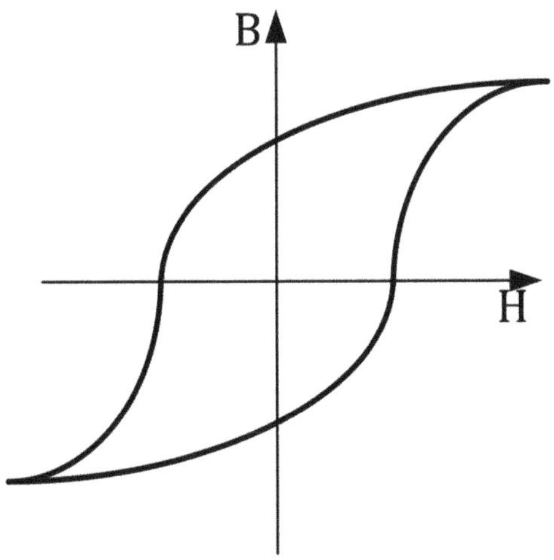

Figure 90. Relative direction of magnetic and electric field

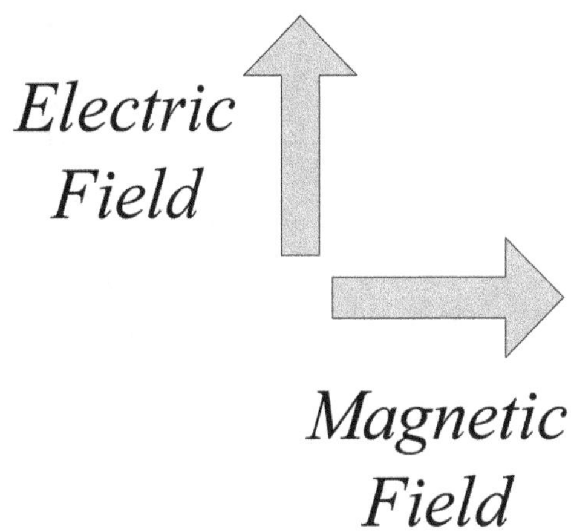

Figure 91. Equivalent circuit of a real transformer

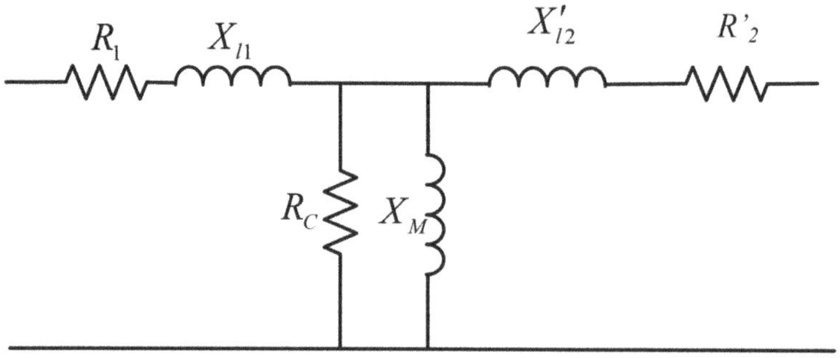

Various Types

Transformers can step up voltage and step down current from generators to high voltage transmission lines, and step down voltage/step up current to local distribution circuits or industrial customers. The step-up transformer is used to increase the secondary voltage relative to the primary voltage. The step-down transformer is used to decrease the secondary voltage relative to the primary voltage. Transformer can be single and three phase, high and low frequency. Three phase transformer and an autotransformer is shown in Figures 92, 93, 94, 95, 96, and 97.

Figure 92. Phasor diagram of a three phase transformer

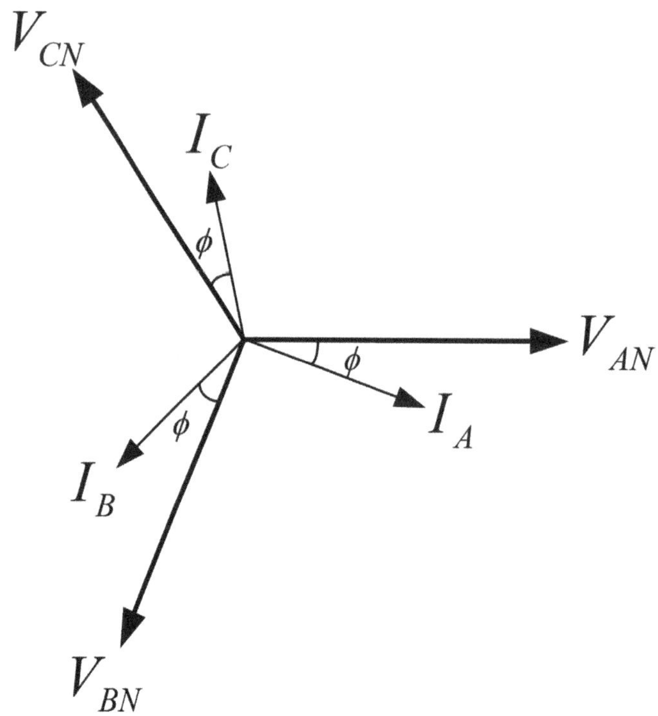

Figure 93. A three phase transformer consists of three single phase transformers

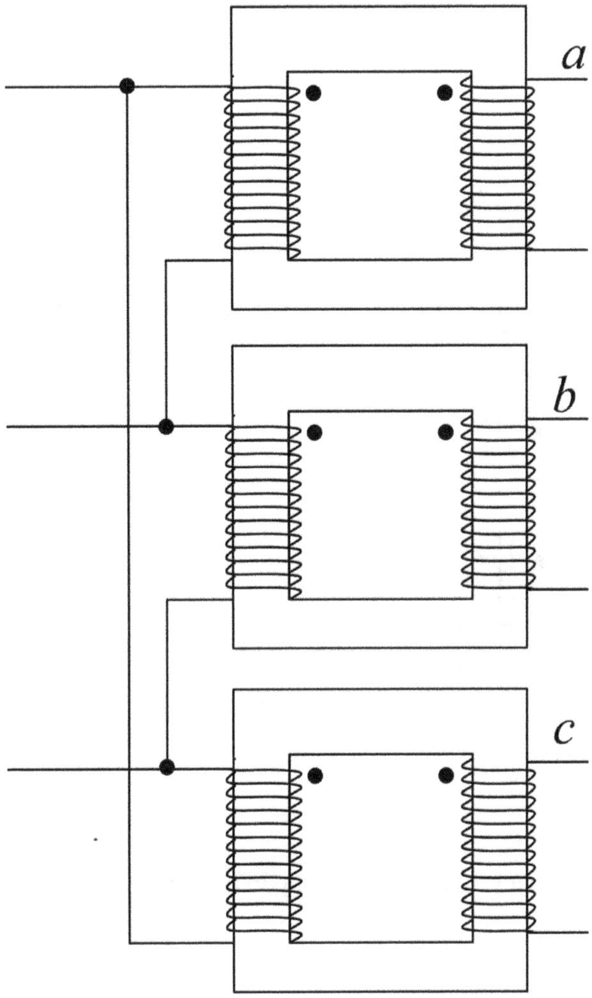

Figure 94. A three phase transformer, Source: Seraj Co. (with permission)

Figure 95. relation between line and phase voltages in a Y connected three phase transformer

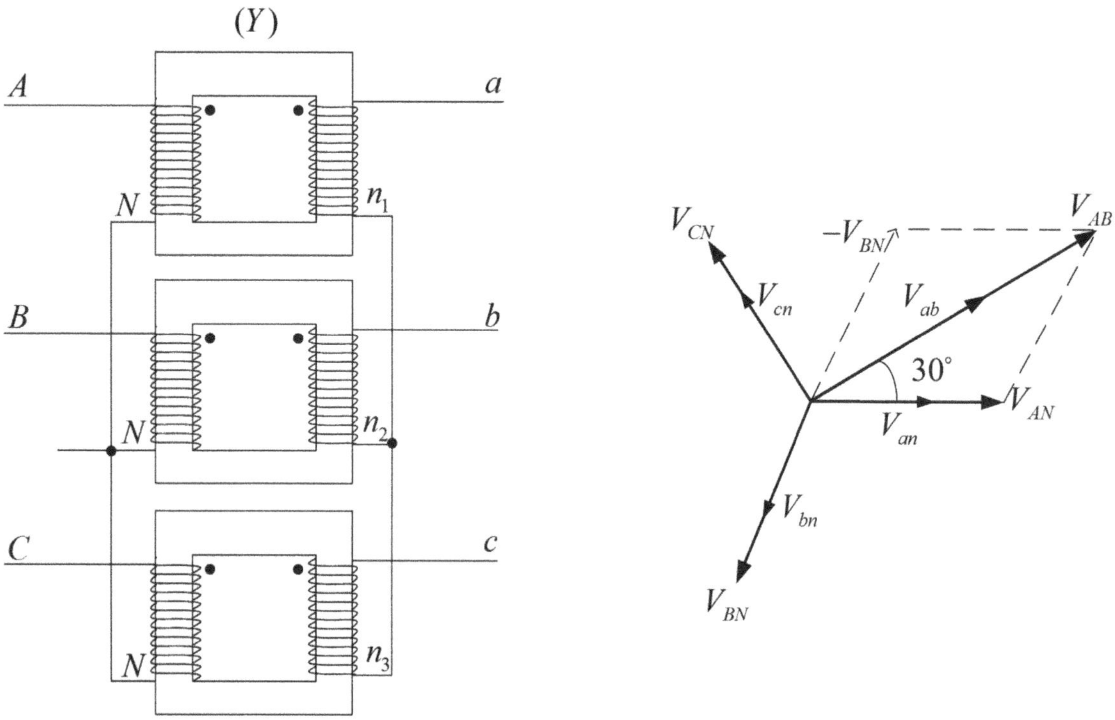

Features

Lack of any rotating element leads to high efficiency in transformers.

Application Example: Power Transmission

Transformers are used to increase voltage before transmitting electrical energy over long distances through wires. Wires have resistance which loses energy through joule heating at a rate corresponding to square of the current. By transforming power to a higher voltage transformers enable economical transmission of power and distribution. Consequently, transformers have shaped the electricity supply industry, permitting generation to be located remotely from points of demand.

Application Example: Impedance Heating

Impedance heating is perfectly suited for pipe maintenance temperatures. The pipe becomes the heater when low AC voltage is applied to it. The applied voltage is very low. The pipe is the heater and therefore, there is no heater to burn out. Figures 98 and 99 show a simplified diagram of a heating system and its performance.

Figure 96. Three phase transformer: Conceptual connection (up), Simplified diagram (middle), Three winding type (down)

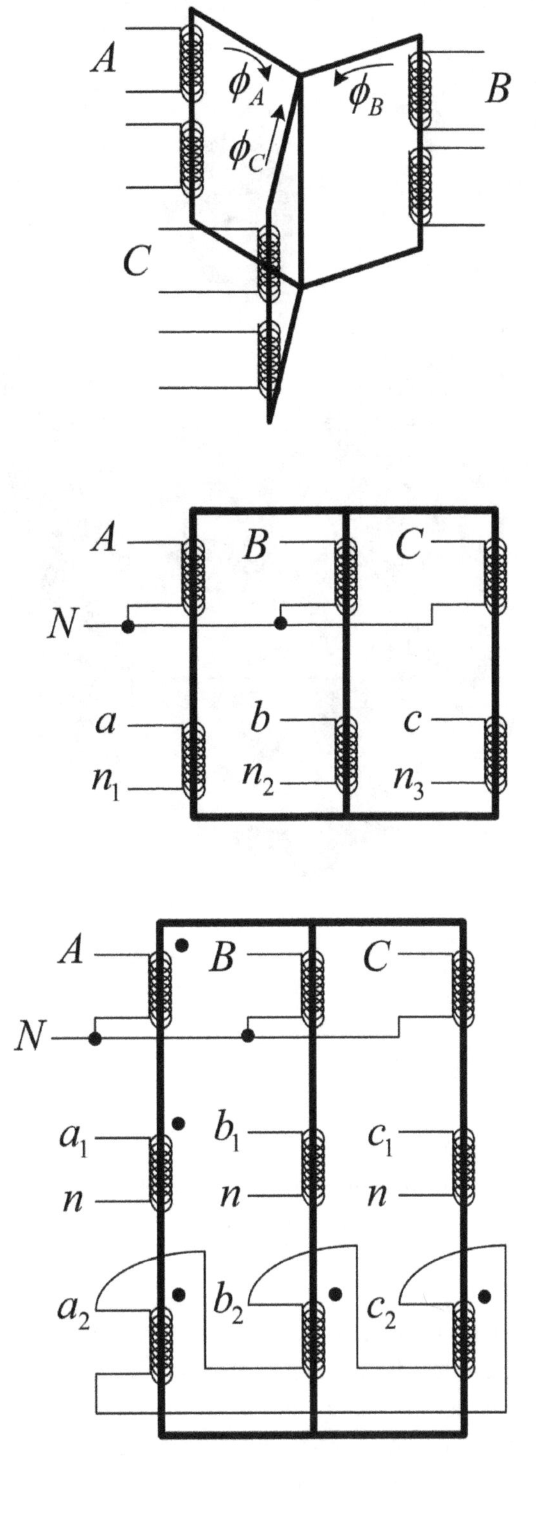

Figure 97. A three phase autotransformer

SUMMARY AND CONCLUSION

In this chapter, a brief introduction about importance of power electronic converters and electrical machines was presented. Operation principles of these systems were described. This chapter is used to support the next chapter dealing with failure mechanisms in these equipments. The results of this chapter are summarized as follows.

1. Power electronic converters are electronic circuits used for power conditioning. They may be named as static power converters because they do not include any rotary part. Power conditioning is performed to achieve higher efficiency and better performance in electrical energy conversion process. Power conditioning can be carried out with linear or switching electronic circuits. The main category of power electronic converters utilize solid state switches for various applications. These switches are the key parts of the converters for proper operation. As the power loss of the

Figure 98. Application of transformer in pipeline impedance heating

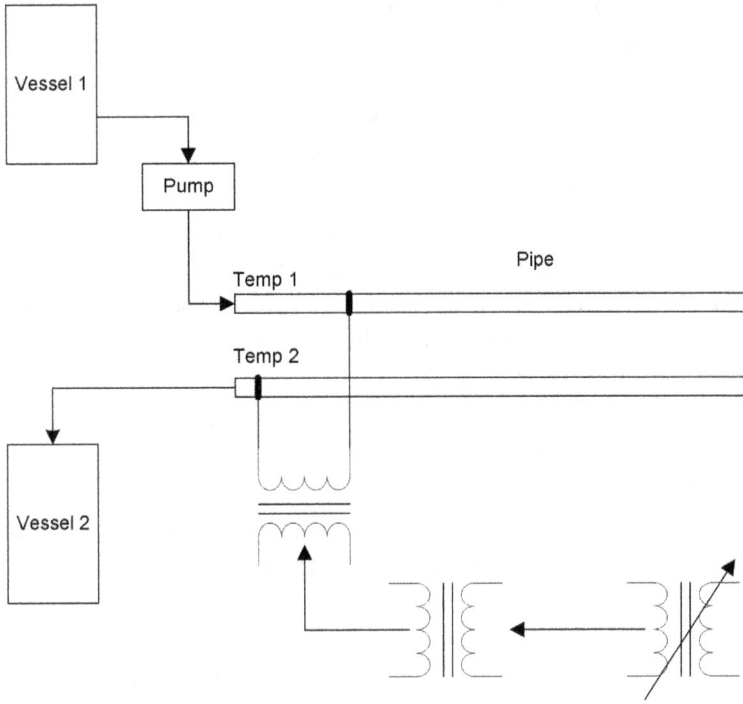

Figure 99. Temperature profile in pipeline impedance heating

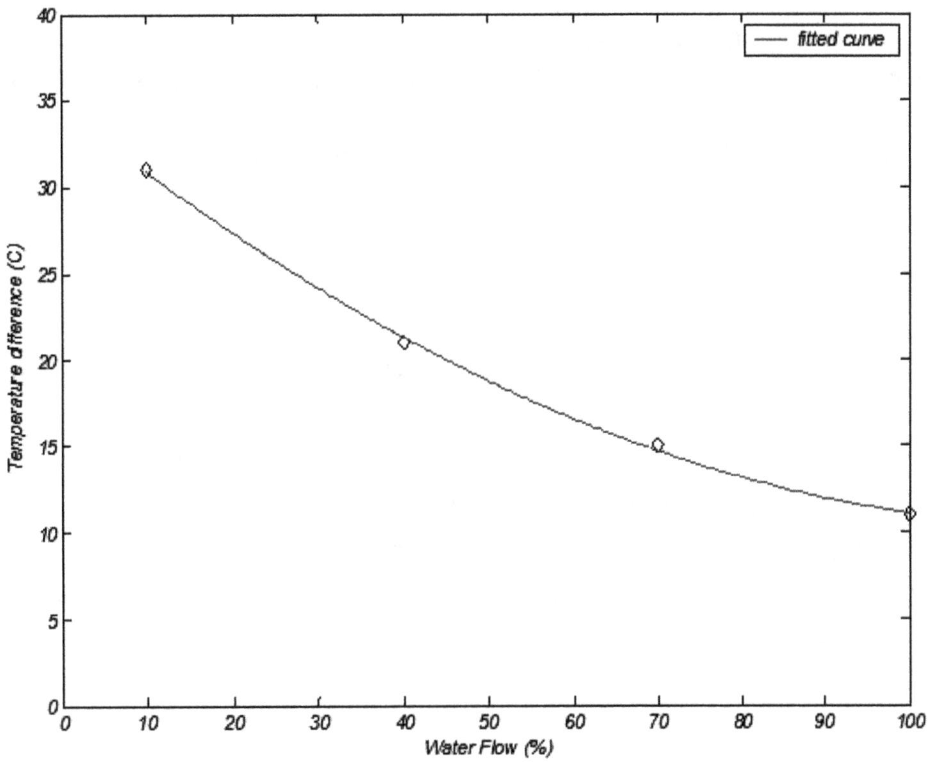

switch is very low, these types of power electronic converters have high efficiency. It is important today because loss minimization is a well-accepted method for reducing the electric energy demand. Four different types of switching power electronic converters are used in conversion of AC to DC, AC to AC, DC to AC and DC to DC. All of these converters can be isolated by transformers or are not isolated. Some power electronic converters are faced with power system from input terminal (for example: rectifiers) or output terminals (for example: inverters). Thus, they are affected by problems occurring in the power system.

2. Rotary electrical machines are divided into two main categories: AC machines and DC machines. DC machines were historically used earlier than AC machines. They show better controlling performance and are widely used in precise control processes like robots. There are two main types of AC machines: synchronous and asynchronous machines. Synchronous generators are the essential part in electrical networks for generating electric power. Synchronous motors as BLDC and SRM find wide usage during recent years for their excellent performance. Electrical machines have a more mechanical structure than power electronic converters.
3. The main difference between static converters and electrical machines as rotary converters from reliability point of view is the mass of active parts associated with power conditioning. This leads to better performance of rotary power converters in harsh environment.

The last result of the current chapter implies the necessity for including both static and rotary electric power converters in a single book. Because they are complement from reliability point of view. The static power converters have high performance and long life in normal condition. But they are very sensitive to harsh transient conditions. On the other hand, rotary converters have better stability in unwanted conditions. We could add some other reasons for necessity to include both static and rotary converters in this book. As a difference, rotary power converters deal with electromechanical power conversion while static converters are fully electrical circuits. Thus, mechanical functions of rotary converters can not be carried out by static converters. As a similarity, power electronic converters are merged to rotary machine todays.

In the next chapter, we describe the fault mechanisms in power converters based on the presented background in chapter 1.

REFERENCES

Abdi, E., McMahon, R., Malliband, P., Shao, Mathekga, M.E., Tavner, P., … Tatlow, M. (2013). Performance analysis and testing of a 250 kW medium-speed brushless doubly-fed induction generator. *IET Renewable Power Generation, 7*(6), 63638.

Abusara, M. A., & Sharkh, S. M. (2013). Design and Control of a Grid-Connected Interleaved Inverter. *IEEE Transactions on Power Electronics*, *28*(2), 748–764. doi:10.1109/TPEL.2012.2201505

Barati, F., & Oraee, H. (2010). Vector model utilization for nested-loop rotor Brushless Doubly-Fed Machine analysis, control and simulation. In *Proceedings of 1st Power Electronic & Drive Systems & Technologies Conference* (pp. 295–301). IEEE. doi:10.1109/PEDSTC.2010.5471806

Barati, F., Shao, S., Abdi, E., Oraee, H., & McMahon, R. (2011). Generalized Vector Model for the Brushless Doubly-Fed Machine With a Nested-Loop Rotor. *IEEE Transactions on Industrial Electronics*, *58*(6), 2313–2321. doi:10.1109/TIE.2010.2064279

Bilgin, B., Emadi, A., & Krishnamurthy, M. (2013). Comprehensive Evaluation of the Dynamic Performance of a 6/10 SRM for Traction Application in PHEVs. *IEEE Transactions on Industrial Electronics*, *60*(7), 2564–2575. doi:10.1109/TIE.2012.2196015

Boglietti, A., El-Refaie, A. M., Drubel, O., Omekanda, A. M., Bianchi, N., Agamloh, E. B., & Bartolo, J. B. et al. (2014). Electrical Machine Topologies: Hottest Topics in the Electrical Machine Research Community. *IEEE Industrial Electronics Magazine*, *8*(2), 18–30. doi:10.1109/MIE.2013.2294077

Cipriano, E., Jacobina, C. B., da Silva, E. R. C., & Rocha, N. (2012). Single-Phase to Three-Phase Power Converters: State of the Art. *IEEE Transactions on Power Electronics*, *27*(5), 2437–2452. doi:10.1109/TPEL.2011.2175751

Collins, E. R. (1992). Torque and slip behavior of single-phase induction motors driven from variable-frequency supplies. *IEEE Transactions on Industry Applications*, *28*(3), 710–715. doi:10.1109/28.137461

Gerada, D., Mebarki, A., Brown, N. L., Gerada, C., Cavagnino, A., & Boglietti, A. (2014). High-Speed Electrical Machines: Technologies, Trends, and Developments. *IEEE Transactions on Industrial Electronics*, *61*(6), 2946–2959. doi:10.1109/TIE.2013.2286777

Gorginpour, H., Jandaghi, B., Oraee, H., & Abdi, E. (2014). Magnetic equivalent circuit modelling of brushless doubly-fed induction generator. *IET Renewable Power Generation*, *8*(3), 334–346. doi:10.1049/iet-rpg.2012.0383

Gorginpour, H., Oraee, H., & McMahon, R. A. (2013). A Novel Modeling Approach for Design Studies of Brushless Doubly Fed Induction Generator Based on Magnetic Equivalent Circuit. *IEEE Transactions on Energy Conversion*, *28*(4), 902–912. doi:10.1109/TEC.2013.2278486

Hamzeh, M., Karimi, Y., Asadi, E., & Oraee, H. (2014). Design and implementation of a single phase grid-connected PV inverter with a new active damping strategy. In *Proceedings of 5th Power Electronics, Drive Systems and Technologies Conference* (pp.72-77). IEEE doi:10.1109/PEDSTC.2014.6799347

Jang, D. H. (2013). Problems Incurred in a Vector-Controlled Single-Phase Induction Motor, and a Proposal for a Vector-Controlled Two-Phase Induction Motor as a Replacement. *IEEE Transactions on Power Electronics*, *28*(1), 526–536. doi:10.1109/TPEL.2012.2199772

Jassal, A., Polinder, H., & Ferreira, J. A. (2012). Literature survey of eddy-current loss analysis in rotating electrical machines. *IET Electric Power Applications*, *6*(9), 743–752. doi:10.1049/iet-epa.2011.0335

Kaboli, S., Moayedi, A., & Oraee, H. (2008). Application of Random PWM technique for reducing the high frequency harmonics in class-D amplifier. In *Proceedings of 4th IET Conference on Power Electronics, Machines and Drives* (pp.406-410). IEEE doi:10.1049/cp:20080553

Kolar, J. W., & Friedli, T. (2013). The Essence of Three-Phase PFC Rectifier Systems - Part I. *IEEE Transactions on Power Electronics*, *28*(1), 176–198. doi:10.1109/TPEL.2012.2197867

Levi, E., Bojoi, R., Profumo, F., Toliyat, H. A., & Williamson, S. (2007). Multiphase induction motor drives - a technology status review. *IET Electric Power Applications*, *1*(4), 489–516. doi:10.1049/iet-epa:20060342

Motor Reliability Working Group. (1985). Report of Large Motor Reliability Survey of Industrial and Commercial Installations, Part I. *IEEE Transactions on Industry Applications*, *21*(4), 853–864.

Nag, S. S., & Mishra, S. (2014). Current-Fed Switched Inverter. *IEEE Transactions on Industrial Electronics*, *61*(9), 4680–4690. doi:10.1109/TIE.2013.2289907

Nasirian, V., Kaboli, S., & Davoudi, A. (2013). Output Power Maximization and Optimal Symmetric Freewheeling Excitation for Switched Reluctance Generators. *IEEE Transactions on Industry Applications*, *49*(3), 1031042. doi:10.1109/TIA.2013.2253438

Nasirian, V., Kaboli, S., Davoudi, A., & Moayedi, S. (2013). High-Fidelity Magnetic Characterization and Analytical Model Development for Switched Reluctance Machines. *IEEE Transactions on Magnetics*, *49*(4), 1505–1515. doi:10.1109/TMAG.2012.2222427

Oraee, A., Abdi, E., Abdi, S., McMahon, R., & Tavner, P. J. (2015). Effects of Rotor Winding Structure on the BDFM Equivalent Circuit Parameters. *IEEE Transactions on Energy Conversion*, *30*(4), 1660–1669. doi:10.1109/TEC.2015.2432272

Pellegrino, G., Vagati, A., Boazzo, B., & Guglielmi, P. (2012). Comparison of Induction and PM Synchronous Motor Drives for EV Application Including Design Examples. *IEEE Transactions on Industry Applications*, *48*(6), 2322–2332. doi:10.1109/TIA.2012.2227092

Rahman, K. M., & Hiti, S. (2005). Identification of machine parameters of a synchronous motor. *IEEE Transactions on Industry Applications*, *41*(2), 557–565. doi:10.1109/TIA.2005.844379

Song, Y., & Wang, B. (2013). Survey on Reliability of Power Electronic Systems. *IEEE Transactions on Power Electronics*, *28*(1), 59604. doi:10.1109/TPEL.2012.2192503

Sorrentino, E., & Fernández, S. (2011). Comparison of six steady-state models for single-phase induction motors. *IET Electric Power Applications*, *5*(8), 61617. doi:10.1049/iet-epa.2011.0001

Tohidi, S., Oraee, H., Zolghadri, M. R., Shiyi, Sh., & Tavner, P. (2013). Analysis and Enhancement of Low-Voltage Ride-Through Capability of Brushless Doubly Fed Induction Generator. *IEEE Transactions on Industrial Electronics*, *60*(3), 1146–1155. doi:10.1109/TIE.2012.2190955

Tohidi, S., Zolghadri, M. R., Oraee, H., & Oraee, A. (2012). Dynamic modeling of a wind turbine with brushless doubly fed induction generator. In *Proceedings of 3rd Power Electronics and Drive Systems Technology* (pp. 490–494). IEEE. doi:10.1109/PEDSTC.2012.6183379

Wang, X., Zhong, H., Yang, Y., & Mu, X. (2010). Study of a Novel Energy Efficient Single-Phase Induction Motor with Three Series-Connected Windings and Two Capacitors. *IEEE Transactions on Energy Conversion*, *25*(2), 433–440. doi:10.1109/TEC.2009.2039218

Yang, Sh., Bryant, A., Mawby, P., Xiang, D., Ran, L., & Tavner, P. (2011). An Industry-Based Survey of Reliability in Power Electronic Converters. *IEEE Transactions on Industry Applications*, *47*(3), 1441451.

ADDITIONAL READING

Baliga, B. J. (2002). *Fundamentals of Power Semiconductor Devices Springer Science & Business Media. Berlin, Germany*. AP: Technology &Engineering.

Brown, R. E. (2009). *Electric Power Distribution Reliability*. Florida, USA: CRC.

Erickson, R. W. (2001). *Fundamentals of Power Electronics*. NJ, USA: Kluwer Academic. doi:10.1007/b100747

Sheng, W. W., & Colino, R. P. (2005). *Power Electronic Modules*. Florida, USA: CRC.

Strzelecki, R., & Benysek, G. (2008). *Power Electronics in Smart Electrical Energy Networks*. Berlin, Germany: Springer. doi:10.1007/978-1-84800-318-7

KEY TERMS AND DEFINITIONS

Electric Generator: Equipment which is used to convert electrical energy to mechanical energy (reverse operation of a motor).

Electric Motor: Equipment which is used to convert electrical energy to mechanical energy.

Power Electronics: Application of solid state switches for power conditioning.

Reliability: A probabilistic expression for normal operation of any system.

Solid State Power Switch: A solid state device with high capability for blocking high voltage and carrying high current.

Static Power Converter: A name used for power electronic converters.

Chapter 2
Fault Mechanism

ABSTRACT

Static and rotary electric power converters have different structures but there are common failure factors in both of them. In this chapter, causes of failure in electric power converters is described. All of the failure factors which are described in this chapter are catastrophic factors and lead to destructive damage in the systems. Other types of failure without destructive effect on converter like electromagnetic interference will be presented in the next chapters. All descriptions are based on details of operation of the converters which were presented in the previous chapter. Over temperature, over voltage, mechanical forces and environmental effects like humidity are the main factors of failure in systems. Origins of these factors are described in this chapter. Over temperature is a special factor among them because other failure factors finally act as over temperature in failure process of the converters. Since over temperature is the main failure factor in electric power converters, loss model of components in electric power converters are presented in details. In addition, practical technique for measuring power loss is described. Sample industrial examples of damaged equipments due to these failure factors are shown to give a real sense to reader about failure results.

INTRODUCTION: FAILURE OF ELECTRIC POWER CONVERTERS

In this chapter, we present the failure mechanism in electric power converter based on the background of Chapter 1. Figure 1 shows the state of Chapter 2 in the flowchart of the book.

CATASTROPHIC FAILURE

All of investigated failure in this chapter are catastrophic type and lead to damage of converter. There are other types of failure that does not damage the converter are studied in the next chapters as availability concept. A catastrophic failure is a sudden and total failure from which recovery is impossible. Catastrophic failures often lead to cascading systems failure (Tohidi, Zolghadri, Oraee, Tavner, Abdi, & Logan, 2012). The term is most commonly used for structural failures, but has often been extended

DOI: 10.4018/978-1-4666-9429-3.ch002

Figure 1. State of chapter 2 in the flowchart of the book

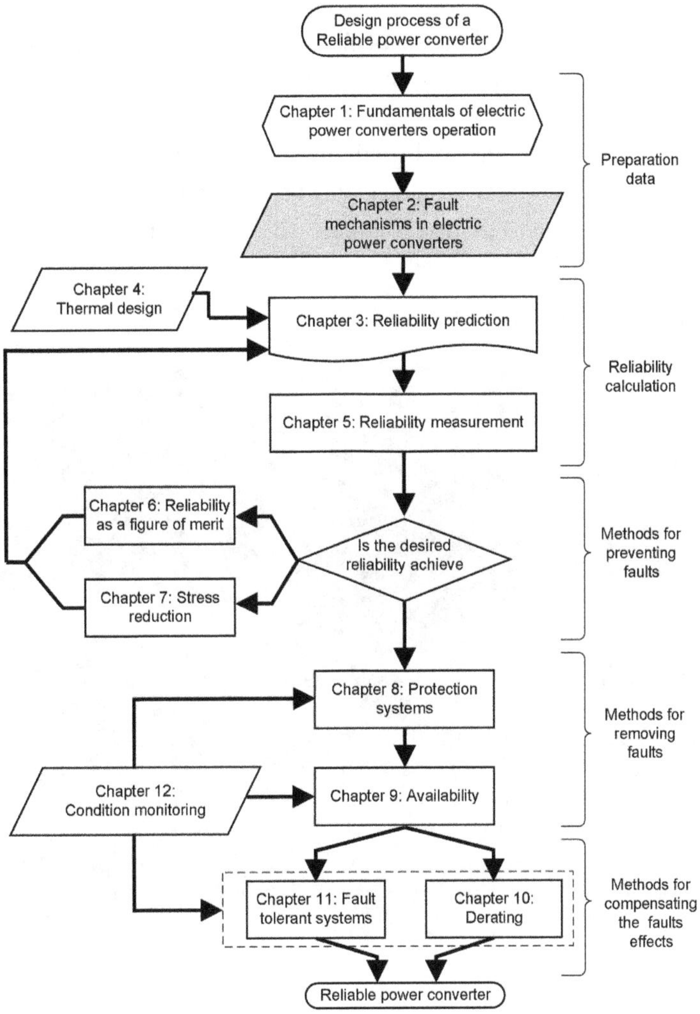

to many other disciplines where total and irrecoverable loss occurs. Such failures are investigated using the methods of forensic engineering (Abdi, Abdi, Oraee, & McMahon, 2015), which aims to isolate the cause or causes of failure.

FAILURE FACTORS

Meaning of failure in this chapter is the type of failure that leads to damage in a converter. Based on this view, four main reasons of failure are:

- Thermal shock
- Over voltage
- Mechanical forces
- Environmental effects.

THERMAL SHOCK

Over temperature is the most important factor of failure in all of systems. Thermal damage is a very important factor in the fault of electric power converters. Figures 2 and 3 show sample damaged converters due to over temperature. There are two scenarios for thermal damage: over temperature and thermal shock. Thermal shock occurs when a thermal gradient causes different parts of an object to expand by different amounts. At some point, this stress can exceed the strength of the material, causing a crack to form. To investigate the failure due to thermal damage, first we studied the origins of heat generation.

Figure 2. A thermal damage in an electronic board

Figure 3. Thermal damage in an integrated circuit

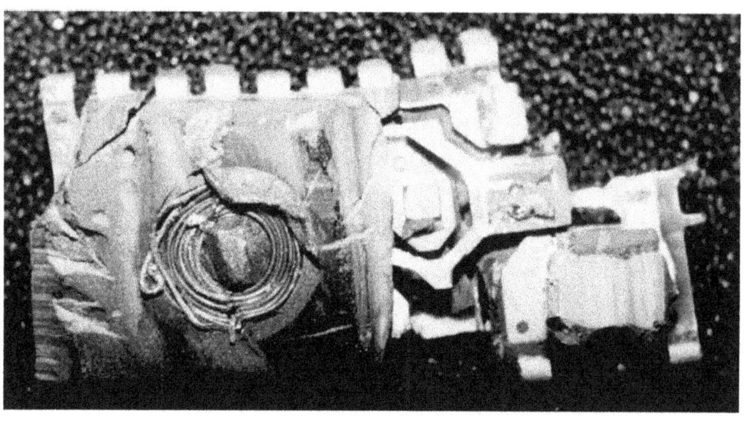

Electric Power Losses in Power Electronics

A power electronic converter consists of solid state active devices and some passive devices. Therefore, the power loss of a power electronic converter can be listed as follows (Yuancheng, Ming, Zhou, & Lee, 2006).

- Power losses in semiconductor devices
- Losses in passive devices

Conduction Losses

Every solid state device is a resistive element, so it dissipates power as current is conducted through the device. For the MOSFETs, these conduction losses are inversely proportional to the size of the MOSFET; the larger the switching transistor, the lower its on resistance and Figure 4 shows effect of temperature rise on the conduction voltage drop of some switches. It can be seen that the on-state voltage drop of some solid state switches increases with temperature and it leads to increase the conduction losses of the switches. Figure 5 shows effect of temperature on conduction area of solid state switches characteristic.

Switching Losses

Another term of losses in solid state devices is switching losses. Switching losses are created as a result of a simultaneous high voltage and current in a solid state device during a transition between the open and closed states.

Figure 4. Effect of temperature on conduction resistance of solid state switches, (a): a MOSFET, Published with permission of International Rectifier, (b): an IGBT, Source: ON Semiconductor Co. (with permission)

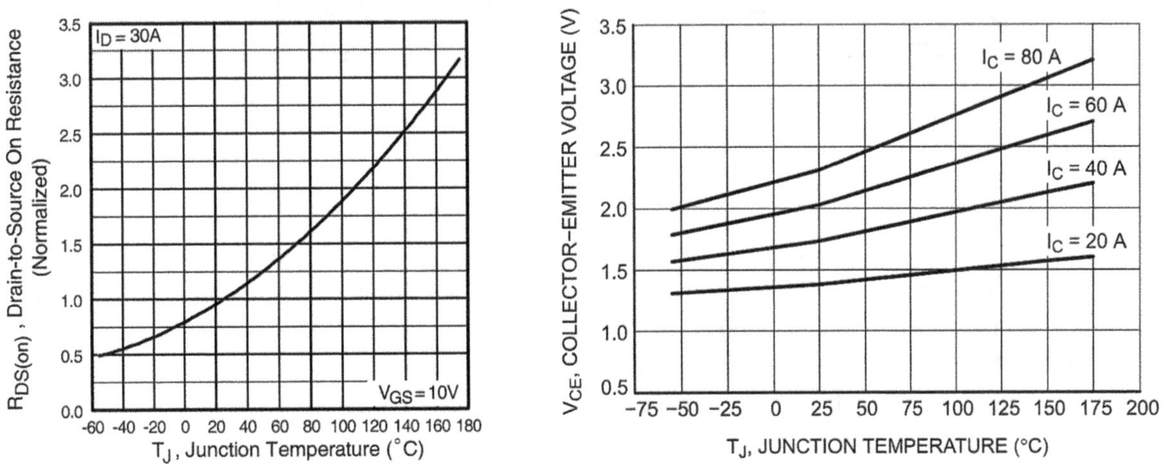

(a) (b)

Figure 5. Effect of temperature on operation of solid state switches, (a) a MOSFET in junction temperature 25ºC, (b): a MOSFET in junction temperature 125ºC, Published with permission of International Rectifier, (c): an IGBT in junction temperature 25ºC, (d): an IGBT in junction temperature 125ºC, Source: ON Semiconductor Co. (with permission)

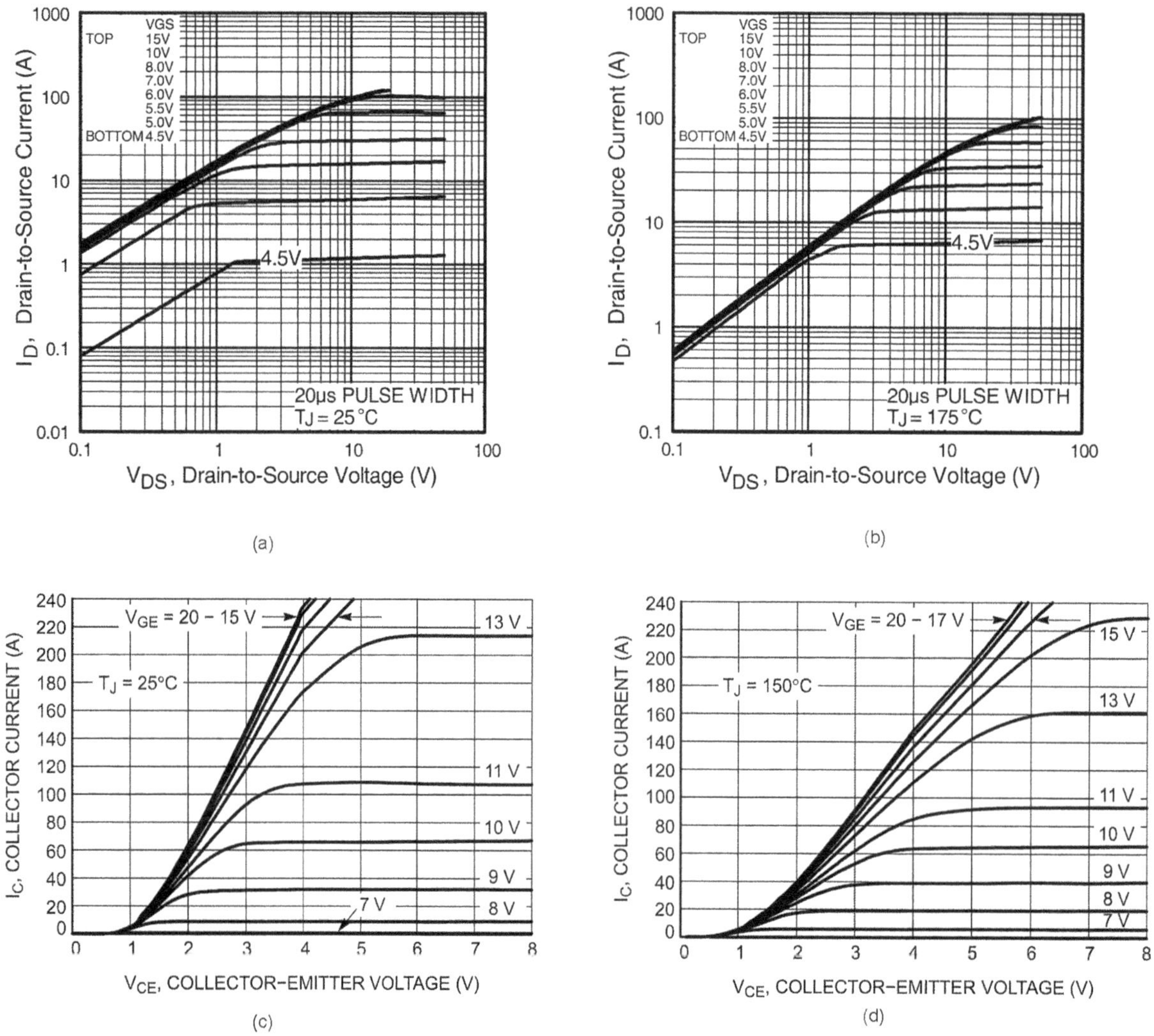

Amount of energy loss in switching interval has a direct relation with the switched voltage and current.

Therefore, we have to capture relevant waveforms in the working model and calculate the switching loss. This task used to involve a bit of computational effort, but with new oscilloscopes the whole math will be done very easily. Figures 6 and 7 show switching loss of an IGBT. Figures 8 and 9 show measurement method of switching loss with TEKTRONIX oscilloscope.

Figure 6. Typical waveforms of voltage and current of a solid state switch during switching process, (a): turn on process, (b): turn off process, Source: ON Semiconductor Co. (with permission)

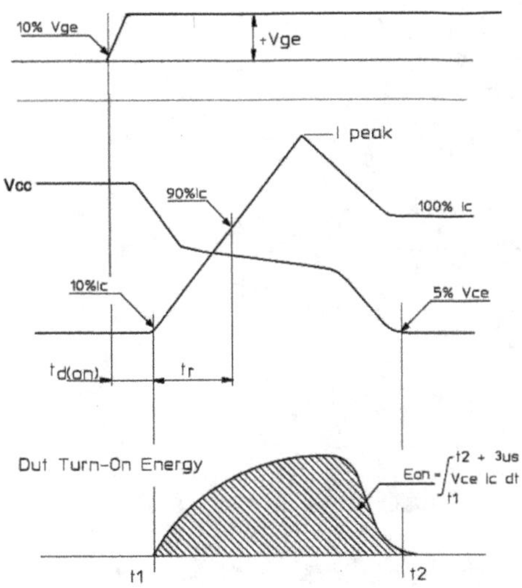

(a)

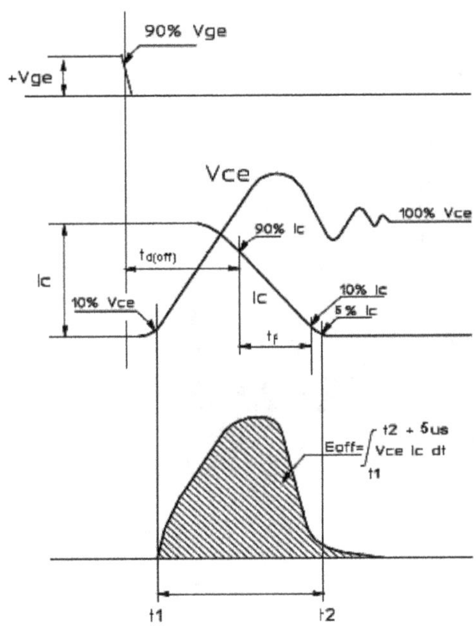

(b)

Figure 7. Dependence of IGBT switching losses to the operating voltage of converter (a), Dependence of IGBT switching losses to the forward current (b), Dependence of IGBT switching losses to junction temperature (c), Source: ON Semiconductor Co. (with permission)

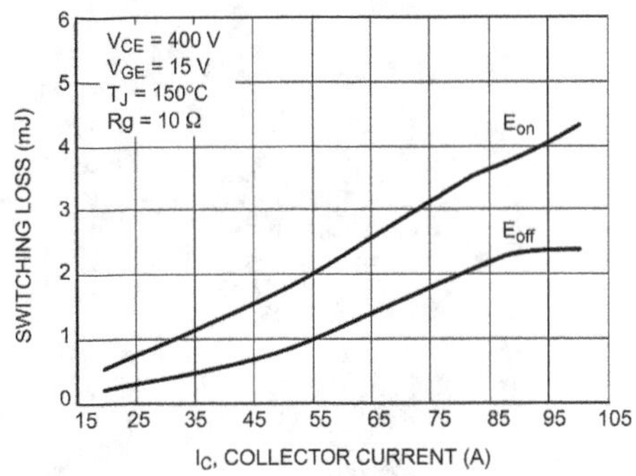

(a)

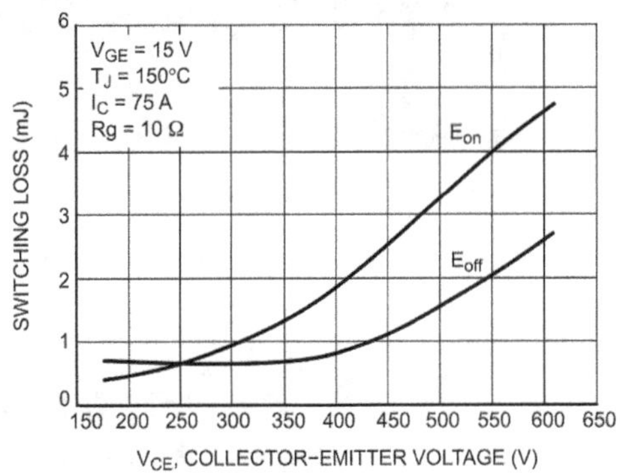

(b)

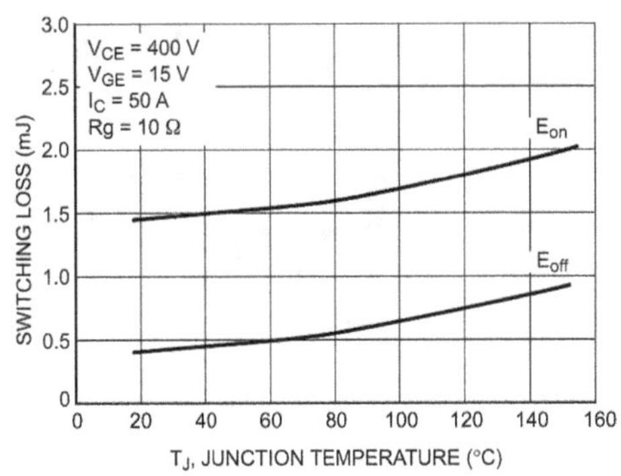

(c)

Figure 8. Calculation of switching losses with TEK oscilloscope

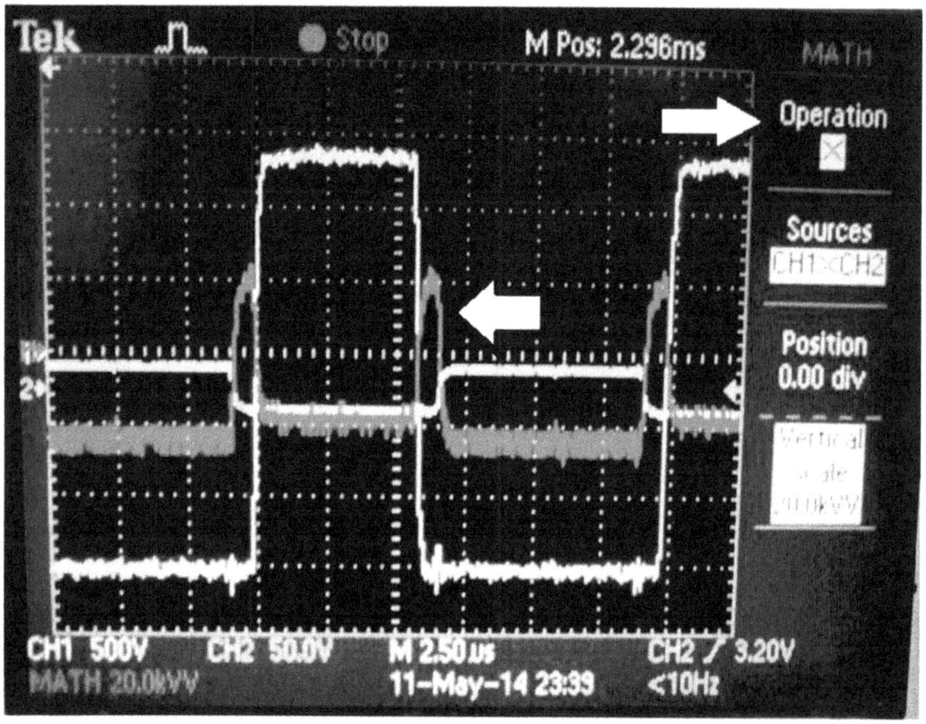

Figure 9. Calculation of switching losses with Math key in TEK oscilloscope

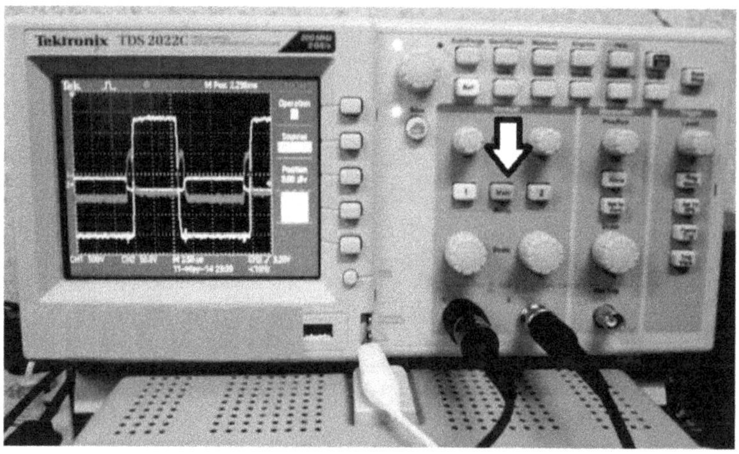

Safe Operating Area (SOA)

For power semiconductor devices, safe operating area is defined as the voltage and current conditions over which the device can be expected to operate without damage. The SOA specification combines the various limitations of the device: maximum voltage, current and power, allowing simplified design of protection circuitry. Often, in addition to the continuous rating, separate SOA curves are plotted for short duration pulse conditions (1 ms pulse, 10 ms pulse, etc.). SOA specifications are useful to the design engineer working on power circuits such as amplifiers and power supplies as they allow quick assessment of the limits of device performance, the design of appropriate protection circuitry, or selection of a more capable device. Figure 10 shows a flyback converter as a converter with inductive load. Figure 11 shows voltage and current waveform of the converter switch during switching interval. Figure 12 shows I-V trajectory of the switch. Figure 13 and 14 show typical SOA for a solid state switch. Figure 15 shows a real SOA for a BJT.

Power Losses in Passive Devices

Pure capacitors and inductors do not dissipate energy; any process which dissipates energy must be treated as one or more resistors in the component model. For discrete electrical circuit components, power loss

Figure 10. A flyback converter

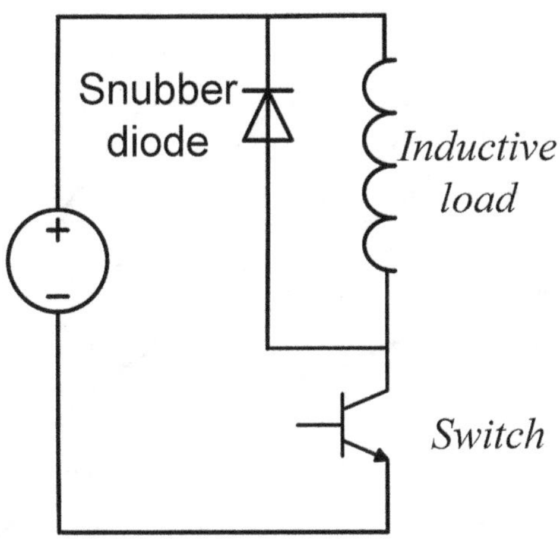

Figure 11. Time domain waveforms during switching

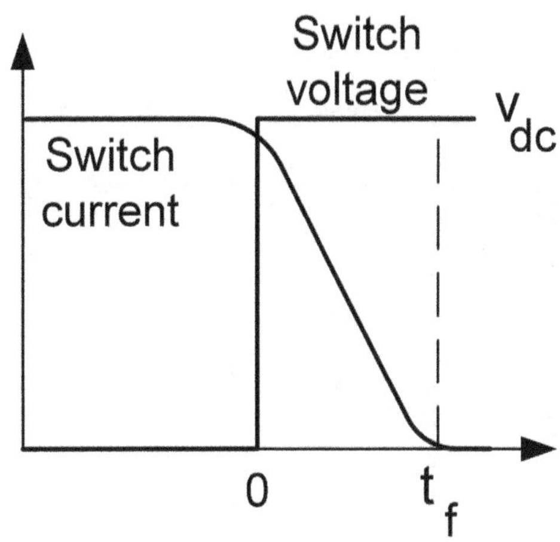

Figure 12. State space trajectory of switch voltage and current during switching without proper snubber

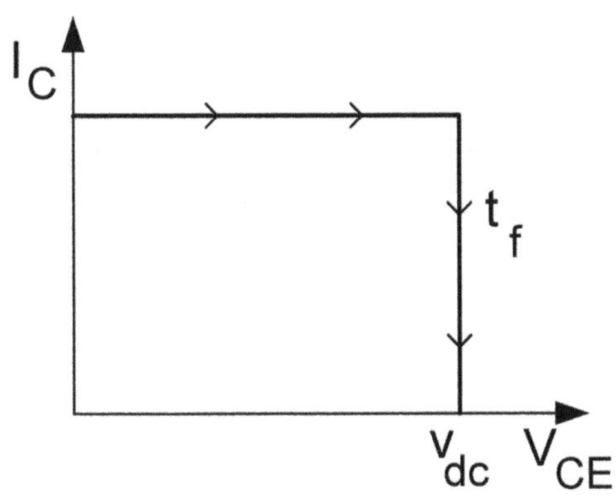

Figure 13. Current and voltage limitations of a switch

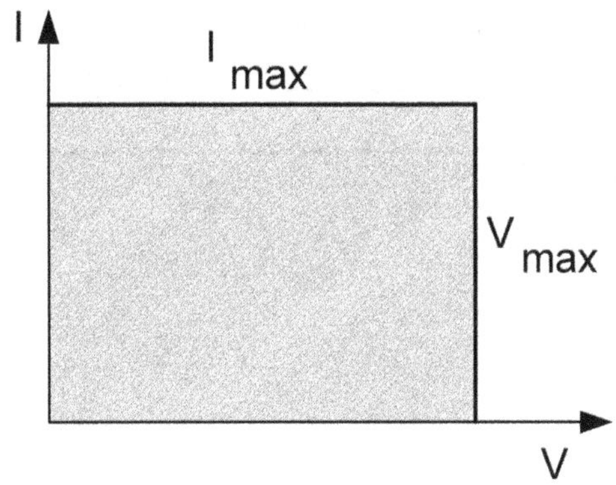

Figure 14. Power limitation of a switch

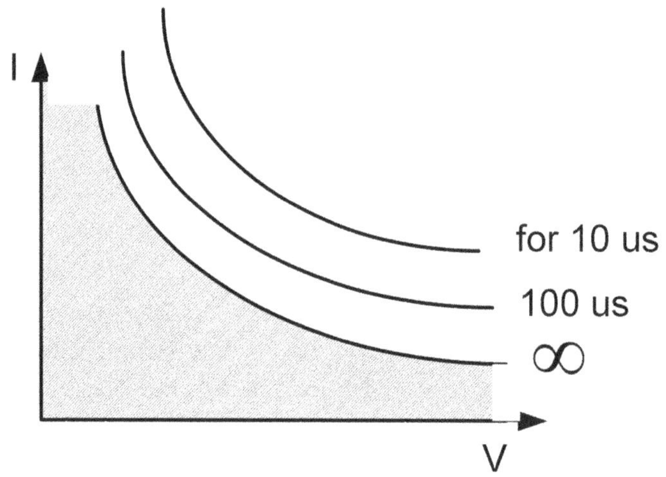

Figure 15. Real active SOA of a BJT Source: ON Semiconductor Co. (with permission)

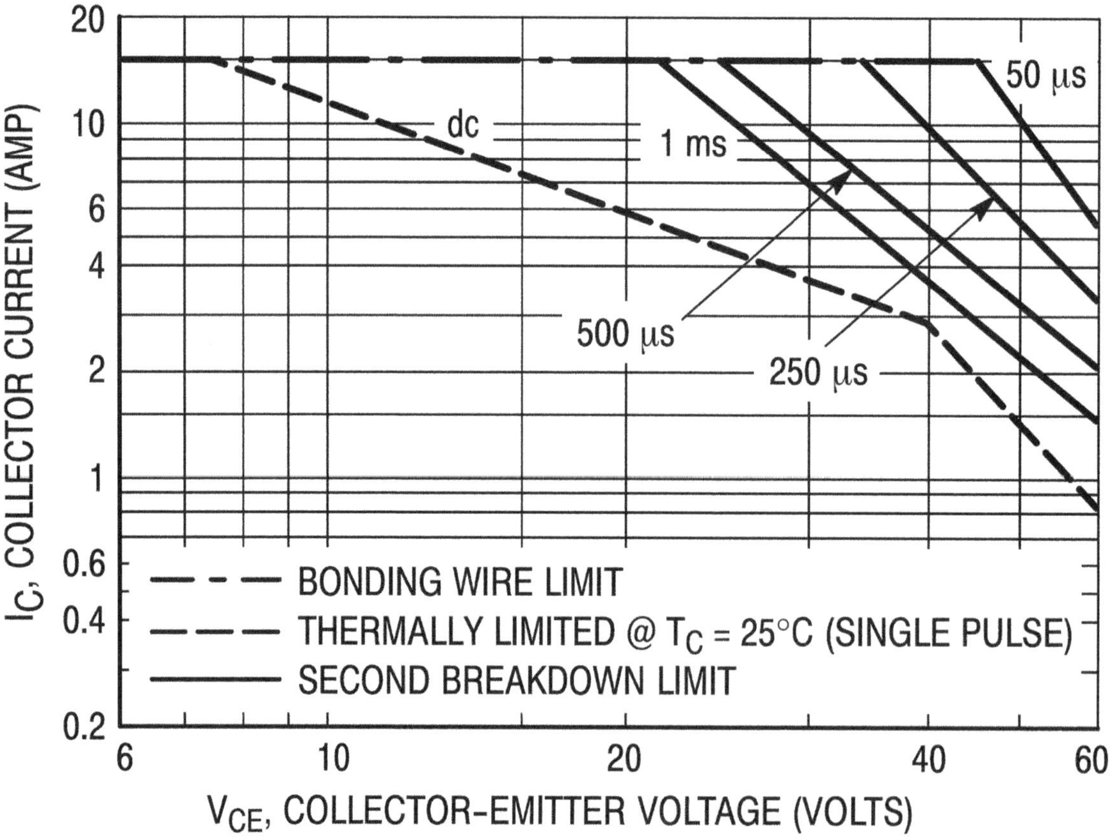

Figure 16. a damaged power resistor

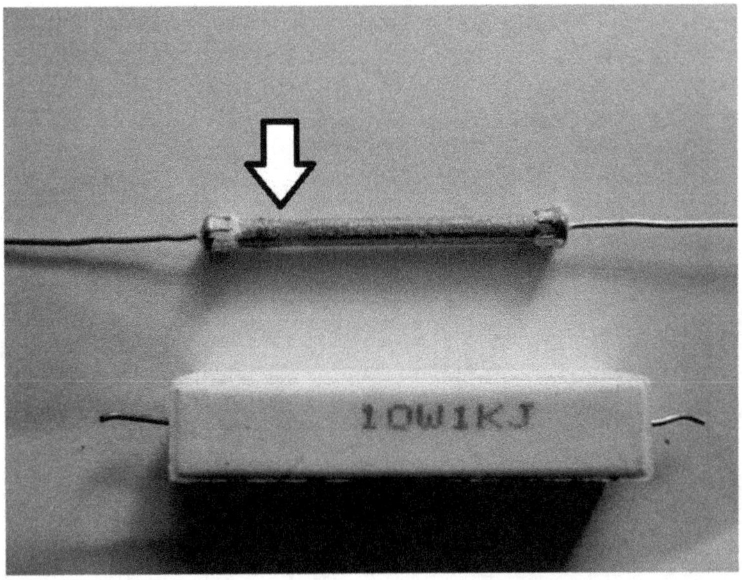

includes inductor winding and core loss and capacitors loss. A capacitor is typically made of a dielectric placed between conductors. The lumped element model of a capacitor includes a lossless ideal capacitor in series with a resistor termed the equivalent series resistance (ESR). The ESR represents losses in the capacitor. Figure 16 shows a damaged resistor because of over current. Figures 17, 18, and 19 shows real power losses in some passive elements.

High Frequency Conductor Losses

Skin effect is the tendency of an alternating electric current to become distributed within a conductor such that the current density is largest near the surface of the conductor, and decreases with greater depths in the conductor. The skin effect causes the effective resistance of the conductor to increase at higher frequencies where the skin depth is smaller, thus reducing the effective cross-section of the conductor. The skin effect is due to opposing eddy currents induced by the changing magnetic field resulting from the alternating current. Figures 20, 21, and 22 show the effect of high frequency current on resistance of a conductor.

Core Losses in Electrical Machines

We start this section with power losses in transformers because other types of electrical machines has the same losses (Hanna, & Schmitt, 2012). A real transformer is typically 95 to 99% efficient, due to several loss mechanisms, including winding resistance, winding capacitance, leakage flux, core losses, and hysteresis loss. Larger transformers are generally more efficient than small units, and those rated for electricity distribution usually perform better than 98%.

As transformer losses vary with load, it is often useful to express these losses in terms of no-load loss, full-load loss, half-load loss, and so on. Hysteresis and Eddy current losses are constant at all load

Figure 17. Real Z-f characteristic of capacitors, (left): aluminum electrolyte capacitor, (right): ceramic capacitor, Source: EPCOS Co. (with permission)

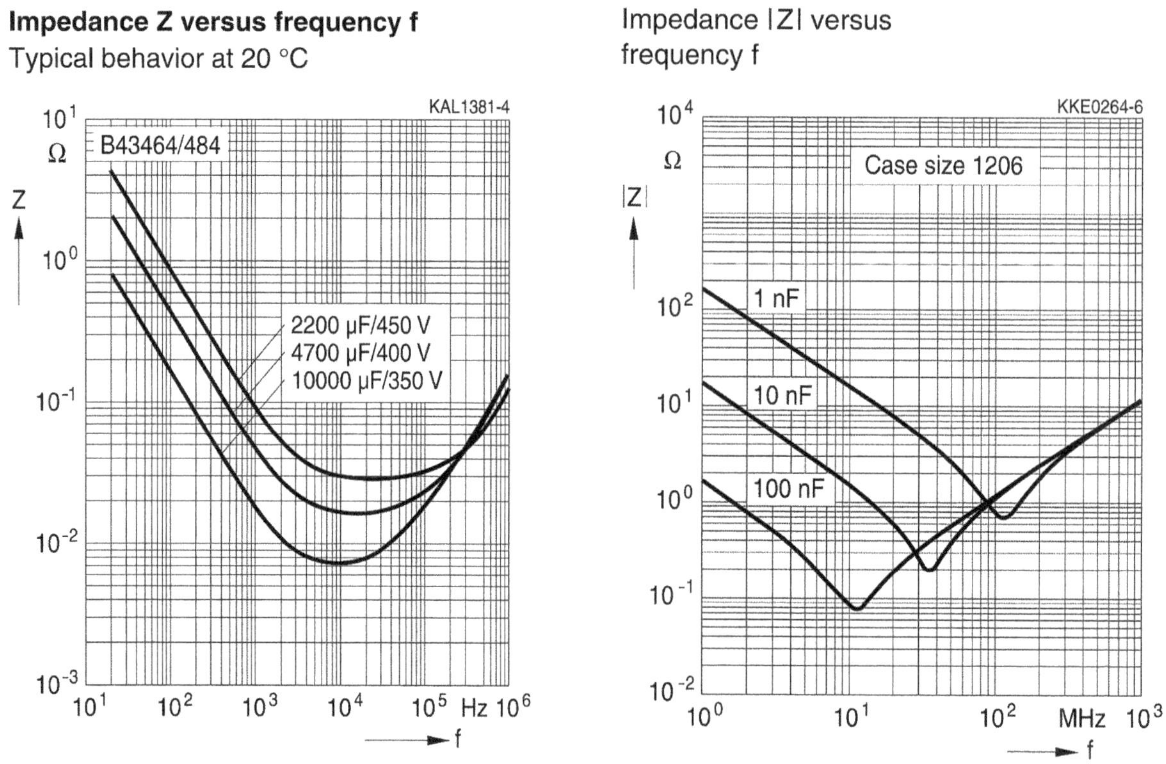

Figure 18. Frequency response of an inductor

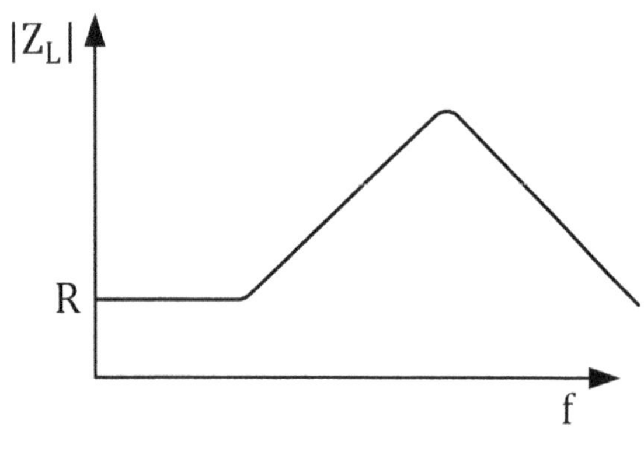

Figure 19. Equivalent circuit of a real capacitor

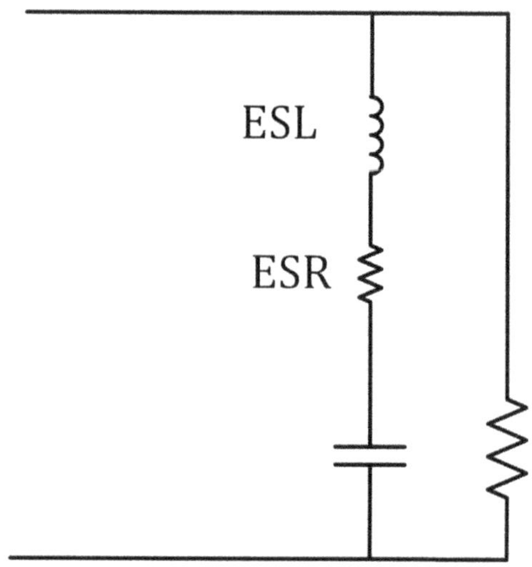

Figure 20. A DC current carrying conductor

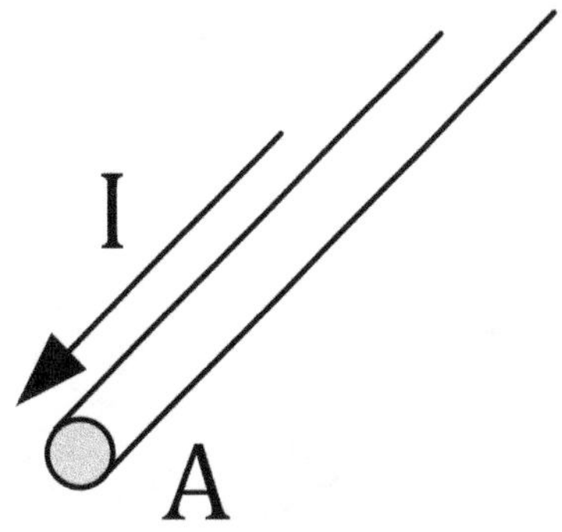

Figure 21. Characteristic of AC resistance of a conductor

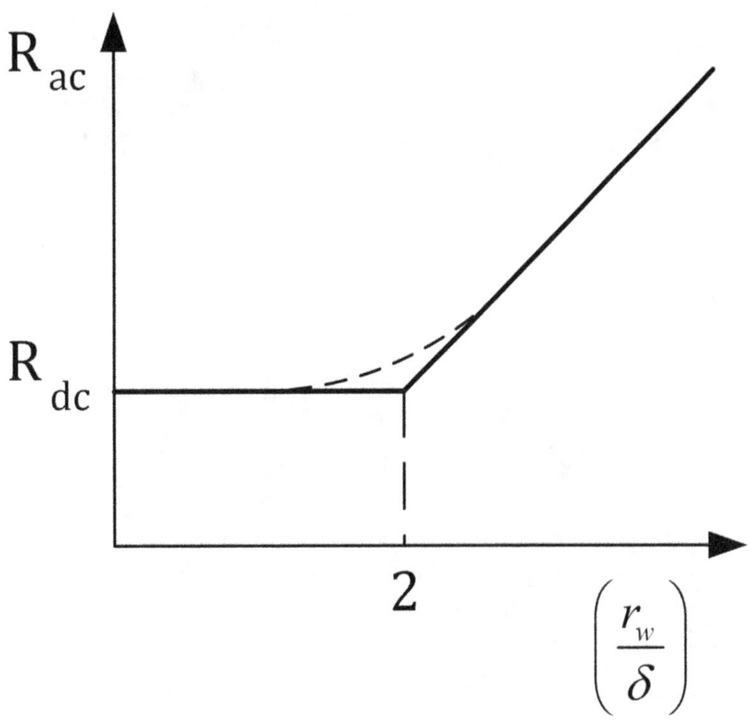

Figure 22. Skin effect in a conductor with AC current

levels and dominate overwhelmingly without load, while variable winding joule losses dominating increasingly as load increases (Gorginpour, Oraee, & Abdi, 2014). Figures 23, 24, 25, and 26 show core sheet of an induction motor.

Figure 23. Separated core sheets of stator and rotor

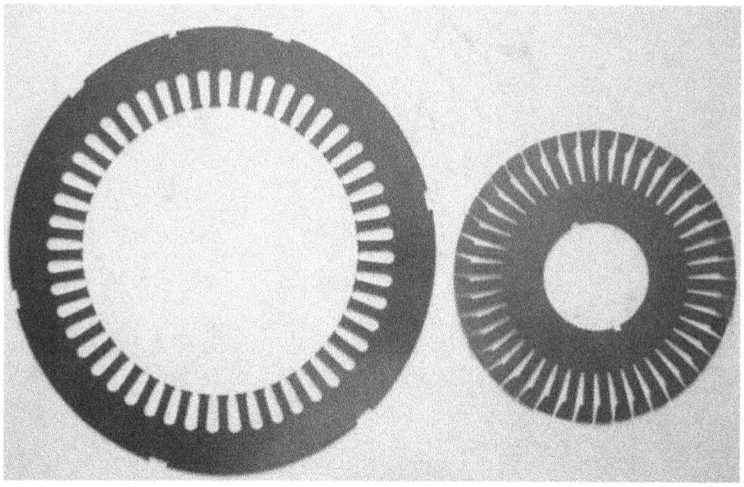

Figure 24. Assembled core sheets of stator and rotor

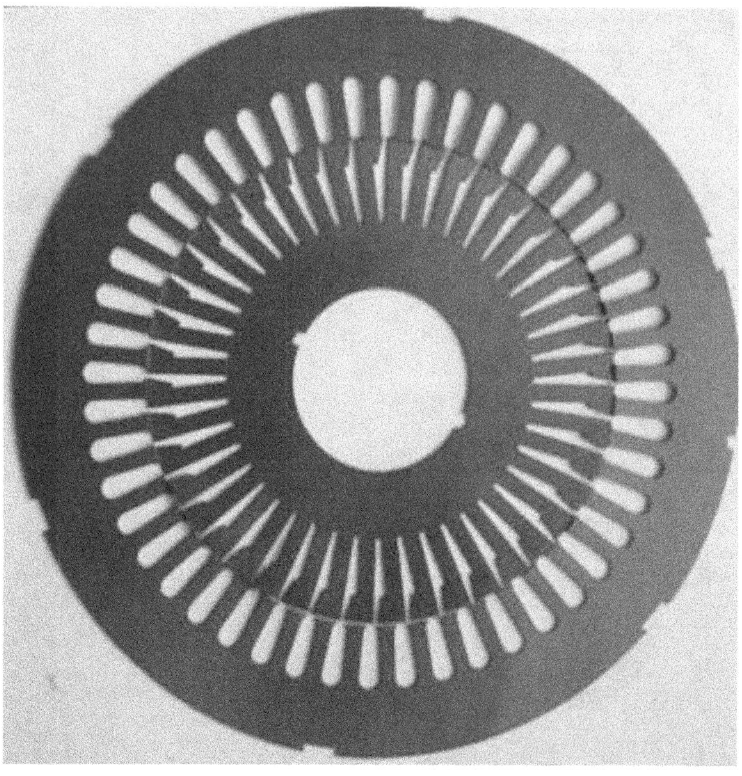

Figure 25. Tooth of stator core

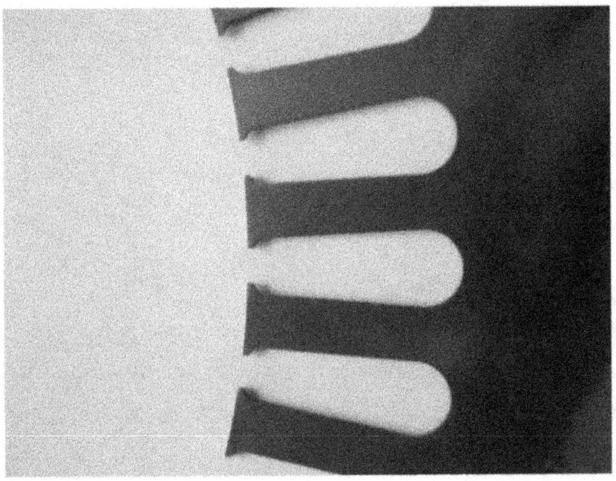

Figure 26. Tooth of rotor core

Hysteresis Losses

Each time the magnetic field is reversed, a small amount of energy is lost due to hysteresis within the core (Hanna, Hiscock, & Klinowski, 2007). Figure 27 shows a real B-H curve for a ferrite. Figure 28 shows a minor loop in B-H curve of a steel sheet. Figures 29 and 30 show B-H minor loop for an inductor core with DC current.

Figure 27. Real B-H Hysteresis loop in a ferrite core, Source: EPCOS Co. (with permission)

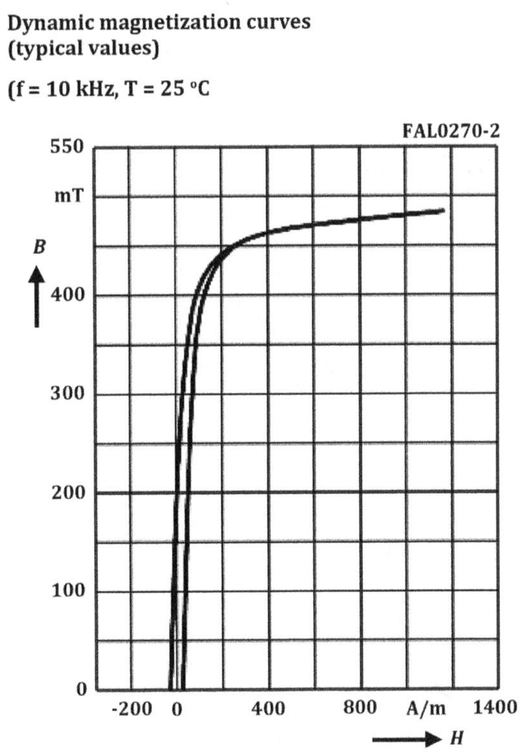

Figure 28. B-H Hysteresis loop

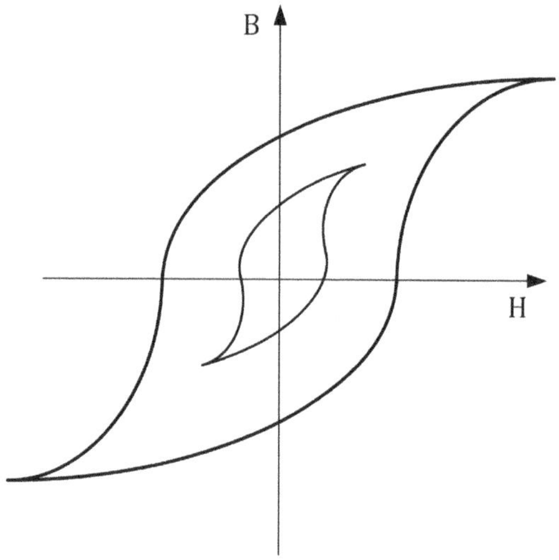

Figure 29. Voltage and current waveforms in a DC inductor

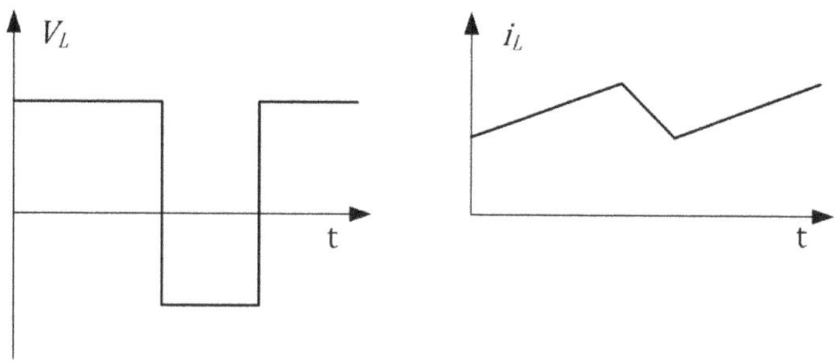

Figure 30. Local B-H loop for DC inductor

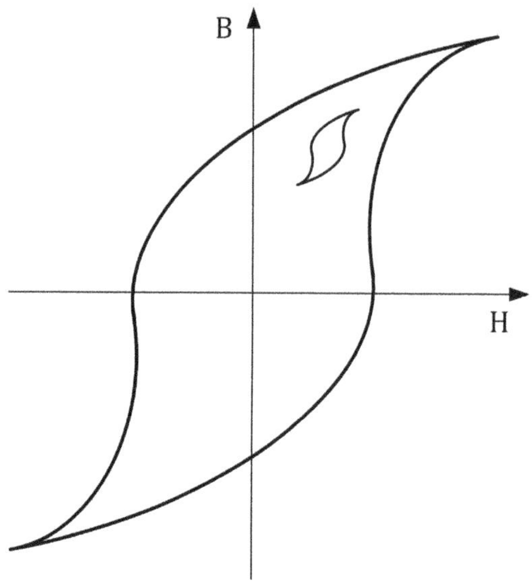

Eddy Current Losses

Ferromagnetic materials are also good conductors and a core made from such a material also constitutes a single short-circuited turn throughout its entire length. Eddy currents therefore circulate within the core in a plane normal to the flux, and are responsible for resistive heating of the core material. The eddy current loss is a function of the square of supply frequency. Eddy current losses can be reduced by making the core of a stack of plates electrically insulated from each other, rather than a solid block; all transformers operating at low frequencies use laminated or similar cores.

Lamination

Laminated magnetic cores are made of thin, insulated iron sheets, lying, as much as possible, parallel with the lines of flux. Using this technique, the magnetic core is equivalent to many individual magnetic circuits, each one receiving only a small fraction of the magnetic flux. Because eddy currents flow around lines of flux, the laminations prevent most of the eddy currents from flowing at all, restricting any flow to much smaller, thinner and thus higher resistance regions. From this, it can be seen that the thinner the laminations, the lower the eddy currents. Figures 31 and 32 show lamination of transformer core.

Figure 31. Simplified diagram for laminated core

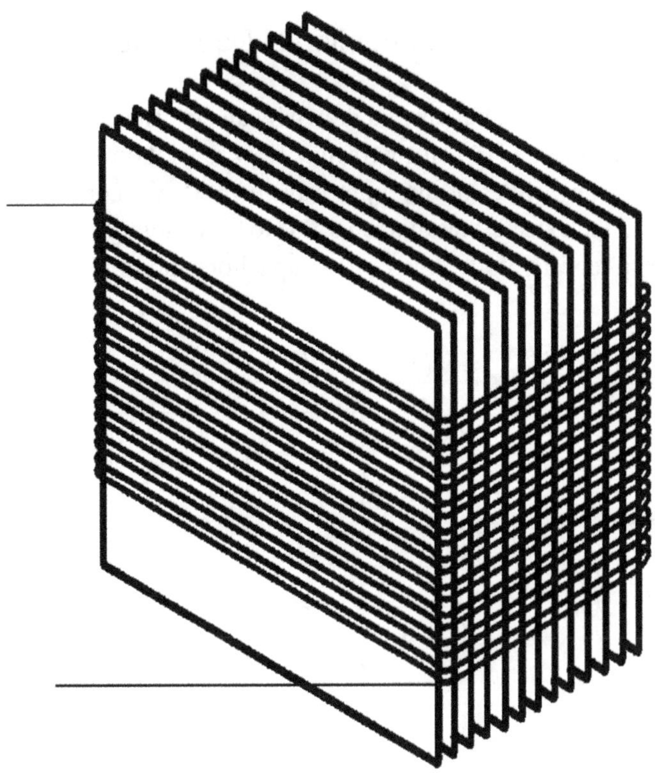

Figure 32. A real transformer with laminated core

Stray Load Loss

The stray load losses represent a not negligible term in the power balance of industrial induction motors and these losses can not be directly measured with targeted experimental tests.

Example: Heat Generation in Transient Interval Due to Converter Inrush Current

Inrush current is the instantaneous input current drawn by an electrical device when first turned on. Alternating current electric motors and transformers may draw several times their normal full-load current when first energized, for a few cycles of the input waveform. Power converters also often have inrush currents much higher than their steady state currents, due to the charging current of the input capacitance. Figures 33 and 34 show a rectifier with large output capacitors and the converter input stage current.

Figure 33. A single phase rectifier with capacitive filter

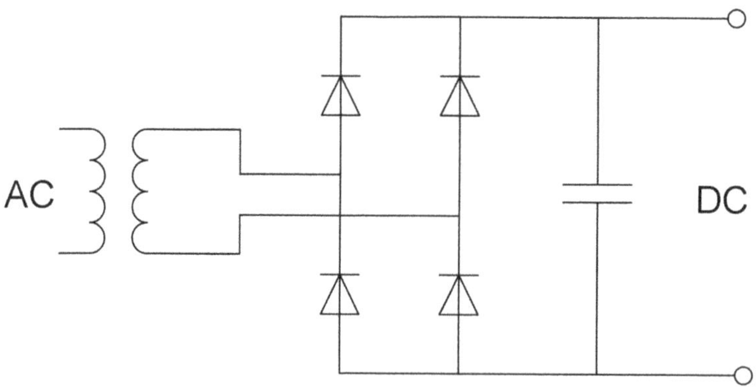

Figure 34. Inrush current of rectifier with large output capacitor

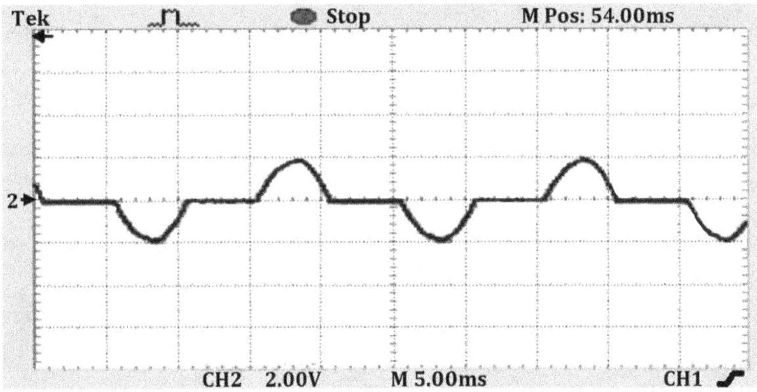

Figure 35. Inrush current in switching of a transformer, (a): circuit diagram, (b), input voltage

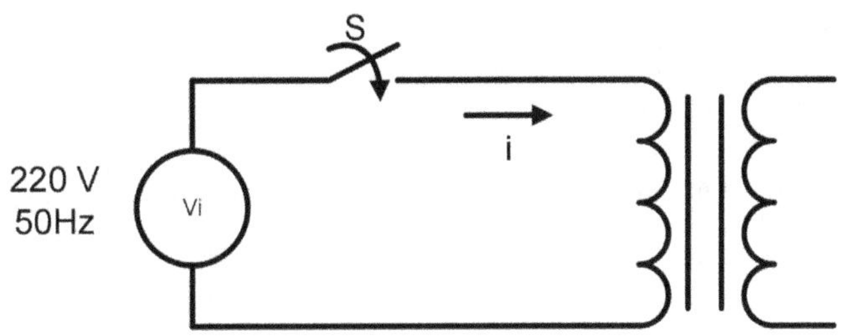

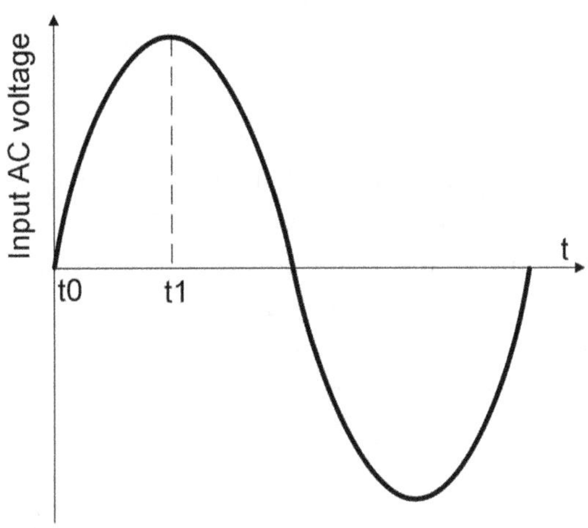

Example: Heat Generation in Transient Interval Due to Transformers Inrush Current

In a transformer, when power is applied, the residual field will cause a high inrush current until the effect of the remanent magnetism is reduced (Chiesa, & Høidalen, 2010). Figures 35 and 36 show inrush current of a transformer.

Figure 36. various types of Inrush current in switching of a transformer, (a) positive current with high transient amplitude because of switching at t_0, (b) positive current with low transient amplitude because of switching after t_0, (c), negative current with high transient amplitude, (d): quasi-symmetric inrush current

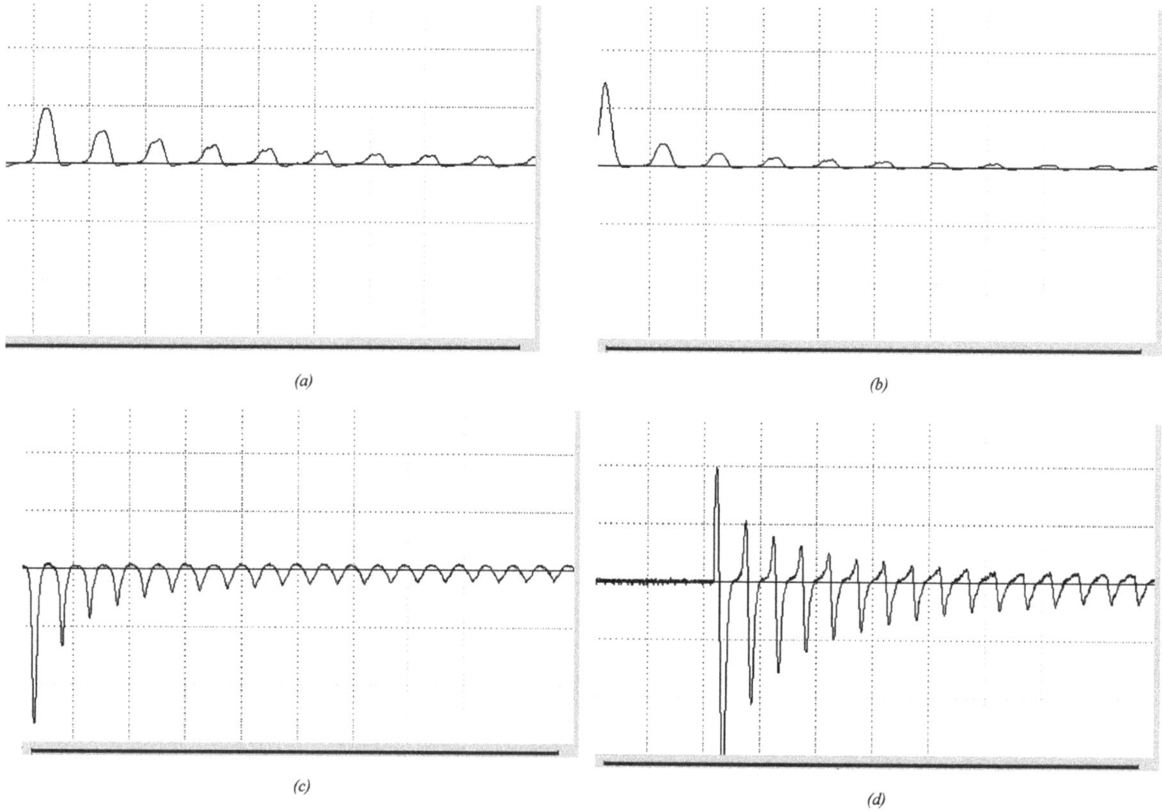

A resistor in series with the line can be used to limit the current charging input capacitors. However, this approach is not very efficient, especially in high power devices, since the resistor will have a voltage drop and dissipate some power. Inrush current can also be reduced by inrush current limiters. Negative temperature coefficient thermistors are commonly used in switching power supplies, motor drives and audio equipment to prevent damage caused by inrush current. Another option, particularly for high voltage circuits, is to use a pre-charge circuit. The circuit would support a current limited precharge mode during the charging of capacitors, and then switch to an unlimited mode for normal operation when the voltage on the load is 90% of full charge. Figure 37 shows a rectifier with a soft starter.

Example: Heat Generation Due to Shoot-Thorough Over Current in Inverter Leg

The two-switch phase leg consists of a high- and a low-side semiconductor switch spanning a DC voltage source. Dead time delay intervals are typically added to the software controlling these two switches in order to avoid shoot-through conditions. Shoot-through conditions occur when both switches are on at the same time and cause a short in the DC voltage source. It is one of the most dangerous failure modes encountered in conventional inverter circuits. Shoot-through failures occur when two series semiconduc-

Figure 37. A rectifier with soft starter

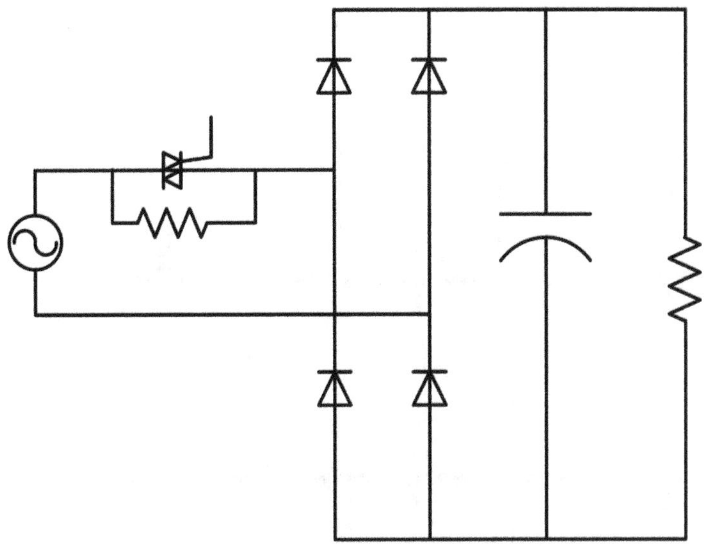

tor switches are inadvertently turned on simultaneously, causing the flow of extremely high currents that destroy the device. Figure 38 shows complement pulses of an inverter leg.

Example: Heat Generation Due to Saturation of Core

Saturation leads to over current in the drive circuit and thermal damage of the switches. Different materials have different saturation levels. For example, high permeability iron alloys used in transformers reach magnetic saturation at 1.6 - 2.2 T, whereas ferrites saturate at 0.2 - 0.5 T. Some amorphous alloys saturate at 1.1.3 T. Figures 39, 40, and 41 show some characteristic of core materials.

Example: Over Current Due to Voltage Unbalance

Voltage unbalance can cause various problems. Three-phase motors may run hotter. Un-balanced voltages have caused malfunctions in variable speed drives. A major cause of unbalanced voltage is unbalanced load current. This imbalance can be due to customer load imbalance, utility system imbalance, external influence or self-induced (Gorginpour, Jandaghi, & Oraee, 2011).

Relatively small voltage imbalances are often achievable close to substations on circuits with well-balanced loads. Farther out on circuits it can be extremely difficult and very expensive to achieve a high level of voltage balance.

An excessive level of voltage imbalance can have serious impacts on induction motors. Induction motors are designed to tolerate a small level of imbalance, but they have to be de-rated if the imbalance is excessive, which is an added cost for the installation. Figure 42 shows unbalance three phase currents that cause to voltage unbalance. Figures 43 and 44 shows two non-sinusoidal current waveform with low and high frequency harmonics. Figure 45 shows the problem of voltage harmonics for core of electric machines. In this figure, the motor operates in very low speed. (Figure 46) Thus, its self-cooling via its fan does not work properly and voltage harmonics causes to over temperature. Figure 47 shows ideal voltage and current waveforms.

Figure 38. Complement gate pulses of an inverter leg

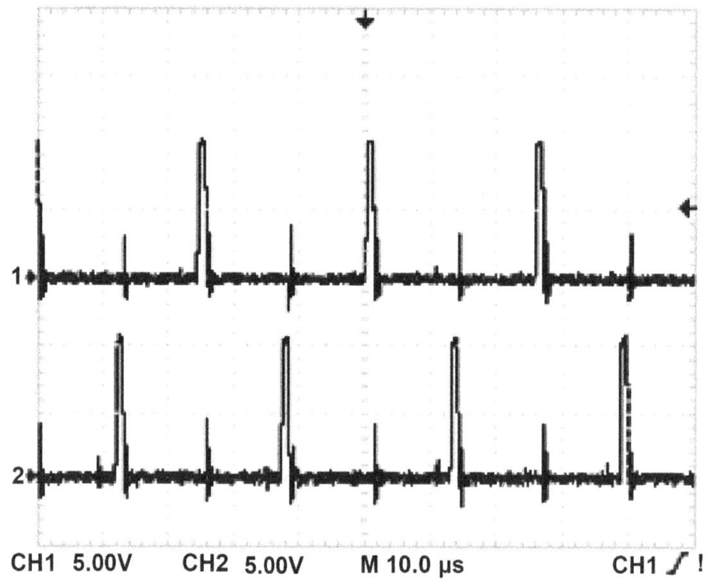

Figure 39. A typical B-H curve for an iron sheet

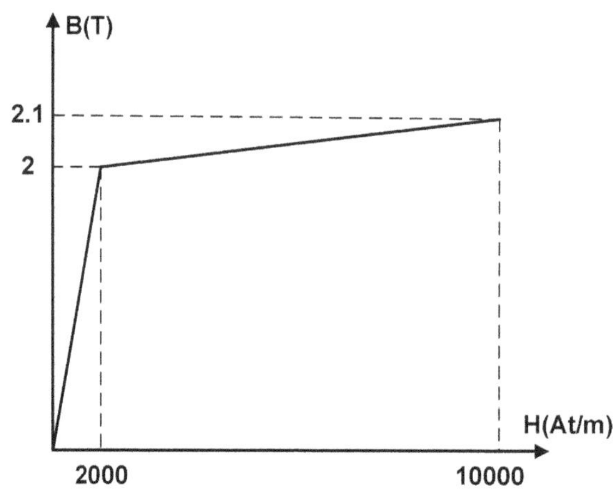

Figure 40. A typical B-H curve for a ferrite

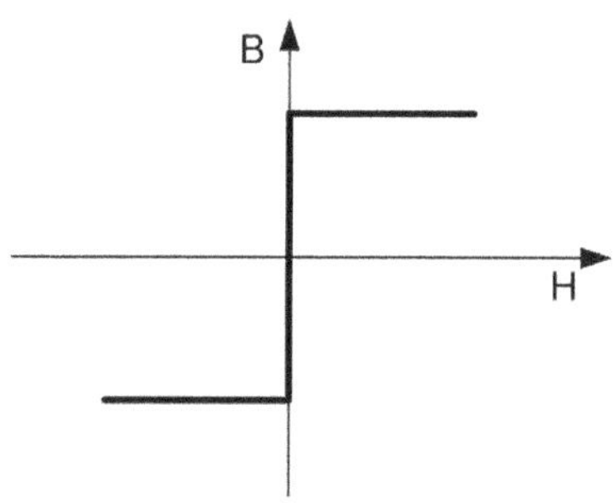

Figure 41. Real properties of a ferrite core, (up): real B-H Hysteresis loop in a ferrite core with sharp saturation, (down): Variation of a ferrite core permeability with temperature, (center): Variation of a ferrite core losses with temperature, Source: EPCOS Co. (with permission)

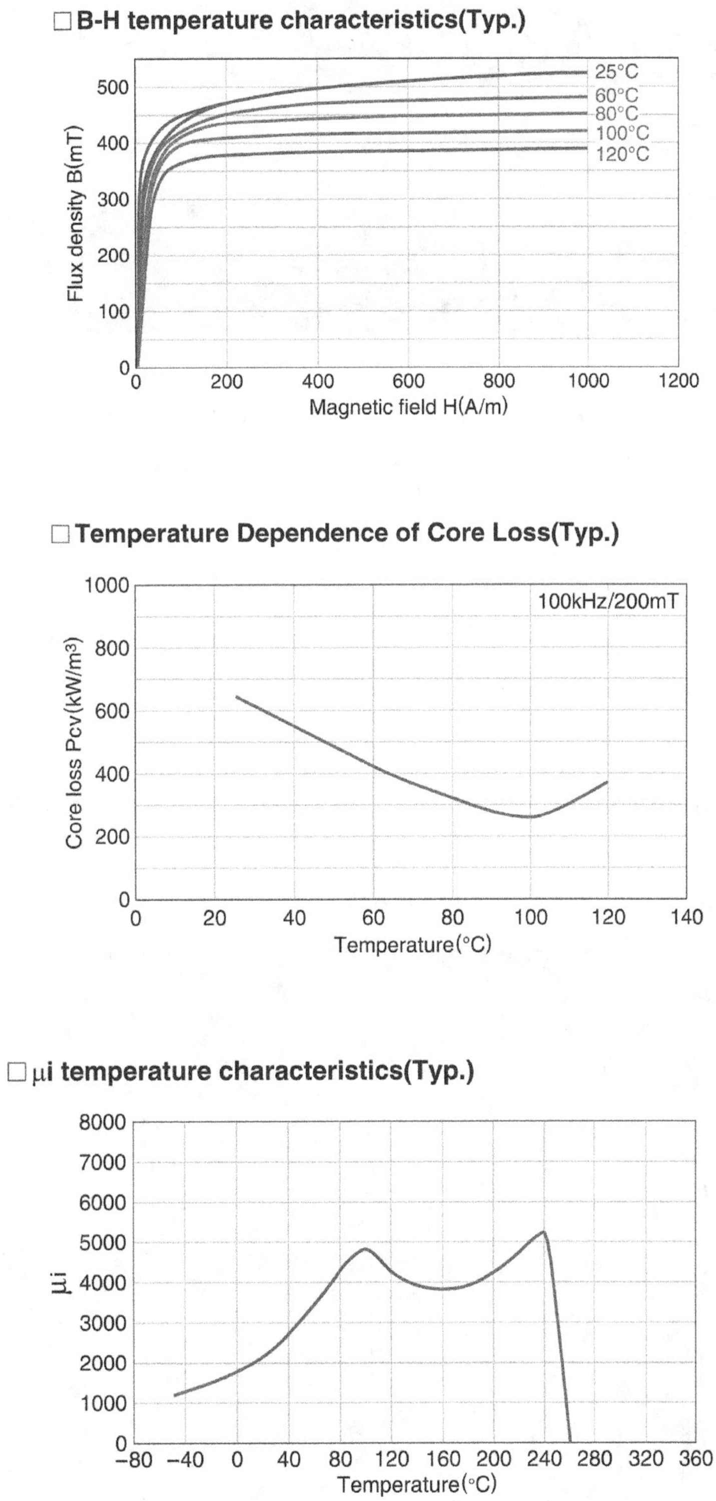

Figure 42. unbalance 3-phase currents

Figure 43. Low frequency distortion in a non-sinusoidal current waveform

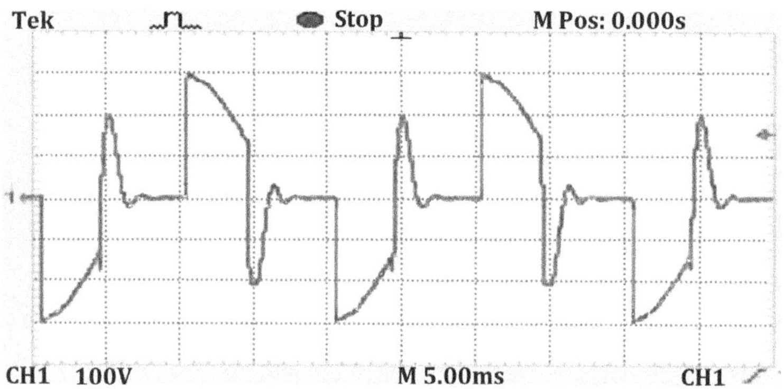

Figure 44. High frequency harmonics in line current

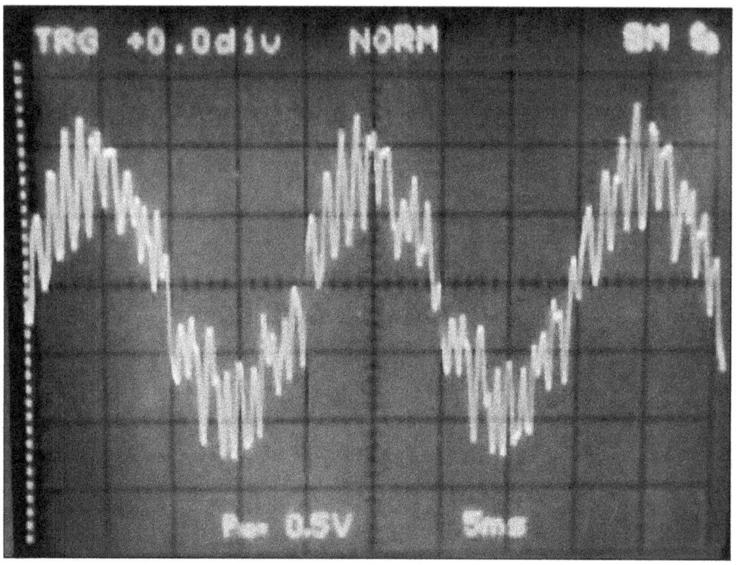

Figure 45. A very low speed motor affected by voltage harmonics

Figure 46. Harmonic analysis of a voltage waveform with TEK oscilloscope

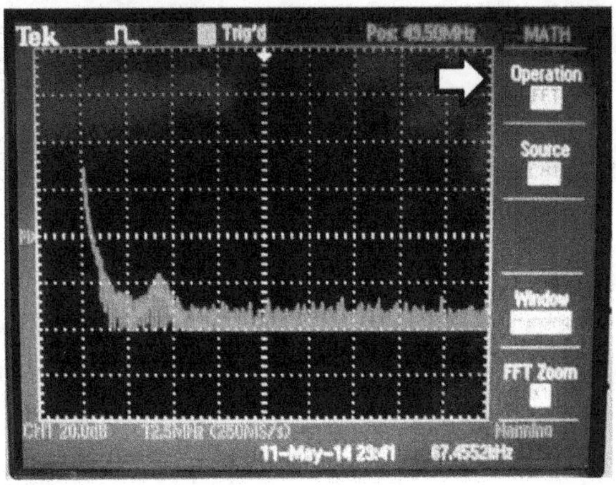

Figure 47. Voltage and current waveforms of an ideal Load: power factor=1 without harmonis

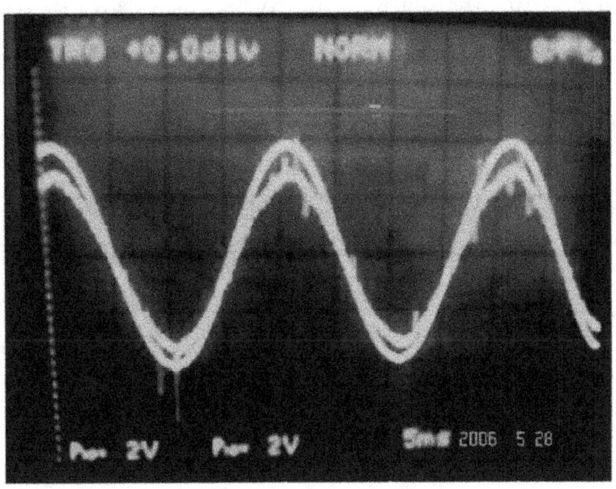

Figure 48. Damaged winding of an air gapped inductor due to electrical breakdown

ELECTRIC BREAKDOWN

Electrical breakdown or dielectric breakdown refers to a rapid reduction in the resistance of an electrical insulator when the voltage applied across it exceeds the breakdown voltage. This results in a portion of the insulator becoming electrically conductive. Under sufficient electrical stress, electrical breakdown can occur within solids, liquids, gases or non-ideal vacuum. Electrical breakdown is often associated with the failure of solid or liquid insulating materials used inside high voltage transformers or capacitors in the electricity distribution grid, usually resulting in a short circuit or a blown fuse. Dielectric breakdown within a solid insulator can permanently change its appearance and properties (Oraee, 2000). Thus, we talk more about breakdown of solid insulators. Figure 48 shows an electric breakdown in winding of an inductor.

Breakdown in Solid Dielectric

In solid dielectrics, the breakdown strength is high. The highest breakdown strength obtained under carefully controlled conditions (Movahed, Oraee, Mirzamani, Rajabi, & Daneshvar, 2010). Dielectrics usually fail at stresses well below the intrinsic strength due usually to one of the following causes.

- electro-mechanical breakdown
- breakdown due to internal discharges
- thermal breakdown

Figures 49, 50, 51, and 52 shows some commonly used insulators.

Heat is generated continuously in electrically stressed insulation by dielectric losses, which is transferred to the surrounding medium by conduction through the solid dielectric. If the heat generated exceeds the heat lost to the surroundings, the temperature of the insulation increases and it damages. Figures 53, 54, and 55 show application of insulator in various electric machines.

Creepage

The creepage distance for insulators is the shortest distance along the insulator surface between the metal parts at each end of the insulator. A proper and adequate creepage distance protects against tracking, a process that produces a partially conducting path of localized deterioration on the surface of an insulating material as a result of the electric discharges on or close to an insulation surface. Tracking that damages the insulating material normally occurs because of one or more of the following reasons:

- Humidity in the atmosphere.
- Presence of contamination.
- Corrosive chemicals.
- Altitude at which equipment is to be operated

Figure 56 and 57 show faults due to creepage in power converters.

Example: Failure of Excitation Systems Due to Electric Breakdown

Any synchronous generator requires exciting the field of the generator, and the field is generally the part that is rotated by the prime mover. Excitation systems have a powerful impact on generator dynamic performance and availability, it ensures quality of generator voltage and reactive power.

Following types are common (Barati, Oraee, Abdi, Shiyi, & McMahon, 2008):

Figure 49. A thick solid isolator

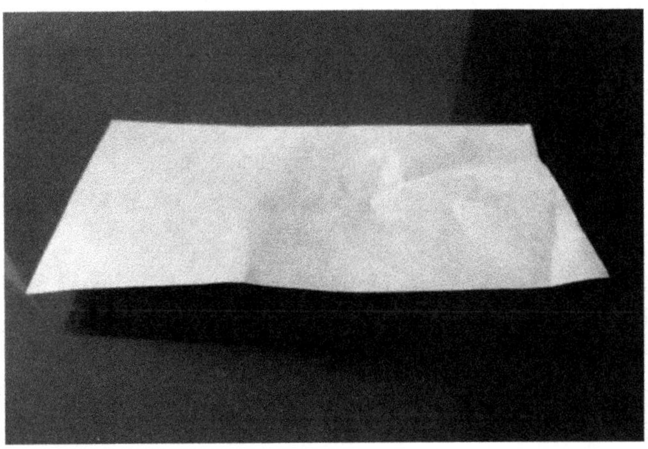

Figure 50. Typical thickness of a thick solid isolator

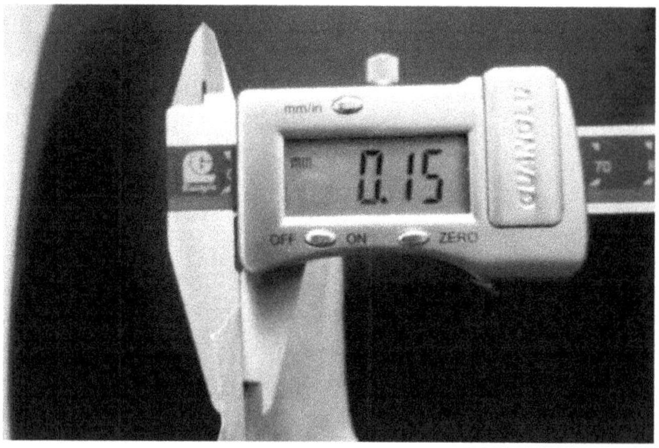

Figure 51. A thin solid isolator

Figure 52. Typical thickness of a thin solid isolator

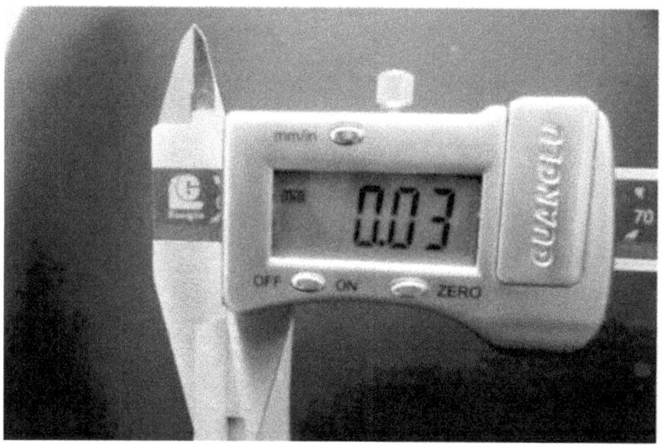

Figure 53. Different layers of a transformer

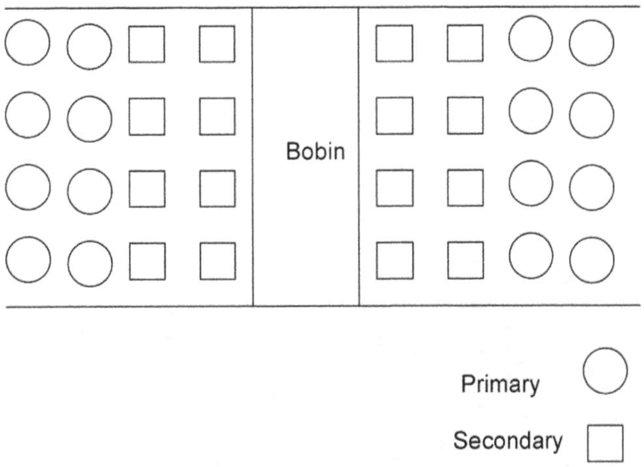

Figure 54. Solid isolators used for isolation between winding of a motor and its core

Figure 55. Solid isolators used for isolation between winding of a generator and its core

Figure 56. Creepage on a PCB

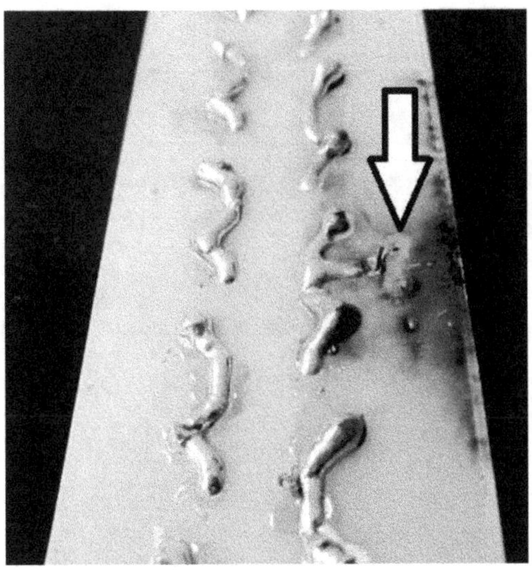

Figure 57. A damaged rotor winding due to electrical breakdown

- Brushless excitation systems, with rotating exciter machines and Automatic Voltage Regulator (AVR), or
- Static excitation systems (SES), feeding rotor directly from thyristor bridges via brushes.

Figures 58, 59, 60, 61, 62, and 63 show excitation system of synchronous generators.

Figure 58. A synchronous generator

Figure 59. Rotary excitation system (down) of a synchronous generator (up)

Figure 60. Static rectifier (up) used for rotor (down) of a synchronous generator

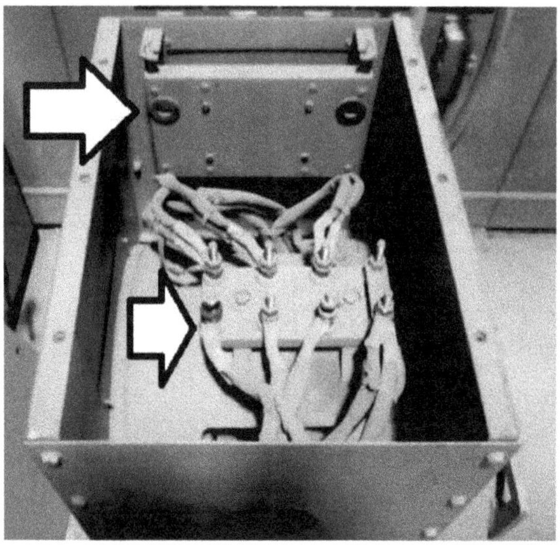

Figure 61. Mounting of a static rectifier used for rotor of a synchronous generator

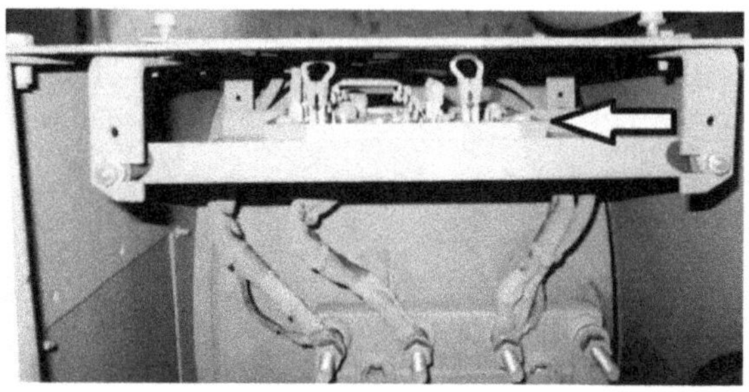

Figure 62. Rotor of a synchronous generator

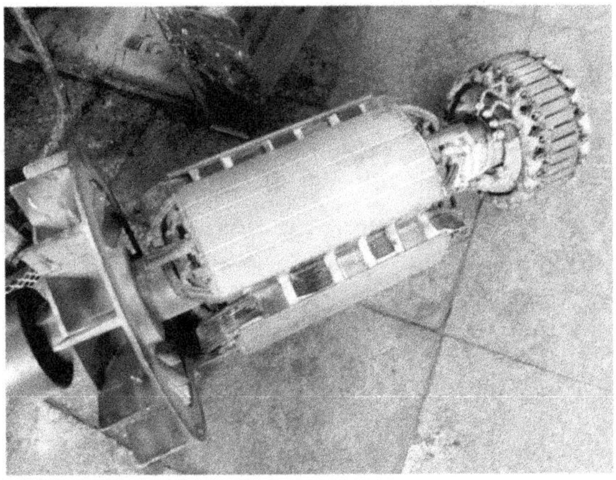

Figure 63. Brushless rotary rectifier synchronous generator

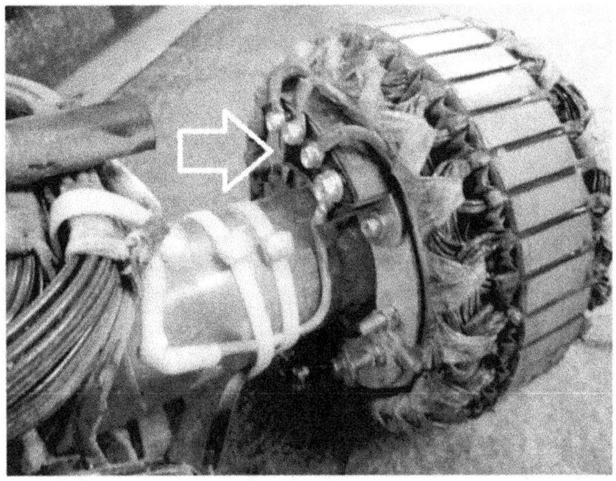

Example: Gate Oxide Breakdown

In MOSFETs and IGBTs, the gate oxide is very thin, so it can only hold a limited voltage. In the datasheets, manufacturers often state a maximum gate to source voltage, around 20 V, and exceeding this limit can result in destruction of the component. Furthermore, a high gate to source voltage reduces significantly the lifetime of the MOSFET, with little to no advantage on reduction of on state resistance of the switch. Figures 64 and 65 show an IGBT failure because of gate oxide breakdown.

Example: Maximum Drain to Source Voltage

Power semiconductors switches have a maximum specified drain to source voltage when turned off, beyond which breakdown may occur. Exceeding the breakdown voltage causes the device to conduct,

Figure 64. A damaged IGBT due to voltage breakdown in its gate

Figure 65. damaged IGBT gate

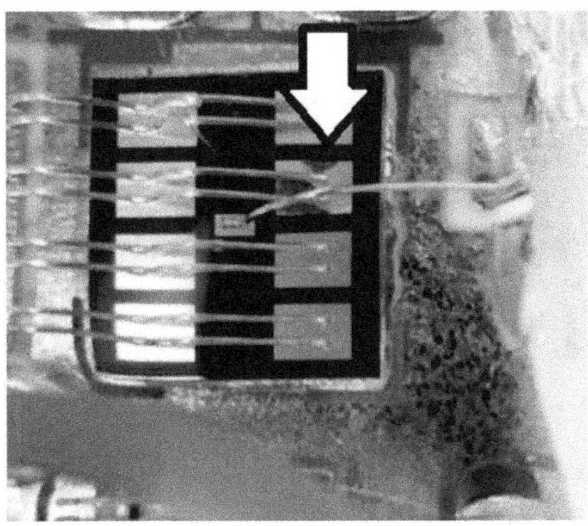

potentially damaging it and other circuit elements due to excessive power dissipation. Figures 66, 67, 68, and 69 shows some voltage spike in power electronic converters.

Example: Electric Breakdown in Capacitor

The capacitor nominal working voltage indicates how much voltage the capacitor can withstand long term. In capacitors, the thickness of the electrolyte element and the insulating material determine the working voltage of the capacitor. Figure 70 shows structure of a capacitor. Figures 71 and 72 show damaged capacitors because of electrical breakdown. An overvoltage will damage the dielectric of a capacitor and may affect the performance and life of a capacitor or cause catastrophic failure (Lahyani, Venet, Grellet, & Viverge, 1998).

Figure 66. Voltage spike on a MOSFET drain voltage

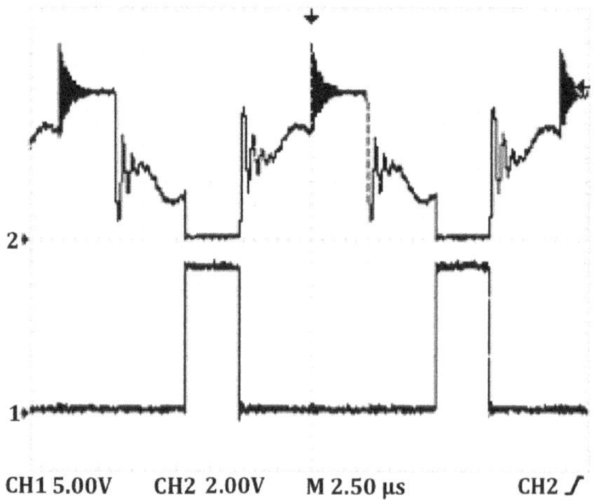

Figure 67. Voltage spike on a DC motor control switch

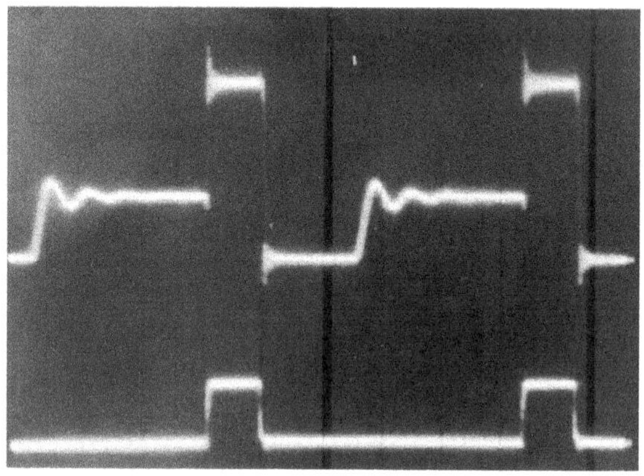

Figure 68. Voltage spike on a diode blocking voltage

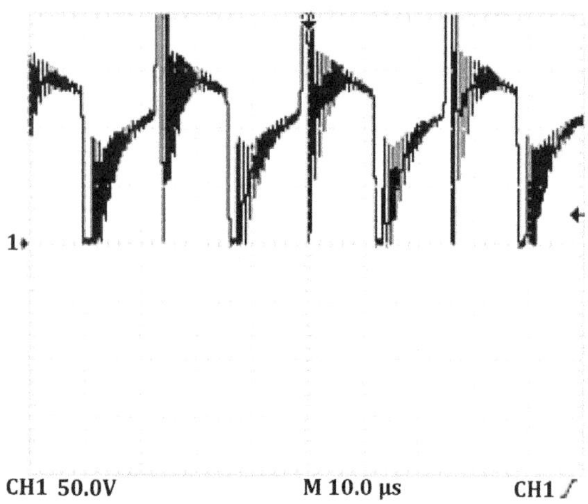

Figure 69. Parasitic capacitor in a transformer

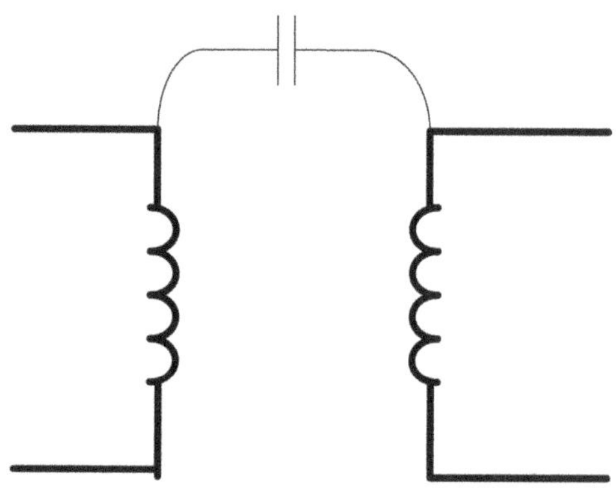

Figure 70. Structure of a capacitor, (a), schematic view, (b) simplified model

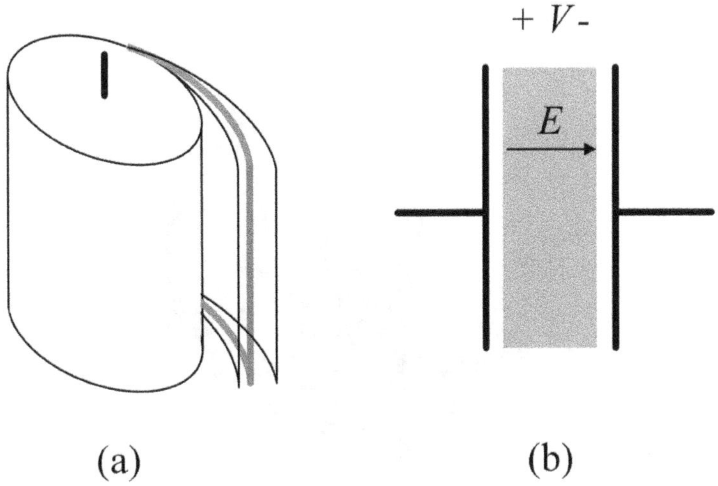

Figure 71. A damaged electrolyte type capacitor with overvoltage

Figure 72. A damaged film type capacitor with overvoltage

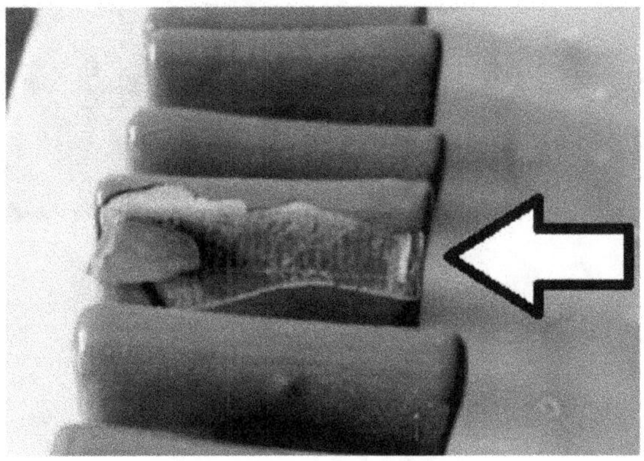

Example: Electric Breakdown in Auxiliary Part

To make rotating magnetic field in stator single phase supply is made like double phase supply with the help of a capacitor and double phase stator winding. Figure 73 shows a single phase induction motor and its capacitor.

Flashover

Undesired or unintended electric arcing can have detrimental effects on electric power transmission, distribution systems and electronic equipment. When an inductive circuit is switched off the current cannot instantaneously jump to zero and a transient arc will be formed across the separating contacts. If a circuit has enough current and voltage to sustain an arc formed outside of a switching device, the

Figure 73. A single phase induction motor with its axillary capacitor

Figure 74. A damaged stator bar of a plant turbogenerator due to flashover

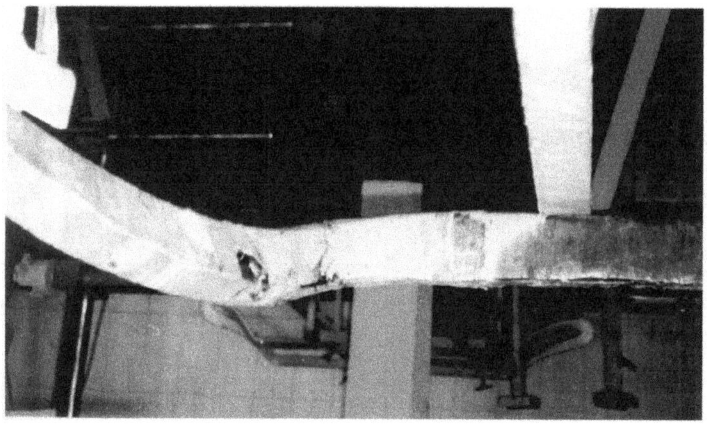

arc can cause damage to equipment such as melting of conductors, destruction of insulation, and fire (Gorgin, Sotoodeh, & Oraee, 2010). An arc flash describes an explosive electrical event that presents a hazard to people and equipment. Figure 74 shows a generator stator bar which is damaged by flashover.

Arcing can also occur when a low resistance channel due to conductive dust or moisture forms between places with different potential. The conductive channel then can facilitate formation of an electric arc. The ionized air has high electrical conductivity approaching that of metals, and can conduct extremely high currents, causing a short circuit and tripping protective devices.

Electric arc over the surface of plastics causes their degradation. A conductive carbon-rich track tends to form in the arc path, negatively influencing their insulation properties. Arcing over some types of printed circuit boards, possibly due to cracks of the traces or the failure of a solder, renders the affected insulating layer conductive as the dielectric is combusted due to the high temperatures involved. This conductivity prolongs the arcing due to cascading failure of the surface.

Example: Transformer Internal Phase-to-Earth and Phase-to-Phase Short-Circuit

This is the most common type of internal fault. Figures 75 and 76 show effect of internal short circuit on a generator stator bar.

Internal phase-to-phase short-circuit must be detected and cleared by:

- 3 fuses on the primary side of the transformer
- An over current relay that trips a circuit-breaker upstream of the transformer

The protection of transformers by transformer-mounted devices, against the effects of internal faults, is provided on transformers by the classical Buchholz mechanical relay. These relays can detect a slow accumulation of gases which results from the arcing of incipient faults in the winding insulation or from the ingress of air due to an oil leak.

Partial Discharge

A Partial Discharge is an electrical discharge or spark that bridges a small portion of the insulation between two conducting electrodes. Partial Discharge activity can occur at any point in the insulation system, where the electric field strength exceeds the breakdown strength of that portion of the insulating material.

Example: Failure in Transformer

For large transformers used in power distribution or electrical substations, the core and coils of the transformer are immersed in oil which cools and insulates. Oil circulates through ducts in the coil and around the coil and core assembly, moved by convection. The oil is cooled by the outside of the tank in small ratings, and in larger ratings an air-cooled radiator is used. Where a higher rating is required, or where the transformer is used in a building or underground, oil pumps are used to circulate the oil and an oil-to-water heat exchanger may also be used. Cast-resin power transformers encase the windings in epoxy resin. These transformers simplify installation since they are dry, without cooling oil, and so require no fire-proof vault for indoor installations. The epoxy protects the windings from dust and corrosive atmospheres. Figure 77 shows dry type transformer.

Example: Carbonized Arc

When arc-over occurs, it could cause carbonization on the PCB surface. This could result in a permanent short. This is also irreversible damage and would be especially bad if a single high-voltage spike created a permanent short, then any low voltage would still have a low impedance path available. Figures 78 and 79 show carbonized paths.

Figure 75. Voltage difference between two consecutive strands in a generator stator bar

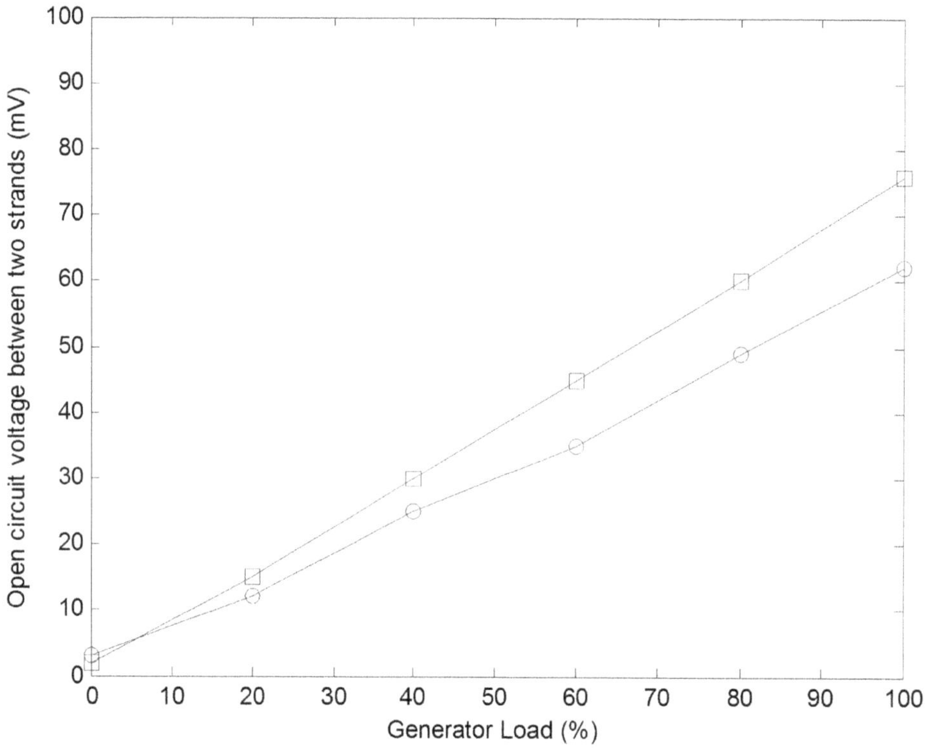

Figure 76. Leakage current between two consecutive strands in a generator stator bar

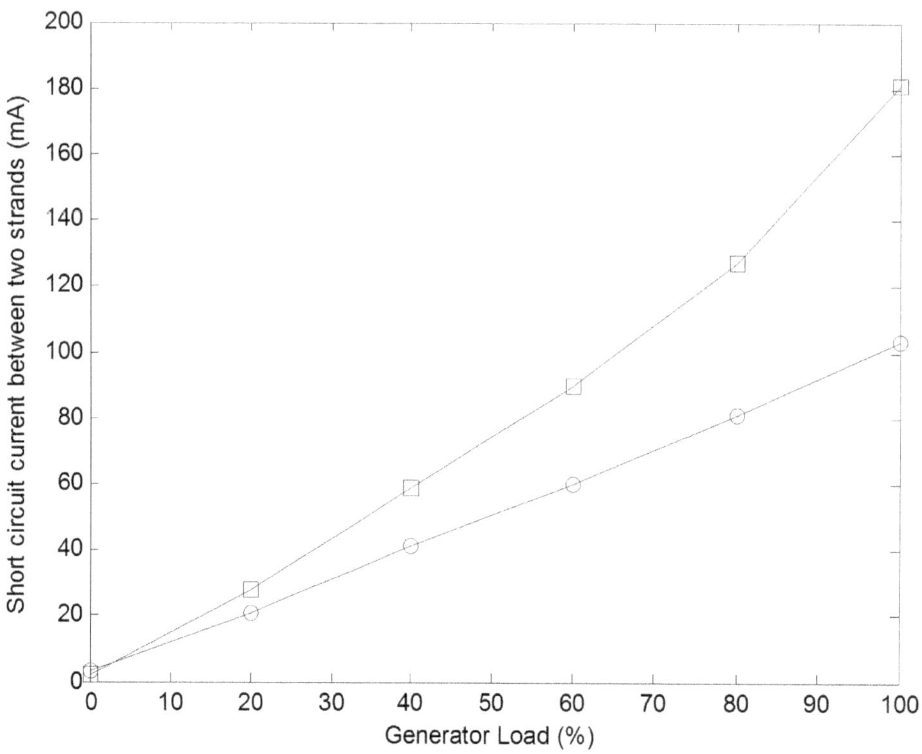

Figure 77. A transformer with resin as insulator

Figure 78. Carbonized path of electrical arc on a PCB

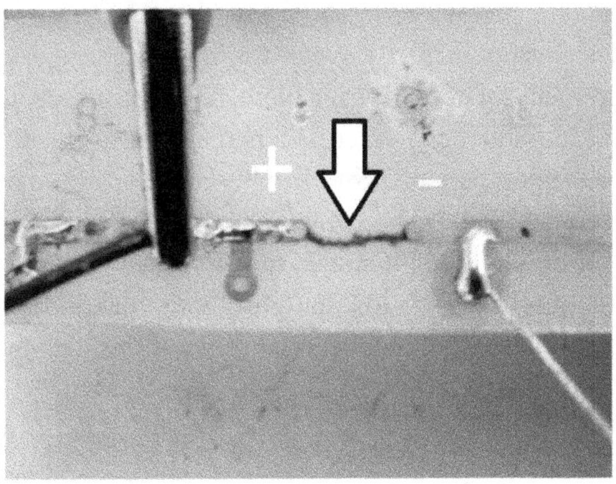

Figure 79. Carbonized path of electrical discharge on a connector box

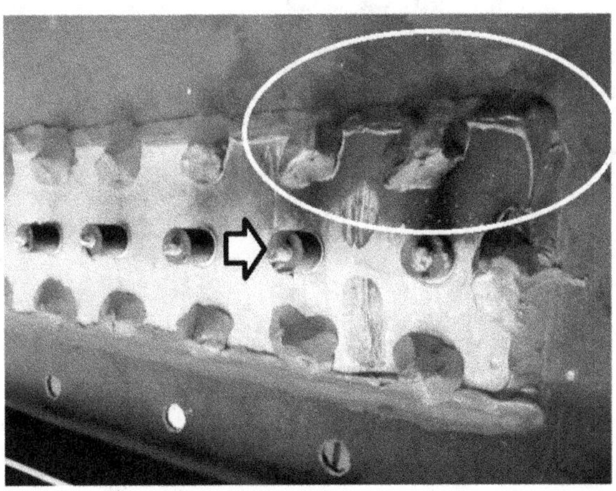

ENVIRONMENTAL FACTORS

Dust

Dust/dirt can accumulate on a board surface, reducing dielectric strength.

Cooling Water

High power systems are often cooled by circulating a liquid called engine coolant through the system, where it is heated, then through a radiator where it loses heat to the atmosphere, and then returned to the system. The coolant is usually water-based, but may also be oil. It is common to employ a water pump to force the coolant to circulate, and also for an axial fan to force air through the radiator. Figure 80 show a water cooling system. Figures 81, 82, 83, and 84 show stator bar of a turbogenerator and its blocked cooling water channels.

Humidity

Humidity is the amount of water vapor in the air. Water vapor is the gaseous state of water and is invisible. There are three main measurements of humidity: absolute, relative and specific. Absolute humidity is the water content of air. Relative humidity, expressed as a percent, measures the current absolute humidity relative to the maximum for that temperature. Specific humidity is a ratio of the water vapor content of the mixture to the total air content on a mass. Many electronic devices have humidity specifications, for example, 5% to 95%. At the top end of the range, moisture may increase the conductivity of permeable insulators leading to malfunction. Too low humidity may make materials brittle or electro-static discharge. A particular danger to electronic items, regardless of the stated operating humidity range, is

Figure 80. A water based cooling system

Figure 81. A stator bar, (a): cooling channels. (b): stator conductor

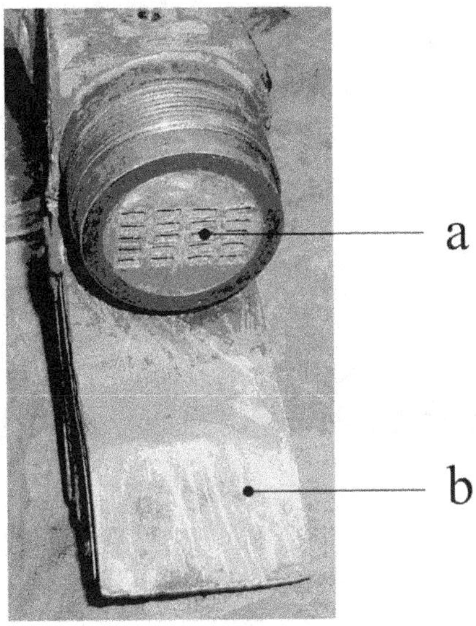

Figure 82. A normal stator bar with cooling channels

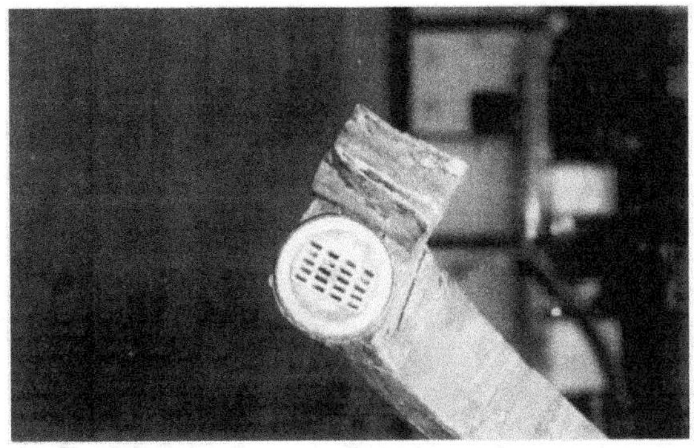

Figure 83. A stator bar with blocked cooling channel

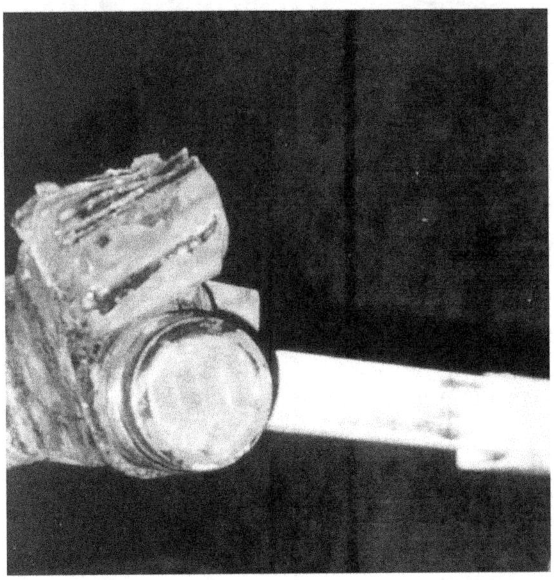

Figure 84. Mechanism of cooling water blockage

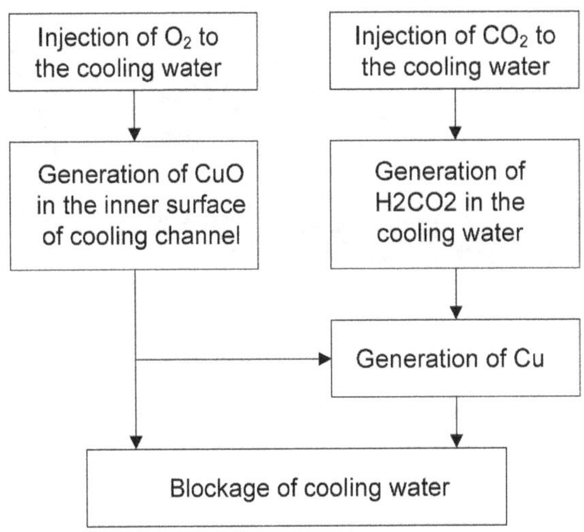

condensation. When an electronic item is moved from a cold place to a warm humid place, condensation may coat circuit boards and other insulators, leading to short circuit inside the equipment. Figures 85, 86, 87, and 88 show effect of humidity on the elements of power converters.

In contrast, a very low humidity level favors the build-up of static electricity, which may result in spontaneous shutdown of computers when discharges occur. In addition, electrostatic discharges can cause dielectric breakdown in solid state devices, resulting in irreversible damage.

Figure 85. A damaged autotransformer due to humidity

Figure 86. A damaged winding due to humidity

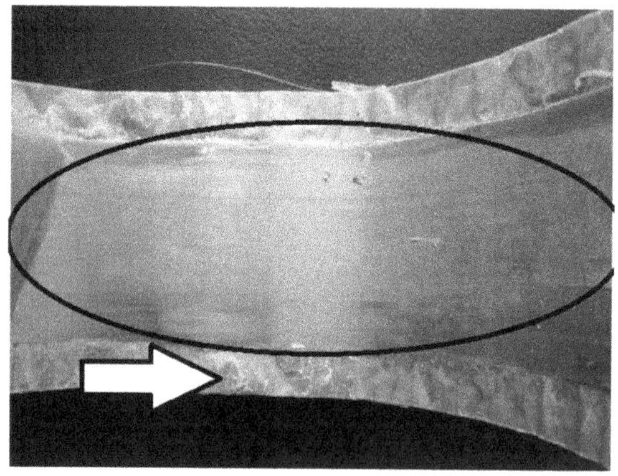

Figure 87. A damaged winding due to humidity

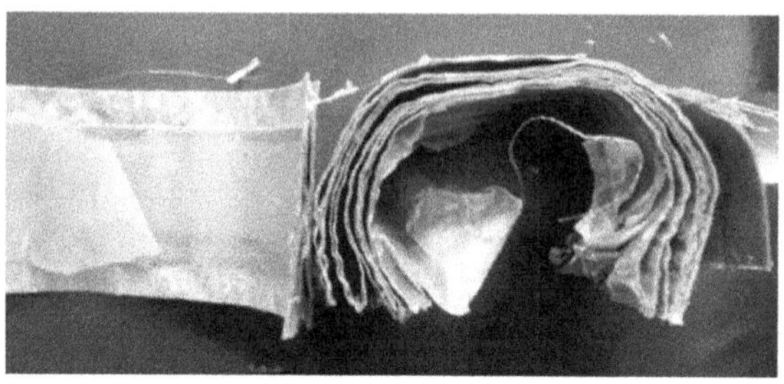

Figure 88. Effect of humidity on stator core

Figure 89 and 90 shows effect of environmental factors on failure of power converters.

MECHANICAL FACTORS

Mechanical Problems in Power Electronic Boards

One of the most likely failure modes is the PCB flexing which can cause the solder joints on the PCB to crack causing intermittent or complete failure of the connection. It is tried to keep the PCB as compact as possible. The smaller the PCB the smaller the overall flex of the board. The user can use a conformal coating on the PCB which will hold surface mount components in place as well as add some stiffness to the PCB.

Figure 89. Effect of humidity of winding isolation

Figure 90. Effect of dust on electric isolation

Mechanical Problems in Rotary Power Converters

Vibration can cause damage to electric motors in several ways. First, it can accelerate bearing failure by causing indentations on the bearing raceways at the ball or roller spacings. Secondly, it can loosen windings and cause mechanical damage to insulation by fracturing, flaking or eroding of the material. Third, the excessive movement it causes can result in lead wires becoming brittle. Fourth and final, it can cause brush sparking at commutators or current collector rings. As a result of these problems, whenever vibration is located in an electric motor, its source should be located quickly and corrected.

Motor Fan

All rotating electrical machines generate heat as a result of the electrical and mechanical losses inside the machine. Cooling is necessary to continuously transfer the heat to a cooling medium, such as the air. The different methods of cooling rotating machines are classified in the standards. Another method to remove heat from electric motors is by providing forced air cooling. This is commonly done by providing an electric fan to blow air over the motor. Force air cooling can reduce the amount of heat transferred into the machine structure and allow the motor to be operated at a higher load point. Figures 91 and 92 shows effect of dust on electric machines. Figure 93 shows effect of humidity on a PCB.

The most common type of AC motor is the totally enclosed fan cooled motor, which is provided with an external forced cooling fan mounted on the non-drive end of the shaft, with cooling ribs running axially along the outer surface of the motor frame. These are designed to keep the air flow close to the surface of the motor along its entire length, thus improving the cooling and self-cleaning of the ribs. An air-gap is usually left between the ribs and the fan cover for this purpose. Figures 94, 95, and 96 show an induction motor and its cooling system.

Figure 91. Effect of dust on a fan

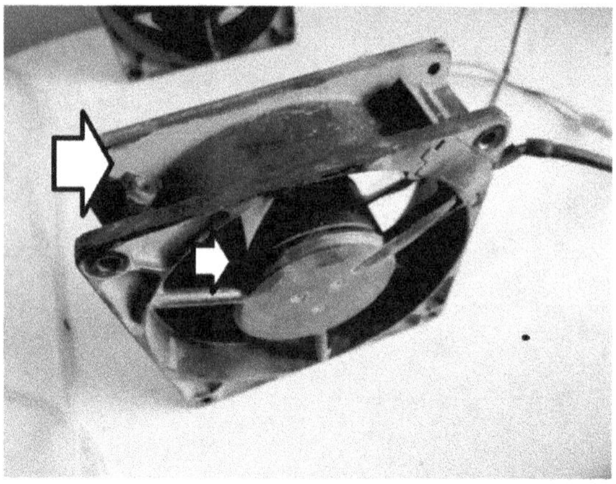

Figure 92. Effect of dust on rotor of a DC generator

Figure 93. Islanding due to humidity in a PCB

Rotor Broken Bar

Broken rotor bar is one of the commonly encountered induction motor faults that may cause serious motor damage to the motor if not detected timely. Rotor windings in squirrel cage induction motors are usually manufactured from aluminum alloy. Larger motors generally have rotors and end-rings fabricated out of these whereas motors with ratings less than a few hundred horsepower generally have die-cast aluminum alloy rotor cages. Replacement of the rotor core in larger motors is costly; therefore, by detecting broken rotor bars early, such secondary deterioration can be avoided. The rotor can be repaired at a fraction of the cost of rotor replacement, not to mention averting production revenue losses due to unplanned downtime. Figure 97 shows a squirrel cage rotor.

Some of the more common secondary effects of broken rotor bars are:

- Broken bars can cause sparking, a serious concern in hazardous areas.
- If one or more rotor bars are broken, the healthy bars are forced to carry additional current leading to rotor core damage from persistent elevated temperatures in the vicinity of the broken bars and current passing through the core from broken to healthy bars.

Figure 94. Rotor of a self-cooled induction motor, right: cooling ducts, left: fan

Figure 95. Stator core (right) and cooling ducts (left) in a self-cooled induction motor

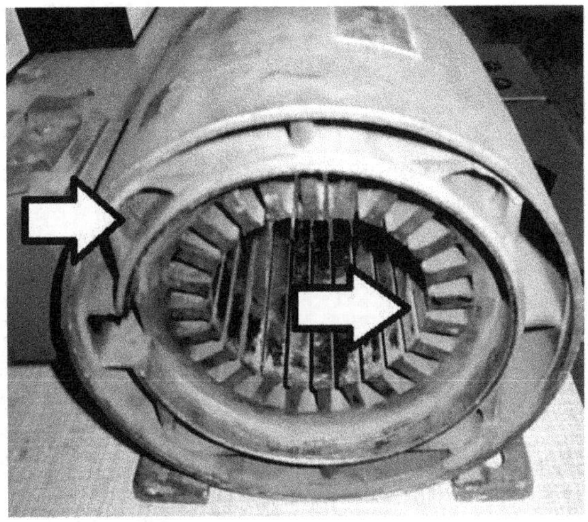

Figure 96. A natural convection cooled induction motor, right: rotor bars, left: end ring

Figure 97. Rotor of an induction motor, up: end ring, right: bearing, left: rotor

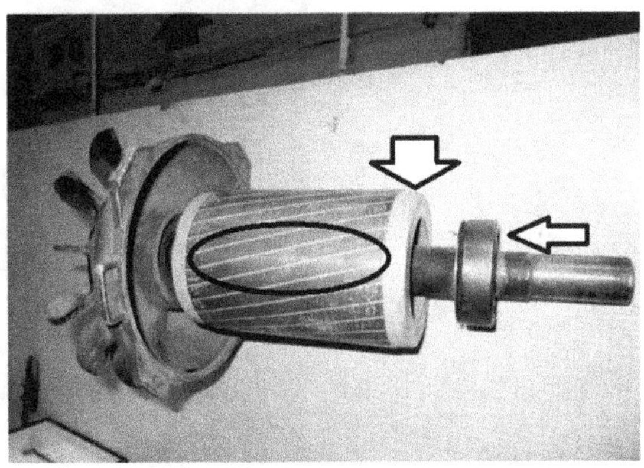

- Broken bars cause torque and speed oscillations in the rotor, provoking premature wear of bearings and other driven components.
- As the rotor rotates at high radial speed, broken rotor bars can lift out of the slot due to centrifugal force and strike against the stator winding causing a catastrophic motor failure.

Rotor Eccentricity

Air gap eccentricity is a mechanical fault with the motor. There are several possible causes for the presence of variances in the distance between a rotor and a stator. Some basic types of air gap eccentricities that can occur are:

- Rotor is eccentric to the axis of rotation
- Stator bore is eccentric
- Rotor and stator are round, but do not have the same axis of rotation
- Rotor and shaft are round, but do not have the same axis

Figures 98, 99, 100, 101, 102, and 103 shows eccentricity failure in various electric motors. The following are only a few of the possible causes of an air gap eccentricity:

Figure 98. A typical diagram of an electric machine to demonstrate the position of bearings and rotor

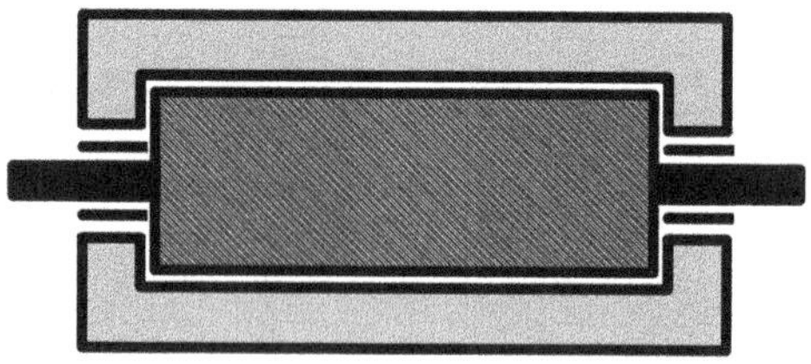

Figure 99. A cooling fan under stress

Figure 100. very thin air gap of a motor

Figure 101. Effect of eccentricity on rotor of an induction rotor

Figure 102. Effect of eccentricity on stator of an induction stator

Figure 103. Three different scenario for eccentricity

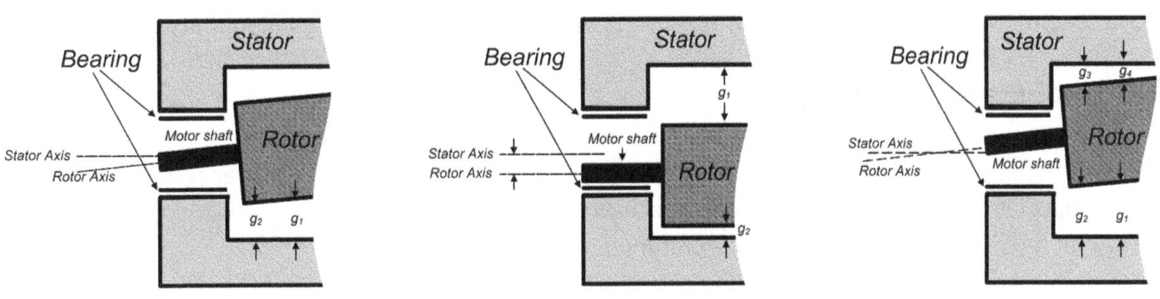

- Improper mounting of the motor to its bedplate can lead to an air gap distortion. A loose or missing bolt allows shifting of the motor's mounting foot during thermal expansion of the frame. This shifting over time could lead to a distortion of the frame and possible distortion of the stator bore.
- During construction of the motor, out-of-roundness of either the rotor or stator will lead to an air gap eccentricity. Industry standards recommend that measurements for total indicated roundness should be performed at different locations along the length of each of these components. Couple these measurements with the circumferences of each component, and depending on the speed and size of the motor, there are recommended tolerances from 5 to 20 percent variation in the air gap.
- Eccentricity can develop due to improper tensioning of drive belts coupled to a motor.
- Distorted end bells, cocked bearings, or a bent shaft will all cause an air gap eccentricity. During the manufacturing of the rotor, uneven mechanical stresses could be introduced into the cage and lamination stack leading to bowing of the completed rotor.
- An air gap eccentricity results in increased levels of vibration due to the uneven magnetic pull it creates between the circumference of the rotor and stator bore. Over time, these elevated levels of vibration can result in excessive movement of the stator winding, which could lead to increased friction and eventually a turn-to-turn, coil-to-coil, or ground fault. Increases in mechanical vibration accelerate bearing failure, which could seize the shaft and overheat the windings or allow additional movement of the shaft leading to a rotor/stator rub. The uneven magnetic stresses applied to the rotor, coupled with the increased vibration, will also contribute to mechanical looseness developing in the rotor assembly. Risk of rotor pull-over increases exponentially with the amount of air gap eccentricity.
- Rotor pull-over is an example of rotor/stator rub that appears as random marks on both the stator bore and surface of the rotor. Pull-over describes the bending of the motor shaft allowing the rotor to come into contact with the stator. Pull-over most often occurs during start up when the magnetic forces pulling on the rotor are greatest. The magnetic pull acting upon the rotor varies as the square of the difference in the air gap. For example, when the air gap at the narrowest point between the rotor and the stator is one half that of the air gap at its widest, the magnetic force at the narrowest will be four times stronger than the widest point.

Brushes

Carbon-brush collectors have successfully transferred current from excitation power sources to the rotating fields of synchronous generators for over a century. The basic principles of current transfer have remained the same over that time, although numerous improvements have been made in brush and collector-ring materials, brush-holder designs, and ventilation arrangements.

Collectors are sometimes overlooked, and failures occur. Such failures can be severe and the resulting forced outage may be long and costly (Abdi, Abdi, Oraee, & McMahon, 2014). Figures 104, 105, 106, 107, 108, and 109 show application of brushes in electric machines and possible failures.

DC Machine Commutator

The commutator ring may be worn or out-of-round, causing some brush float. This can be self-reinforcing as the spark chows out material. Some arcing is inevitable during interrupting significant current. Figures 110, 111, 112, 113, 114, and 115 shows commutator in DC machines.

Figure 104. Brush of an autotransformer

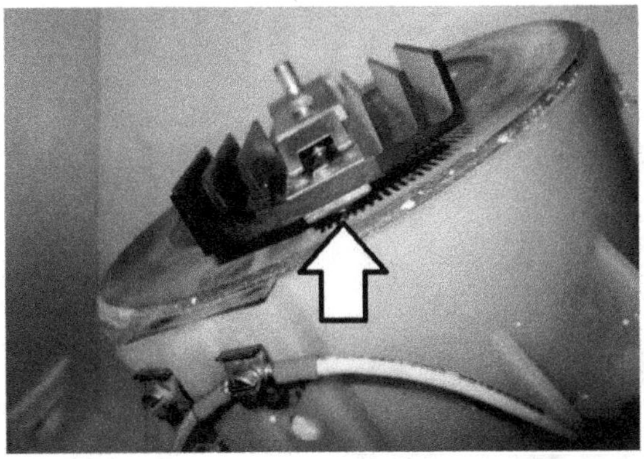

Figure 105. Brushes of a synchronous generator

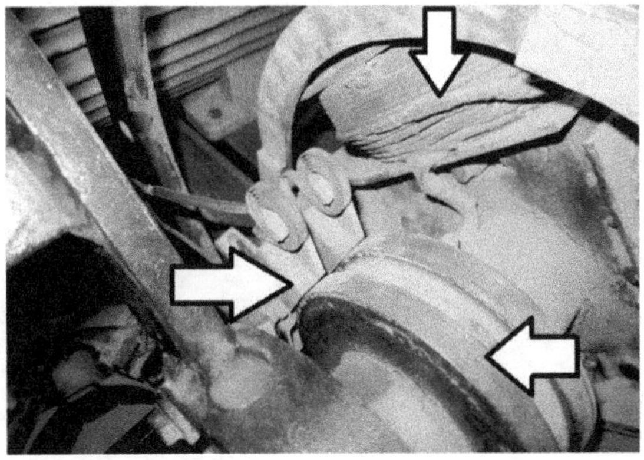

Figure 106. Brushes in rotor of an induction motor

Figure 107. Faulty brush of an autotransformer

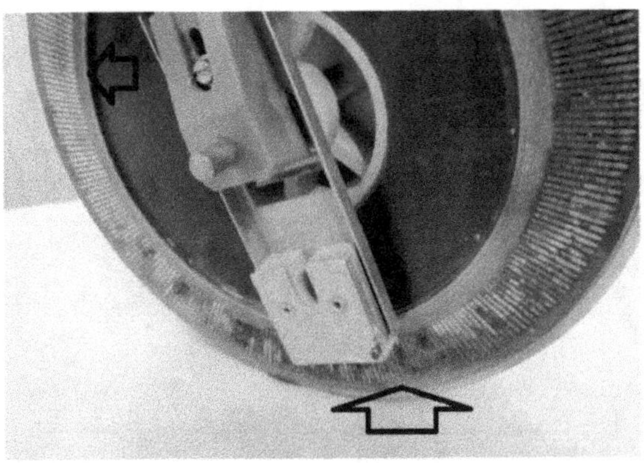

Figure 108. Faulty brush of an autotransformer

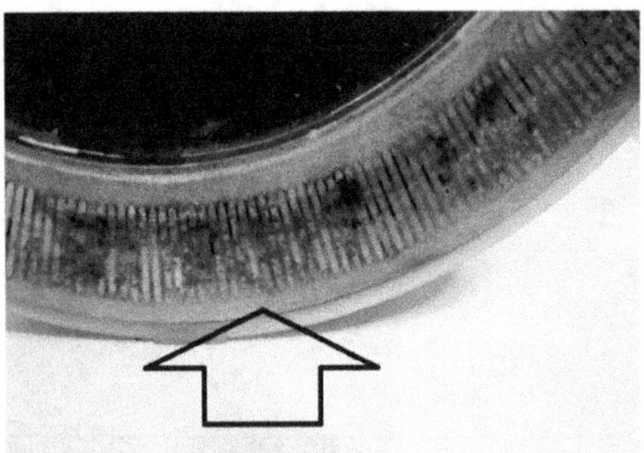

MECHANICAL AUX SYSTEMS

Tacho Generator

A generator specially designed and constructed for speed measuring is called a tachometer or tachogenerator. When not connected to a load resistance, generators will generate voltage roughly proportional to shaft speed (Figure 116). With precise construction and design, generators can be built to produce very precise voltages for certain ranges of shaft speeds, thus making them well-suited as measurement devices for shaft speed in mechanical equipment.

Figure 109. Faulty brush of synchronous generator

Figure 110. Rotor(down) and stator of a DC motor, up: main poles, left: auxiliary poles

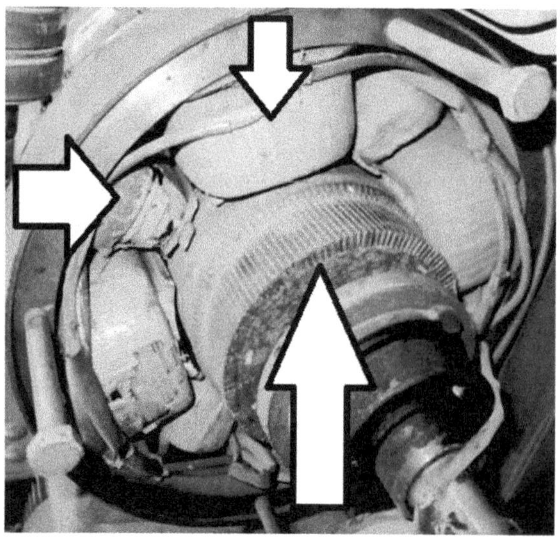

Figure 111. Brush (down) and commutator (up) of a DC motor

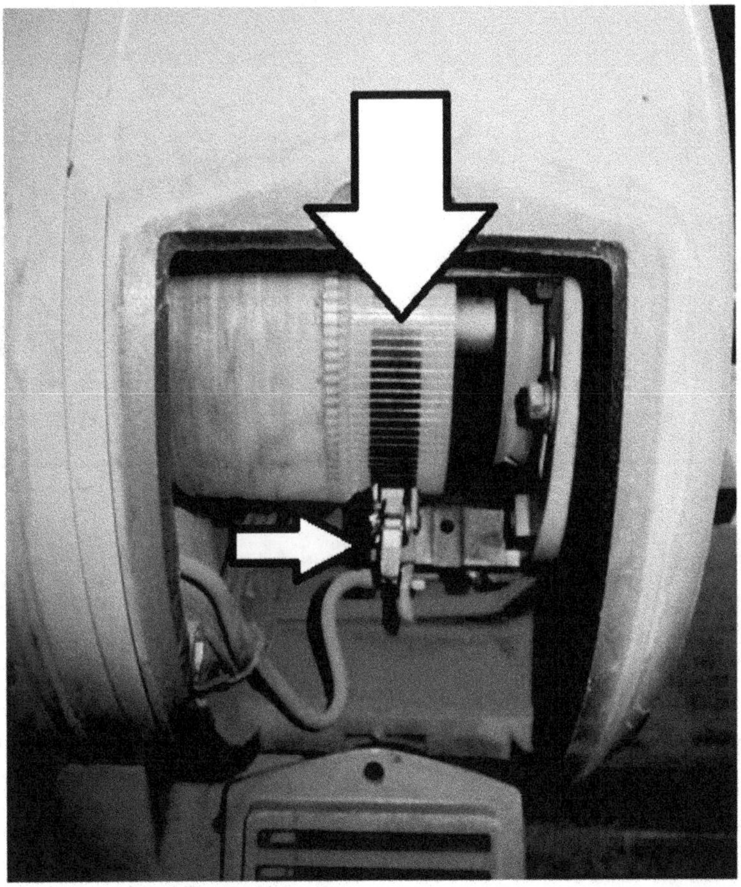

Figure 112. Rotor of a DC motor, left: core, middle: winding, right: commutator

Figure 113. A damaged commutator of a DC generator

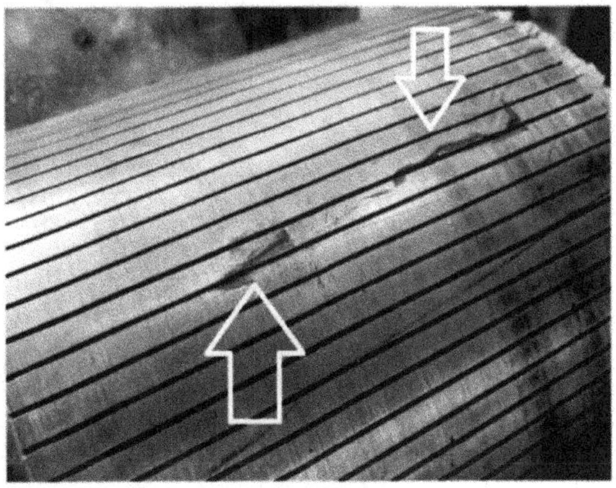

Figure 114. Connection of rotor conductors to commutator of a DC generator

Figure 115. Connection of rotor conductors to commutator of a DC generator

Tachogenerators can also indicate the direction of rotation by the polarity of the output voltage. When a permanent-magnet style DC generator's rotational direction is reversed, the polarity of its output voltage will switch. In measurement and control systems where directional indication is needed, tachogenerators provide an easy way to determine the direction of rotation.

Rotary Encoder

A rotary encoder, also called a shaft encoder, is an electro-mechanical device that converts the angular position or motion of a shaft or axle to an analog or digital code (Figure 117). There are two main types:

Figure 116. Tachogenerator of a generator

Figure 117. A SRM drives with optical encoder

absolute and incremental. The output of absolute encoders indicates the current position of the shaft, making them angle transducers. The output of incremental encoders provides information about the motion of the shaft, which is typically further processed elsewhere into information such as speed, distance, and position. Rotary encoders are used in many applications that require precise shaft unlimited rotation.

Bearings

There are different types of bearings and the required maintenance on them will depend on the type of bearing, operating environment and the motor application. Lubrication is only one of three maintenance tasks involved with motor bearings. Cleaning, removal and replacement are the other tasks. For most types, the sources of bearing failures are:

- Insufficient oil or grease.
- Too much grease causing churning and overheating.
- Worn bearings (i.e., broken balls or rough races, etc.)
- Hot motor or external environment.

Figure 118 shows bearing reliability characteristic.

Figure 118. Effect of torque ripple on reliability

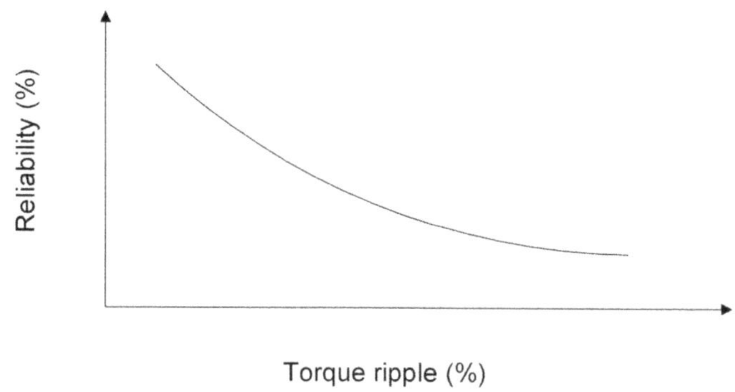

SUMMARY AND CONCLUSION

Some of the main reasons of failure in electric power converters were presented in this chapter. All these factors cause catastrophic damage in the systems. Over temperature, over voltage, mechanical forces and environmental effects are the main factors of failure in power converters which are described in details. The results of this chapter are summarized as follows:

1. Over temperature is the most important failure factor in electric power converters. Most failure factors finally lead to over temperature. Since the power conversion process is not ideal, heat generation is a common problem in all power converters. Of course, this is true for all electrical systems and not only for electric power converters. But, it is a serious problem in these converters because the amount of energy transfer is high and therefore, their energy loss is considerable. Conduction loss in conductors and semiconductors, switching loss in switching devices and core loss in the core of electric machines are the main mechanisms of energy loss in electric power converters. Transient phenomena like inrush currents with high power in a short time also cause over temperature damage. Damage due to over temperature is a long term factor in comparison to damage due to electric break down.
2. Electrical breakdown or dielectric breakdown is a rapid reduction in the resistance of an electrical insulator when the voltage applied across it exceeds a specified voltage. Electric breakdown has a fast effect on the failure of converters. It acts as break down inside the body of insulator or show a creepage on the surface of the insulator. Environmental factors has a direct effect on the creepage type of insulator breakdown. A breakdown in solid insulator usually leads to a permanent damage. Therefore, special attention to this failure factor is very important because low power and low voltage power converters use elements with solid insulator.
3. Mechanical factors affect over long term in comparison to electrical factors. Vibration is an important mechanical factor in failure of rotary power converters. In addition, outside generated vibration causes mechanical damage of leads in electronic elements. In mechanical world, transient condition like mechanical shock occurs similar to transient phenomena in electrical world. Improper mounting of the elements can lead to mechanical damage. Modal analysis is a tool for analyzing the mechanical behavior of structures and elements.

4. Operation in harsh environment accelerates the failure due to the above mentioned factors. Operation in high ambient temperature accelerates the failure due to over temperature. Humidity and dust act by reducing the insulator resistance and cause to electrical breakdown in lower voltages than the designed nominal values.

In the next chapter, we formulate the calculations of reliability based on the concepts which are reviewed in the current chapter. Each of the presented failure factors plays a role in the prediction of power converter reliability. They appear as a part of reliability formula and can be updated to the current condition of converter. In addition, recognition of dominant failure factor help the designers to prevent the predictable faults in power converters. Therefore, the materials which are described in chapter 2 are also useful in the chapters 7 and 8 for reliability improvement.

REFERENCES

Abdi, S., Abdi, E., Oraee, A., & McMahon, R. (2014). Equivalent Circuit Parameters for Large Brushless Doubly Fed Machines (BDFMs). *IEEE Transactions on Energy Conversion*, *29*(3), 706–715. doi:10.1109/TEC.2014.2311736

Abdi, S., Abdi, E., Oraee, A., & McMahon, R. (2015). Optimization of Magnetic Circuit for Brushless Doubly Fed Machines. *IEEE Transactions on Energy Conversion*, *30*(4), 1611–1620. doi:10.1109/TEC.2015.2468063

Barati, F., Oraee, H., Abdi, E., Shiyi, Sh., & McMahon, R. (2008). The Brushless Doubly-Fed Machine Vector Model in the rotor flux oriented reference frame. In *Proceedings of 34th Annual Conference of IEEE Industrial Electronics* (pp.1415-1420). IEEE. doi:10.1109/IECON.2008.4758161

Chiesa, N., & Høidalen, H. K. (2010). Novel Approach for Reducing Transformer Inrush Currents: Laboratory Measurements, Analytical Interpretation and Simulation Studies. *IEEE Transactions on Power Delivery*, *25*(4), 2609–2616. doi:10.1109/TPWRD.2010.2045772

Gorgin, H., Sotoodeh, P., & Oraee, H. (2010). Dynamic model for Brushless Doubly-Fed Machine with stator winding faults. In *Proceedings of 1st Power Electronic & Drive Systems & Technologies Conference* (pp. 316–322). IEEE. doi:10.1109/PEDSTC.2010.5471801

Gorginpour, H., Jandaghi, B., & Oraee, H. (2011). Time and space harmonics in brushless doubly-fed machine. In *Proceedings of 19th Iranian Conference on Electrical Engineering* (pp.1-6). IEEE.

Gorginpour, H., Oraee, H., & Abdi, E. (2014). Calculation of Core and Stray Load Losses in Brushless Doubly Fed Induction Generators. *IEEE Transactions on Industrial Electronics*, *61*(7), 3167–3177. doi:10.1109/TIE.2013.2279357

Hanna, R., & Schmitt, D. W. (2012). Failure Analysis of Induction Motors: Magnetic Wedges in Compression Stations. *IEEE Industry Applications Magazine*, *18*(4), 40–46. doi:10.1109/MIAS.2012.2191340

Hanna, R. A., Hiscock, W., & Klinowski, P. (2007). Failure Analysis of Three Slow-Speed Induction Motors for Reciprocating Load Application. *IEEE Transactions on Industry Applications*, *43*(2), 429–435. doi:10.1109/TIA.2006.889969

Lahyani, A., Venet, P., Grellet, G., & Viverge, P.-J. (1998). Failure prediction of electrolytic capacitors during operation of a switchmode power supply. *IEEE Transactions on Power Electronics*, *13*(6), 1199–1207. doi:10.1109/63.728347

Movahed, S. R., Oraee, H., Mirzamani, S. H., Rajabi, A., & Daneshvar, H. (2010). Estimation of insulation life of inverter- fed induction motors. In *Proceedings of 1st Power Electronic & Drive Systems & Technologies Conference* (pp. 335–339). IEEE. doi:10.1109/PEDSTC.2010.5471797

Oraee, H. (2000). A quantative approach to estimate the life expectancy of motor insulation systems. *IEEE Transactions on Dielectrics and Electrical Insulation*, *7*(6), 790–796. doi:10.1109/94.891990

Tohidi, S., Zolghadri, M. R., Oraee, H., Tavner, P., Abdi, E., & Logan, T. (2012). Performance of the brushless doubly-fed machine under normal and fault conditions. *IET Electric Power Applications*, *6*(9), 621–627. doi:10.1049/iet-epa.2012.0101

Yuancheng, R., Ming, X., Zhou, J., & Lee, F. C. (2006). Analytical loss model of power MOSFET. *IEEE Transactions on Power Electronics*, *21*(2), 310–319. doi:10.1109/TPEL.2005.869743

ADDITIONAL READING

Abraham, A. P., & Prabhakar, B. R. (1992). Effect of humidity and temperature on the DC breakdown of rod-rod and rod-plane gaps. *IEEE Transactions on Electrical Insulation*, *27*(2), 207–213. doi:10.1109/14.135592

Antonino-Daviu, J. A., Riera-Guasp, M., Folch, J. R., & Molina Palomares, M. P. (2006). Validation of a new method for the diagnosis of rotor bar failures via wavelet transform in industrial induction machines. *IEEE Transactions on Industry Applications*, *42*(4), 990–996. doi:10.1109/TIA.2006.876082

Baker, P. C., Judd, M. D., & McArthur, S. D. J. (2010). A frequency-based RF partial discharge detector for low-power wireless sensing. *IEEE Transactions on Dielectrics and Electrical Insulation*, *17*(1), 133–140. doi:10.1109/TDEI.2010.5412011

Bellini, A., Filippetti, F., Franceschini, G., Tassoni, C., & Kliman, G. B. (2001). Quantitative evaluation of induction motor broken bars by means of electrical signature analysis. *IEEE Transactions on Industry Applications*, *37*(5), 1248–1255. doi:10.1109/28.952499

Benbouzid, M. E. H., & Kliman, G. B. (2003). What stator current processing-based technique to use for induction motor rotor faults diagnosis? *IEEE Transactions on Energy Conversion*, *18*(2), 238–244. doi:10.1109/TEC.2003.811741

Bhuvaneswari, G., & Mahanta, B. C. (2009). Analysis of Converter Transformer Failure in HVDC Systems and Possible Solutions. *IEEE Transactions on Power Delivery*, *24*(2), 814–821. doi:10.1109/TPWRD.2009.2014271

Blewitt, W. M., & Gurwicz, D. I. (2008). Reduction of power MOSFET losses in hard-switched converters. *Electronics Letters*, *44*(18), 1088–1089. doi:10.1049/el:20081414

Blodt, M., Granjon, P., Raison, B., & Rostaing, G. (2008). Models for Bearing Damage Detection in Induction Motors Using Stator Current Monitoring. *IEEE Transactions on Industrial Electronics*, *55*(4), 1813–1822. doi:10.1109/TIE.2008.917108

Bonnett, A. H., & Soukup, G. C. (1988). Analysis of rotor failures in squirrel-cage induction motors. *IEEE Transactions on Industry Applications*, *24*(6), 1124–1130. doi:10.1109/28.17488

Bonnett, A. H., & Soukup, G. C. (1992). Cause and analysis of stator and rotor failures in three-phase squirrel-cage induction motors. *IEEE Transactions on Industry Applications*, *28*(4), 921–937. doi:10.1109/28.148460

Brekken, T. K. A., & Mohan, N. (2007). Control of a Doubly Fed Induction Wind Generator Under Unbalanced Grid Voltage Conditions. *IEEE Transactions on Energy Conversion*, *22*(1), 129–135. doi:10.1109/TEC.2006.889550

Chen, Y. H., & Cheng, P. T. (2011). Flux Estimation Techniques for Inrush Current Mitigation of Line-Interactive UPS Systems. *IEEE Transactions on Industry Applications*, *47*(2), 901–911. doi:10.1109/TIA.2010.2103913

Cusido, J., Romeral, L., Ortega, J. A., Rosero, J. A., & Garcia Espinosa, A. (2008). Fault Detection in Induction Machines Using Power Spectral Density in Wavelet Decomposition. *IEEE Transactions on Industrial Electronics*, *55*(2), 633–643. doi:10.1109/TIE.2007.911960

da Silva, A. M., Povinelli, R. J., & Demerdash, N. A. O. (2008). Induction Machine Broken Bar and Stator Short-Circuit Fault Diagnostics Based on Three-Phase Stator Current Envelopes. *IEEE Transactions on Industrial Electronics*, *55*(3), 1310–1318. doi:10.1109/TIE.2007.909060

del Perugia, C., Findlay, R. D., & Stranges, N. (2006). Skin Effect Factor in the Bar Extension of Large Two-Pole Induction Motors by Three-Dimensional Finite-Element Simulations. *IEEE Transactions on Magnetics*, *42*(10), 3404–3406. doi:10.1109/TMAG.2006.879711

Dey, D., Chatterjee, B., Chakravorti, S., & Munshi, S. (2010). Cross-wavelet transform as a new paradigm for feature extraction from noisy partial discharge pulses. *IEEE Transactions on Dielectrics and Electrical Insulation*, *17*(1), 157–166. doi:10.1109/TDEI.2010.5412014

Dissado, L. A. (2002). Understanding electrical trees in solids: From experiment to theory. *IEEE Transactions on Dielectrics and Electrical Insulation*, *9*(4), 483–497. doi:10.1109/TDEI.2002.1024425

Dorrell, D. G., Thomson, W. T., & Roach, S. (1997). Analysis of airgap flux, current, and vibration signals as a function of the combination of static and dynamic airgap eccentricity in 3-phase induction motors. *IEEE Transactions on Industry Applications*, *33*(1), 24–34. doi:10.1109/28.567073

Escobar, G., Valdez, A. A., Leyva-Ramos, J., & Mattavelli, P. (2007). Repetitive-Based Controller for a UPS Inverter to Compensate Unbalance and Harmonic Distortion. *IEEE Transactions on Industrial Electronics*, *54*(1), 504–510. doi:10.1109/TIE.2006.888803

Farag, A. S., Wang, C., Cheng, T. C., Zheng, G., Du, Y., & Hu, L. et al. (1998). Failure analysis of composite dielectric of power capacitors in distribution systems. *IEEE Transactions on Dielectrics and Electrical Insulation*, *5*(4), 583–588. doi:10.1109/94.708276

Hu, Q., Shu, L., Jiang, X., Sun, C., Zhang, S., & Shang, Y. (2011). Effects of air pressure and humidity on the corona onset voltage of bundle conductors. *IET Generation. Transmission & Distribution*, *5*(6), 621–629. doi:10.1049/iet-gtd.2010.0560

Idir, N., Weens, Y., & Franchaud, J.-J. (2009). Skin effect and dielectric loss models of power cables. *IEEE Transactions on Dielectrics and Electrical Insulation*, *16*(1), 147–154. doi:10.1109/TDEI.2009.4784562

(2012).. . *IGBT and Diode Loss Estimation Under Hysteresis Switching.*, *27*(3), 1044–1048.

Jeong, S. G., & Choi, J. Y. (2002). Line current characteristics of three-phase uncontrolled rectifiers under line voltage unbalance condition. *IEEE Transactions on Power Electronics*, *17*(6), 935–945. doi:10.1109/TPEL.2002.805580

Jiabing, H., Yikang, H., Lie, X., & Williams, B. W. (2009). Improved Control of DFIG Systems During Network Unbalance Using PI–R Current Regulators. *IEEE Transactions on Industrial Electronics*, *56*(2), 439–451. doi:10.1109/TIE.2008.2006952

Junhao, L., Quanwei, H., Xuefeng, Z., Xiu, Y., Yongfen, Y., & Yanming, Y. (2011). Partial-Discharge Characteristics of Free Spherical Conducting Particles Under AC Condition in Transformer Oils. *IEEE Transactions on Power Delivery*, *26*(2), 538–546. doi:10.1109/TPWRD.2010.2103575

Kaczer, B., Degraeve, R., De Keersgieter, A., Van de Mieroop, K., Simons, V., & Groeseneken, G. (2002). Consistent model for short-channel nMOSFET after hard gate oxide breakdown. *IEEE Transactions on Electron Devices*, *49*(3), 507–513. doi:10.1109/16.987123

Kaufhold, M., Aninger, H., Berth, M., Speck, J., & Eberhardt, M. (2000). Electrical stress and failure mechanism of the winding insulation in PWM-inverter-fed low-voltage induction motors. *IEEE Transactions on Industrial Electronics*, *47*(2), 396–402. doi:10.1109/41.836355

Kaufhold, M., Borner, G., Eberhardt, M., & Speck, J. (1996). Failure mechanism of the interturn insulation of low voltage electric machines fed by pulse-controlled inverters. *IEEE Electrical Insulation Magazine*, *12*(5), 9–16. doi:10.1109/57.537190

Kini, P. G., Bansal, R. C., & Aithal, R. S. (2007). A Novel Approach Toward Interpretation and Application of Voltage Unbalance Factor. *IEEE Transactions on Industrial Electronics*, *54*(4), 2315–2322. doi:10.1109/TIE.2007.899935

le Roux, W., Harley, R. G., & Habetler, T. G. (2007). Detecting Rotor Faults in Low Power Permanent Magnet Synchronous Machines. *IEEE Transactions on Power Electronics*, *22*(1), 32328. doi:10.1109/TPEL.2006.886620

Lee, C. Y. (2011). Effects of unbalanced voltage on the operation performance of a three-phase induction motor. *IEEE Transactions on Energy Conversion*, *14*(2), 20208.

Lee, J. C., Chen, I. C., & Hu, Ch. (1988). Modeling and characterization of gate oxide reliability. *IEEE Transactions on Electron Devices*, *35*(12), 2268–2278. doi:10.1109/16.8802

Li, L., Tang, J., & Liu, Y. (2015). Partial discharge recognition in gas insulated switchgear based on multi-information fusion. *IEEE Transactions on Dielectrics and Electrical Insulation*, *22*(2), 1080–1087. doi:10.1109/TDEI.2015.7076809

Miller, H. C. (1989). Surface flashover of insulators. *IEEE Transactions on Electrical Insulation, 24*(5), 765–786. doi:10.1109/14.42158

Nakagawa, T. (2007). *Shock and Damage Models in Reliability Theory*. Berlin, Germany: Springer.

Nandi, S., Bharadwaj, R. M., & Toliyat, H. A. (2002). Performance analysis of a three-phase induction motor under mixed eccentricity condition. *IEEE Transactions on Energy Conversion, 17*(3), 39399. doi:10.1109/TEC.2002.801995

Ordaz-Moreno, A., de Jesus Romero-Troncoso, R., Vite-Frias, J. A., Rivera-Gillen, J. R., & Garcia-Perez, A. (2008). Automatic Online Diagnosis Algorithm for Broken-Bar Detection on Induction Motors Based on Discrete Wavelet Transform for FPGA Implementation. *IEEE Transactions on Industrial Electronics, 55*(5), 2193–2202. doi:10.1109/TIE.2008.918613

Pechtmichael, G. (2004). Parts Selection and Management. John Wiley & Sons Ltd.

Piccin, R., Mor, A., Morshuis, P., Girodet, A., & Smit, J. (2015). Partial discharge analysis of gas insulated systems at high voltage AC and DC. *IEEE Transactions on Dielectrics and Electrical Insulation, 22*(1), 218–228. doi:10.1109/TDEI.2014.004711

Qi, B., Li, Ch., Xing, Zh., & Wei, Zh. (2014). Partial discharge initiated by free moving metallic particles on GIS insulator surface: Severity diagnosis and assessment. *IEEE Transactions on Dielectrics and Electrical Insulation, 21*(2), 766–774. doi:10.1109/TDEI.2013.003585

Rahmani, S., Mendalek, N., & Al-Haddad, K. (2010). Experimental Design of a Nonlinear Control Technique for Three-Phase Shunt Active Power Filter. *IEEE Transactions on Industrial Electronics, 57*(10), 3364–3375. doi:10.1109/TIE.2009.2038945

Rajagopalan, S., Aller, J. M., Restrepo, J. A., Habetler, T. G., & Harley, R. G. (2007). Analytic-Wavelet-Ridge-Based Detection of Dynamic Eccentricity in Brushless Direct Current (BLDC) Motors Functioning Under Dynamic Operating Conditions. *IEEE Transactions on Industrial Electronics, 54*(3), 1410–1419. doi:10.1109/TIE.2007.894699

Rodriguez, D., Gorur, R. S., & Hansen, P. M. (2010). Effect of humidity on the breakdown characteristics of air in non-uniform fields at 30 kHz. *IEEE Transactions on Dielectrics and Electrical Insulation, 17*(1), 45–52. doi:10.1109/TDEI.2010.5412001

Roy, M., Nelson, J. K., MacCrone, R. K., Schadler, L. S., Reed, C. W., Keefe, R., & Zenger, W. (2005). Polymer nanocomposite dielectrics-the role of the interface. *IEEE Transactions on Dielectrics and Electrical Insulation, 12*(4), 629–643. doi:10.1109/TDEI.2005.1511089

Saarinen, K., Frisk, L., & Ukkonen, L. (2012). Effects of Cycling Humidity on the Performance of RFID Tags With ACA Joints. *IEEE Transactions on Reliability, 61*(1), 23–31. doi:10.1109/TR.2011.2182223

Sergeant, P., Dupre, L., Vandenbossche, L., Garshelis, I., & Tollens, S. (2008). Numerical Model for the Drag Force Method to Evaluate Hysteresis Loss. *IEEE Transactions on Magnetics, 44*(6), 84845. doi:10.1109/TMAG.2008.916550

Sune, J. (2001). New physics-based analytic approach to the thin-oxide breakdown statistics. *IEEE Electron Device Letters, 22*(6), 296–298. doi:10.1109/55.924847

Tinga, T. (2013). *Principles of Loads and Failure Mechanisms*. Berlin, Germany: Springer. doi:10.1007/978-1-4471-4917-0

Tseng, H.-T., & Chen, J.-F. (2012). Voltage compensation-type inrush current limiter for reducing power transformer inrush current. *IET Electric Power Applications*, *6*(2), 101–110. doi:10.1049/iet-epa.2011.0151

Umemura, T., Nakamura, S., Hikita, M., Maeda, T., & Higashiyama, M. (2013). Partial discharges of small-air-gap in cast-resin insulation system. *IEEE Transactions on Dielectrics and Electrical Insulation*, *20*(1), 255–261. doi:10.1109/TDEI.2013.6451365

Verdelho, P., & Marques, G. D. (1997). An active power filter and unbalanced current compensator. *IEEE Transactions on Industrial Electronics*, *44*(3), 321–328. doi:10.1109/41.585829

von Jouanne, A., & Banerjee, B. (2001). Assessment of voltage unbalance. *IEEE Transactions on Power Delivery*, *16*(4), 78790. doi:10.1109/61.956770

Wang, X., & You, Ch. (2013). Effect of humidity on negative corona discharge of electrostatic precipitators. *IEEE Transactions on Dielectrics and Electrical Insulation*, *20*(5), 1720–1726. doi:10.1109/TDEI.2013.6633702

Xiong, Y., Sun, Sh., Jia, H., Shea, P., & Shen, Z. J. (2008). New Physical Insights on Power MOSFET Switching Losses. *IEEE Transactions on Power Electronics*, *24*(2), 525–531. doi:10.1109/TPEL.2008.2006567

Xu, Y., Tolbert, L. M., Kueck, J. D., & Rizy, D. T. (2010). Voltage and current unbalance compensation using a static var compensator. *IET Power Electronics*, *3*(6), 977–988. doi:10.1049/iet-pel.2008.0094

Yong, L. (2012). *Power Electronic Packaging*. Berlin, Germany: Springer.

Yoshida, M., Kojima, H., Hayakawa, N., Endo, F., & Okubo, H. (2011). Evaluation of UHF method for partial discharge measurement by simultaneous observation of UHF signal and current pulse waveforms. *IEEE Transactions on Dielectrics and Electrical Insulation*, *18*(2), 425–431. doi:10.1109/TDEI.2011.5739446

Zarri, L., Mengoni, M., Tani, A., Serra, G., & Casadei, D. (2010). Minimization of the Power Losses in IGBT Multiphase Inverters with Carrier-Based Pulsewidth Modulation. *IEEE Transactions on Industrial Electronics*, *57*(11), 3695–3706. doi:10.1109/TIE.2010.2041737

Zhang, P., Du, Y., Habetler, T. G., & Lu, B. (2011). A Survey of Condition Monitoring and Protection Methods for Medium-Voltage Induction Motors. *IEEE Transactions on Industry Applications*, *47*(1), 34–46. doi:10.1109/TIA.2010.2090839

KEY TERMS AND DEFINITIONS

Electric Field: A quantity for characterization of electric force.
Heat: A form of energy that affect as a failure factor in this chapter.
Humidity: An index for the amount of water in air.
Vibration: Mechanical movement in a system.

Section 2
Reliability Calculation

Chapter 3
Reliability Prediction

ABSTRACT

It is important to have an expectation about useful life of a system before its construction or even its remaining useful life during its operation. Reliability prediction is a tool for this goal. Reliability is the probability of performing adequately to achieve the desired aim of the system. In this chapter, probability calculation is used to predict failure rate of the converter. The formulation of these calculations are based on the concepts of failure factors which were described in the previous chapter. Some detailed examples are presented to show the power of probability tool for analyzing the behavior of complex systems. This chapter covers the methods for reliability calculation from component to system level. Some standards of reliability are presented. One can use the information from a reliability prediction to guide design decisions throughout the development cycle. MIL-HDBK-217 is described in details as a well-known standard for reliability prediction at component level. Reliability modeling is introduced for calculating the reliability at system level. Difference between system block diagram and reliability model is presented. The reliability models of various static and rotary power converters are expressed. Some examples are presented to demonstrate the procedure of calculations for a simple converter with its auxiliary components. This chapter gives a quantitative view to reader about evaluation of reliability and it can be used in the next chapters for reliability improvement.

INTRODUCTION: RELIABILITY PREDICTION

Now, we know the reasons for failure in power converters. In this chapter, we try to predict the chance of failure. The state of this chapter in the flowchart of the book is shown in Figure 1. Reliability predictions provide a quantitative basis for evaluating the power converters reliability.

The information obtained from a reliability prediction is used to guide design decisions throughout the development cycle. When an initial design concept is proposed, a reliability prediction can indicate the design feasibility from a reliability standpoint. For example, the designer might have a requirement of a 200,000 hr MTBF for a power converter. If the predicted value is 35,000 hr, the current design concept may not be feasible. With modifying the design concept or revising the requirement, a predicted value of 500,000 hr can give confidence in design concept. (Figure 2).

DOI: 10.4018/978-1-4666-9429-3.ch003

Figure 1. State of chapter 3 in the flowchart of the book

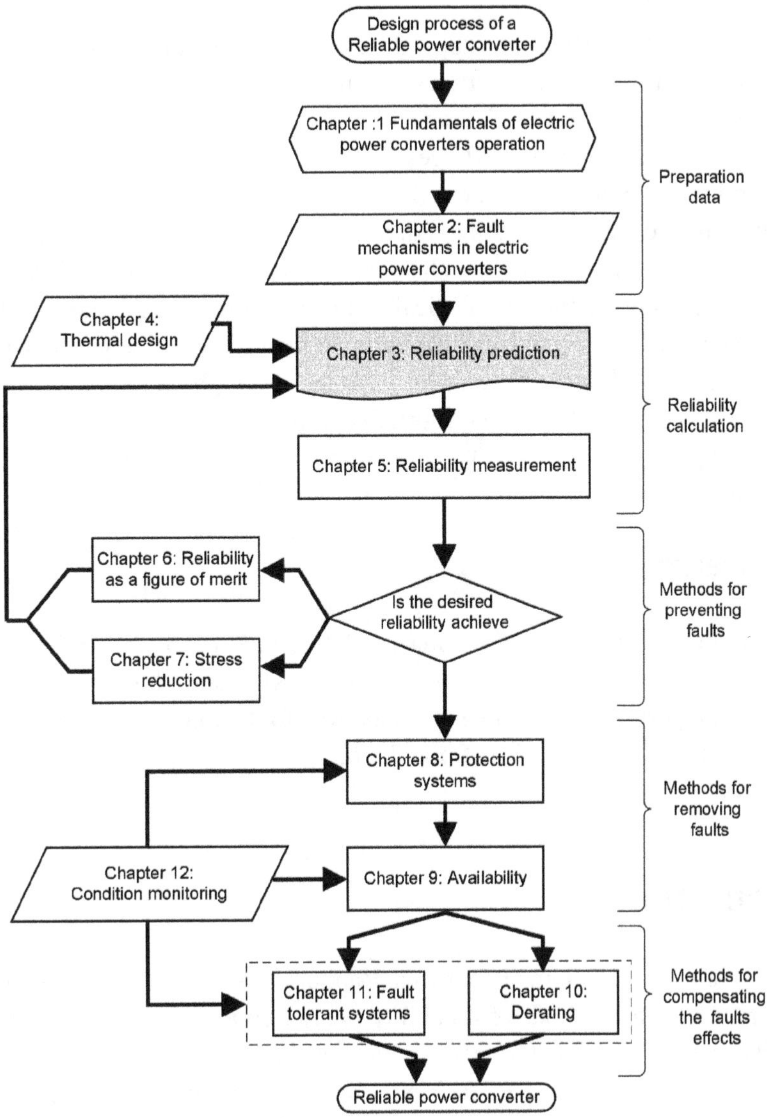

Figure 2. Comparison between failure in electrical (solid) and mechanical (dashed) devices

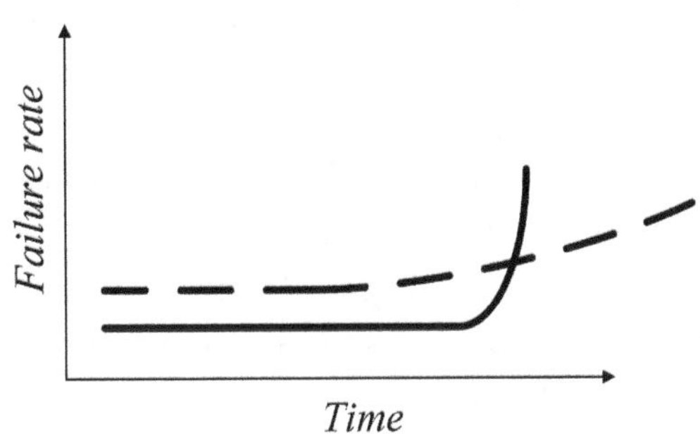

Reliability Prediction Methodology

There are reliability prediction techniques depending on the knowledge about design. As more details of the design are known, more accurate methods become available. These methods use part failure rate models, which predict the failure rates of parts based on various part parameters, such as technology, complexity, package type, quality level, and stress levels 3.

Predictive methods attempt to predict the reliability of a part based on some model typically developed through empirical studies and/or testing. An attempt is made to identify critical variables such as materials, application environmental and mechanical stresses, application performance requirements, duty cycle and manufacturing techniques. Typically, a base failure rate for the component is assigned, and this is multiplied by factors for each critical variable identified. Some predictive models assume a constant failure rate over the lifetime of a product. This ignores higher failure rates typically seen at the beginning and end of component life, infant mortality, and wear-out, respectively. Predictive methods can provide a relatively accurate reliability estimate in cases where good studies have been done to analyze field failures.

Reliability Comparison

Predictive methods are also useful in providing a relative ranking of reliability between alternative designs, but the absolute reliability numbers (or failure rates) obtained with these methods will rarely be indicative of real-life performance. This is the best way to demonstrate the real reliability level of a product. It requires a statistically significant population in the field, and a reasonably long time in the field (Javadian, & Kaboli, 2013).

PROBABILISTIC TOOL

Fault occurrence is a relatively random phenomena. Randomness means lack of pattern or predictability in events. Therefore, essential methods of reliability prediction are based on probability analysis. The fields of probability and statistics use formal definitions of randomness. In statistics, a random variable is an assignment of a numerical value to each possible outcome of an event space. This association facilitates the identification and the calculation of probabilities of the events. Random variables can appear in random sequences. A random process is a sequence of random variables describing a process whose outcomes do not follow a deterministic pattern, but follow an evolution described by probability distributions. These and other constructs are extremely useful in probability theory and the various applications of randomness (Sutanto, Outhred, & Lee, 1989). Probabilistic design is a discipline within engineering design. It deals primarily with the consideration of the effects of random variability upon the performance of an engineering system during the design phase. Typically, these effects are related to quality and reliability. Thus, probabilistic design is a tool that is mostly used in areas that are concerned with quality and reliability. Probability theory is applied in everyday life in risk assessment and in trade on financial markets (Bak-Jensen, Bech, Bjerregaard, & Jensen, 1999). Figure 3 shows application of probabilistic tool for analyzing the flash over.

Figure 3. Application of probability in flash over analysis IEC publication number and edition Copyright © year of publication IEC Geneva, Switzerland. www.iec.ch

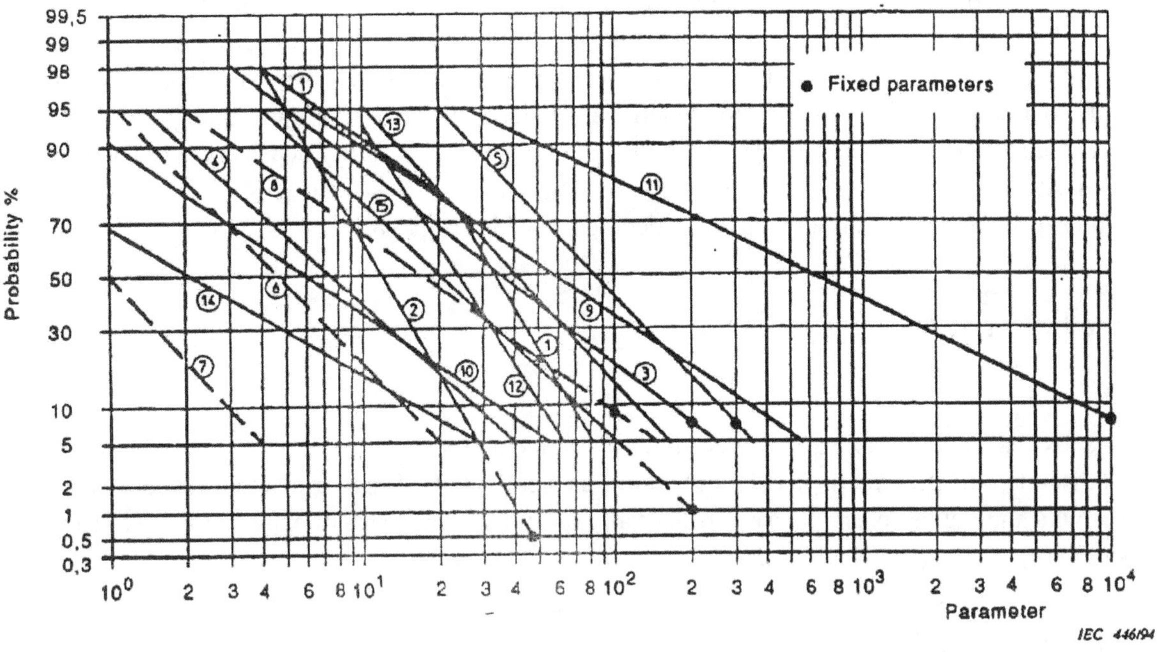

Parameter	Scale of abscissa	Positive flash	First positive stroke	Negative flash	First negative stroke	Negative subsequent strokes
I	kA		(3) •		(1)	(2) •
Q_f	C	(5) •		(4)		
Q_s	C		(8) •		(6)	(7)
W/R	kJ/Ω		(11) •		(9)	(10)
$(\Delta i/\Delta t)_{max}$	kA/μs		(14)		(12)	(13)
$(\Delta i/\Delta t)_{30\%/90\%}$	kA/μs					(15) •

Random Variables

A random variable x is, in the simplest term, a variable which randomly takes some values in a specific range (Melo, Pereira, & Leite da Silva, 1993). In probability field, a random variable is expressed by its statistical moments. The probability with which different values are taken by a random variable is quantified by the probability distribution function F(x), defined by F(x)=Pr(x≤x), that is, the probability that random variable x has a value less than or equal to x. The probability density function (PDF) p(x) is the derivative given by p(x)=dF(x)/dx.

The expectation of a random variable x, denoted as E{x}, is given by:

$$\bar{X} = E\{\mathrm{x}\} = \int_{-\infty}^{+\infty} xp(x)dx \tag{1}$$

This is also called the mean value of x, or the first moment of x.

A discrete random process may be thought of as a collection of real or discrete sequences of time, any of which might be observed on any trial of an experiment. The mean or expected value of a random process x[n] at the time index n is defined as:

$$\bar{x}[n] = E\{x[n]\} = \sum_{i=1}^{n} x_i p_i \tag{2}$$

The autocorrelation of a discrete random process between two sample times indices n_1 and n_2 is defined as:

$$R_{xx}[n_1, n_2] = E\{x[n_1]x^*[n_2]\} \tag{3}$$

Where * is the conjugate of x. Equation (3) is the engineering definition for autocorrelation, as first suggested by Wiener. The autocorrelation function of a random process is the appropriate statistical average that will be concerned with characterizing random signals in the time domain. A random process is wide-sense stationary (WSS) if its mean is constant for all time indices (i.e. independent of time) and its autocorrelation depends only on time index difference m = n_1-n_2. Therefore, autocorrelation of a WSS discrete random process x[n] is given by

$$R_{xx}[n_1, n_1 - m] = E\{x[n+m]x^*[n]\} \tag{4}$$

The power spectral density of a discrete random process is defined as the discrete time Fourier transform (DFT) of the autocorrelation sequence.

$$P_{xx}(f) = \sum_{m=-\infty}^{+\infty} R_{xx}[n_1, n_1 - m]\, e^{-j2\pi f m} \tag{5}$$

Example: Probabilistic Tool for Harmonic Estimation

Amplitude and order of voltage harmonics fed into the motor are a measure of the motor harmonic loss and also electromagnetic interference (Kaboli, & Zolghadri, 2003). The voltage spectrum of DTC drive has not previously been investigated systematically because of difficulties that arise from its nonlinear nature. There are only a few papers studying the harmonic spectrum of DTC.

Probabilistic methods are appropriate tools for analyzing the complex processes because they can eliminate the dependence of the process on time. Although the result of a probabilistic analysis is an estimation of the real response, but, it is the only systematic analyzing tools used for many complex processes such as radar studies and atomic phenomenon.

In this section, a probabilistic method is used to study the voltage spectrum of a direct torque controlled induction motor drive. However, this method is a general technique which can be used for any other type of machine controlled by DTC (Kaboli, Vahdati-Khajeh, & Zolghadri, 2006). In the method presented, the inverter output voltage vector is considered as a discrete random process with seven possible values. Using, the voltage vectors chain which we explain later, the autocorrelation function of voltage vectors is obtained and its power spectral density (PSD) is calculated. The effect of torque and flux hysteresis bands, machine parameters and inverter dc-link voltage on the harmonic spectrum is investigated. DTC block diagram is shown in Figure 4. A three phase inverter with its respective voltage vectors is shown in Figure 5. Selection of voltage vectors in the DTC loop is a nonlinear function of both stator flux and torque errors. The modeling of the voltage vector chain with a random process is the basic idea used for analyzing the voltage spectrum in DTC. Origin of this idea comes from where the acoustic noise radiation of DTC has been compared with a random PWM (RPWM) scheme and it is shown that they are very similar. In addition, comparison between characteristics of RPWM and DTC in simulation shows that the selection of voltage vectors has a semi-random behavior. For a sampling time index n, the inverter output voltage vector, U[n], gets one of the eight possible voltage vectors in a two-level three-phase inverter. Thus, U[n] may be considered as a complex discrete random process with seven possible values (six non-zero voltage vector and one zero voltage vector). Figure 6 shows a statistical diagram about probability of selection of different vectors. The probability with which non-zero voltage vectors are taken by random variable is the same and is generally different from the probability of zero vectors.

The mean value of U[n] is calculated as follows.

$$\begin{aligned}\bar{\mathbf{U}}[n] &= E\{\mathbf{U}[n]\} = \sum_{i=0}^{7}\mathbf{U}_i p(\mathbf{U}_i[n]) \\ &= \Pr 1\times 0 + \Pr 2\sum_{i=1}^{6}\mathbf{U}_i + \Pr 1\times 0 = 0\end{aligned} \tag{6}$$

Therefore, the mean value of U[n] is constant and independent of time. The autocorrelation of U[n] is expressed by:

$$R_{UU}[n,n-m] = E\{U[n]U^*[n-m]\} \tag{7}$$

Figure 4. DTC block diagram

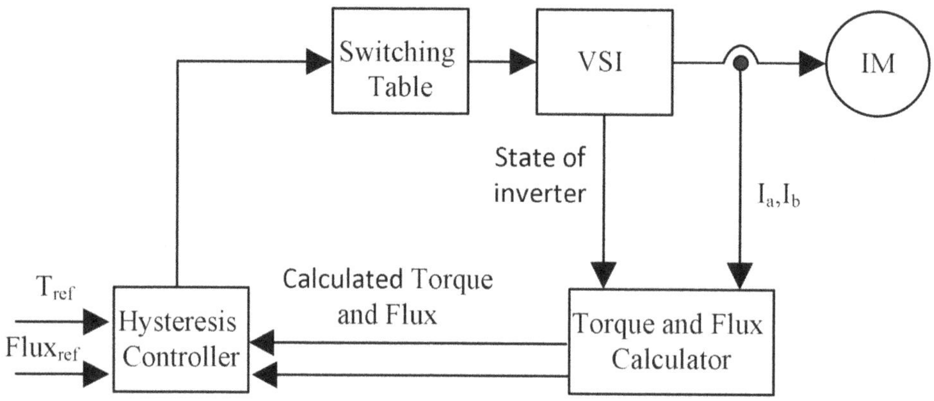

Figure 5. Respective voltage vectors of three phase inverter

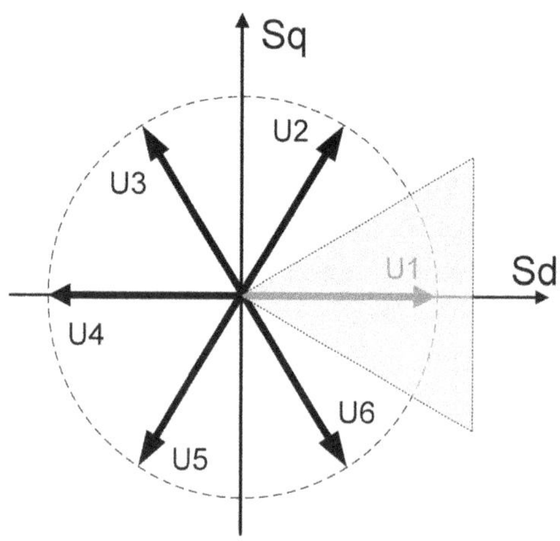

Figure 6. Time diagram of DTC variables

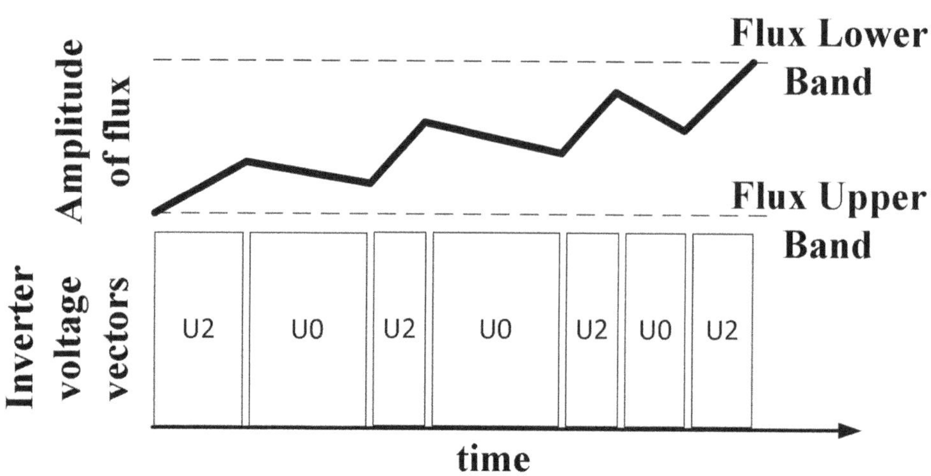

Because of symmetry, the conjugate of each vector (U) and product of two of them (U_i U_j) remain in the same set. Therefore, the product value U[n]U[n-m] takes one of the seven values (six non-zero vectors and two zero vectors). Thus, the above mentioned equation can be rewritten as the following:

$$\begin{aligned} R_{UU}[n,n-m] &= \sum_{i=0}^{7} \mathbf{U}_i p\left(\mathbf{U}[n]\mathbf{U}^*[n-m]=\mathbf{U}_i\right) \\ &= 0 \times p\left(\mathbf{U}[n]\mathbf{U}^*[n-m]=\mathbf{U}_0\right) \\ &\quad + \sum_{i=1}^{6} \mathbf{U}_i p\left(\mathbf{U}[n]\mathbf{U}^*[n-m]=\mathbf{U}_i\right) \\ &\quad + 0 \times p\left(\mathbf{U}[n]\mathbf{U}^*[n-m]=\mathbf{U}_7\right) \\ &= \sum_{i=1}^{6} \mathbf{U}_i p\left(\mathbf{U}[n]\mathbf{U}^*[n-m]=\mathbf{U}_i\right) \end{aligned} \tag{8}$$

If ψ[n] is in the k_{th} sector (k=1 to 6) then the possible choices for U[n] are determined. In this study, the solution A is used to control the inverter. However, this is not a limiting assumption and every switching table can be analyzed using this algorithm. Therefore, U[n] takes three values in each sector: U_0, U_{k+1} and U_{k+2}. Using Bayes' law, the probability of voltage vectors product (U[n]U*[n-m]) can be given by:

$$\begin{aligned} & p\left(\mathbf{U}[n]\mathbf{U}^*[n-m]=\mathbf{U}_i\right)=p\left(\mathbf{U}[n]\mathbf{U}^*[n-m]=\mathbf{U}_i/\mathbf{U}[n]=\mathbf{U}_0\right)p\left(\mathbf{U}[n]=\mathbf{U}_0\right) \\ & +p\left(\mathbf{U}[n]\mathbf{U}^*[n-m]=\mathbf{U}_i/\mathbf{U}[n]=\mathbf{U}_{k+1}\right)p\left(\mathbf{U}[n]=\mathbf{U}_{k+1}\right) \\ & +p\left(\mathbf{U}[n]\mathbf{U}^*[n-m]=\mathbf{U}_i/\mathbf{U}[n]=\mathbf{U}_{k+2}\right)p\left(\mathbf{U}[n]=\mathbf{U}_{k+2}\right), \text{for } i=1 \text{ to } 6 \end{aligned} \tag{9}$$

or

$$\begin{aligned} & p\left(\mathbf{U}[n]\mathbf{U}^*[n-m]=\mathbf{U}_i\right)=p\left(\mathbf{U}[n]\mathbf{U}^*[n-m]=\mathbf{U}_i/\mathbf{U}[n]=\mathbf{U}_{k+1}\right)\times p_1 \\ & +p\left(\mathbf{U}[n]\mathbf{U}^*[n-m]=\mathbf{U}_i/\mathbf{U}[n]=\mathbf{U}_{k+2}\right)\times p_2, \text{for } i=1 \text{ to } 6 \end{aligned} \tag{10}$$

where p_1 and p_2 are the probabilities of selection of U_{k+1} and U_{k+2}, respectively (p(U[n]=U_{k+1}) and p(U[n]=U_{k+2})). The probabilities indicated in equation (10) depend on the machine parameters, and torque hysteresis band and flux hysteresis band. Figure 8 illustrates the typical waveforms of torque and stator flux by DTC. It is assumed that the sampling time is short enough to keep the torque and flux errors within their hysteresis bands. Increasing and decreasing slopes of torque and flux waveforms vary from one sampling period to the other one. In this section, the slopes are considered to be constant. The effect of variation of these slopes and long sampling period on the voltage spectrum is discussed in the next section. The increasing and decreasing slope of torque waveform are expressed in equations (11-a) and (11-b).

$$S_T^+ = \frac{3}{2}\frac{P}{2}\frac{1}{\sigma L_s}(\frac{L_m}{L_r})^2(\frac{\sqrt{3}}{\pi}V_{dc}\left|\psi_s\right| - \omega_m\left|\psi_s\right|^2) \text{ (11-a)}$$

$$S_T^- = -\frac{3}{2}\frac{P}{2}\frac{1}{\sigma L_s}(\frac{L_m}{L_r})^2\omega_m\left|\psi_s\right|^2 \text{ (11-b)}$$

Neglecting the stator resistance, the increasing and decreasing slopes of flux waveform can be presented as equation (12):

$$S_F^+ = -S_F^- = \frac{1}{3}V_{dc} \tag{12}$$

The relation between the times indicated in Figure 8, motor parameters and hysteresis bands of torque and flux controller are expressed in (13).

$$t_{rT} = \frac{2\Delta T}{S_T^+} \text{ (13-a)}$$

$$t_{fT} = \frac{2\Delta T}{S_T^-} \text{ (13-b)}$$

$$t_{rF} = \frac{2\Delta\psi}{S_F^+} \text{ (13-c)}$$

$$t_{fF} = \frac{2\Delta\psi}{S_F^-} \text{ (13-d)}$$

Thus, the components are calculated as follows.

A. p(U[n]=U_0) A zero vector is used to decrease the torque. Therefore, p(U[n]=U_0) can be described as:

$$p(\mathbf{U}[n] = \mathbf{U}_0) = p(T\downarrow) = \frac{t_{fT}}{t_{rT} + t_{fT}} \tag{14}$$

B. p_1 = p(U[n]=U_{k+1}) U_{k+1} will be selected during the rising of torque when the flux is also increasing.

$$p(\mathbf{U}[n] = \mathbf{U}_{\mathbf{k+1}}) = p(T\uparrow) \times p(\left|\grave{\mathbf{E}}_{\mathbf{s}}\right|\uparrow) = \frac{t_{rT}}{t_{rT} + t_{fT}} \times \frac{t_{rF}}{t_{rF} + t_{fF}} \tag{15}$$

C. $p_2 = p(U[n]=U_{k+2})$ U_{k+2} is used to increase the torque while decreasing the flux.

$$p(\mathbf{U}[n] = \mathbf{U}_{k+2}) = p(T\uparrow) \times p(|\dot{\mathbf{E}}_s|\downarrow) = \frac{t_{rT}}{t_{rT} + t_{fT}} \times \frac{t_{fF}}{t_{rF} + t_{fF}} \tag{16}$$

D. $p(U[n]\ U[n-m] = U_i / U[n]=U_{k+1})$ and $p(U[n]\ U[n-m] = U_i / U[n]=U_{k+2})$

The terms $p(U[n]\ U[n-m] = U_i / U[n]=U_{k+1})$ and $p(U[n]\ U[n-m] = U_i / U[n]=U_{k+2})$ are conditional probabilities. They depend on the drive condition and the state of U at the n_{th} sampling interval (U[n]) and $(n-m)_{th}$ sampling interval (U[n-m]). These terms can be simplified as in (17):

$$p\left(\mathbf{U}[n]\mathbf{U}^*[n-m] = \mathbf{U}_i / \mathbf{U}[n] = \mathbf{U}_{k+1}\right) = p\left(\mathbf{U}[n-m] = \mathbf{U}_{k+2-i} / \mathbf{U}[n] = \mathbf{U}_{k+1}\right) \quad \text{(17-a)}$$

$$p\left(\mathbf{U}[n]\mathbf{U}^*[n-m] = \mathbf{U}_i / \mathbf{U}[n] = \mathbf{U}_{k+2}\right) = p\left(\mathbf{U}[n-m] = \mathbf{U}_{k+3-i} / \mathbf{U}[n] = \mathbf{U}_{k+2}\right) \quad \text{(17-b)}$$

Calculation of these terms is done in two stages. At first, consider two consecutive sampling interval (i.e. m=1). 9 possibilities, as shown in Figure 7, exist for the next sampling interval voltage vector. The detailed calculations are presented in the appendix.

These elements construct the transition matrix H as shown in Figure 9.

$$H = \begin{bmatrix} H_{00} & H_{01} & H_{02} \\ H_{10} & H_{11} & H_{12} \\ H_{20} & H_{21} & H_{22} \end{bmatrix} \tag{18}$$

Now, consider m>1 and there are some sampling intervals between U[n] and U[n-m]. Figure 9 shows the voltage chain and the transition matrix that describes the relation between consecutive vectors and their probabilities. There are some patches from a specific voltage in $(n-m)_{th}$ sampling interval to the other voltage vector in n_{th} sampling interval. The probability of this patch is equal to the multiplication of probabilities which are in this direction and are defined by H. If U[n] and U[n-m] are located in the same sector, the probability $p(U[n-m] = U_{k+1} / U[n]=U_{k+1})$ is equal to the summation of probabilities for all patches from voltage vector U_{k+1} in the sampling interval n-m to the voltage vector U_{k+1} in the sampling interval n. For example, if m=2 the probability $p(U[n-m] = U_{k+1} / U[n]=U_{k+1})$ is calculated as:

$$p\left(\mathbf{U}[n-2] = \mathbf{U}_{k+1} / \mathbf{U}[n] = \mathbf{U}_{k+1}\right) = H_{11}H_{11} + H_{10}H_{01} + H_{12} + H_{21} \tag{19}$$

This is equal to the (2,2) element in the matrix S which is shown in Figure 6.

Figure 7. Different probability for various voltage vectors

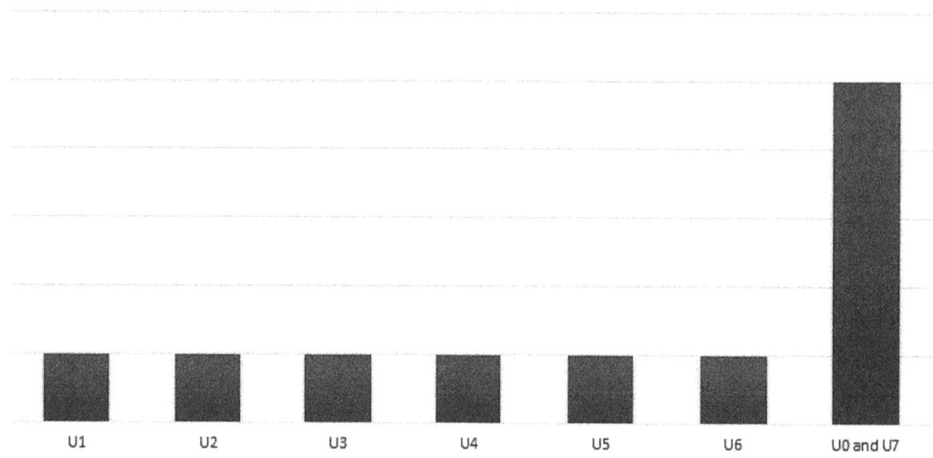

Figure 8. One stage random process

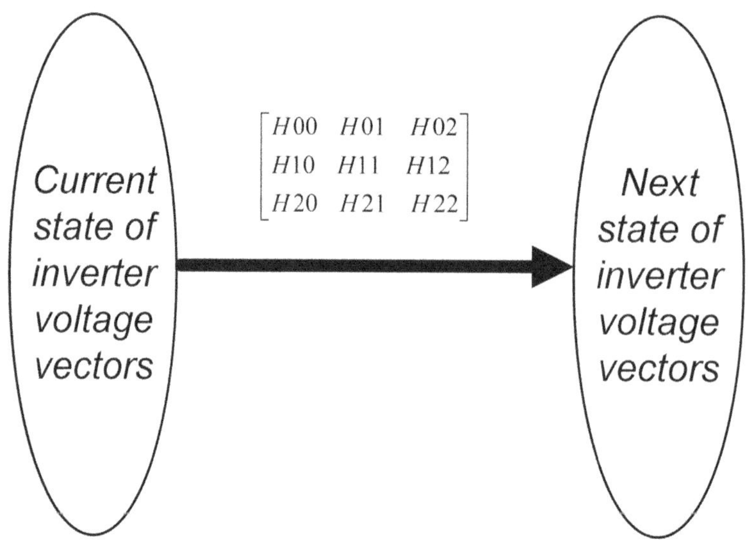

Figure 9. Random process of DTC

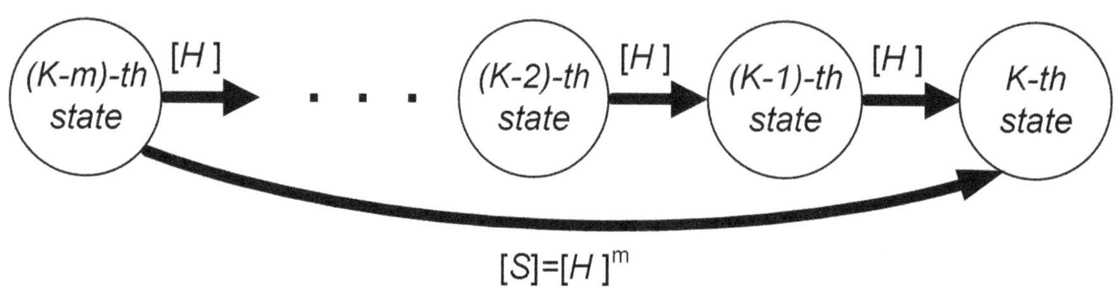

$$S = \begin{bmatrix} S_{11} & S_{12} & S_{13} \\ S_{21} & S_{22} & S_{23} \\ S_{31} & S_{32} & S_{33} \end{bmatrix} \tag{20}$$

If U[n-m] locates in the $(k-1)_{th}$ sector, U_{k+1} for k_{th} sector is equal to U_{k+2} for $(k-1)_{th}$ sector. Therefore, the probability p(U[n-m] $=U_{k+1}$ /U[n]=U_{k+1}) is calculated by summation of the probabilities for all patches from U_{k+2} in the sampling interval n-m to U_{k+1} in the sampling interval n. This is equal to the (3,2) element in the matrix [S]. Index k in the voltage vectors of the probability p(U[n-m] $=U_{k+1}$ /U[n]=U_{k+1}) is referred to the k_{th} sector. When the index m increases and enters into the other sectors (for example sector k') the index k should be converted to that reference. Because U_0 and U_{k+1} and U_{k+2} are used as U_0 and $U_{k'+1}$ and $U_{k'+2}$ in that sector.

$$p\left(\mathbf{U}[n]\mathbf{U}^*[n-m] = \mathbf{U}_i / \mathbf{U}[n] = \mathbf{U}_{k+1}\right) = \begin{cases} S_{22} & \text{in the sector k} \\ S_{32} & \text{in the sector k-1} \\ 0 & \text{in the sector k-2} \\ 0 & \text{in the sector k-3} \\ 0 & \text{in the sector k-4} \\ 0 & \text{in the sector k-5} \end{cases}, \text{ for } i = 1 \tag{21}$$

$$\begin{aligned} & R_{UU}[n,n-m] = \\ & S_{22}p_1 + \frac{1}{2}S_{23}p_2 + \frac{1}{2}S_{32}p_1 + S_{33}p_2 + j(\frac{\sqrt{3}}{2}S_{23}p_2 - \frac{\sqrt{3}}{2}S_{32}p_1), \text{for sector k} \\ & \frac{1}{2}S_{22}p_1 - \frac{1}{2}S_{23}p_2 + S_{32}p_1 + \frac{1}{2}S_{33}p_2 + j(\frac{\sqrt{3}}{2}S_{22} + \frac{\sqrt{3}}{2}S_{23}p_2 - \frac{\sqrt{3}}{2}S_{32}p_1), \text{for sector k} - 1 \\ & -\frac{1}{2}S_{22}p_1 - S_{23}p_2 + \frac{1}{2}S_{32}p_1 - \frac{1}{2}S_{33}p_2 + j(\frac{\sqrt{3}}{2}S_{22} + \frac{\sqrt{3}}{2}S_{23}p_2 + \frac{\sqrt{3}}{2}S_{32}p_1), \text{for sector k} - 2 \\ & -\frac{1}{2}S_{22}p_1 + \frac{1}{2}S_{23}p_2 - S_{32}p_1 - \frac{1}{2}S_{33}p_2 + j(-\frac{\sqrt{3}}{2}S_{22} - \frac{\sqrt{3}}{2}S_{23}p_2 + \frac{\sqrt{3}}{2}S_{32}p_1), \text{for sector k} - 4 \\ & \frac{1}{2}S_{22}p_1 + S_{23}p_2 - \frac{1}{2}S_{32}p_1 + \frac{1}{2}S_{33}p_2 + j(-\frac{\sqrt{3}}{2}S_{22} - \frac{\sqrt{3}}{2}S_{23}p_2 - \frac{\sqrt{3}}{2}S_{32}p_1), \text{for sector k} - 5 \end{aligned} \tag{22}$$

This equation shows that the autocorrelation function of voltage vectors is independent of time origin because the elements of (22) are independent of time origin. Therefore, R_{UU}[n, n-m] can be replaced with R_{UU}[n, n-m]. So the amplitude of voltage power spectrum as a function of frequency can be obtained by the norm of P_U(f) (||P_U(f)||).

Figure 10 can be used to compare the analytical voltage spectrum and practical results. It can be seen that the analytical method can predict the specific harmonics in the spectrum. In this case, some of

Figure 10. Comparison between simulation and analytic results

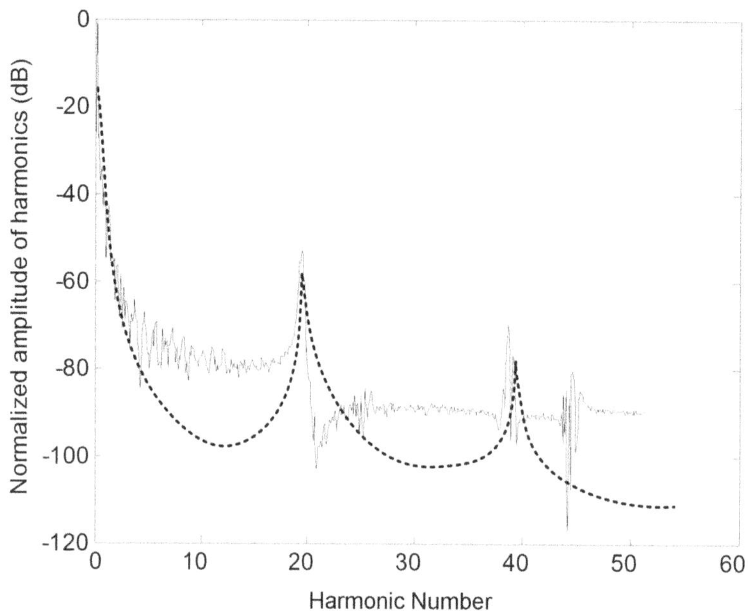

the probabilities used in the theoretical study tend to zero. Due to the fast jumping of torque and large practical sampling time, selection of two consecutive nonzero voltage vectors does not occur. Thus, the values of H_{11}, H_{22}, H_{21} and H_{12} should be replaced by zero.

It can be seen that even if the error in amplitude of the high frequency harmonics is high, this method can calculate the location of these harmonics precisely. The error in the harmonic amplitude estimation is not unusual and it has seen in the other methods for approximation of power spectrum.

RELIABILITY AND PROBABILITY

Reliability is the probability of performing adequately to achieve the desired aim of the system. This can be mentioned as a time dependent equation. Reliability concept has more importance in specific applications such as space and military equipment that mission can hardly be replaced or be done by another system instead of the failed part. In order to improve the system reliability, different researches have been done and several methods have been introduced.

Concept of Failure Rate

Many of these methods need information about failure rate of the system. Each system contains a number of components. One method to enhance reliability is improvement of components reliability. This goal may be achieved by component specific derating or by improvement of components' specifics. In this level, failure rate of the components should be modeled properly. The other method is using redundant or fault tolerant system in which after partial fault, the rest of the system can work adequately to achieve the goal of whole system. In this level, systematic analysis of reliability is necessary. For example, a systematic analysis indicates that one problem of using redundant system is to load current balancing. Hence, different studies have been done in current sharing and current control. The other method that may be useful is comparison of various possible topologies or different operation conditions to choose a proper state to achieve the goal of design. Because power electronic systems usually are not available to be repaired or maintained, maintenance can rarely be used in power electronic systems. Also this method is not meaningful for each power electronic component. For example only a fan can be maintained to work properly and doing its duty for cooling heat sink while for a MOSFET switch or film capacitor, maintenance do not have a practical meaning 6.

Application of Failure Rate

In researches, failure rate of some components have been reported to be higher than others. Electrolytic capacitor and switch are the power electronic components which are more exposed to be failed based on empirical reports for reliability evaluation. As a general approach, component stress reduction can enhance reliability of component.

There is a method for dynamic and time dependent reliability assessment named Markov Chain that in power system analysis has been used for many years. In this method, by using redundancy component, it will be more important that failing mode of component, how damages whole converter operation. For example, if there are 2 parallel output capacitors, open circuit or short circuit failing of one capacitor, has different effects on system reliability.

Rather than theoretical reliability evaluation with standards such as MIL-HDBK-217, there are some accelerated or aging tests that manufacturers use them to evaluate the reliability of their products. In this method, at each test, one or some parameters of environmental conditions are stressed more than typical state to reduction of test time less than real state. Then, there are some determined relations between this accelerated test results and typical condition results that are used to finding failure rates and reliability evaluation. This is the subject of chapter 5.

Time Variation of Failure Rate in a Power Converter

Failure process in a converter can be divided into three intervals. The infant mortality period is characterized by an initially high failure rate. This is normally the result of poor design, the use of substandard components, or lack of adequate controls in the manufacturing process. When these mistakes are not caught by quality control inspections, an early failure is likely to result. Early failures can be eliminated from the customer by "burn in" during which time the equipment is operated at stress levels equal to the intended actual operating conditions. The equipment is then released for actual use only when it has passed through the "burn-in" period.

The useful life period, is characterized by an essentially constant failure rate. This is the period dominated by chance failures. Chance failures are those failures that result from strictly random or chance causes. They cannot be eliminated by either lengthy burn-in periods or good preventive maintenance practices. Equipment is designed to operate under certain conditions and up to certain stress levels. When these stress levels are exceeded due to random unforeseen or unknown events, a chance failure will occur. While reliability theory and practice is concerned with all three types of failures, its primary concern is with chance failures, since they occur during the useful life period of the equipment. The time when a chance failure will occur cannot be predicted; however, the likelihood or probability that one will occur during a given period of time within the useful life can be determined by analyzing the equipment design. If the probability of failure chance is too great, either design changes must be introduced or the operating environment made less severe.

In the wearout period, is characterized by an as a result of equipment deterioration due to age or use. For example, mechanical components such as transmission bearings will eventually wear out and fail, regardless of how well they are made. Early failures can be postponed and the useful life of equipment extended by good design and maintenance practices. The only way to prevent failure due to wearout is to replace or repair the deteriorating component before it fails.

Mathematical Expression of Reliability

MTTF is mean time which is expected the component or system to work adequately before failure. It can be deduced that MTTF is equal to the integral of reliability of component as

$$MTTF = \int_0^\infty R(t)\,dt \tag{23}$$

MTTR is mean time which takes to repair the failed component that in electronic reliability evaluations is generally negligible with respect to MTTF. MTBF is the mean time between failure that is summation of MTTF and MTTR, which can be assumed to equal to MTTF.

In general, the relation between reliability and failure rate can be deduced as

$$R(t) = e^{-\int_0^t \lambda(\tau)\,d\tau} \tag{24}$$

MTBF will be derived as

$$R(t) = e^{-\lambda t} \tag{25}$$

$$MTBF = \frac{1}{\lambda} \tag{26}$$

In a DC-DC converter, there are some components such as switch, diode, inductor, capacitor and controller IC, which are in series or parallel relation as reliability point of view. For evaluation of converter reliability, failure rate of components must be determined. Then the converter should be analyzed systematically and the reliability model of the converter should be obtained.

RELIABILITY MODELS

There are various modeling approach in power electronics (Poshtan, Kaboli, & Mahdavi, 2006). Reliability modelling is the process of predicting or understanding the reliability of a component or system prior to its implementation (Richardeau, & Pham, 2013). Reliability design begins with the development of a model. Reliability and availability models use block diagrams and fault trees to provide a graphical means of evaluating the relationships between different parts of the system. These models may incorporate predictions based on failure rates taken from historical data. While the predictions are often not accurate in an absolute sense, they are valuable to assess relative differences in design alternatives.

Series Models

The simplest and perhaps most commonly occurring configuration in reliability mathematical modeling is the series configuration. The successful operation of the system depends on the proper functioning of all the system components. In this configuration, a component failure represents total system failure. There are two approaches for component failure expression. One is that the failure of any one component is statistically independent of the failure or success of any other. This is usually the case for most practical purposes. If this is not the case, then conditional probabilities must be used, which increase the complexity of the calculations.

In series configuration, the system failure rate with exponential distribution for the reliability function is the sum of the individual component failure rates and the system mean life.

For a series system, total failure rate is a summation of failure rate of all components and system reliability is multiplication of components reliability:

$$\lambda_{system} = \sum \lambda_{component} \tag{27}$$

$$R_{system}(t) = \prod R_{component}(t) = e^{-\sum \lambda t} \tag{28}$$

Parallel Models

The next most commonly occurring configuration encountered in reliability mathematical modeling is the parallel configuration. For this case, all of the components would have to fail for system failure. Thus, parallel configuration, or the use of redundancy, is one of the design procedures used to achieve extremely high system reliability, greater than the individual component reliabilities. Of course, this is a very simple concept, which becomes more complicated in actual practice. Redundant equipment can be active ("on-line") or turned off ("standby"), some redundant units can repaired without shutting down the system, others can not, and the number of repair crews can vary. All these factors must be considered in formulating appropriate reliability models.

For a parallel system, the system reliability cannot be expressed by one exponential term but the unreliability of system is multiplication of components unreliability:

$$Q_{system}(t) = 1 - R_{system}(t) = \prod \left(1 - R_{component}(t)\right) \quad (29)$$

For a more complex system, the system should be divided in to simple subsystems.

Figures 11 to 14 show different reliability models for various electrical connection (Jian-Ping & Thompson, 2005). Figures 11 and 12 show a parallel connection of two transistor in a power electronic converter. Current rating of each transistor in Figure 11 is more than total nominal current. Therefore, the reliability model of this connection is parallel. In Figure 12, current rating of each transistor is less that total nominal current and it leads to series reliability model. Similar result is obtained with series electrical connection of switches as shown in Figures 13 and 14.

Figure 15 shows reliability model of an electric motor. Proper operation of both stator and rotor is necessary for correct operation of the motor. Therefore, the reliability model of an electric motor is generally series connection of stator reliability and rotor reliability. Similarly, both winding and core of stator and rotor must operate properly and it leads to series reliability model as shown in Figure 16. Reliability model can be achieved from various point of view. For example, reliability model of an electric motor can be considered as a combination of electrical parts and mechanical parts instead of division into stator and rotor as shown in Figure 17. Reliability model gives a clear view to designer and operator

Figure 11. Parallel reliability model for two parallel connected switches

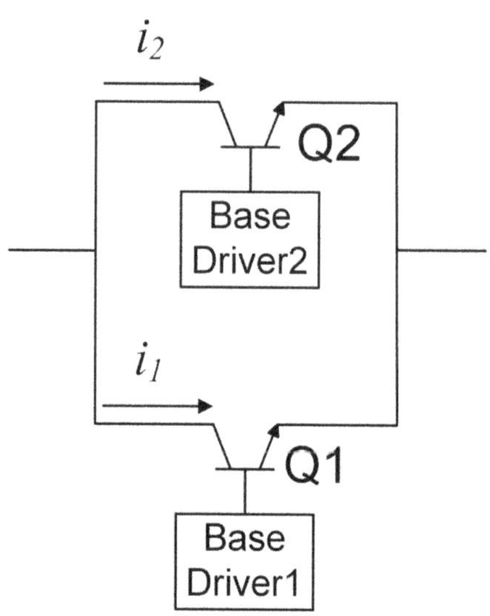

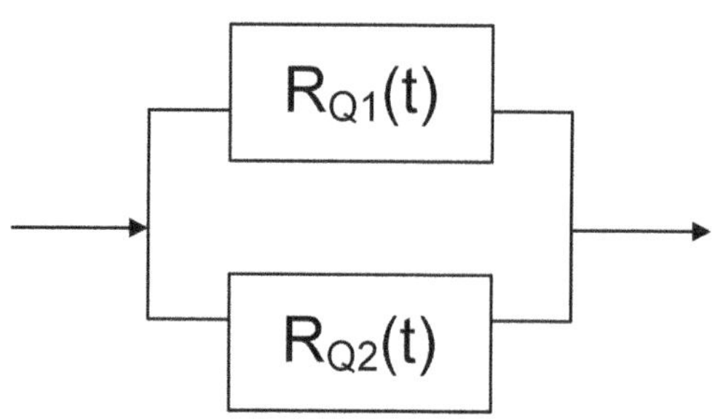

Figure 12. Series reliability model for two parallel connected switches

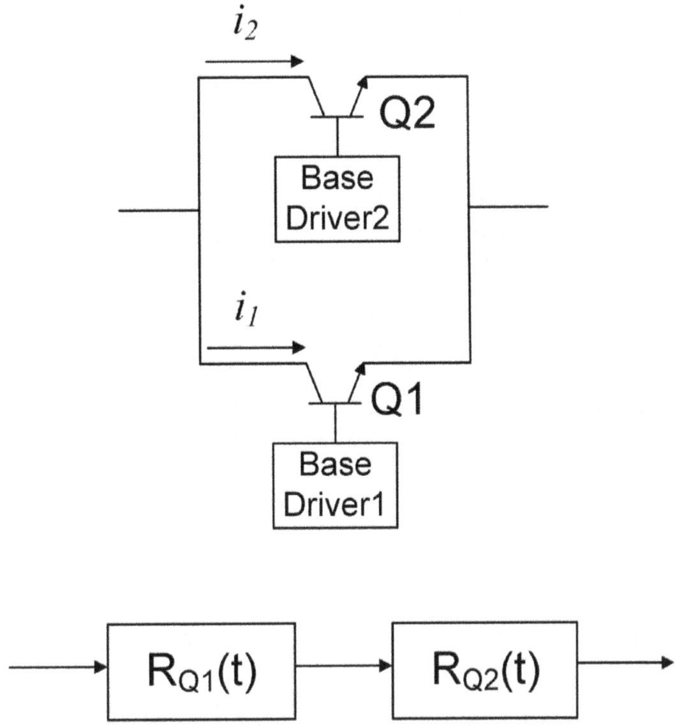

Figure 13. Parallel reliability model for two Series connected switches

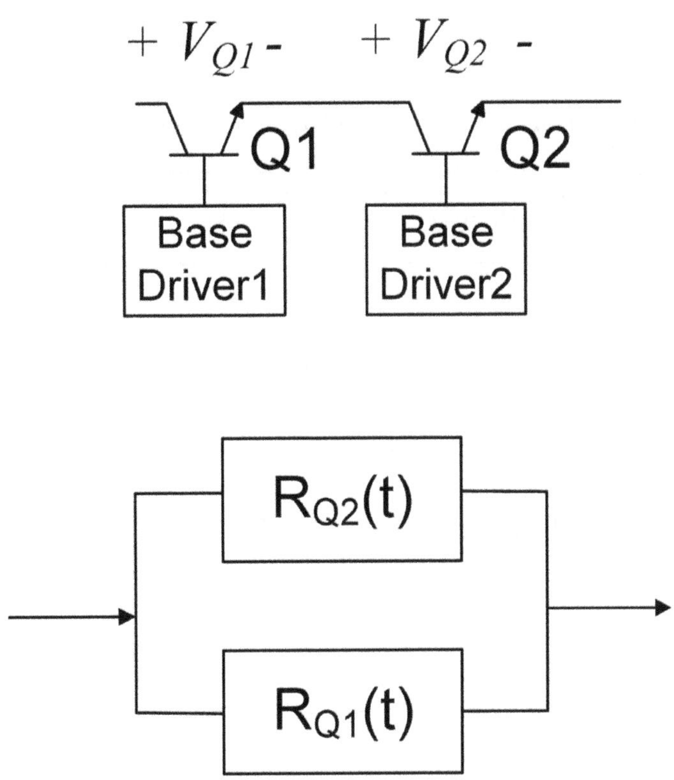

Figure 14. Series reliability model for two Series connected switches

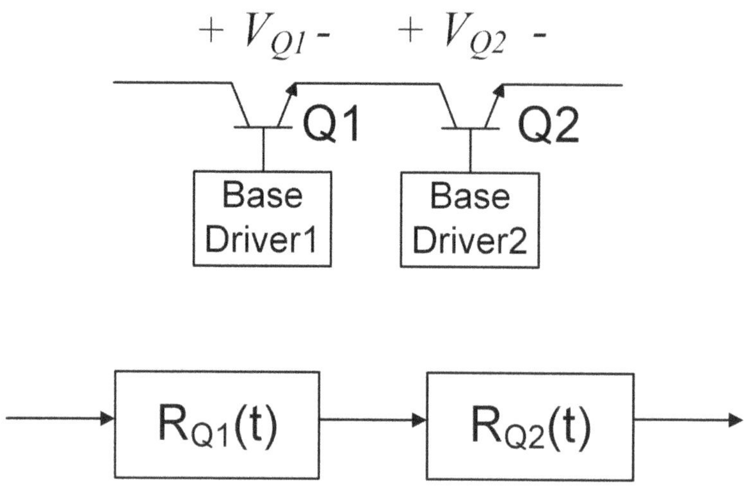

Figure 15. Reliability model of an electric machine based on physically separated sections

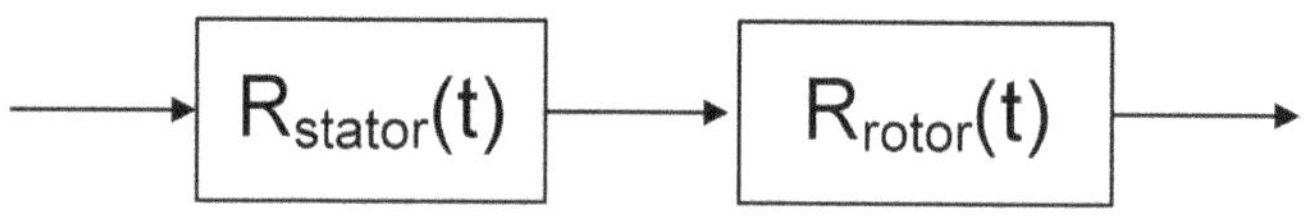

Figure 16. Reliability model of an electric machine based on separated sections with different nature

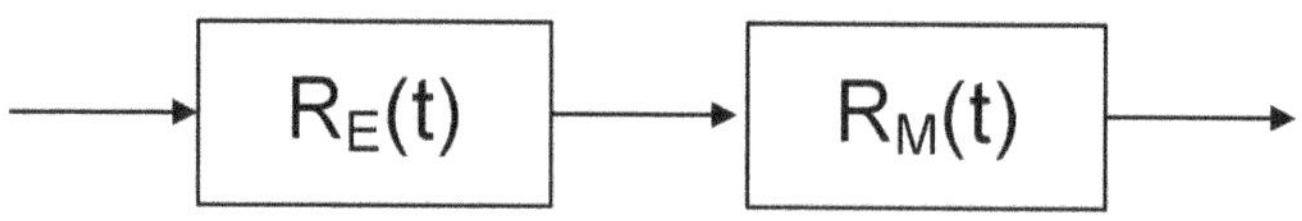

Figure 17. Different methodology for Reliability modeling of an electric machine

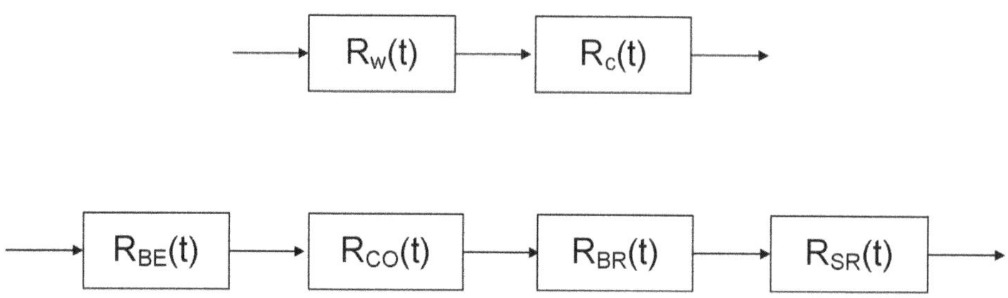

Figure 18. Reliability model of a DC rotor

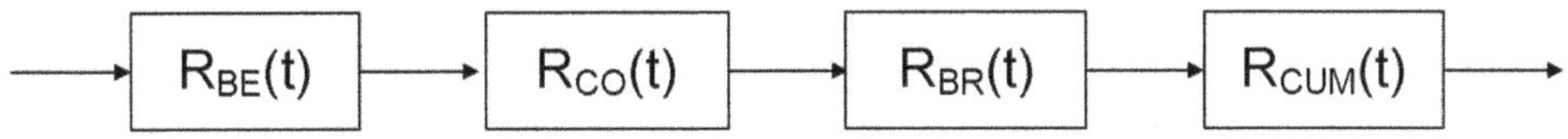

Figure 19. Reliability model of a sysnchronous rotor

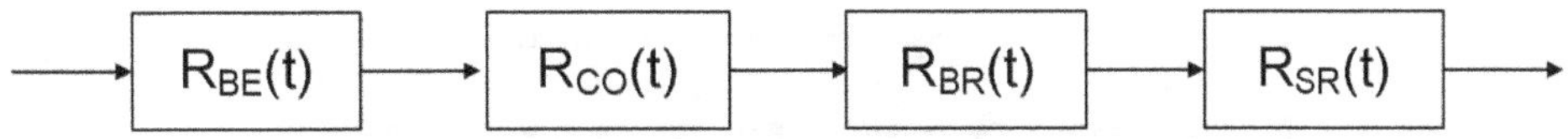

Figure 20. Electrical reliability model of a SRM rotor

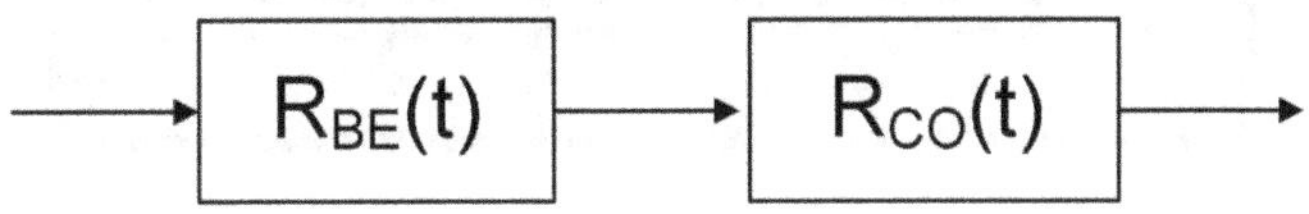

about the problem of converter. A comparison between reliability model of a SRM (Figure 18) and a DC motor (Figure 19) roughly shows a SRM has a robust structure because it does not many parts used in DC motor and its reliability model is shorter than that found in DC motor. Figures 20 and 21 show reliability models of electrical parts of a SRM rotor and a synchronous generator.

COMPONENTS RELIABILITY EVALUATION

In system or even sub-system level, the reliability prediction process uses some powerfull mathematical tools like Markov Chain analysis and etc. But, as we close to component level, practical data finds more benefits in comparison to analytic tools. For part level predictions, two separate fields of investigation are common:

- The physics of failure approach uses an understanding of physical failure mechanisms involved, such as mechanical crack propagation or chemical corrosion degradation or failure;
- The parts stress modelling approach is an empirical method for prediction based on counting the number and type of components of the system, and the stress they undergo during operation.

Herein, we present one of the commonly used tools for calculation of component failure rate.

Figure 21. Reliability model of electrical section of a SRM rotor

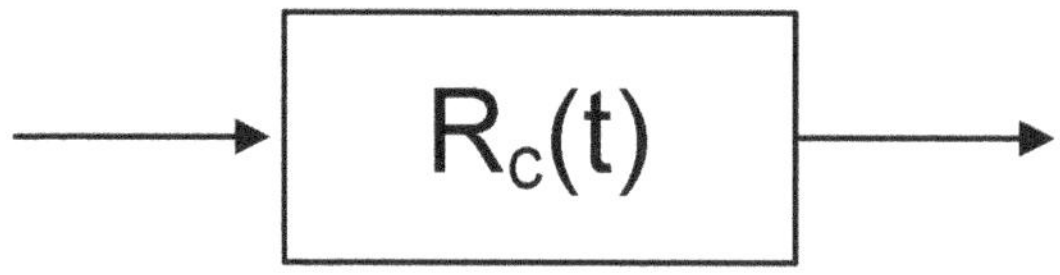

MIL-HDBK-217

Maintaining reliability and providing reliability engineering is an essential need with modern electronic systems. Reliability engineering for electronic equipment requires a means for a quantitative baseline, or a reliability prediction analysis. There are some technical standards for numerical evaluation of power electronic components failure rate (Bhuiyan, & Yazdani, 2010). One of the most widely used reliability prediction methodology handbook is MIL-HDBK-217, the Military Handbook for "Reliability Prediction of Electronic Equipment". In this chapter, MIL-HDBK-217 is used for numerical evaluation of reliability 14.

Structure of the Standard

The MIL-HDBK-217 handbook contains failure rate models for the various part types used in electronic systems, such as ICs, transistors, diodes, resistors, capacitors, relays, switches, connectors, etc. These failure rate models are based on the best field data that could be obtained for a wide variety of parts and systems; this data is then analyzed, with many simplifying assumptions thrown in, to create usable models. The purpose for developing this handbook was to establish and maintain consistent and uniform methods for estimating the inherent reliability of military electronic equipment and systems.

Methodology of the Standard

MIL-217 uses part stress analysis method for characterizing the components reliability. The Part Stress Analysis method is used the majority of time and is applicable when the design is near completion and a detailed parts list, or BOM, plus component stresses are available. By component stresses, the standard is referring to the actual operating conditions such as environment, temperature, voltage, current and power levels applied. The MIL-217 standard groups components or parts by major categories and then has subgroups within the categories. An example is a "fixed electrolytic (dry) aluminum capacitor" is a subcategory of the "capacitor" group. Each component or part category and its subgroups have a unique formula or model applied to it for calculating the failure rate for that component or part. The failure rate formulas include a base failure rate, for the category and subgroup selected. The base failure rates apply to components and parts operating under normal environmental conditions, with power applied, performing the intended functions, using base component quality levels and operating at the design stress levels. The standard then applies many correction factors, to the base failure rates in order to factor in the actual operating conditions, environment and stress levels. Base failure rates are adjusted by applying the correction factors to the underlying equation or model provided for each component category 15 (Figure 22).

Thermal Considerations

As it was shown in Chapter 2, temperature is one of the most important parameters affected the failure rate of a component. Ambient and operating temperatures have a major impact on the failure rate prediction results of electronic equipment, especially equipment involving semiconductors and integrated circuits. The MIL-217 standard requires an input of ambient temperatures and more definitive data required for the calculation of junction temperatures in semiconductors and microcircuits.

A thermal analysis should be a part of the design and reliability analysis process for electronic equipment. Therefore, we dedicate a separated chapter for thermal study.

Electric Field Considerations

The voltage stress factor accounts for the acceleration of failure mechanisms associated with abnormally high supply voltages.

Figure 22. different values of quality factor in MIL-217

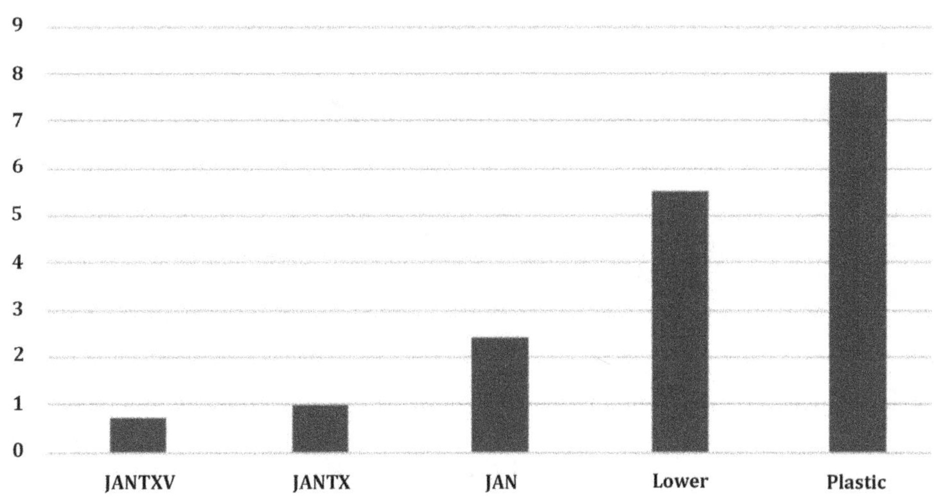

Mechanical Considerations

The design quality of the component utilized has a direct effect on the part failure rate and appears in the models as a correction factor. Many of the components covered by the MIL-217 specification are available in several quality levels and each has an associated correction factor.

Environmental Considerations

Environmental stress is of major concern in establishing the failure rate for components and parts included in a system. Environmental stresses can be quite different from one application environment to another and can subject the equipment to a controlled environment with constant temperature and humidity, or an environment with rapid temperature changes, high humidity, high vibration and high acceleration (Figures 23, 24, 25, 26, 27, 28, 29, and 30).

Example: Failure Rate for Various Parts

Switch

In MIL-HDBK-217 equation is determined for switch failure rate:

Figure 23. A comparison between temperature factor in MIL-217 between diode (dashed) and MOSFET (solid)

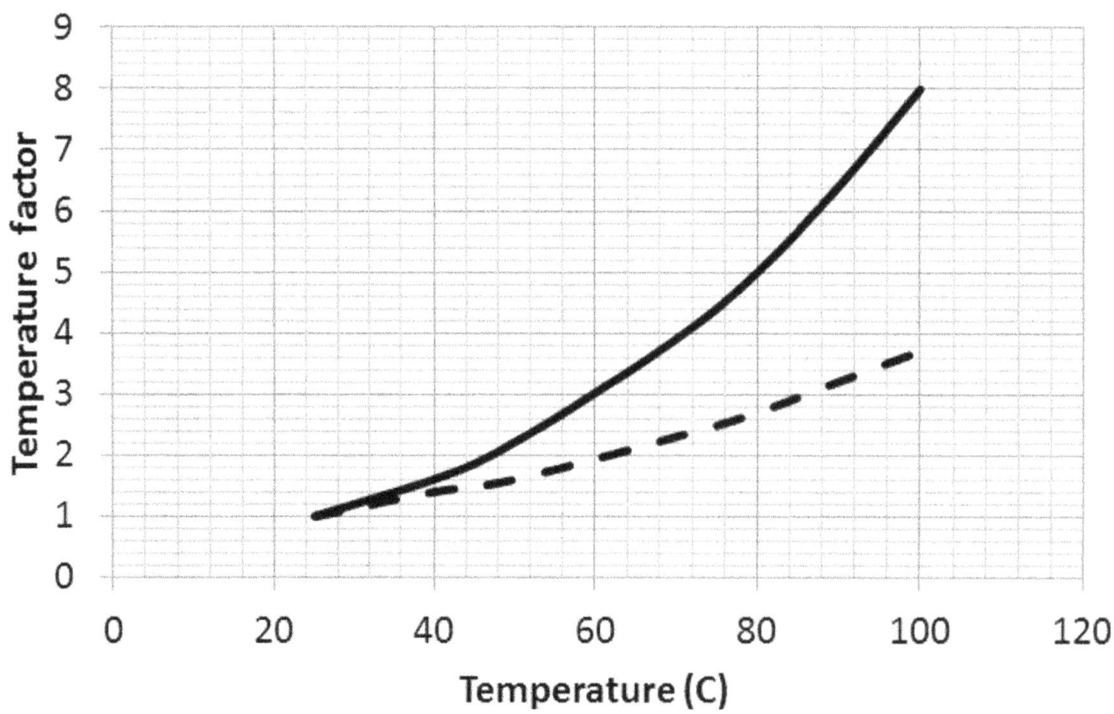

Figure 24. A comparison between temperature factor in MIL-217 among diode and MOSFET and thyristor

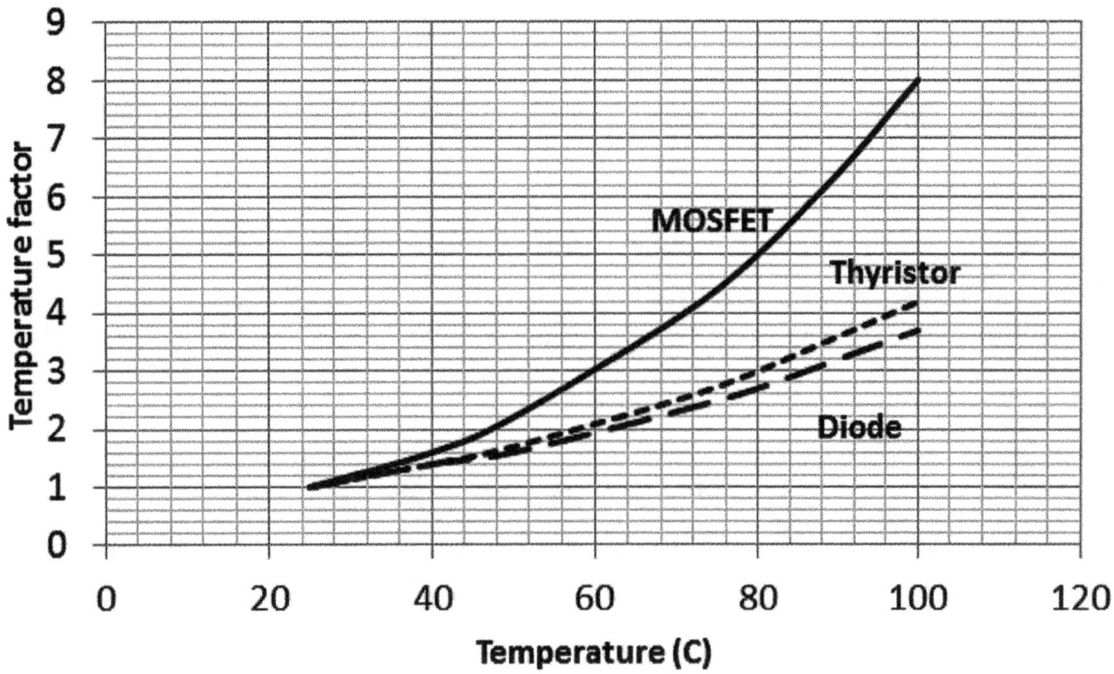

Figure 25. A comparison between temperature factor in MIL-217 between different switches

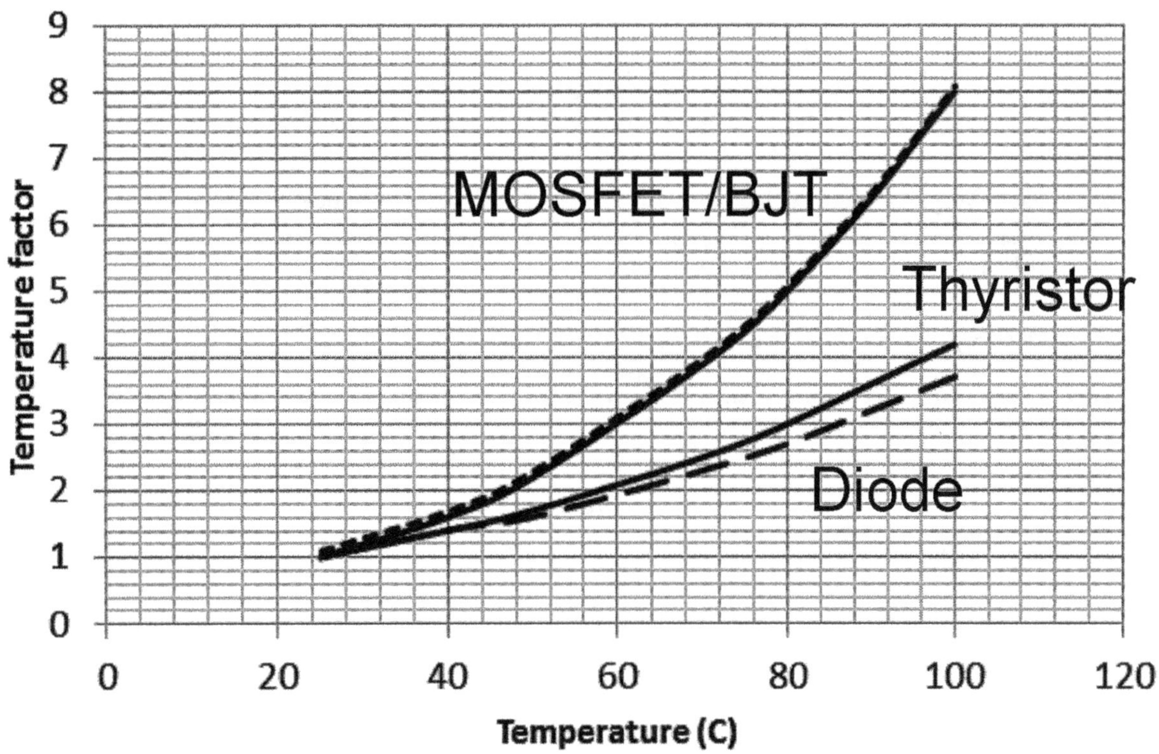

Figure 26. different values of environmental factor in MIL-217

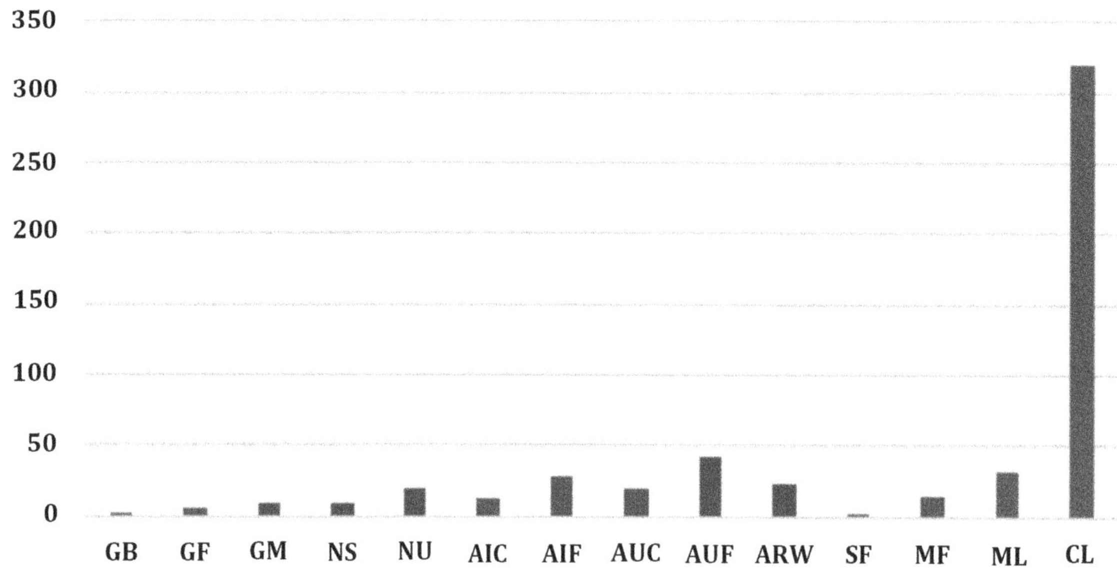

Figure 27. different values of voltage factor for diode in MIL-217

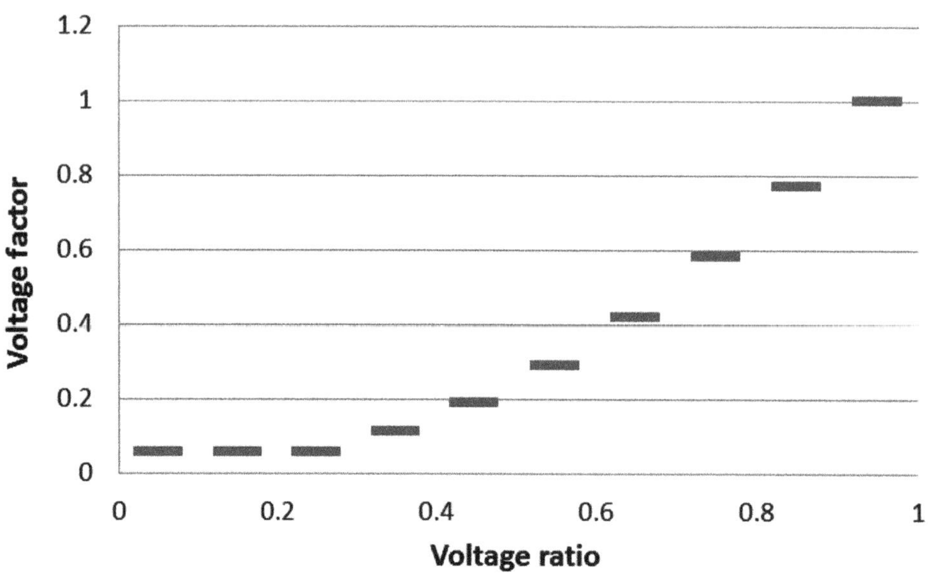

Figure 28. a comparison between values of voltage factor for diode (left) and BJT (right) in MIL-217

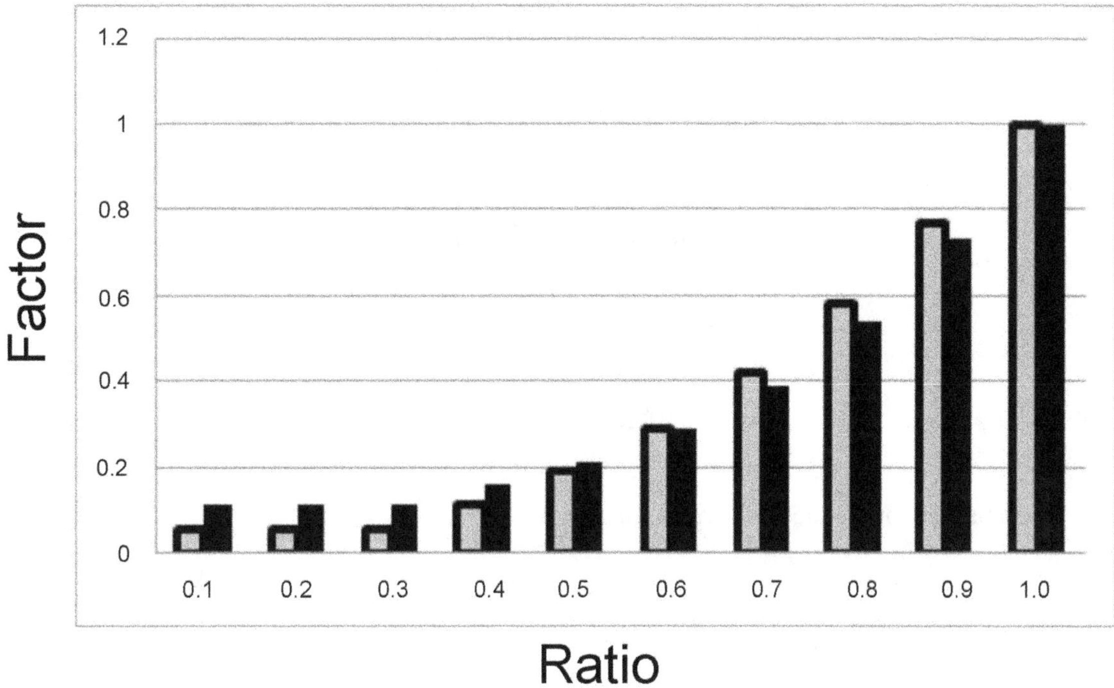

Figure 29. different values of factor for a capacitor in MIL-217

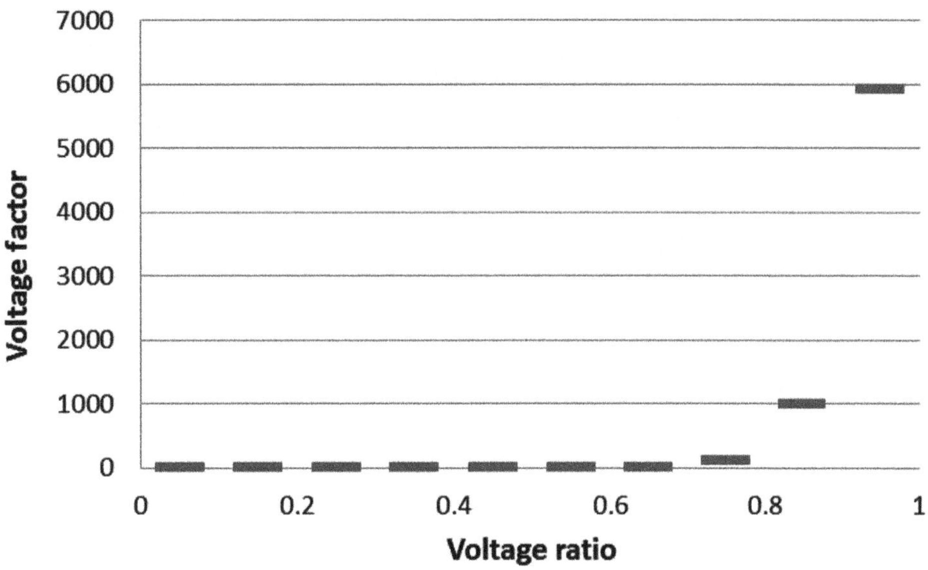

Figure 30. A step down buck converter

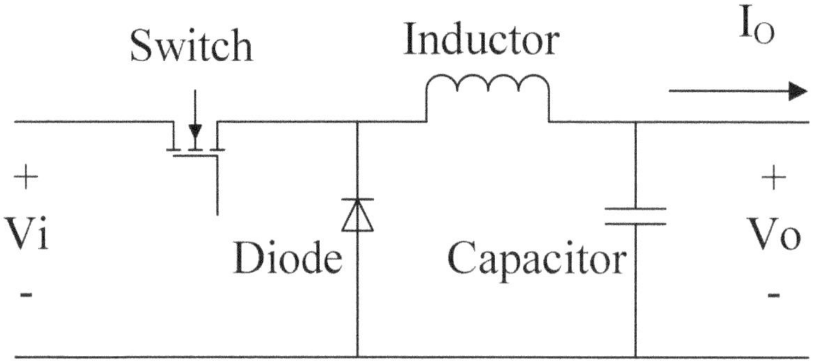

$$\lambda_p = \lambda_b \times \pi_T \times \pi_A \times \pi_Q \times \pi_E \quad Failure\,/\,10^6\,Hours$$

π_T is temperature factor that deduces from equation:

$$\pi_t = \exp\left\{-1925 \times \left(\frac{1}{T_j + 273} - \frac{1}{298}\right)\right\}$$

T_j is junction temperature that is related to MOSFET power losses. So, for determination of π_T, one important step is MOSFET power losses evaluation. Reduction of power losses result to decrement of π_T and therefore failure rate comes down. π_Q that is quality factor is determined by packaging and covering method that for commercial applications it is "lower" quality and is equal to 5.5. Other quality factors that have less value are for specific applications such as military and aerospace. For example, using JANTX package quality which has π_Q equal to 1.0 results to lower switch failure rate.

Diode

Equation 10 expresses diode failure rate

$$\lambda_p = \lambda_b \times \pi_T \times \pi_S \times \pi_C \times \pi_Q \times \pi_E \quad Failure\,/\,10^6\,Hours$$

π_T has relation to diode power losses. Lower power losses cause smaller π_T and failure rate.

$$\pi_t = \exp\left\{-3091 \times \left(\frac{1}{T_j + 273} - \frac{1}{298}\right)\right\}$$

π_S is electrical stress factor that is related to rated and applied voltage. Equation 12 indicates this relation. Whatever the portion of applied voltage to rated voltage becomes smaller, this factor gets smaller too. Hence, selection of diode with higher rated voltage, results to lower failure rate. But, there is one important fact that should be considered for diode selection. In fact, choosing diodes which have larger rated voltage may have larger forward voltage that can increase diode conduction loss and it affects π_T. Therefore, optimizing between different parameters is required to choose an appropriate diode.

$$\pi_S = \left(\frac{Applied\ voltage}{Rated\ voltage}\right)^{2.43}$$

Inductor

For inductor, the following equation is used to calculate failure rate

$$\lambda_p = \lambda_b \times \pi_T \times \pi_Q \times \pi_E \quad Failure / 10^6\, Hours$$

λ_b is equal to 0.00003 for inductors which have fixed inductance.
π_T is temperature factor that can be obtained:

$$\pi_T = \exp\left(\frac{-0.11}{8.617\times10^{-5}}\left(\frac{1}{T_{HS}+273}-\frac{1}{298}\right)\right)$$

Where:

$$T_{HS} = T_A + \frac{1.1\times11.5\times P_{Loss}}{W_L^{0.6766}}$$

As it is clear in equations 115, inductor failure rate has direct relation with power losses. However, inductor failure rate is much lower than other components and has not been any reports that inductor failure rate makes problem for reliability of power electronic converters.

Capacitor

The other component is capacitor that equation 16 expresses failure rate for it:

$$\lambda_p = \lambda_b \times \pi_T \times \pi_C \times \pi_V \times \pi_Q \times \pi_E \; Failure / 10^6\, Hours$$

π_T is determined by ambient temperature:

$$\pi_T = \exp\left\{\frac{0.15}{8.617\times10^{-5}}\times\left(\frac{1}{298}-\frac{1}{T_A+273}\right)\right\}$$

$$\pi_C = C^{0.09} \quad (C\ is\ in\ \mu f\ unit)$$

As it is presented, greater rated voltage leads to smaller π_V which results in lower failure rate. Consequently, one method to failure rate reduction is using capacitor which has higher rated voltage.

$$S = \frac{Applied\ voltage\ to\ capacitor}{Rated\ voltage}$$

$$\pi_V = (\frac{S}{0.6})^5 + 1$$

Controller IC

In MIL-HDBK-217, section 5 there are some standard formulas to determine ICs' failure rate. But it is too complicated to use them. In some websites of manufacturers there are reliability accelerated test results. Component failure rate can be deduced from these results. As an example, one of these tests is HTOL that is based on Arrhenius acceleration model. In this test, component will be tested in temperature which is stressed more than normal use condition. Acceleration factor is defined and then failure rate will be evaluated. λ which is evaluated is upper bound failure rate that is expected to be measured. Thus, for different ICs, failure rate is evaluated by results of tests which are done by manufacturers or with FIT data that is presented directly in websites of manufacturers.

$$A_T = \exp\left\{\frac{E_a}{K_B}\left[\frac{1}{T_{use}} - \frac{1}{T_{stress}}\right]\right\}$$

$$\lambda = \frac{\chi^2(\gamma, 2 \times Y + 2)}{2 \times N \times A_T \times t}$$

Example: Buck Converter

In this section, a Buck converter is considered to be investigated as reliability point of view with assessing switching frequency influence on power losses of mentioned components that can affect the reliability of those components and total converter reliability (Javadian, & Kaboli, 2013).

In the following, evaluation of power losses of each component and the possible procedures to losses reduction are discussed.

MOSFET Switch Power Losses

MOSFET losses are the summation of conduction and switching losses. Conduction losses are product of terminal current and voltage during on-period of MOSFET which can be estimated. Switching losses are product of terminal current and voltage during transitions between on-period and off-period for MOSFET which can be estimated.

$$P_{conduction} = R_{on} \times I_{rms}^2$$

$$P_{switching,CCM} = \frac{1}{2} V_{off} \times \mathrm{I}_{on} \times (t_{on} + t_{off}) \times f_s$$

Choosing a MOSFET which simultaneously has lower R_{on}, turn-on and turn-off time means to have lower power losses that causes to smaller failure rate.

Diode Power Losses

Diode losses consist of conduction and switching losses. Conduction loss is because of product of forward voltage dropped on diode during on-state and diode current that can be evaluated.

$$P_{conduction} = V_f \times (1 - D) \times \mathrm{I}_o$$

Switching loss is caused by reverse recovery loss related to P-N junction during transition of turn-off time. Turn-on loss can be negligible. There is one precise method in for evaluation of diode turn-off loss. But using that method needs a lot of data and estimation. Therefore, in this section Pspice simulation with Pspice model of used diode which is MUR460 has been considered to determination of E_{off}. This E_{off} has been measured equal to 3.34 µJ. Thus, switching loss of free-wheel diode for proposed Buck converter can be determined.

$$P_{switching} = \mathrm{E}_{off} \times f_s$$

Choosing a Diode which simultaneously has lower V_f and lower reverse recovery loss means to have lower power losses that causes to smaller failure rate.

Inductor Power Losses

Inductor losses are caused by magnetic core loss and copper loss of inductor wiring. There have been some studies for core loss during non-sinusoidal operation that are much complicated whereas in this paper, magnetic core loss during CCM operation is estimated to be simplified.

$$P_{core} = 1.5 \times K \times f_s^{\alpha} \times \Delta \mathrm{B}^{\beta} \times V_e$$

Because of ripple in inductor current, copper losses consist of two parts named AC and DC loss. DC loss is caused by DC component of inductor current while AC loss is caused by inductor ripple current. This ripple has a main component frequency which is equal to switching frequency and the other components are harmonics of f_s. DC loss is based on interaction between DC current and R_{dc} that can be measured in situation which conducts just DC current through the inductor wiring. AC loss is the result of interaction between AC components of inductor current that are calculated and related values of R_{ac} which are estimated by equation 29. Total copper losses are deduced.

$$b_h = -\frac{(-1)^h \times \Delta i_{peak-peak}}{h^2 \times (D - D^2) \times \pi^2} \times \sin\left[h \times (1-D) \times \pi\right]$$

$$I_{rms,h} = \frac{b_h}{\sqrt{2}}$$

$$R_{ac,h} = \frac{l}{\sigma \times A_h} = \frac{l_{cu}}{\sigma_{cu} \times [\pi r_{wire}^2 - \pi(r_{wire} - \delta_h)^2]}$$

$$\delta_h = \frac{0.066}{\sqrt{h \times f_s}}$$

$$P_{copper} = R_{dc} \times I_{dc}^2 + \sum_{h=1}^{\infty}(R_{ac,h} \times I_{rms,h}^2)$$

Capacitor Power Loss

It should be noted just ambient temperature is involved in capacitor failure rate. Nevertheless, it is reasonable that increasing in capacitor temperature which can be occurred because of ESR loss, affect the temperature of the air near the component. Thus, the capacitor loss during choosing an appropriate capacitor should be taken into account. Hence, capacitor loss is presented in table 1 and table 2:

$$P_{loss,capacitor} = ESR \times I_{rms.ripple}^2$$

Example of System Reliability Estimation: MOSFET Driver Comparison with Reliability Consideration

One of methods to choose optimum reliability state of system is comparison of different topologies that can be used to achieve the design goal. For the aim of this section, that is Buck Converter design, two

Table 1. Application factor of the switch

Application Factor – π_A	
Application (P_r, Rated Output Power)	π_A
Linear Application (P_r < 2w)	1.5
Small Signal Switching	0.7
Power FETs(Non-linear, $P_r \geq$2w)	
2w $\leq P_r$ < 5w	2.0
5w $\leq P_r$ < 50w	4.0
50w $\leq P_r$ < 250w	8.0
$P_r \geq$ 250w	10

Table 2. Packaging factor

Quality Factor – π_Q	
Quality	π_Q
JANTXV	0.70
JANTX	1.0
JAN	2.4
Lower	5.5
Plastic	8.0

options for types of MOSFET switch are available. One option is N-channel MOSFET and the other is P-channel MOSFET. Since the N- channel MOSFET source in the buck topology is not connected to circuit ground or constant voltage reference, so MOSFET gate driver has more complexity than the other topologies in which MOSFET source is connected to circuit ground. 4 topologies for Buck converter are presented in Figures 31, 32, 33, and 34. Using N-channel MOSFET needs isolation between MOSFET gate signal and MOSFET source pin. This isolation, in Figure 33 is established by bootstrap technique which exists in IR2113 and in Figure 34 is established by using optocoupler. But using P-channel MOSFET does not need this kind of isolation. Because P-channel MOSFET source will be connected to input voltage and consequently voltage of MOSFET source is constant and there is not problem for providing gate signal without isolation. It should be noted that there may be some other methods for gate isolation but in this section just these methods are considered.

In Figures 31 and 32 MOSFET switch is P-channel type. When current passes through R_1, voltage drop between gate and source goes up and MOSFET turns on. This method can be used only for input voltage range under 50 volts. Because of higher input voltage, power losses of R_1 and R_2 will be a problem and increment of R_1 and R_2 resistance to reduction of power losses, results to bad MOSFET switching performance. In Figure 2, auxiliary switch is used to enhance current injection to MOSFET gate which lead to better performance. But in Figure 32, controller IC supply the whole current needed for P-channel MOSFET drive.

In Figure 33, MOSFET is N-channel type and IR2113 driver which is designed to use bootstrap technique to charging gate capacitor is used. In Figure 34, optical isolation between controller and driver is designed to separate signal and power zones. ICL7667 has a floating ground point that needs an isolated power supply. There are some DC power supplies that because of isolation between input and output zones are proper for this propose.

After requirements that were obtained before, whole converter reliability can be evaluated. Proposed Buck converter specifications are presented in Table 3. Operating point is located in CCM.

Figure 31. Reliability model of Buck converter

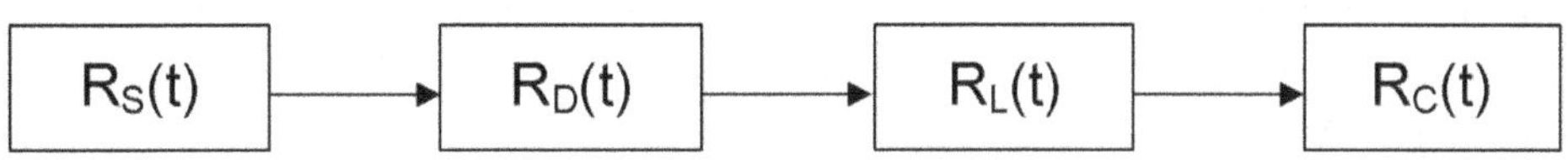

Figure 32. Buck converter with P-channel MOSFET without auxiliary switch

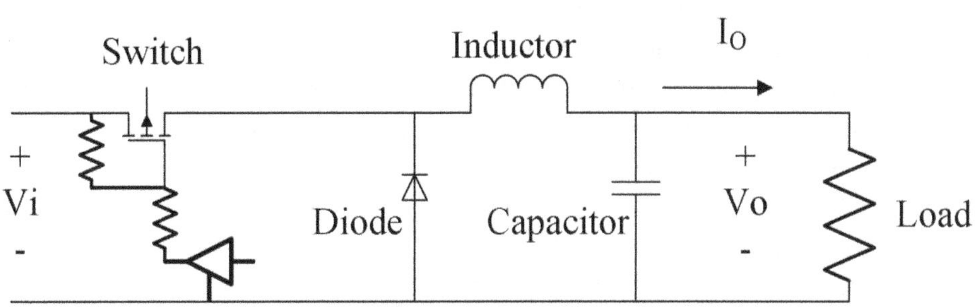

Figure 33. Buck converter with P-channel MOSFET with auxiliary switch

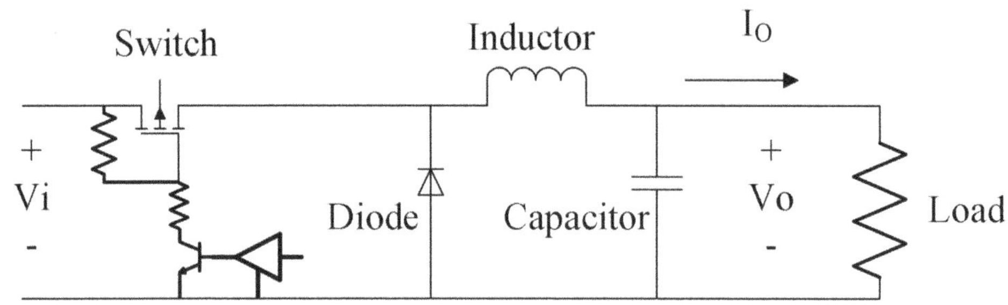

Figure 34. Buck converter with N-channel MOSFET and boot strap driver

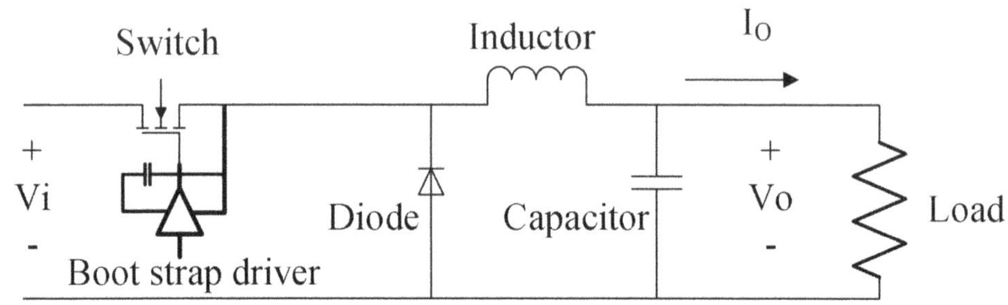

Table 3. Proposed buck converter specification

P_o	V_{in}	V_o	L	C	D	F_s
60W	30V	15V	375uH	1.5uF	0.5	50KHz

In each of 4 converter topologies that are presented, it is assumed that different drivers do not have important effect on converter component operation just on MOSFET switch operation and switch loss. One important difference between N-channel and P-channel MOSFET is input capacitor values. Input capacitor in products of some manufacturer has been assessed and it is deduced that the average of input capacitor in P-channel MOSFETs is more than N-channel MOSFETs. Whatever input capacitor is higher, charge interval will takes more time and consequently MOSFET switching loss becomes higher. Higher loss results to higher switch failure rate. (Figure 35 and 36)

In Figure 37 failure rate of different drivers which are used in Figures 38 are shown. These failure rates are calculated based on data of manufacturers for the failure rate of ICs and failure rate prediction of MIL-HDBK-217 for diodes, capacitors and resistors which are used in driver circuits. Details of driver circuits can be seen in application note which are presented in websites of manufacturers. As it is seen in Figure 37, drivers' failure rate for P-channel MOSFET are lower than N-channel MOSFET drivers.

For reliability comparison between 4 topologies that are presented, 2 switches of types P-channel and N-channel are selected to use datasheet parameters and spice simulation to calculate loss and then evaluate switch failure rate. It should be noted that to use available spice models, 2 switches of IRF are selected for spice simulation. IRF540N and IRF5210 are switches which used in converter simulation. Other power electronic components are same and gate signal is ideal. In spice simulation, power loss of

Figure 35. Buck converter with N-channel MOSFET and optical isolation

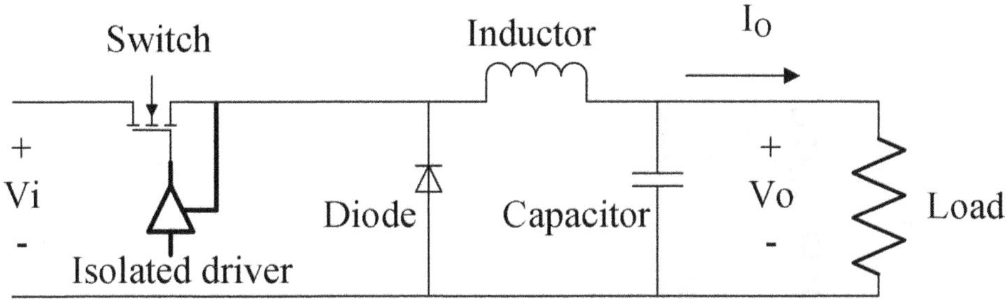

Figure 36. Power losses in switch driver for various drive topologies

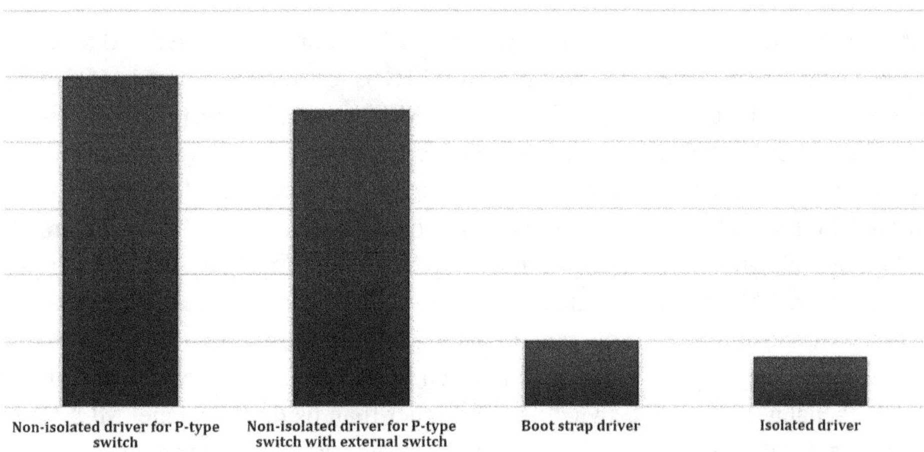

Figure 37. A comparison between failure rates of different BUCK converter topologies

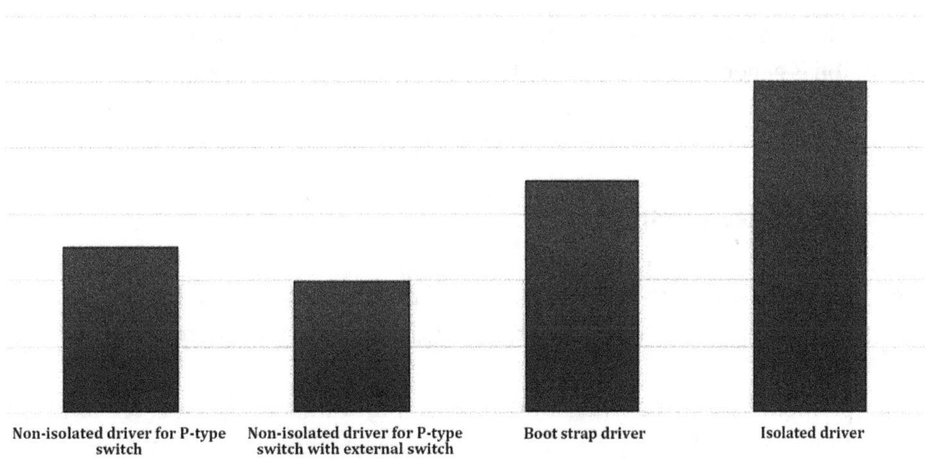

Figure 38. Total Buck Converter failure rate

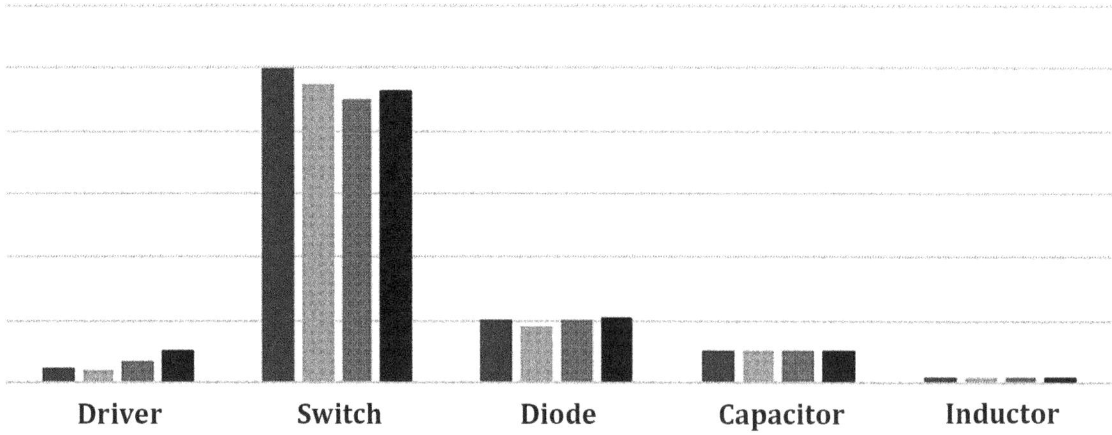

P-channel MOSFET is about 0.99 watt but power loss of N-channel MOSFET is 0.67 watt. Therefore, it is evident that switch failure rate for N-channel MOSFET is less than P-channel MOSFET. As already mentioned, resistance of R_1 and R_2 have affect on MOSFET switching performance. In 4 cases, spice simulation for P-channel MOSFET with varied R_1 and R_2 were done. Whatever these resistors have more resistance, power loss of these resistors are less but switching performance drops more down and power loss of main MOSFET switch increases that results to higher switch failure rate. In Figure 38 power loss of R_1, R_2 and main P-channel MOSFET switch in 4 cases of varied resistors of driver are shown. Also, failure rate of P-channel MOSFET is evaluated. For each of R_1 and R_2 equal to 15 ohm, power loss of switch is 1.475 watt and failure rate of switch equals to 4.15 failures per million hours. In this case, power loss of driver is very high to reduce main switch failure rate. However, the overall failure rate of converter using P-channel MOSFET switch is still higher than using N-channel MOSFET switch and IR2113 driver. The difference of total failure rate in the cases that the resistance of R_1 and R_2 are greater than 15 ohm will be greater. Consequently, it is obvious that using N-channel MOSFET results to lower failure rate for whole converter failure rate.

Other Standards

Fides is a guide allowing estimated reliability calculation for electronic components and systems. The reliability prediction is generally expressed in FIT or MTBF. FIDES Methodology is based on the physics of failures and supported by the analysis of test data, field returns and existing modeling.

SUMMARY AND CONCLUSION

In this chapter, probability calculation was used for reliability prediction. Reliability models from component to system levels were described. The results of this chapter are summarized as follows.

1. Reliability predictions provide a quantitative basis for evaluating product reliability. You can use the information from a reliability prediction to guide design decisions throughout the development

cycle. When an initial design concept is proposed, a reliability prediction can indicate the design feasibility from a reliability standpoint. Even though these early stage predictions are based on limited design information and are approximate at best, they can give direction to your design decisions, which may be critical to product success. Reliability design begins with the development of a system model. Reliability is defined in terms of probability. Probabilistic design is a discipline within engineering design. It deals primarily with the consideration of the effects of random variability upon the performance of an engineering system during the design phase. Typically, these effects are related to quality and reliability. Thus, probabilistic design is a tool that is mostly used in areas that are concerned with quality and reliability.
2. Test of samples.
3. One of the most widely used reliability prediction methodology handbook is MIL-HDBK-217, the Military Handbook for "Reliability Prediction of Electronic Equipment". MIL-HDBK-217 is published by the Department of Defense. It contains failure rate models for numerous electronic components. These models may incorporate predictions based on failure rates taken from historical data. Various parts of a converter has different rate of failure. A failure in a system occurres in the part with highest rate of failure.
4. Series and parallel models are two main types of reliability models of a converter. In series model, the successful operation of the system depends on the proper functioning of all the system components. A component failure represents total system failure. Parallel configuration, or the use of redundancy, is one of the design procedures used to achieve extremely high system reliability, greater than the individual component reliabilities.

The presented tools in this chapter are used for reliability re-evaluation of power converter during the operation process or optimal design process which are presented in the next chapters. Before that, in view of the considerable importance of temperature in all reliability models, we dedicate chapter 4 to this topic.

REFERENCES

Bak-Jensen, B., Bech, J., Bjerregaard, C. G., & Jensen, P. R. (1999). Models for probabilistic power transmission system reliability calculation. *IEEE Transactions on Power Systems*, *14*(3), 1166–1171. doi:10.1109/59.780952

Bhuiyan, F. A., & Yazdani, A. (2010). Reliability assessment of a wind-power system with integrated energy storage. *IET Renewable Power Generation*, *4*(3), 211–220. doi:10.1049/iet-rpg.2009.0070

Javadian, V., & Kaboli, S. (2013). Reliability assessment of some high side MOSFET drivers for buck converter. In *Proceedings of 3rd International Conference on Electric Power and Energy Conversion Systems* (pp.1-6). IEEE. doi:10.1109/EPECS.2013.6713092

Kaboli, S., Vahdati-Khajeh, E., & Zolghadri, M. R. (2006). Probabilistic voltage harmonic analysis of direct torque controlled induction motor drives. *IEEE Transactions on Power Electronics*, *21*(4), 1041–1052. doi:10.1109/TPEL.2006.876838

Kaboli, S. H., & Zolghadri, M. R. (2003). Flux harmonic spectrum processing of direct torque controlled induction motor. In *Proceedings of the 2003 International Symposium on Circuits and Systems* (pp. III-383 - III-386). IEEE. doi:10.1109/ISCAS.2003.1205036

Li, J.-P., & Thompson, G. (2005). A method to take account of inhomogeneity in mechanical component reliability calculations. *IEEE Transactions on Reliability*, *54*(1), 159–168. doi:10.1109/TR.2004.837702

Melo, A. C. G., Pereira, M. V. F., & Leite da Silva, A. M. (1993). A conditional probability approach to the calculation of frequency and duration indices in composite reliability evaluation. *IEEE Transactions on Power Systems*, *8*(3), 1118–1125. doi:10.1109/59.260886

Poshtan, M., Kaboli, S., & Mahdavi, J. (2006). On the Suitability of Modeling Approaches for Power Electronic Converters. In *Proceedings of IEEE International Symposium on Industrial Electronics* (pp.1486-1491). IEEE. doi:10.1109/ISIE.2006.295691

Richardeau, F., & Pham, T. T. L. (2013). Reliability Calculation of Multilevel Converters: Theory and Applications. *IEEE Transactions on Industrial Electronics*, *60*(10), 4225–4233. doi:10.1109/TIE.2012.2211315

Sutanto, D., Outhred, H. R., & Lee, Y. B. (1989). Probabilistic power system production cost and reliability calculation by the Z-transform method. *IEEE Transactions on Energy Conversion*, *4*(4), 559–566. doi:10.1109/60.41714

ADDITIONAL READING

Al Badawi, F. S., & AlMuhaini, M. (2015). Reliability modelling and assessment of electric motor driven systems in hydrocarbon industries. *IET Electric Power Applications*, *9*(9), 605–611. doi:10.1049/iet-epa.2015.0089

Bazzi, A. M., Dominguez-Garcia, A., & Krein, P. T. (2012). Markov Reliability Modeling for Induction Motor Drives under Field-Oriented Control. *IEEE Transactions on Power Electronics*, *27*(2), 534–546. doi:10.1109/TPEL.2011.2168543

Brown, R. E. (2009). *Electric Power Distribution Reliability*. Florida, USA: CRC.

Chan, F., & Calleja, H. (2011). Reliability Estimation of Three Single-Phase Topologies in Grid-Connected PV Systems. *IEEE Transactions on Industrial Electronics*, *58*(7), 2682689. doi:10.1109/TIE.2010.2060459

Choi, U. M., Jeong, H. G., Lee, K. B., & Blaabjerg, F. (2012). Method for Detecting an Open-Switch Fault in a Grid-Connected NPC Inverter System. *IEEE Transactions on Power Electronics*, *27*(6), 2726–2739. doi:10.1109/TPEL.2011.2178435

De León-Aldaco, S. E., Calleja, H., Chan, F., & Jiménez-Grajales, H. R. (2013). Effect of the Mission Profile on the Reliability of a Power Converter Aimed at Photovoltaic Applications-A Case Study. *IEEE Transactions on Power Electronics*, *28*(6), 2998–3007. doi:10.1109/TPEL.2012.2222673

Finkelstein, M. (2008). *Failure Rate Modelling for Reliability and Risk*. Berlin, Germany: Springer.

Gill, P. (2009). *ELECTRICAL POWER EQUIPMENT MAINTENANCE AND TESTING*. Florida, USA: CRC.

Goel, A., & Graves, R. J. (2006). Electronic system reliability: Collating prediction models. *IEEE Transactions on Device and Materials Reliability*, *6*(2), 258–265. doi:10.1109/TDMR.2006.876570

Harb, S., & Balog, R. S. (2013). Reliability of Candidate Photovoltaic Module-Integrated-Inverter (PV-MII) Topologies-A Usage Model Approach. *IEEE Transactions on Power Electronics*, *28*(6), 3019–3027. doi:10.1109/TPEL.2012.2222447

Harb, S., Mirjafari, M., & Balog, R. S. (2013). Ripple-Port Module-Integrated Inverter for Grid-Connected PV Applications. *IEEE Transactions on Industry Applications*, *49*(6), 2692–2698. doi:10.1109/TIA.2013.2263783

Holtz, J. (2005). Sensorless Control of Induction Machines-With or Without Signal Injection. *IEEE Transactions on Industrial Electronics*, *53*(1), 7–30. doi:10.1109/TIE.2005.862324

Johansson, J., & Leisner, P. (2012). Prognostics of thermal fatigue failure of solder joints in avionic equipment. *IEEE Aerospace and Electronic Systems Magazine*, *27*(4), 16–24. doi:10.1109/MAES.2012.6203714

Kanekawa, N., Ibe, E. H., Suga, T., & Uematsu, Y. (2011). *Dependability in Electronic Systems*. Berlin, Germany: Springer. doi:10.1007/978-1-4419-6715-2

Khosroshahi, A., Abapour, M., & Sabahi, M. (2015). Reliability Evaluation of Conventional and Interleaved DC–DC Boost Converters. *IEEE Transactions on Power Electronics*, *30*(10), 5821–5828. doi:10.1109/TPEL.2014.2380829

Lisnianski, A., & Frenkel, I. (2012). *Recent Advances in System Reliability*. Berlin, Germany: Springer. doi:10.1007/978-1-4471-2207-4

Myers, A. (2010). *Complex System Reliability*. Berlin, Germany: Springer. doi:10.1007/978-1-84996-414-2

Nakagawa, T. (2005). *Maintenance Theory of Reliability*. Berlin, Germany: Springer.

Nakagawa, T. (2008). *Advanced Reliability Models and Maintenance Policies*. Berlin, Germany: Springer.

Testa, A., De Caro, S., & Russo, S. (2012). A Reliability Model for Power MOSFETs Working in Avalanche Mode Based on an Experimental Temperature Distribution Analysis. *IEEE Transactions on Power Electronics*, *27*(6), 3093100. doi:10.1109/TPEL.2011.2177279

Wang, H., Khambadkone, A. M., & Yu, X. (2010). Control of Parallel Connected Power Converters for Low Voltage Microgrid-Part II: Dynamic Electrothermal Modeling. *IEEE Transactions on Power Electronics*, *25*(12), 2971–2980. doi:10.1109/TPEL.2010.2087394

Xu, J., Zhao, P., & Zhao, Ch. (2016). Reliability Analysis and Redundancy Configuration of MMC with Hybrid Submodule Topologies. *IEEE Transactions on Power Electronics*, *31*(4), 2720–2729. doi:10.1109/TPEL.2015.2444877

Yang, Y., Ruan, X., Zhang, L., He, J., & Ye, Zh. (2014). Feed-Forward Scheme for an Electrolytic Capacitor-Less AC/DC LED Driver to Reduce Output Current Ripple. *IEEE Transactions on Power Electronics*, *29*(10), 5508–5517. doi:10.1109/TPEL.2013.2293353

Yang, Z., & Raafat, H. (1992). A statistical method for obtaining the factors in electronic-component reliability-prediction models. *IEEE Transactions on Reliability*, *41*(4), 554–557. doi:10.1109/24.249583

KEY TERMS AND DEFINITIONS

Mean Time Between Failures (MTBF): A Well-known parameter which is used in reliability calculation.

Probability: A mathematical tool for analysis of complex systems.

Reliability Model: A model of probability for reliability calculation.

Useful Life: Time interval of system normal operation.

Chapter 4
Thermal Analysis

ABSTRACT

Over temperature is one of the main reasons of failure in electric power converters. In addition, some of other failure factors such as dielectric breakdown act as over temperature in damaging process of a converter. In the previous chapter, it is emphasized that temperature factor is a key index in reliability calculation. Unlike fully electrical variables, thermal calculations require details of geometry of the system and its environment. In this chapter, thermal analysis as the most important factor in failure of converters is presented. Two main approaches for this goal are presented: numerical and lumped mode. Principles of these methods are described with various examples and a comparison is presented. Basic principles of thermal modeling are described and concept of sample node is explained. Methods for thermal management of an electric power converter are described. These methods are at both component and system levels and contain various heat transfer mechanisms like conduction and convection. Theoretical methods and practical considerations for heat sink selection and proper mounting are presented. Thermal insulation classes and various standards related to thermal management are expressed. Industrial samples are presented to show application of theoretical topics in real world.

INTRODUCTION: FAILURES DUE TO THERMAL PROBLEMS

This chapter is considered as a support for reliability calculation as it is shown in Figure 1. Over temperature is one of the main reasons of failure in the systems. In the opposite of other failure factors (for example: dielectric breakdown), over temperature calculation is not performed easily as the analysis of other failure factors are taken place. Thus, in this chapter, temperature calculations of electrical machines and power electronic converters are explained.

METHOD OF HEAT TRANSFER

There are three modes of heat transfer from a source with higher temperature to a region of lower temperature.

DOI: 10.4018/978-1-4666-9429-3.ch004

Figure 1. State of chapter 4 in the flowchart of the book

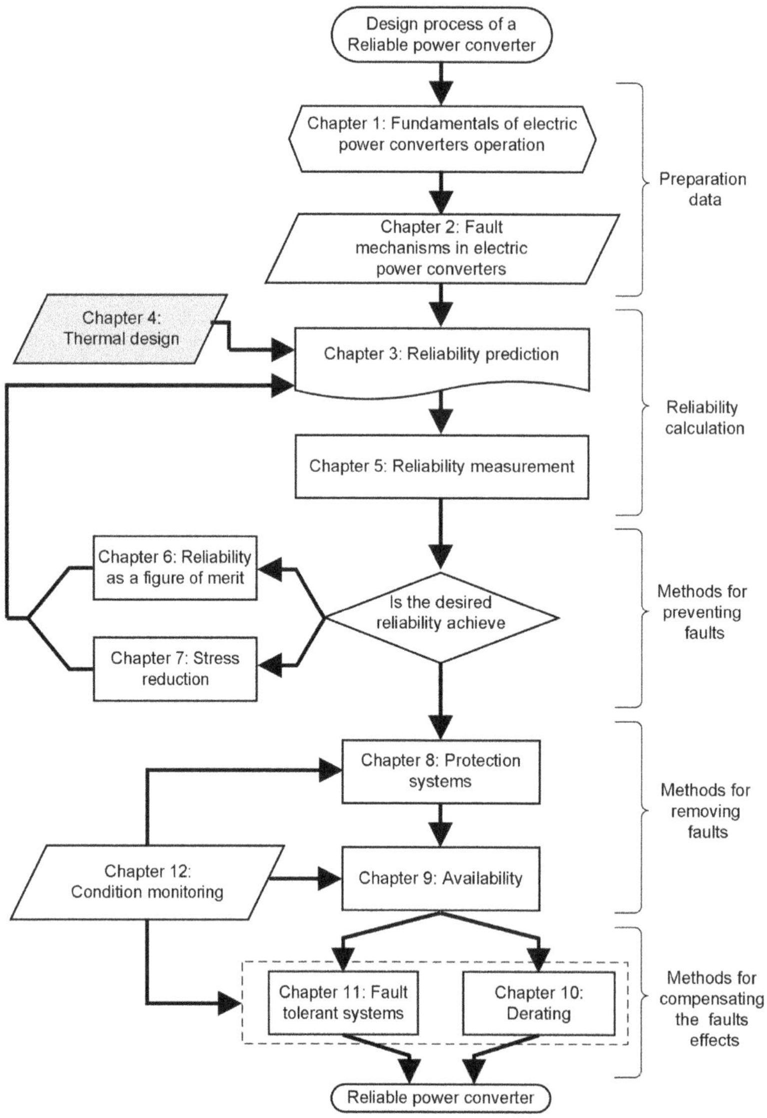

Conduction

Conduction heat transfer occurs when there is a temperature gradient in a system. Energy is transferred from the high temperature region to the low temperature region. The energy transferring requires a material medium to occur. The heat transfer rate is proportional to the temperature gradient in the direction of the heat flow, the area and a property of the system.

Convection

Another means of heat transfer is convection. In convection heat transfer, simply the heat is transferred away from the hot or cold surface. There are a number of concerns to arrive at a adequate analytical treatment of the problem. For example, for the case of a hot plate in front of fan, the velocity at which the air blows over the hot plate obviously influences the rate of heat transfer.

Radiation

The third means of heat transfer is radiation. In contrast to the mechanism of conduction and convection, where energy transfer through a material medium is involved, heat may also be transferred through region where a perfect vacuum exists.

THERMAL ANALYSIS USING FINITE ELEMENT METHOD

In mathematics, the finite element method (FEM) is a numerical technique for finding approximate solutions to boundary value problems for differential equations (Gorginpour, Jandaghi, & Oraee, 2011). It uses the calculation methods to minimize an error function and produce a stable solution. FEM encompasses all the methods for connecting many simple element equations over many small subdomains, named finite elements, to approximate a more complex equation over a larger domain.

Structure of the Calculation

A typical work out of the method is dividing the domain of the problem into a collection of subdomains, with each subdomain represented by a set of element equations to the original problem, systematically recombining all sets of element equations into a global system of equations for the final calculation (Boglietti, Cavagnino, Staton, Shanel, Martin, & Mejuto, 2009). The global system of equations has known solution techniques, and can be calculated from the initial values of the original problem to obtain a numerical answer. The subdivision of a whole domain into simpler parts has several advantages:

- Accurate representation of complex geometry
- Inclusion of dissimilar material properties
- Easy representation of the total solution
- Capture of local effects.

Numerical methods provide a prediction of heat flows. What this means is that it will give a visual or post-processed result of a simulation but the quantitative or absolute accuracy of the result is sensitive to the inclusion and accuracy of the appropriate parameters 1. This method is more appropriate for precise modeling to optimization (Wei, Ghosh, Velusamy, Sankaranarayanan, Skadron, & Stan, 2006). Numerical methods can give an insight into patterns that are difficult, expensive or impossible to study using experimental methods. Experiments can give a quantitative description of flow phenomena using measurements for one quantity at a time, at a limited number of points and time instances. If a full scale model is not available or not practical, scale models or dummy models can be used. The experiments can have a limited range of problems and operating conditions. Simulations can give a prediction of flow phenomena using software for all desired quantities, with high resolution in space and time and virtually any problem and realistic operating conditions.

Methodology

In the first step, the element equations are simple equations that locally approximate the original complex equations to be studied, where the original equations are often partial differential equations. The process is to construct an integral of the inner product of the residual and the weight functions and set the integral to zero. In simple terms, it is a procedure that minimizes the error of approximation by fitting trial functions into the partial differential equations. The residual is the error caused by the trial functions, and the weight functions are polynomial approximation functions that project the residual.

In the second step, a global system of equations is generated from the element equations through a transformation of coordinates from the subdomains' local nodes to the domain's global nodes. This spatial transformation includes appropriate orientation adjustments as applied in relation to the reference coordinate system. The process is often carried out by FEM software using coordinate data generated from the subdomains.

Advantages

FEA is a good choice for analyzing problems over complicated domains, when the domain changes, when the desired precision varies over the entire domain, or when the solution lacks smoothness (Staton, Boglietti, & Cavagnino, 2005). For instance, in a frontal crash simulation it is possible to increase prediction accuracy in "important" areas like the front of the car and reduce it in its rear (thus reducing cost of the simulation). Another example would be in numerical weather prediction, where it is more important to have accurate predictions over developing highly nonlinear phenomena rather than relatively calm areas. FEM allows detailed visualization of where structures bend or twist, and indicates the distribution of stresses and displacements. FEM software provides a wide range of simulation options for controlling the complexity of both modeling and analysis of a system. Similarly, the desired level of accuracy required and associated computational time requirements can be managed simultaneously to address most engineering applications. FEM allows entire designs to be constructed, refined, and optimized before the design is manufactured.

This powerful design tool has significantly improved both the standard of engineering designs and the methodology of the design process in many industrial applications. The introduction of FEM has substantially decreased the time to take products from concept to the production line. It is primarily through improved initial prototype designs using FEM that testing and development have been accelerated. In summary, benefits of FEM include increased accuracy, enhanced design and better insight into critical design parameters, virtual prototyping, fewer hardware prototypes, a faster and less expensive design cycle, increased productivity, and increased revenue.

Figures 2 and 3 show a typical electronic board. To investigate the temperature rise, this beard is analyzed by FEM. Figures 4, 5, 6, and 7 shows the results of this analysis. Some hot spots are seen in the board and should be thermally managed. A similar study can be done on an electric machine as shown in Figure 8.

Figure 2. A PCB board (front)

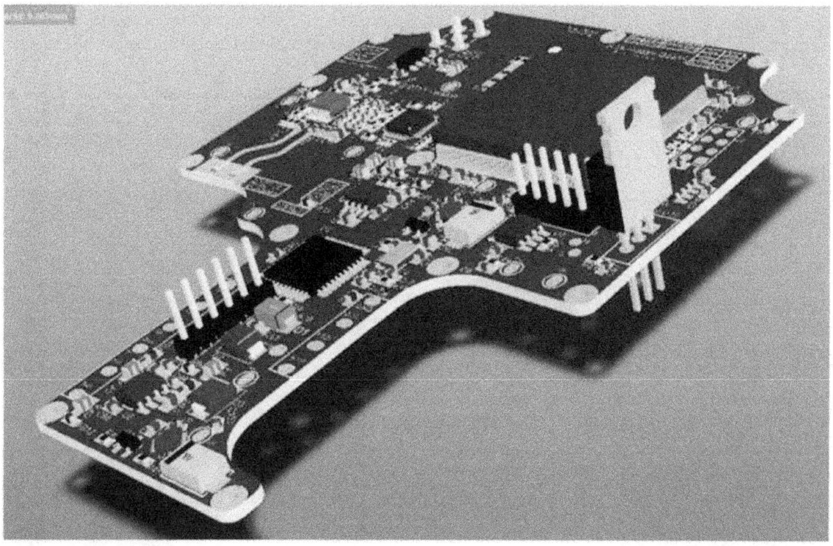

Figure 3. A PCB board (back)

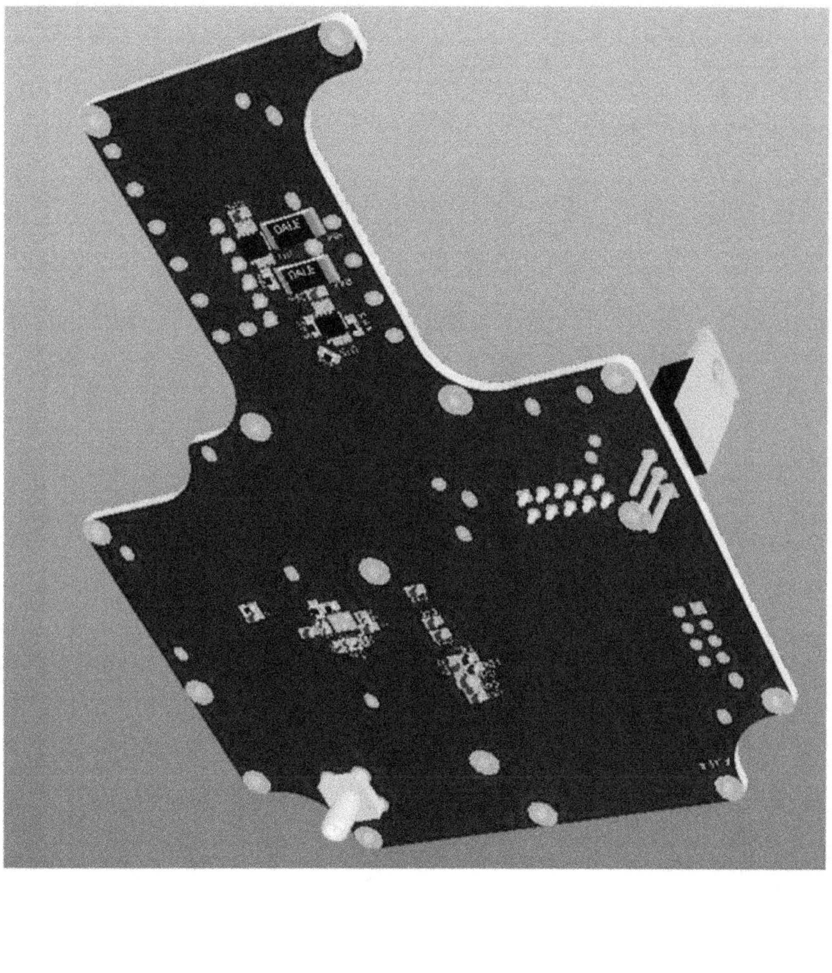

Figure 4. Temperature profile (Front)

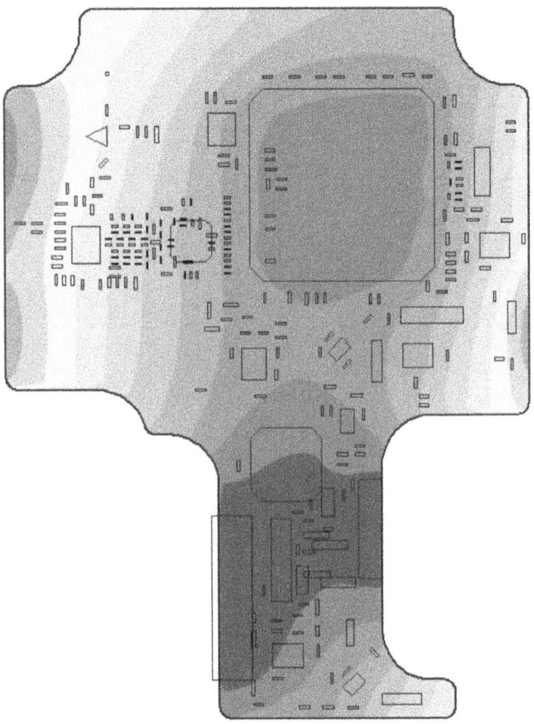

Figure 5. Temperature profile of components

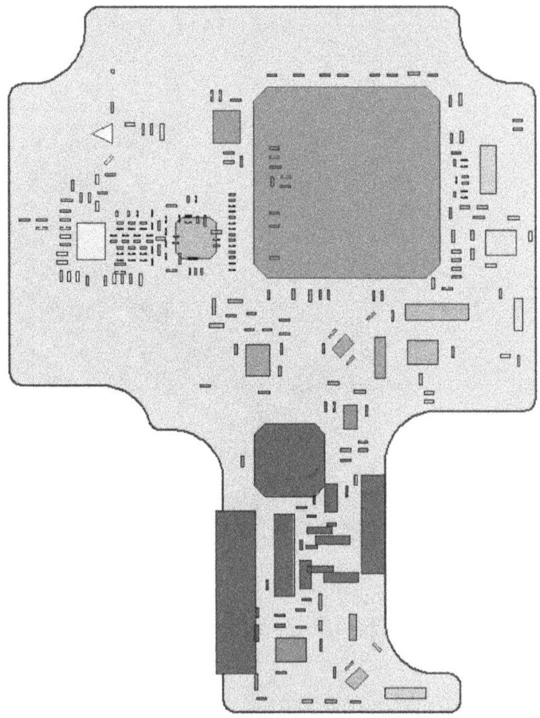

Figure 6. Temperature profile (back)

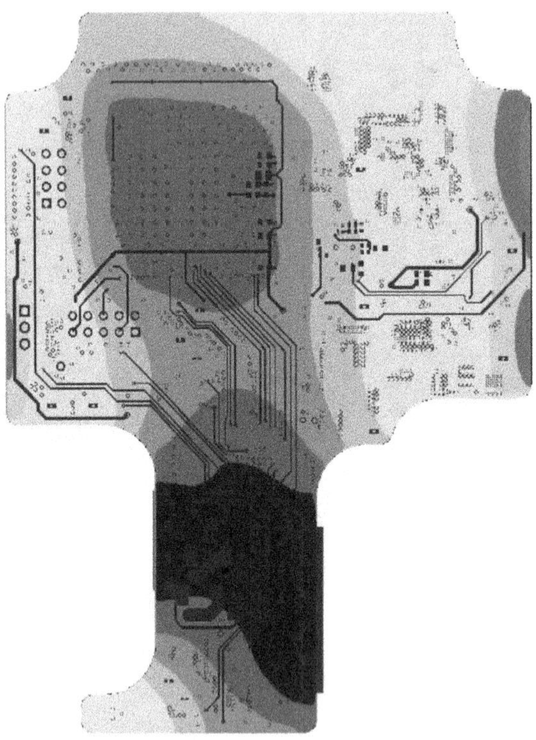

Figure 7. Temperature profile (components)

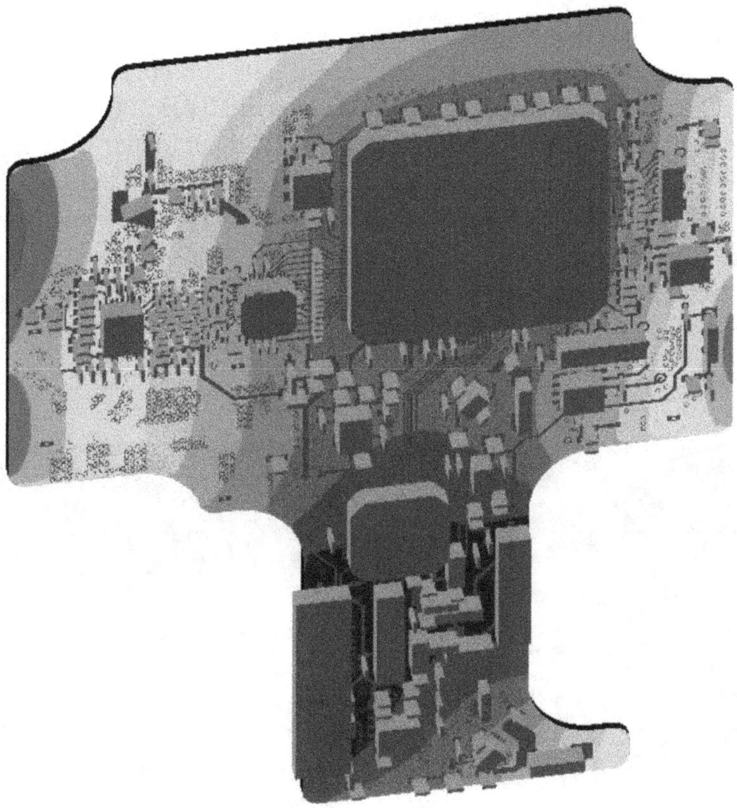

Figure 8. Temperature profile of a rotor (3-D)

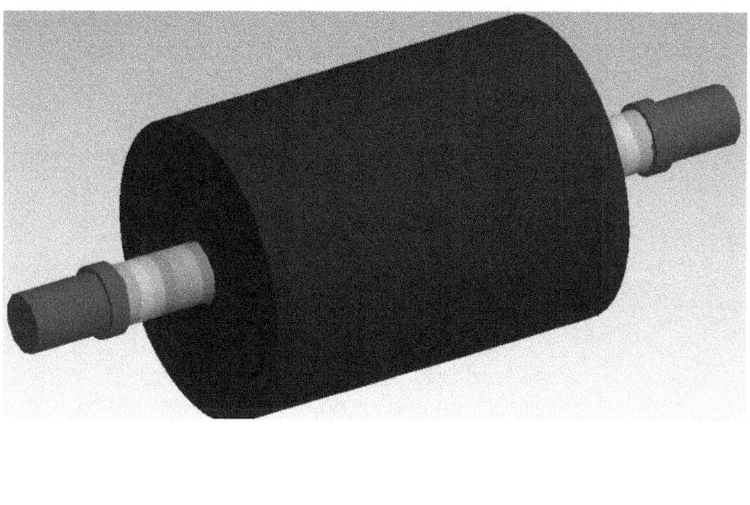

Example: FEM Analysis of Toroid Transformer

To show application of the above mentioned numerical method, a toroid transformer is analyzed using numerical software. The simulation is run using FLUENT software. The geometry of the transformer is set up using GAMBIT software and then it is imported to the FLUENT software to solve the Equation 5.

A large cylinder is placed symmetrically around the transformer to play the role of surroundings. In order to reduce the number of required meshes, and also to reduce the computational time, it is preferred to model only one forth (90 degrees) of the transformer instead of its full circular shape. This should not cause serious problem because the actual heat transfer through the symmetry surface at steady state is zero. Figure 9 shows the transformer section which is imported into FEM software.

Figure 9. Transformer section in a FEM based software

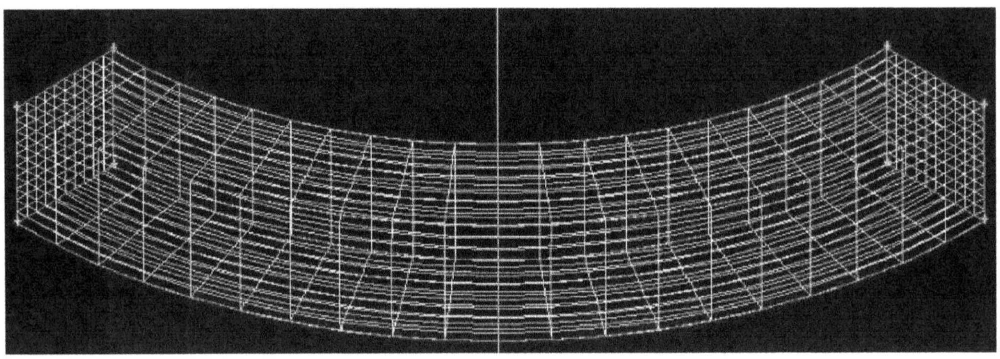

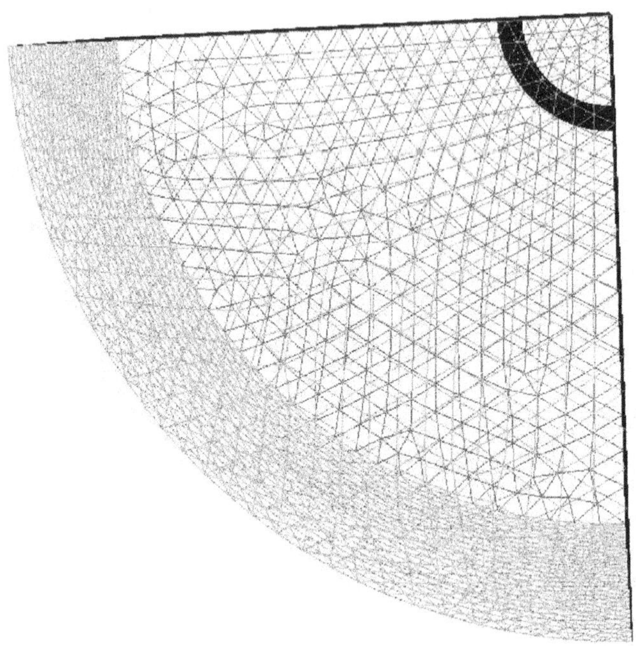

Figure 10. Velocity vectors

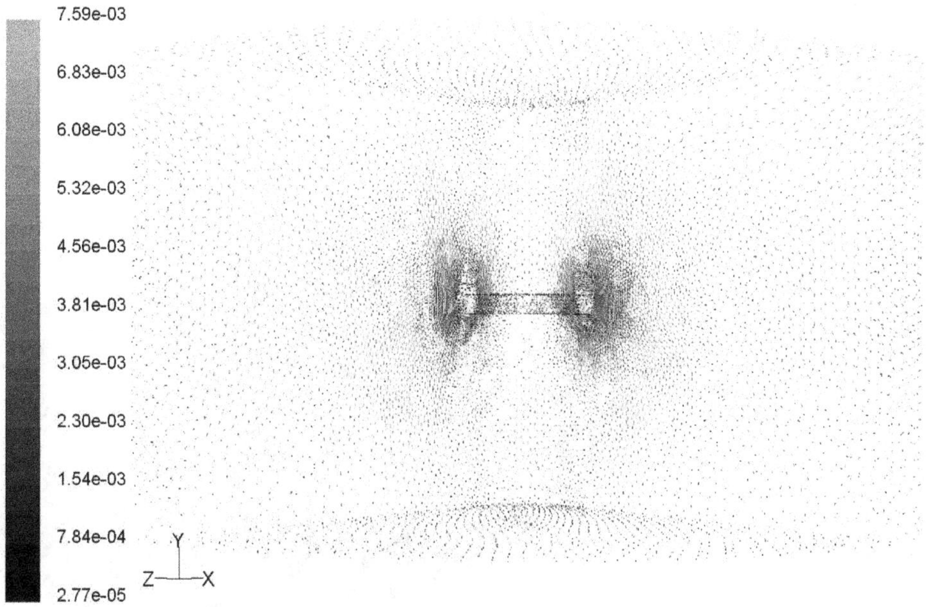

Once all the required parameters are defined, the model can be simulated. Simulation of course needs initial values for say, velocities. Since the heat convection mode is natural, it is reasonable to set the initial values of velocities to zero. In order to simulate the thermal behavior of the model, FLUENT solves the continuity equation and energy balance simultaneously . As FLUENT iterates, it can show the residuals and hence, it can be easily observed if the solution diverges or converges.

Since FLUENT solves the equations for all of parameters velocity, temperature, pressure, density, etc, the plots for any of these parameters can be obtained. The velocity vectors and pressure along a surface crossing origin are shown in Figure 10 and Figure 11, respectively.

The results show the symmetry that would be expected around the surface of the core of the transformer.

Lumped Thermal Model

FEM techniques for analyzing the temperature profile are usually accurate tools and yield the temperature of many nodes in the studied system. However, these methods usually have a long run time and they suffer from divergence problems. Especially in the optimization process with iterative solving techniques, it is an important limitation. Other alternative technique for analyzing the temperature is usage of a thermal model. A thermal model is a summarized model of the one which is used in FEM. In this summarized model, the number of nodes is limited to a few nodes. Therefore, the set of thermal equation become small and it can be solved rapidly. Thus, the run time decreases and the temperature profile is achieved in a reasonable time interval (Gorginpour, Oraee, & McMahon, 2014). The cost paid for this short convergence time is decreasing in the accuracy of the solutions. Regarding to these characteristics, thermal models are used in the design process to reach to an acceptable design and then the accuracy of the design is raised with a FEM based analysis.

Figure 11. Contours of pressure

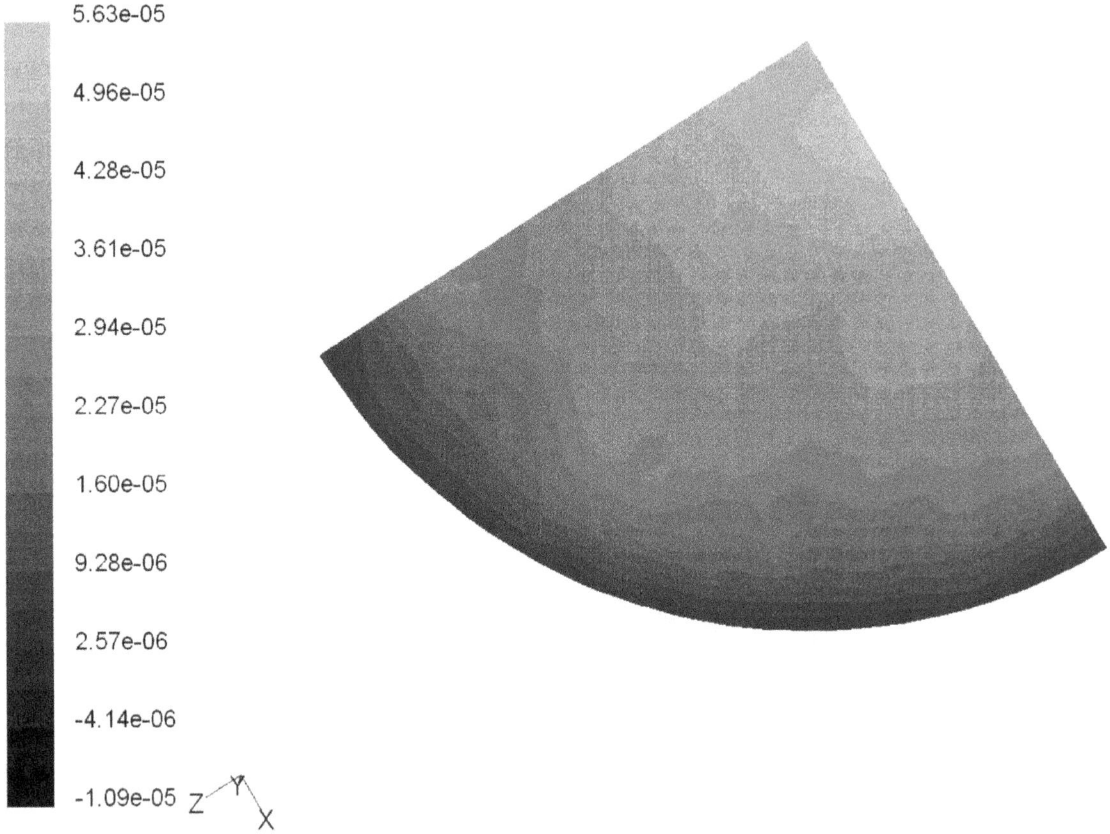

Thermal models facilitate the design and analysis of cooling systems. The underlying idea of such models is the formal similarity between heat transfer and electric circuits. Thermal quantities and their electrical counterparts are listed in Table 1.

Figure 12 shows a typical heat transfer problem, in which a lossy system dissipates some energy into its ambient. The thermal equivalent for this typical problem is shown in Figure 13.

In this circuit, T_A and T_e are the ambient and system temperatures, respectively; P_{loss} is the power loss in the system and R_{th} is the thermal resistance of the heat transfer path. Using this equivalent circuit the following relationship can be readily obtained.

$$T_e - T_A = P_{loss}.R_{th} \tag{1}$$

Table 1. Thermal and electrical equivalent pairs

Thermal Quantity	Electrical Quantity
Heat current (power), P(W)	Current, I(A)
Temperature, T(°K)	Voltage, V(V)
Thermal resistense, R(°K/W)	Resistance, R(Ω)
Thermal capacity, C(J/°K)	Capacitance, C(F)

Figure 12. A simple heat transfer layout

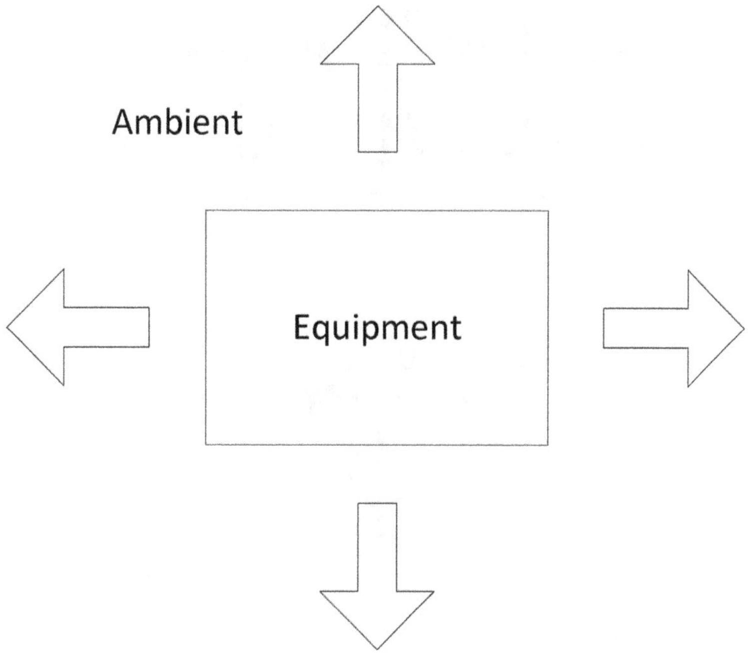

Figure 13. Thermal equivalent circuit of the simple heat transfer layout

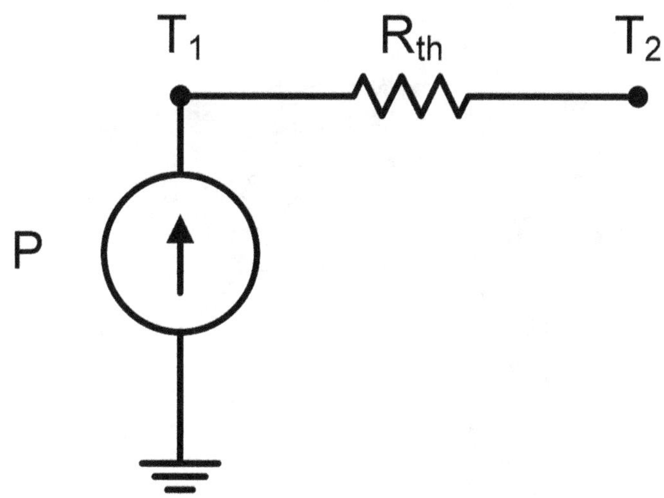

It is worth noting that the thermal capacity is not considered in this study, as it does not affect the steady state temperature.

Figure 14 shows the equivalent thermal model of stator of an electric motor in two levels (Boglietti, Cavagnino, Lazzari, & Pastorelli, 2003). In advanced level, the stator is divided into some parts and a node is considered for each part. In simple level. All of stator is considered as a unique part. A more precise model is shown in Figure 15.

Here, we continue the method with some different type examples.

Figure 14. Two various methods of thermal equivalent circuit

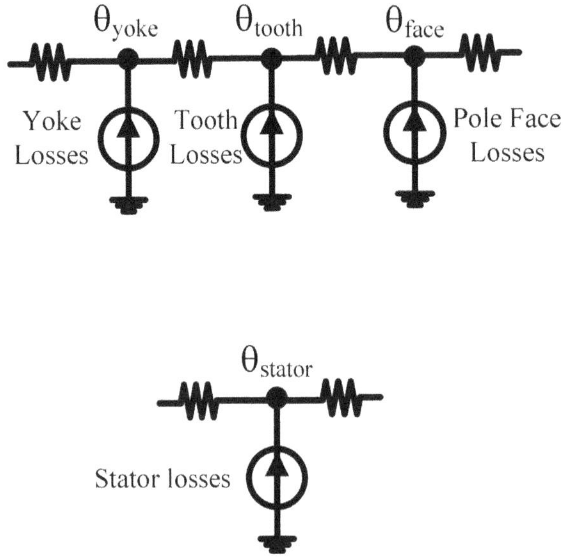

Figure 15. Thermal model of n induction motor

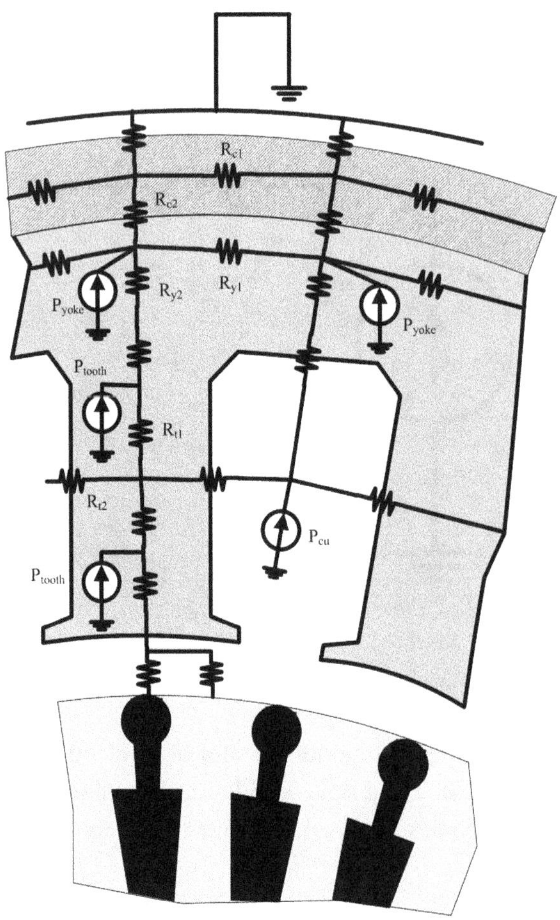

Thermal Modeling with Symmetry

In this section, a method for analysis of thermal behavior of toroidal transformers is proposed. This machine was analyzed in previous section with FEM. Regarding the symmetry in its structure, its thermal model is simple. The method enables prediction of temperature distribution within the transformers (Swift, Molinski, & Lehn, 2001).

A toroidal transformer has a ring shaped core with a circular cross section. Primary and secondary windings are wound on such core. Figure 16 shows this structure and its vertical cross section.

As the whole setup has a polar symmetry, there is no circular heat transfer (Thorough direction1). On the other hand, the heat generated within the core will flow through core iron, winding copper and surrounding air, consecutively (Thorough direction 2).

Figure 17 shows the horizontal cross section of toroidal transformer. A polar symmetry also exists in direction 1 in this figure. Therefore, we will start thermal modeling along the radial axis.

Figure 16. Toroidal transformer and its cutoff cross section

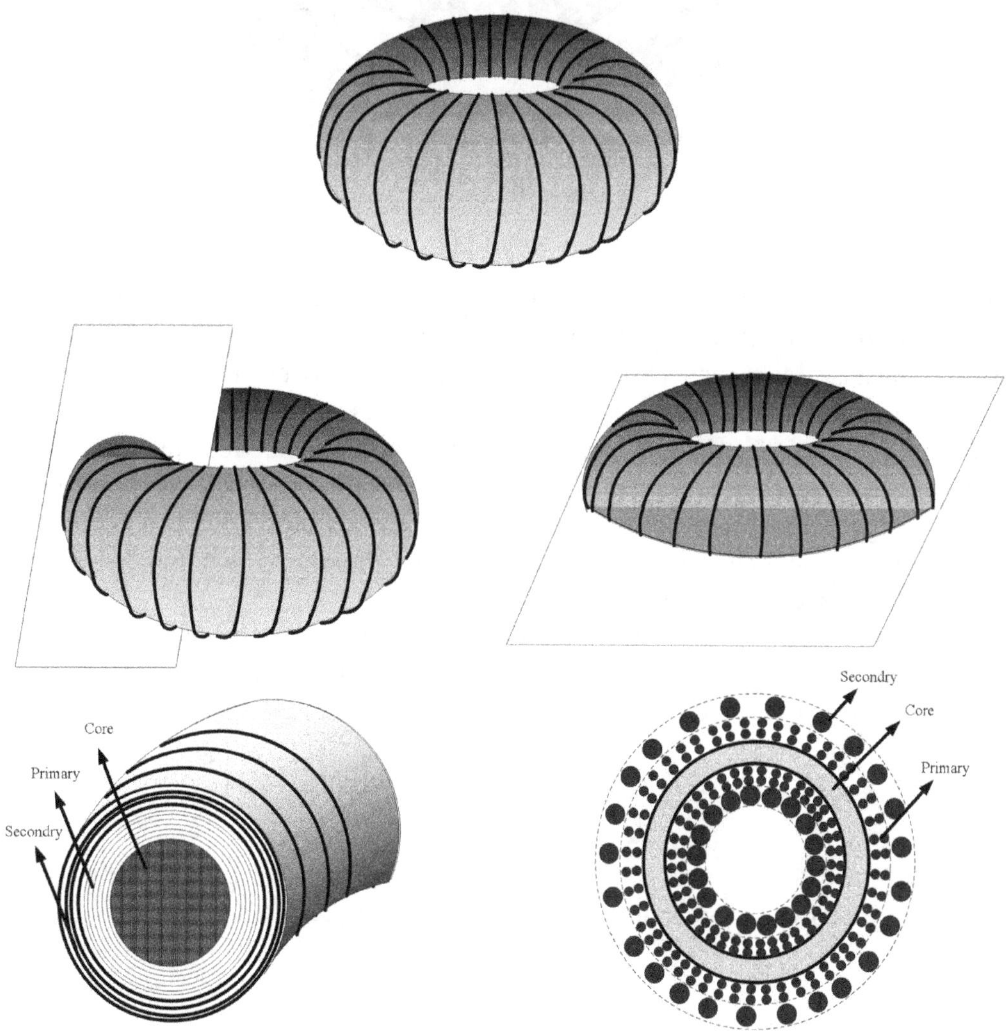

Figure 17. Horizontal cross section of toroidal transformer

As mentioned earlier, thermal modeling requires a transition from a distributed phenomenon to a lumped model. To do so, a differential element of a form of a sector is cut out from the transformer, the view angle of which is φ, as shown in Figure 17. In the following section, heat transfer behavior of this element is examined and its heat sources and thermal resistances are computed 2,3.

Repeatability is the main feature of differential element, i.e., the whole structure is made up of several such differential elements placed one next to another. As shown in Figure 17, the differential element comprises m wires of primary winding in each layer, n wires of secondary winding in each layer and a section of core. As windings are uniformly distributed around the core, ratio n/m is constant regardless of where the differential element is taken from 9.

Heat transfer of a wire to ambient is occurred by convection because a wire has low contact area with its surroundings and the rate of heat transfer through conduction is poor. Heat transfer equation of a single wire through convection obeys equation (2).

$$P_{loss} = hA\Delta T \tag{2}$$

where P_{loss} is the power dissipation of wire, A is its cross section area and ΔT is the temperature difference between wire and its ambient. In this equation constant h is presented by as below.

Figure 18. Differential element

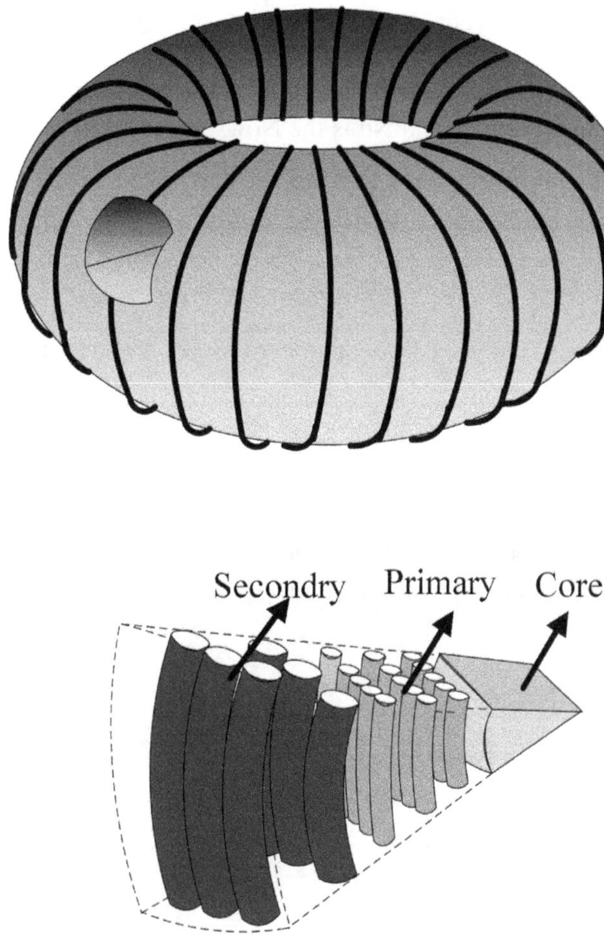

Figure 19. View of differential element from 3 dimensions

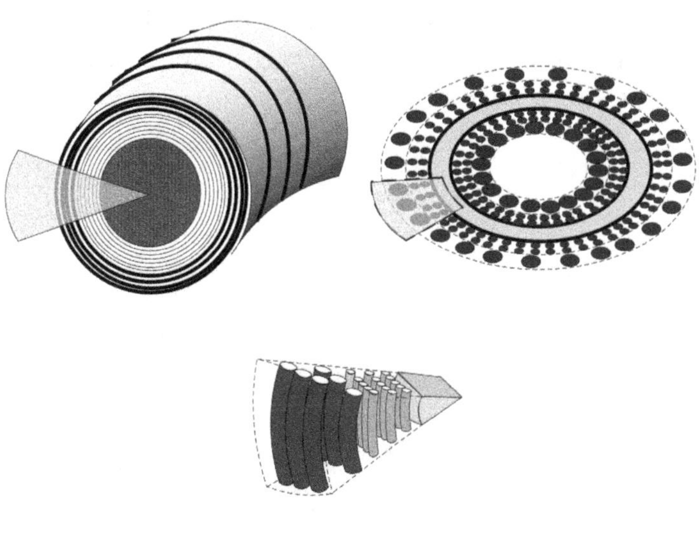

$$h = N_u \frac{k}{d} \tag{3}$$

where d is hydraulic diameter that is equal to wire diameter, k is convection constant of ambient and is equal to 0.026 $W/m.\ ^\circ C$ in this study, and Nu is the Nowselt constant. This constant is expressed as follows 4.

$$N_u = m\left(Ra\right)^c. \tag{4}$$

where m and c are two constants and Ra is Riely number that is presented in the following equation 4.

$$Ra = 0.7\frac{g\beta\Delta T d^3}{v^2} \tag{5}$$

In this equation g is gravity acceleration and equals to $9.8 m/s^2$, v is a constant and is equal to $0.0000156 m^2/s$ and β is equal to the following relation 4.

$$\beta = (\Delta T / 2 + 273)^{-1} \tag{6}$$

Therefore, thermal resistance in convection is equal to the following equation.

$$R_{th} = \frac{1}{hA} \tag{7}$$

In this way, thermal equivalent circuit of differential element for a toroidal transformer is shown in Figure 7. It is assumed that this transformer has one primary and secondary layer. The differential element is related to m wires in primary and n wires in secondary and for sector angle φ (see Figure 17). Current sources P_p and Ps are the power loss in primary and secondary windings, respectively. If the number of primary winding turns is equal to N_p and for secondary winding is equal to Ns, then these current sources can be expressed as follows.

$$P_p = (P_{cup} / N_p)\frac{\varphi}{2\pi} \tag{8}$$

$$P_s = (P_{cus} / N_s)\frac{\varphi}{2\pi} \tag{9}$$

where Pcup and Pcus are the total copper loss in primary and secondary windings, respectively.

In calculation of thermal resistance R_p and R_s, the temperature difference ΔT is equal to temperature difference between two neighborhood point. For example in calculation of R_p the temperature difference is equal to T_{np}-T_{nc}

The current source P_c is the core loss in differential element and is expressed by

$$P_c = P_{ctotal} \frac{\varphi}{2\pi} \frac{m}{N_p} \tag{10}$$

and the core thermal resistance R_c is equal to the following equation

$$R_c = \frac{l}{\lambda A_c} \tag{11}$$

where l is radial of core, A_c is area of core in this element and λ is conductivity of core.

Figure 20 shows the thermal equivalent circuit.

Figure 20. Thermal equivalent circuit for the differential element

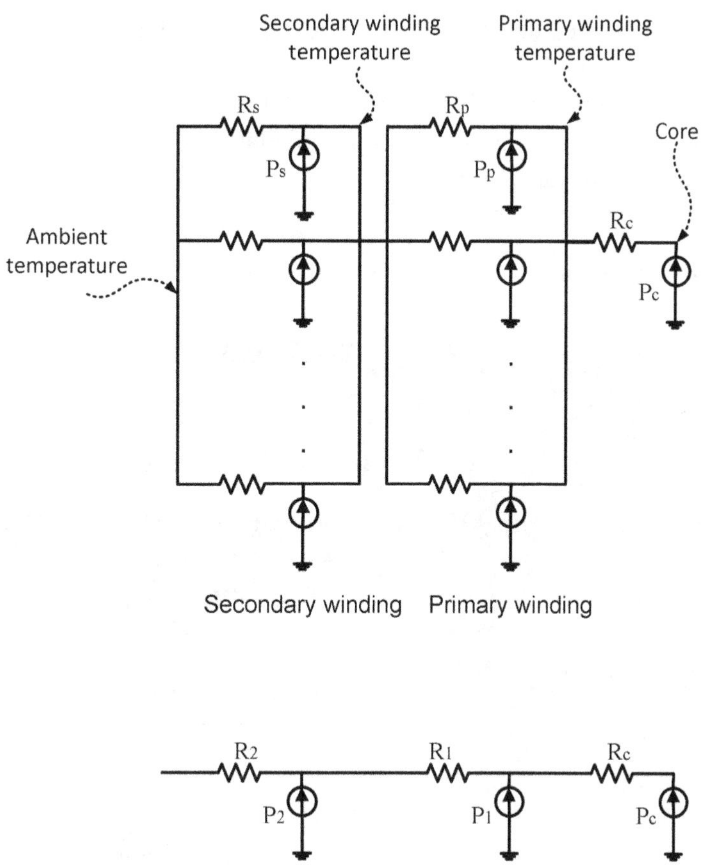

Table 2. Charactristics of transformers used in experimental setup

#	Vi/Vo	Pn (W)	N1	D1 (mm)	N2	D2 (mm)
1	220/24	50	1725	0.28	212	0.80
2	220/24	200	1085	0.60	127	1.80
3	220/24	500	505	0.95	58	2.80

To verify theoretical prediction and simulation results, a test setup has been prepared. This setup is constructed by means of a toroidal transformer, several temperature sensors, which have been put in different layers of transformer windings, a load and some instrumentation devices.

The temperature sensor used is AD590. It is a temperature-controlled current source with high immunity against noise, whose output signal can be transmitted through long wires. Another important feature of AD590 is its small size, which enables putting it in winding layers.

To verify results for a wide range of rated power and size of transformers, three transformers with different sizes and rated powers have been employed in this experiment. The characteristics of these transformers are shown in Table 2.

The photographs of setup are shown in Figure 21. The transformers, the transmission wires and the instrumentation devices are clearly observed.

Tables 3 and 4 show the theoretical predictions and results of experiments, respectively. As can be seen, the experimental results and theoretical calculations are in well agreement. Results show that the core of transformer is its hotspot.

The difference between the measured and calculated values can be due to the fact that the winding wires of transformers in experimental setup had a rectangular cross section, which causes non-uniform heat distribution and transfer in various directions.

Figure 21. Experimental setup for temperature measurement

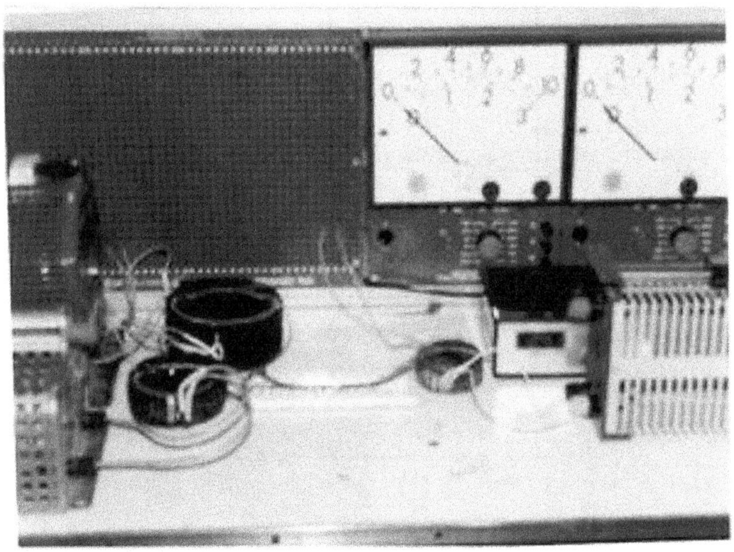

Table 3. Experimental results of temperature in different points of transformers

#	Temperature in the middle of primary winding (° C)	Temperature between primary and secondry windings (° C)	Temperature in the outer layer of secondry windings (° C)
1	65	61	57
2	68	62	58
3	73	68	61

Table 4. Theoretical results of temperature in different points of transformers

#	Temperature in the middle of primary winding (° C)	Temperature between primary and secondry windings (° C)	Temperature in the outer layer of secondry windings (° C)
1	61	59	53
2	63	59	55
3	70	65	62

The simulation and experimental results show the core of transformer is its hotspot and the bottom layer of primary winding which is close to the core is affected by the most thermal stress.

Thermal Modeling Without Symmetry

Example: Thermal Modeling of Stator Bar

The stator winding in a large electric machine as shown in Figure 22 must be able to carry the rated current without exceeding specified thermal limits. A medium material like water is usually used to transfer the heat generated in stator bar. Therefore, it is necessary to place a number of channels for cooling water inside the stator bar. On the other hand, an AC current carrying conductor embedded in a narrow slot in a magnetic material drives magnetic flux around itself and this alternating leakage flux induces alternating voltages along the length of the strips. If solid conductors are used, these voltages would cause circulating currents around the bar, resulting in unacceptable eddy current loss and heating. In order to minimize this effect, the conductor is divided into strips lightly insulated and arranged in a number of stacks along the bar width (Figures 23 and 24). The strips are transposed along the length of the bar using the Roebel method. Figures 25 and 26 show a typical view of this method. In this figure three different cross sectional areas of stator bar along its length are shown. Some of these strands are hollow type to pass the cooling water. In this example, a thermal model is presented considering the displacement of conductors. The proposed model is a thermal equivalent circuit with structure variability in the length of stator bar because of conductor displacement. The network structure is different for each transposition step, but does not change between two successive transposition steps. Therefore, the thermal model consists of different electric networks connected by lateral thermal resistances. Variations of equivalent network affect hot spot of stator bars.

The developed thermal model enables determination of temperature of copper strands in the bar and water temperature at the bar outlet caused by losses dissipated in bar, heat transfer and the number of blocked hollow strands. In order to determine the temperature of the copper strands, the length of the bar

Figure 22. A turbogenerator cutaway

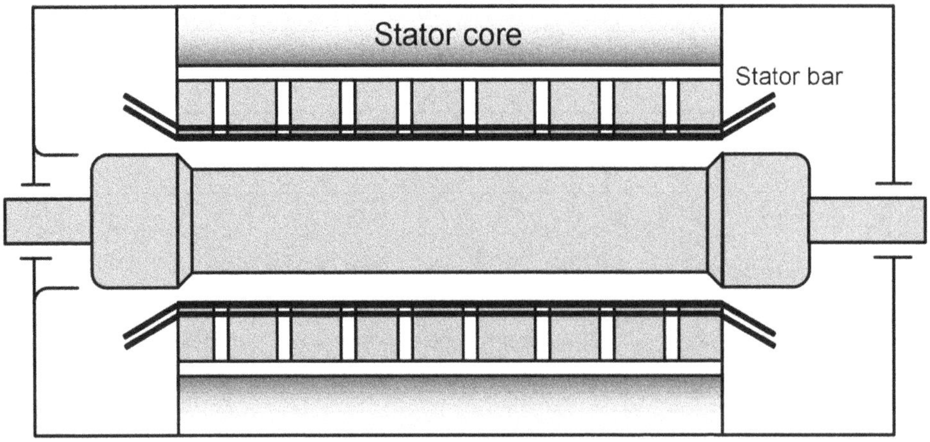

Figure 23. The cross sectional area of a stator bar

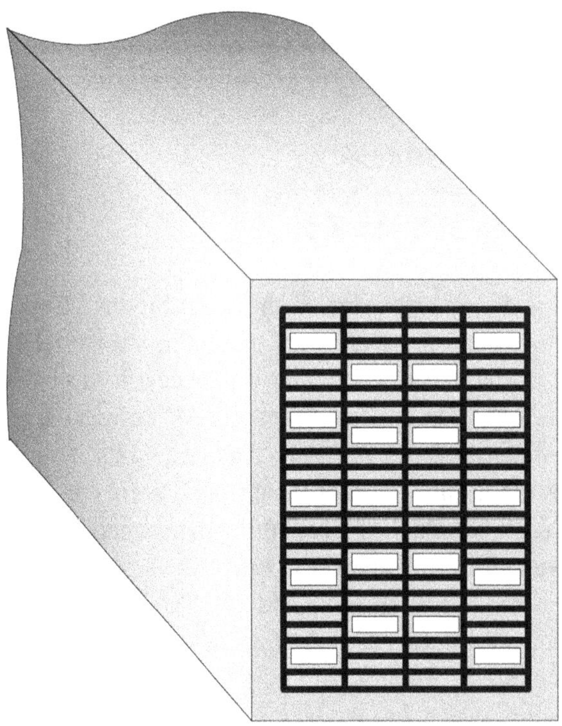

is divided into a number of parts equal in number with transposition steps. The cross sectional area of the bar is fixed between two successive transposition steps and a thermal equivalent circuit is derived for each cross section. In thermal equivalent circuits, node voltages, current sources and resistances represent node temperature, heat loss and thermal resistance, respectively. Therefore, the complete equivalent circuit of stator bar consists of different networks which are connected by lateral thermal resistances. These lateral resistances model water temperature rise in hollow conductors along the length of the bar 12.

Figure 24. Cutaway of a stator bar, (a): insulator holder of bar, (b): insulator between column, (c): copper strand, (d): hollow strand with cooling water channel, (e): insulator of strand

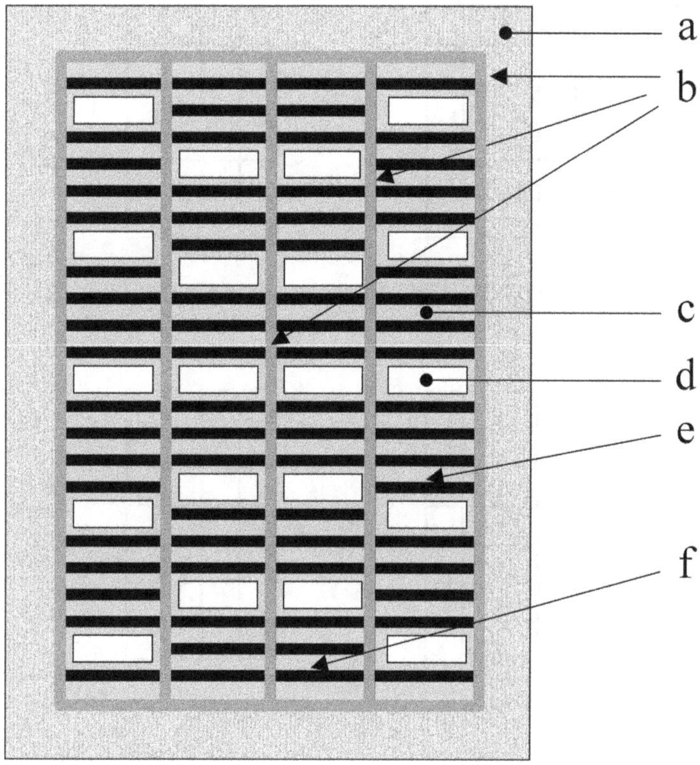

Figure 25. Transposition of stator bar according to the Roebel method

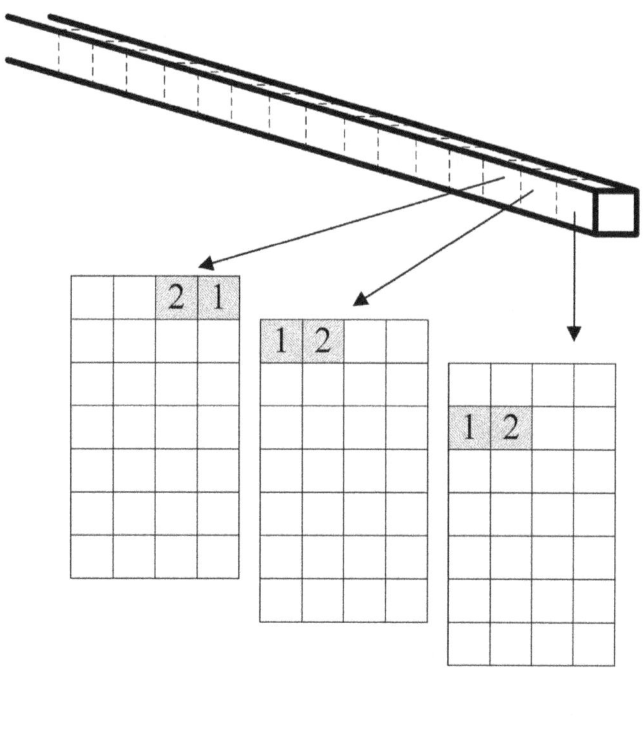

Figure 26. Various pattern in transposition of stator bar according to the Roebel method

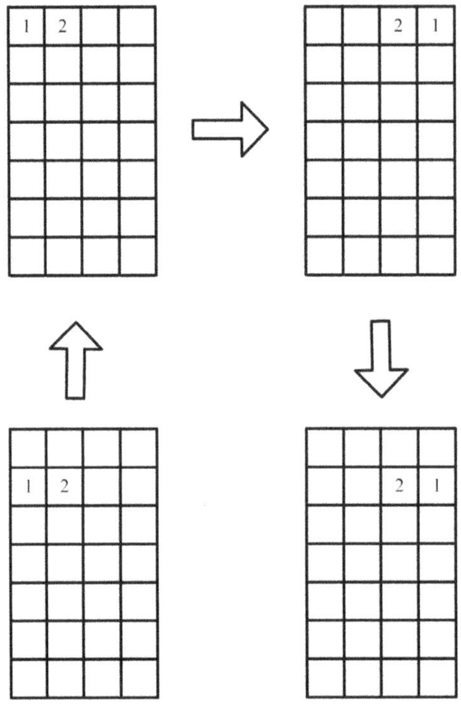

Figure 27 shows the thermal equivalent circuit of a normal and hollow conductor in the bar, respectively.

$$g_1 = \frac{1}{R_1}, R_1 = 2R_{cu1} + 2R_{ins1} + R_{ins2} \tag{12}$$

$$g_2 = \frac{1}{R_2}, R_2 = R_{cu1} + R_{ins1} + R_{ins2} + R_{ins3} + R_{cu2} \tag{13}$$

$$g_3 = \frac{1}{R_3}, R_3 = 2R_{cu2} + 2R_{ins3} + R_{ins2} \tag{14}$$

$$g_4 = \frac{1}{R_4}, R_4 = 2R_{cu3} + 2R_{ins4} \tag{15}$$

Figure 27. Thermal equivalent circuit of a hollow and filled conductor

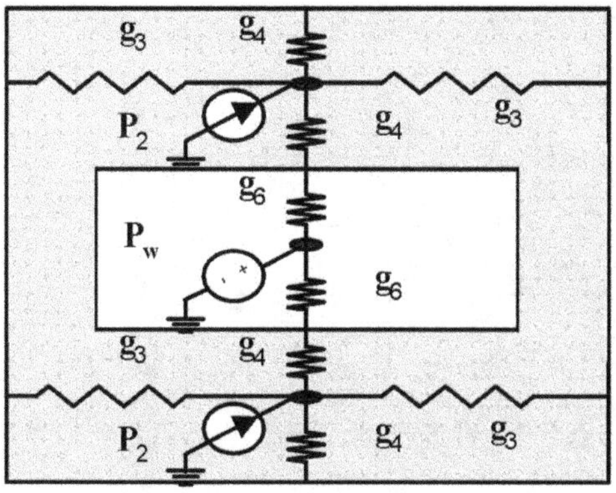

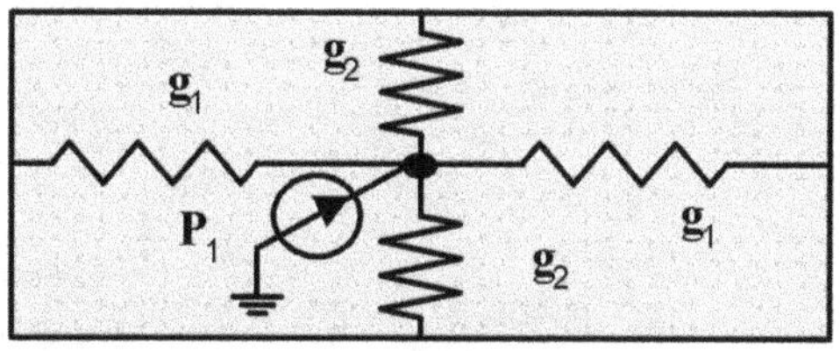

$$g_5 = \frac{1}{R_5}, R_5 = R_{cu3} + R_{ins4} + R_{ins5} + R_{cu4} \tag{16}$$

$$g_6 = \frac{1}{R_6}, R_6 = R_w + R_{cu4} \tag{17}$$

where

$R_{cu1:}$ conduction thermal resistance between two normal conductors along vertical direction.
$R_{cu2:}$ conduction thermal resistance between two hollow conductors along vertical direction.
$R_{cu3:}$ conduction thermal resistance between two normal conductors along horizontal direction.
$R_{cu4:}$ conduction thermal resistance between two hollow conductors along horizontal direction.
$R_{ins1:}$ conduction thermal resistance of normal conductor insulation.
$R_{ins2:}$ conduction thermal resistance of hollow conductor insulation.
$R_{ins3:}$ conduction thermal resistance of insulation between two columns.
$R_{ins4:}$ conduction thermal resistance of bar insulation.
$R_{w:}$ convection thermal resistance of water channel.

Figure 28 shows the thermal equivalent circuit for the bar cross section. In this network, water temperature at hollow conductors is considered constant. Therefore, voltage sources are used to represent water temperature at hollow conductors in each cross section and are shown as their Norton equivalent (P_0). Other current sources are solid conductor copper loss (P_1) and hollow conductor copper loss (P_2), which are expressed based on total loss of the generator.

The voltages at network nodes (which represent node temperatures) are calculated by solving the following system of equations:

$$[Y].[V] = [I]$$

Figure 28a. Thermal equivalent circuit of a cross section

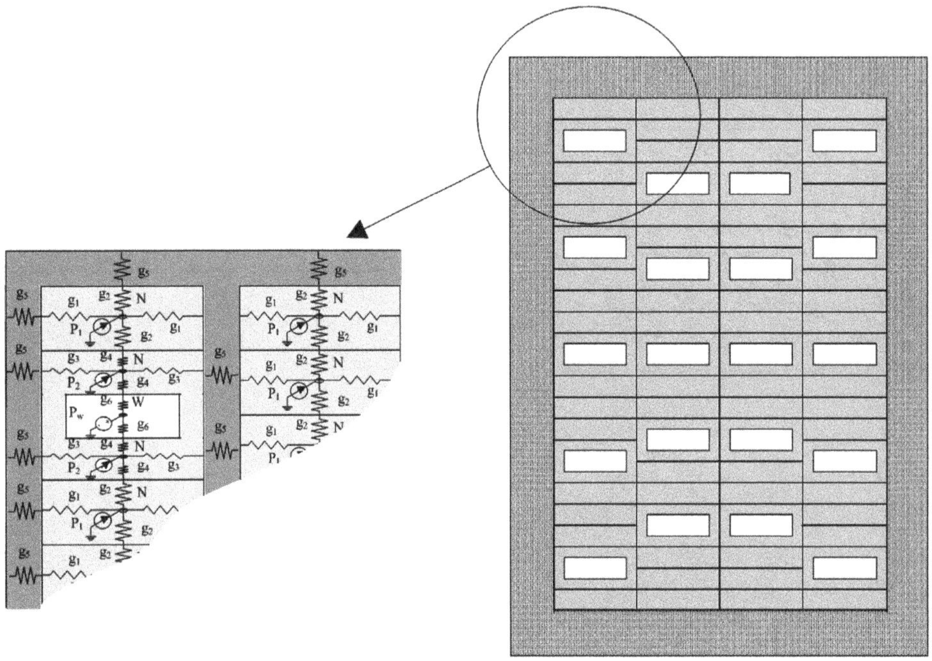

where $[Y]$ is the admittance matrix constructed from the network topology.

To calculate the temperature at different points along the bar, all thermal equivalent circuits of cross sections should be processed. Since it is assumed that water temperature is constant between two successive transposed steps, it is necessary to consider water temperature rise between two successive cross sections. Therefore, water temperature in a cross section is expressed as follows:

$$T_{i+1} - T_i = P_i \frac{N_u}{d}$$

where T_i and T_{i+1} represent water temperature in the i and i+1 cross sections, respectively, P_i represents heat power entering water channels in the i cross section, N_u is the Nouselt constant and d is hydraulic diameter.

Having calculated water temperature rise between two successive cross sections, further process in subsequent cross sections will be possible. Figure 29 shows two consecutive sections of the bar.

Tables 5 and 6 show the electrical and thermal specifications of the stator bar used in simulation, respectively.

Figures 30, 31, and 32 show the thermal equivalent of stator and rotor of the generator.

Figures 33, 34, 35, and 36 show the result of this analysis. Figure 33 shows the temperature of various nodes in the bar. It can be seen that there is a symmetry in this result. Figure 34 show FEM analysis of bar and it verify this symmetry. Figure 35 show the temperature rise of the cooling water in the bar. Figure 36 shows a comparison between simulation analysis and experimental results.

Figure 28b. Completed thermal equivalent circuit of a cross section

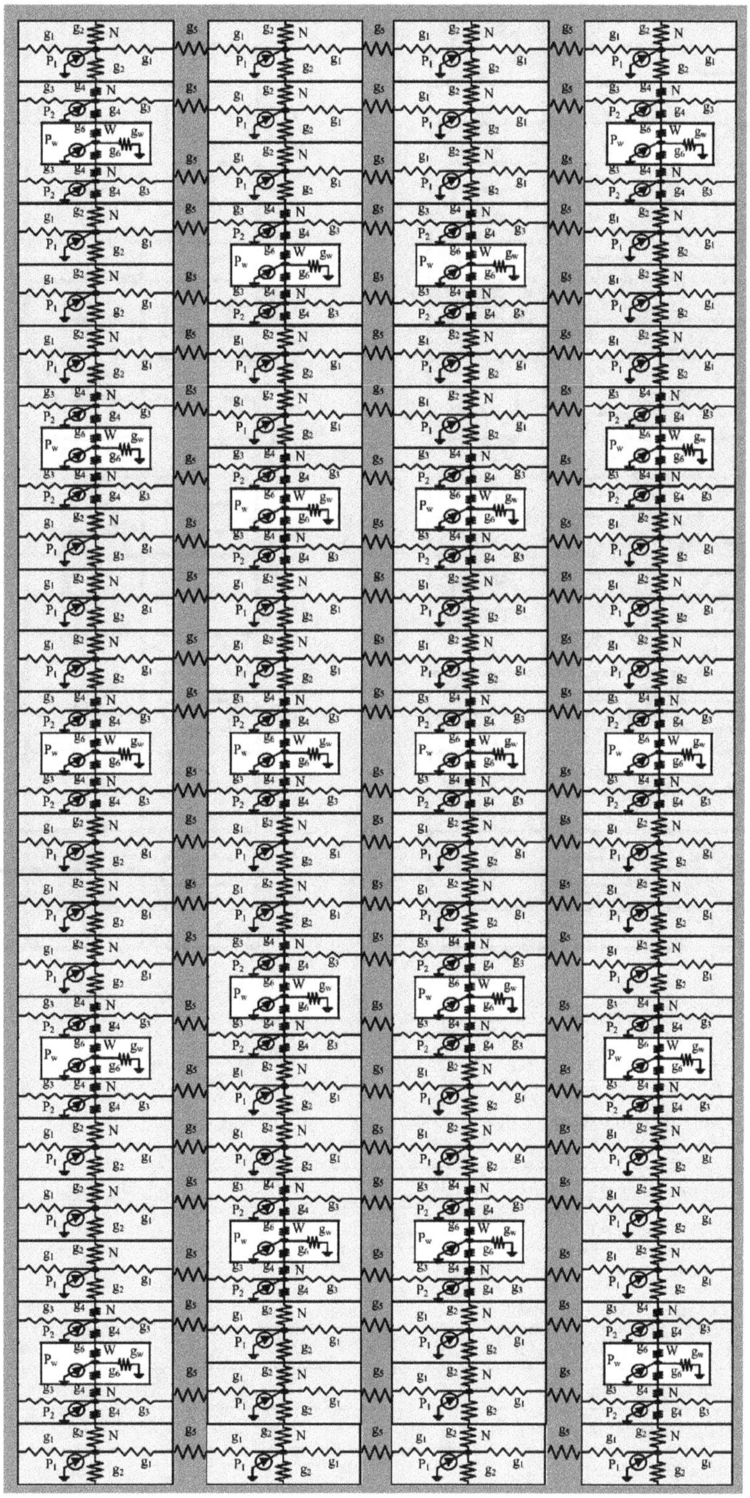

Figure 29. Thermal equivalent models of Two Consecutive segment of stator bar

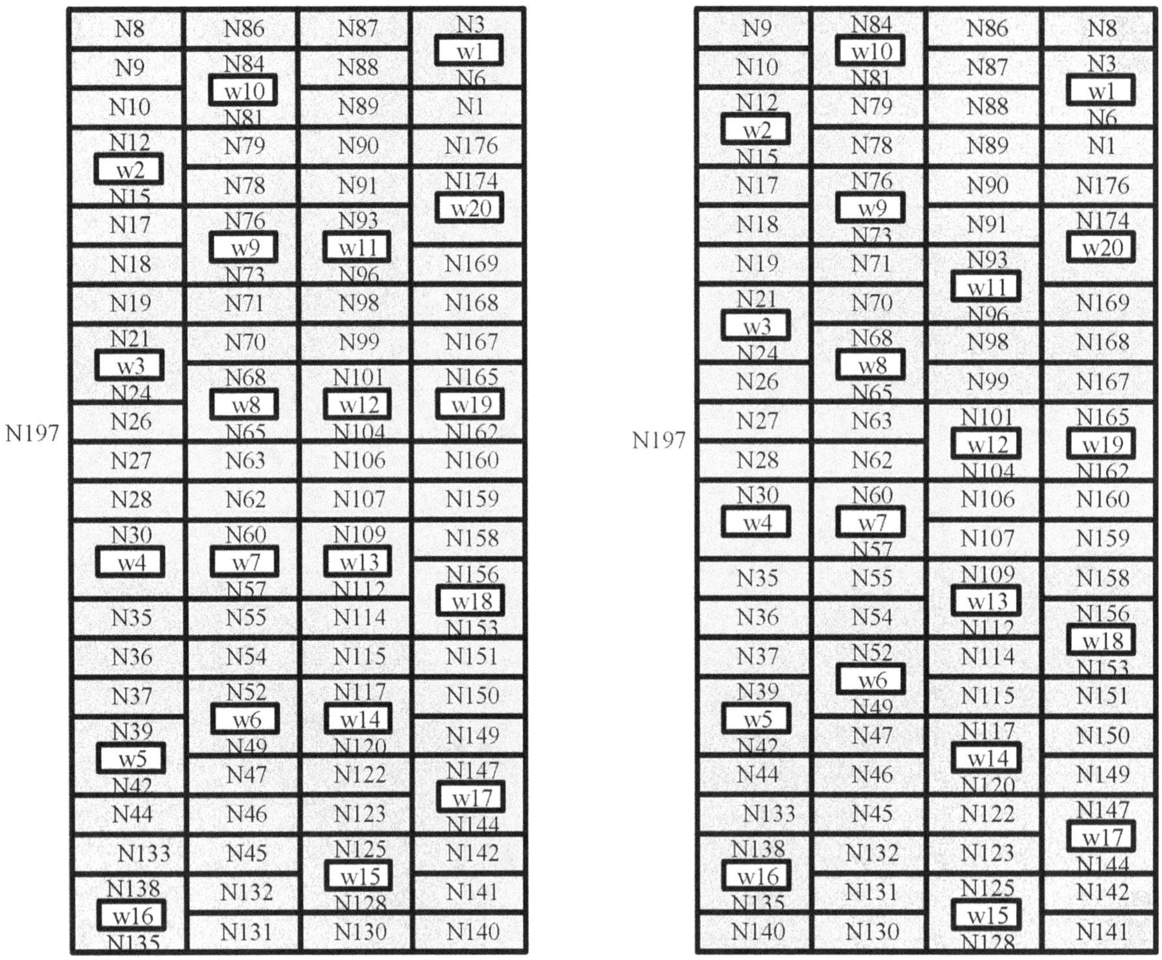

Table 5. Electrical specifications of stator bar

Parameter	Value
Rated current (A)	9150
Length (m)	7
Number of strands per column	19
Number of columns	4
Number of hollow conductors	20
Number of transpose nodes	38
Transposition angle (°)	360

Table 6. Thermal specifications of stator bar

Parameter	Value
Strand insulation	Paper
Column insulation	Mica
Cooling material	Water
Inlet water flow (m/min)	24
Inlet water temperature (ºC)	40

Figure 30. Thermal model of stator and rotor cores

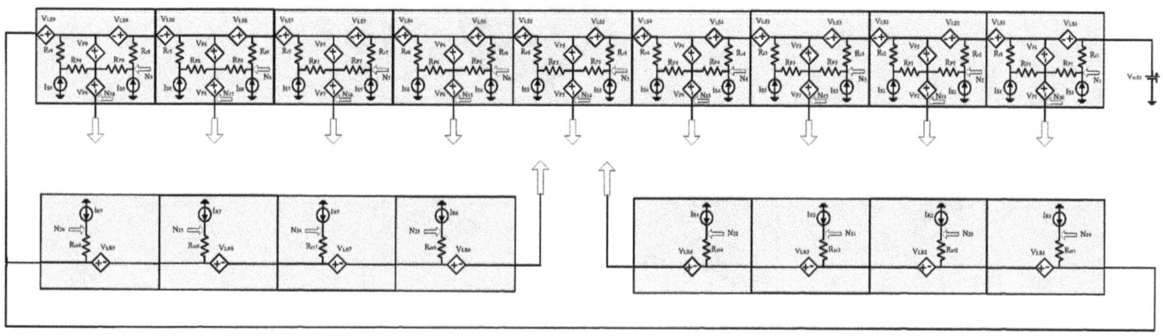

Figure 31. Thermal model of stator bars

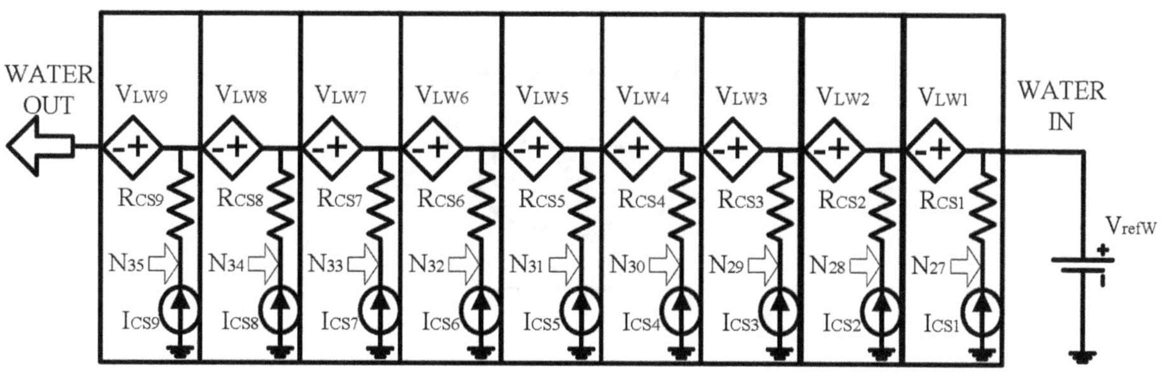

INSULATION CLASS

The electrical insulation system for wires used in generators, electric motors, transformers, and other wire-wound electrical components is divided into different classes by temperature and temperature rise. The electrical insulation system is sometimes referred to as insulation class or thermal classification. The different classes are defined by NEMA, Underwriters Laboratories (UL), and IEC standards 17.

The maximum hot-spot operating temperature is reached by adding the rated ambient temperature of the machine (often 40°C), a temperature rise, and a 10 °C hot-spot allowance. Electrical machines are usually designed with an average temperature below the rated hot-spot temperature to allow for accept-

Figure 32. Thermal model of rotor

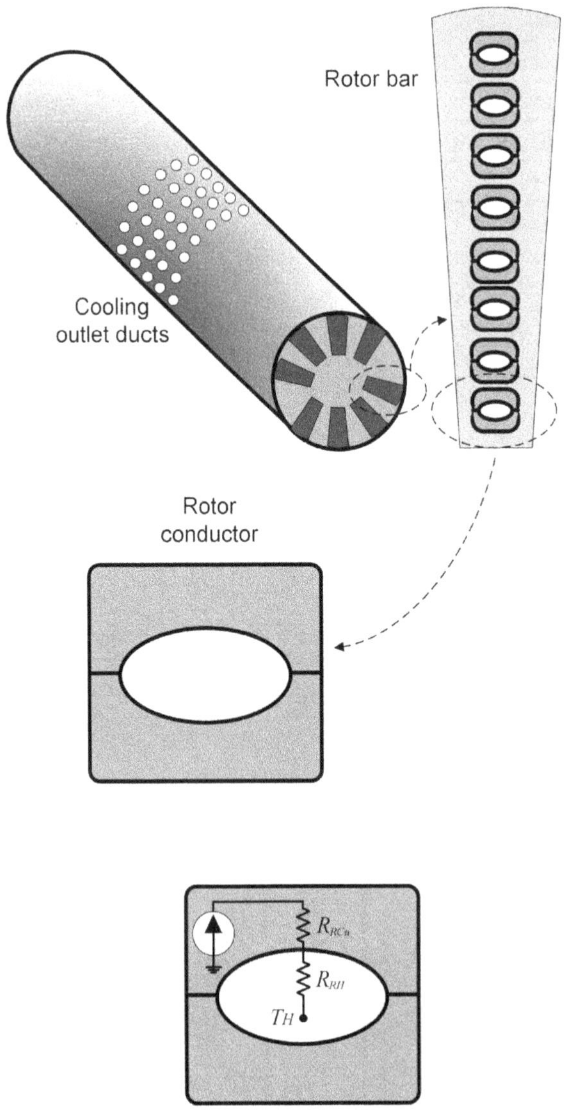

able life. Insulation does not suddenly fail if the hot-spot temperature is reached, but useful operating life declines rapidly; a rule of thumb is a halving of life for every 10 °C.

Older editions of standards listed materials to be used for the various temperature classes. Modern editions of standards are proscriptive, only indicating that the insulation system must provide acceptable life at the specified temperature rise.

In large machines, different systems may be used according to the predicted temperature rise of the machine; for example, in large hydroelectric generators, stator windings may be Class B but the more difficult to cool rotor winding may be Class F.

Figure 33. Temperature profile of stator bar resulted from thermal model

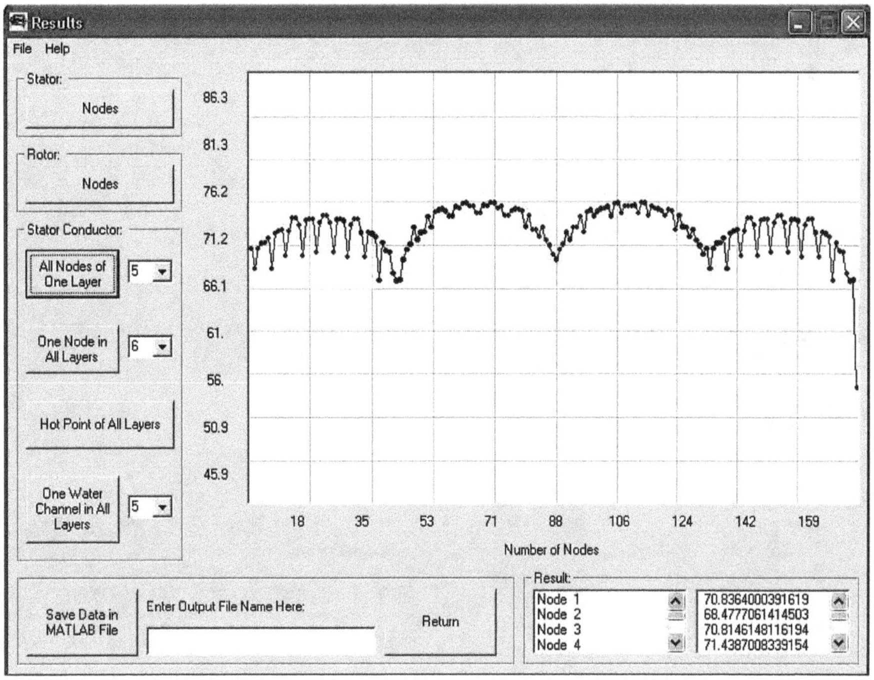

Figure 34. Temperature profile of stator bar resulted from numerical model

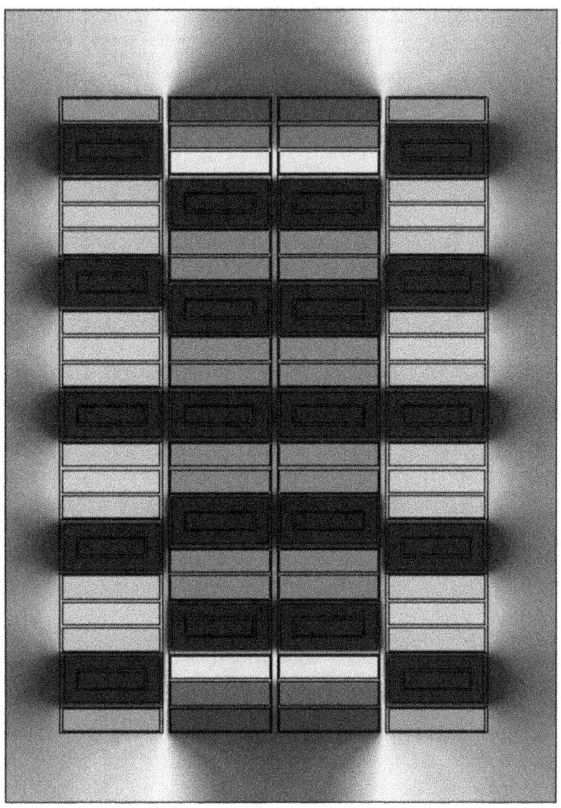

Figure 35. Temperature profile of cooling water resulted from thermal model

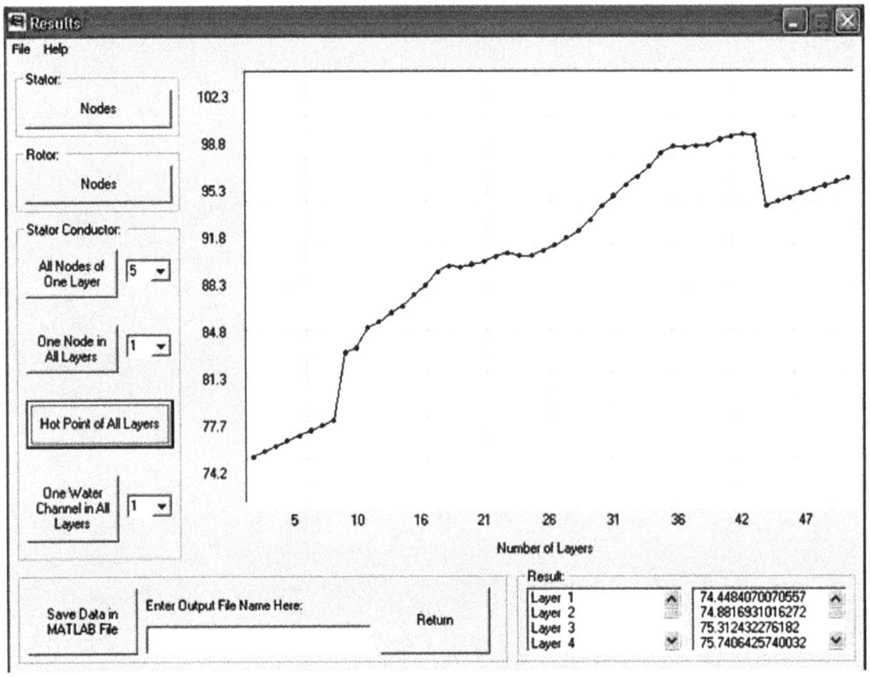

Figure 36. Temperature of various nodes on stator bar

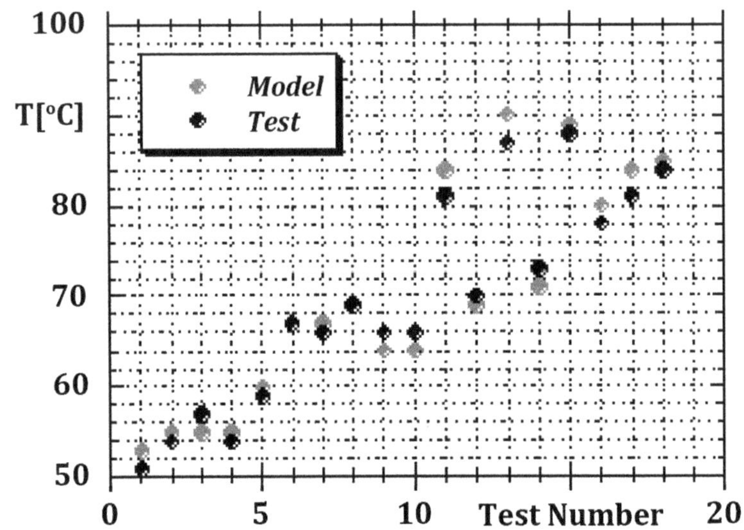

A critical factor in the reduced life of electrical equipment is heat. The type of insulation used in a motor depends on the operating temperature that the motor will experience. Average insulation life decreases rapidly with increases in motor internal operating temperatures.

Table 7.

NEMA Letter Class	Maximum Hot Spot Temperature Allowed
A	105°C
B	130°C
F	155°C
H	180°C
N	200°C
R	220°C

NEMA has established safe maximum operating temperatures for motors based on an average 20,000 hour lifetime. These maximum temperatures are the sum of the ambient and maximum temperature rise ratings of the motor.

There are four NEMA insulation classes based thermal endurance of the system for maximum temperature rating purposes. These are listed on the motor's nameplate and are either A, B, F, or H. These codes indicate the maximum temperature the motor insulation can withstand without failure.

Class A insulation was the standard insulation used on older U Frame motors between 1952 and 1964. Since 1964, T Frame motors use class B insulation as the standard insulation. Most common fractional horsepower motors use either insulation class A or B.

Class B is used on most integral horsepower motors. Classes F and H are generally used for NEMA motor designs which are special applications. Do not confuse the NEMA insulation classes with the NEMA motor designs which are also given by letters.

NEMA insulation ratings assume the motor is operating within its rated ambient temperature. Ambient temperature is the air temperature surrounding the motor and is also indicated on the nameplate.

Motors should be replaced by motors with the same or higher insulation class to avoid reductions in motor life and nuisance tripping of the motor overload device. Each 10 Degree C rise above the motor's rating can reduce motor life by one-half.

SUMMARY AND CONCLUSION

Over temperature is one of the main reasons of failure in the systems. It has a dominant role among the failure factors which were described in chapter 2. In this chapter, thermal management and modeling of an electric power converter was described. The main topics of this chapter are summarized as follows:

1. There are three modes of heat transfer from a source with higher temperature to a region of lower temperature. Among them, conduction heat transfer is the most important method in electric power converter. As they usually work at a low temperature, radiation method is not very effective. On the other hand, natural convection based methods are not effective and forced convection cooling methods are not very reliable because of using active part in cooling process.
2. A theoretical thermal model is normally used as a first order estimate. Thermal analysis can be carried out by numerical analysis or lumped models. Numerical method is not fast but very accurate. FEM is a tool for numerical thermal analysis. FEM is best understood from its practical application

and it is applied in engineering as a computational tool for performing engineering analysis. It includes the use of mesh generation techniques for dividing a complex problem into small elements, as well as the use of software program coded with FEM algorithm. Lumped models are very fast and can be used for optimization. Their accuracy is less than numerical methods. Thermal models facilitate the design and analysis of cooling systems. The underlying idea of such models is the formal similarity between heat transfer and electric circuits.

3. Proper heat sink helps a lossy system to work in a lower temperature than natural convection condition. Forced air and water cooling can be used for temperature control of the systems with high power losses.

In this point, we can predict the temperature and the reliability of a designed power converter or remaining useful life of an operating converter. However, there are many unwanted factors which can not be modeled in reliability calculations. In addition, there are many human errors in implementation of the power converters. Therefore, it is necessary to test the stability of a product, especially an electric power converter. In the next chapter, we discuss reliability measurement.

REFERENCES

A fundamental approach to transformer thermal modeling. (2001). *Theory and Equivalent Circuit, 16*(2),171-175.

Boglietti, A., Cavagnino, A., Lazzari, M., & Pastorelli, M. (2003). A simplified thermal model for variable-speed self-cooled industrial induction motor. *IEEE Transactions on Industry Applications, 39*(4), 945–952. doi:10.1109/TIA.2003.814555

Boglietti, A., Cavagnino, A., Staton, D., Shanel, M., Mueller, M., & Mejuto, C. (2009). Evolution and Modern Approaches for Thermal Analysis of Electrical Machines. *IEEE Transactions on Industrial Electronics, 56*(3), 871–882. doi:10.1109/TIE.2008.2011622

Gorginpour, H., Jandaghi, B., & Oraee, H. (2011). Finite element analysis of Brushless Doubly-Fed Machine under stator winding faults. In *Proceedings of 2nd Power Electronics, Drive Systems and Technologies Conference* (pp.169-174). IEEE doi:10.1109/PEDSTC.2011.5742411

Gorginpour, H., Oraee, H., & McMahon, R. A. (2014). Electromagnetic-Thermal Design Optimization of the Brushless Doubly Fed Induction Generator. *IEEE Transactions on Industrial Electronics, 61*(4), 1710–1721. doi:10.1109/TIE.2013.2267705

Staton, D., Boglietti, A., & Cavagnino, A. (2005). Solving the More Difficult Aspects of Electric Motor Thermal Analysis in Small and Medium Size Industrial Induction Motors. *IEEE Transactions on Energy Conversion, 20*(3), 620–628. doi:10.1109/TEC.2005.847979

Wei, H., Ghosh, S., Velusamy, S., Sankaranarayanan, K., Skadron, K., & Stan, M. R. (2006). HotSpot: A compact thermal modeling methodology for early-stage VLSI design. *IEEE Transactions on Very Large Scale Integration (VLSI). Systems, 14*(5), 501–513.

ADDITIONAL READING

Hefner, A. R. (1994). A dynamic electro-thermal model for the IGBT. *IEEE Transactions on Industry Applications*, *30*(2), 39405. doi:10.1109/28.287517

Holman, S. (1996). *Heat transfer*. Wiley.

Maloberti, O., Gimeno, A., Ospina, A., Friedrich, G., El Kadri Benkara, K., & Charbonnier, L. (2014). Thermal Modeling of a Claw-Pole Electrical Generator: Steady-State Computation and Identification of Free and Forced Convection Coefficients. *IEEE Transactions on Industry Applications*, *50*(1), 279–287. doi:10.1109/TIA.2013.2270222

Remsburg, R. (1998). *ADVANCED THERMAL DESIGN OF ELECTRONIC EQUIPMENT*. NY, USA: Chapman & Hall. doi:10.1007/978-1-4419-8509-5

Senturk, O. S., Helle, L., Munk-Nielsen, S., Rodriguez, P., & Teodorescu, R. (2011). Converter Structure-Based Power Loss and Static Thermal Modeling of The Press-Pack IGBT Three-Level ANPC VSC Applied to Multi-MW Wind Turbines. *IEEE Transactions on Industry Applications*, *47*(6), 2505–2515. doi:10.1109/TIA.2011.2168557

Sheng, K. (2000). A review of IGBT models. *IEEE Transactions on Power Electronics*, *15*(6), 1250–1266. doi:10.1109/63.892840

Staton, D. A., & Cavagnino, A. (2008). Convection Heat Transfer and Flow Calculations Suitable for Electric Machines Thermal Models. *IEEE Transactions on Industrial Electronics*, *55*(10), 3509–3516. doi:10.1109/TIE.2008.922604

Tan, K. K., Huang, S. N., & Lee, T. H. (2002). Robust adaptive numerical compensation for friction and force ripple in permanent-magnet linear motors. *IEEE Transactions on Magnetics*, *38*(1), 221–228. doi:10.1109/20.990111

Trigeol, J.-F., Bertin, Y., & Lagonotte, P. (2006). Thermal modeling of an induction machine through the association of two numerical approaches. *IEEE Transactions on Energy Conversion*, *21*(2), 31323. doi:10.1109/TEC.2005.859964

Wrobel, R., Mellor, P. H., & Holliday, D. (2011). Thermal Modeling of a Segmented Stator Winding Design. *IEEE Transactions on Industry Applications*, *47*(5), 2023–2030. doi:10.1109/TIA.2011.2161741

Xingcun, T. C. (2011). Advanced Materials for Thermal Management of Electronic Packaging. Berlin, Germany: Springer.

KEY TERMS AND DEFINITIONS

Finite Element Method: A numerical technique for calculation of the system operating points.
Heat Sink: A radiator for facilitating heat transfer from a lossy object.
Lumped Model: A model used for calculation of selected points of a complex system.
Thermal Management: Temperature monitoring and control of all parts of a system.
Thermal Modeling: An equivalent electric network for temperature calculation.

Chapter 5
Reliability Measurement

ABSTRACT

After calculation of reliability, the system is constructed. It is important to measure the calculated useful life of the system. Reliability measurement tests can also be used for the converter operating in service to estimate the remaining useful life of the converter. In this chapter, various methods of tests for this goal are presented. The main approach is accelerated aging test that reduce the time needed for failure in a system. In this method, the device is tested under conditions beyond its defined nominal specifications. Limits for this harsh condition is determined based on the calculations which are presented in chapters 3 and 4. If a problem occurs in implementing and operating process of the converter, accelerated aging tests decrease the time to failure. Theoretical concept of accelerated aging tests is described. Standard tests of electric power converters are presented. Equipment and test chambers for standard tests are explained. These tests contain all four various failure factors which are presented in chapter 2. Sample industrial examples are presented to demonstrate the procedure of the tests. Some accelerated aging tests may lead to destruction of the converter. Difference between destructive and nondestructive tests is presented. Sample devices after accelerated aging tests are shown. Measuring devices for system parameter identification are introduced. Various types of tests are expressed in details for some of the most important tests like electric withstand tests.

INTRODUCTION

In Chapter 2, we reviewed the most important failure factors of an electric power converter. However, there are still many unknown failure factors that they become important in special cases. For example, one of the loss factors in electrical machines was stray load loss. This is an unknown factor and cannot be modeled precisely. In many cases, it is a small percent of failure factors. However, it may be dominant factor for the motors operating in special conditions. Theoretical analysis cannot consider all factors about reliability of a converter (Lu, & Rudy, 2001). Therefore, we need to test the reliability of a converter after implementation and before popular allocation. We study the commonly used methods for reliability testing in the current chapter. Figure 1 shows the state of this chapter in the flowchart of the book. But, here is a problem. Assuming an exponential distribution, a reasonable assumption for electronic com-

DOI: 10.4018/978-1-4666-9429-3.ch005

Figure 1. State of chapter 5 in the flowchart of the book

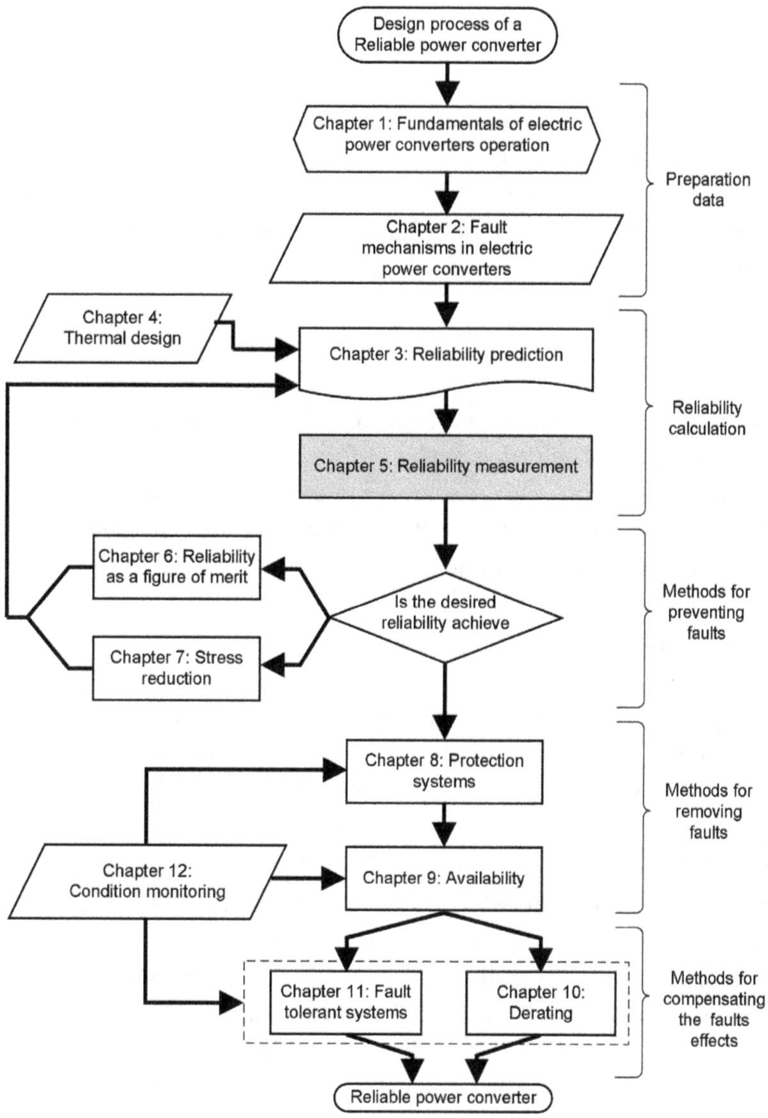

ponents, the specific prediction for the product is simply a matter of determining operating hours and types of failures expected. The failure rate of the product can be determined from the following equation:

Failure Rate = Number of Failures / Operating Time

The advantage of predicting from field and test data is that the reliability results can be accurately determined including the associated uncertainty of the estimate (Wong, Albrecht, Allan, Billinton, Chen, Fong, Haddad, Li, Mukerji, Patton, Schneider, Shahidehpour, & Singh, 1999). The disadvantage is the difficulty of obtaining and assessing accurate field and test data. In addition, more important problem is the time interval that is necessary for fault occurrence.

EFFECT OF TEST ON EQUIPMENT

In reliability tests, at each test one or some parameters of environmental conditions are stressed more than typical state to reduction of test time less than real state. Then, there are some determined relations between this accelerated test results and typical condition results that are used to finding failure rates and reliability evaluation. One of the main concerns in these tests is this question: Does the over stress test have improper effect on the equipment? This question become more important when we want to evaluate the reminder useful life of a converter (Saha, 2003). In this case, the reliability tests must have no destructive effect on the equipment.

Destructive Tests

In destructive testing, tests are carried out to the specimen's failure, in order to understand a specimen's structural performance or material behaviour under different loads. These tests are generally much easier to carry out, yield more information, and are easier to interpret than nondestructive testing. Destructive testing is most suitable, and economic, for objects which will be mass-produced, as the cost of destroying a small number of specimens is negligible 2.

Non-Destructive Tests

Nondestructive testing is a wide group of analysis techniques used in science and industry to evaluate the properties of a material, component or system without causing damage. Because NDT does not permanently alter the article being inspected, it is a highly valuable technique that can save both money and time in product evaluation, troubleshooting, and research (Thebaud, Woirgard, Zardini, Azzopardi, Briat, & Vinassa, 2003). Common NDT methods include ultrasonic, magnetic-particle, liquid penetrant, radiographic, remote visual inspection (RVI), eddy-current testing, and low coherence interferometry. NDT is commonly used in forensic engineering, mechanical engineering, electrical engineering, civil engineering, systems engineering, aeronautical engineering, medicine, and art.

To show the difference between DPA and NDT an insulator is a good example. Figure 2 shows the nonlinear I-V characteristic of a typical isolator. If the test voltage causes to nonlinear breakdown of the isolator, a DPA occurs. If the applied is less than the voltage for breakdown, the isolator is not damaged during the test as shown in Figure 3.

Arrhenius Method

Common way to determine a life stress relationship are

- Arrhenius model
- Eyring model
- Inverse power law model
- Temperature–humidity model
- Temperature non-thermal model

Figure 2. V-I characteristics of an electric insulator

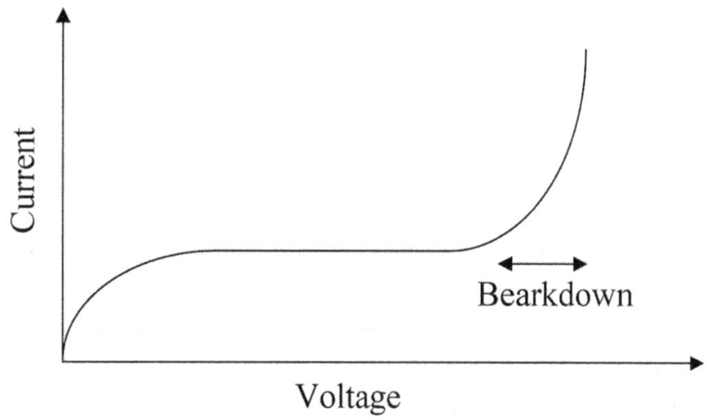

Figure 3. Different between NDT and DT in a electric withstand test

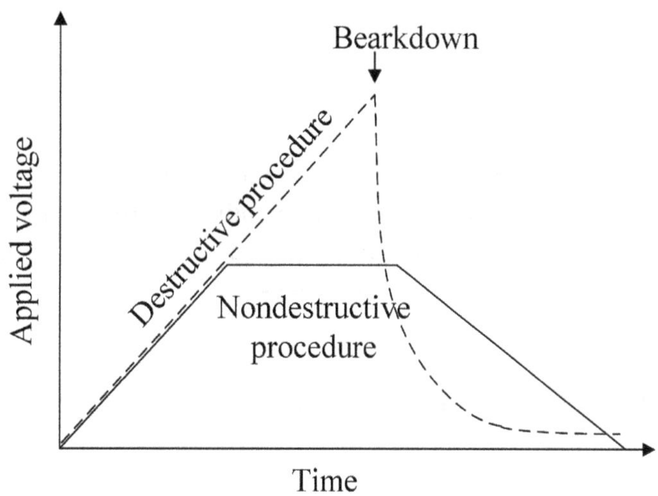

Arrhenius' equation is a simple but remarkably accurate formula for the temperature dependence of reaction rates. Arrhenius provided a physical justification and interpretation for the formula. Currently, it is best seen as an empirical relationship. It can be used to model the temperature variation of diffusion coefficients, population of crystal vacancies, creep rates, and many other thermally-induced processes/ reactions.

A historically useful generalization supported by Arrhenius' equation is that, for many common chemical reactions at room temperature, the reaction rate doubles for every 10 degree Celsius increase in temperature.

Both the Arrhenius activation energy and the rate constant k are experimentally determined, and represent macroscopic reaction-specific parameters that are not simply related to threshold energies and the success of individual collisions at the molecular level. Consider a particular collision (an elementary reaction) between molecules A and B. The collision angle, the relative translational energy, the internal (particularly vibrational) energy will all determine the chance that the collision will produce a product molecule AB. Macroscopic measurements of E and k are the result of many individual collisions with differing collision parameters. To probe reaction rates at molecular level, experiments are conducted under near-collisional conditions and this subject is often called molecular reaction dynamics.

Accelerated Testing

Life data analysis involves analyzing times-to-failure data obtained under normal operating conditions in order to quantify the life characteristics of a product, system or component. For many reasons, obtaining such life data (or times-to-failure data) may be very difficult or impossible. The reasons for this difficulty can include the long life times of today's products, the small time period between design and release, and the challenge of testing products that are used continuously under normal conditions. Given these difficulties and the need to observe failures of products to better understand their failure modes and life characteristics, reliability practitioners have attempted to devise methods to force these products to fail more quickly than they would under normal use conditions. In other words, they have attempted to accelerate their failures.

Advantages of Accelerated Testing

The purpose of accelerated life testing (ALT test) is to induce field failure in the laboratory at a much faster rate by providing a harsher, but nonetheless representative, environment. In such a test, the product is expected to fail in the lab just as it would have failed in the field—but in much less time. The main objective of an accelerated test is either of the following 4:

- To discover failure modes
- To predict the normal field life from the high stress lab life

Various Types of Tests

An Accelerated testing program can be broken down into the following steps:

- Define objective and scope of the test
- Collect required information about the product
- Identify the stress(es)
- Determine level of stress(es)
- Conduct the accelerated test and analyze the collected data.

Manufacturers follow specific industry standards when creating electrical products. To ensure their creations meet those standards, the products must undergo automated or manual testing processes to ensure the specifications are met. Testing also is a way to uncover glitches and problems with a design before it's integrated into a larger project. Testing allows engineers to identify the difference between what they expect their creations to do and what they actually accomplish.

Acceptance and Qualification Tests

Acceptance testing and qualification testing are engineering processes that products undergo before they are deemed safe and ready to use. Cables, electrical wiring and software templates that run electrical panels are subject to a realm of rigorous tests to meet industry and government safety standards, such as those set by the Insulated Cable Engineers Association or the American Society for Testing and Materials. There typically is a range of tests that occur between the qualification of an electrical product and its final acceptance.

Typically the first step in the testing process is a materials test to ensure each piece of the finished electrical product meets or exceeds expected standards. You can't expect a cable or wire to perform effectively and pass the qualification process with substandard materials. The next step is the qualification testing phase that determines whether the overall design is credible. Again, this is an important phase to complete before you even begin manufacturing, because a flawed design will result in a flawed finished product

Before an electrical cable or wire can pass the acceptance testing stage, it goes through exhaustive field testing. Field tests usually include testing while the cable or wire is still on the spool, how it passes the initial installation process and how it interfaces with the existing structures into which it will eventually be used. The cable or software system is run through a wide range of scenarios to test its strength and sustainability under a variety of circumstances. The final stage involves acceptance testing, which is performed after installation and field testing are complete. The product runs through its normal application. The acceptance testing process primarily is performed to test the installation and ferret out any bugs. Acceptance testing should not exceed the capabilities designed into the product.

Industry standards in the electrical industry guide the qualification testing process and usually include aging tests as well as comparisons to similar products. Manufacturers of different cables or software platforms must comply with a set of industry qualification testing standards. Once the product passes

Figure 4. Short circuit protection test of a power supply

Figure 5. Functional temperature rise test of a converter

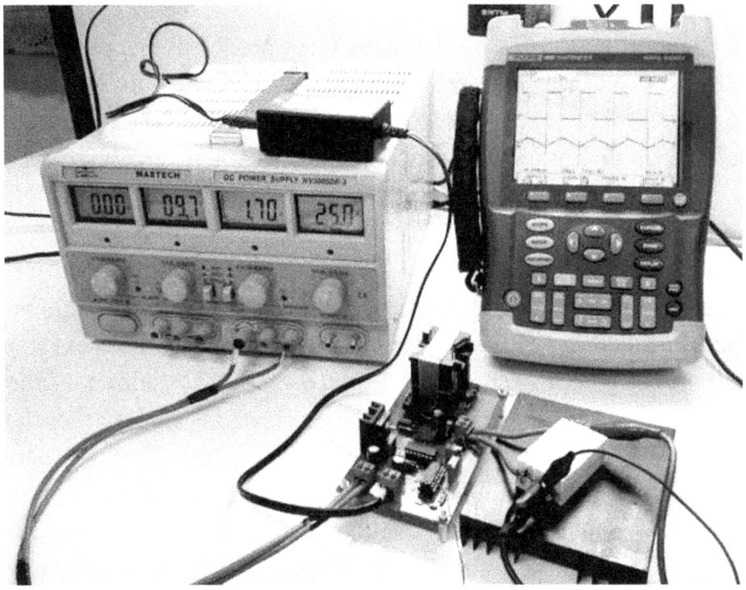

the materials and qualification tests, it can then go to production. Production testing occurs periodically during and after the manufacturing process. Again, industry standards are used to measure the viability of the product as it goes through production. Similarly, in the software industry, each step of a software application is assessed as it's built.

Figure 6 shows standard test sheet for permissible defectives for qualification approval test. This is a test sheet for a capacitor and includes many various tests. The number of permissible defectives are different in various tests.

Figure 4 shows short circuit test of a high voltage power supply. In this test, the output of the power supply is connected to a load with very low impedance by a contactor. Figure 5 shows temperature rise test of a converter in normal condition.

Test Chambers

An environmental chamber is an enclosure used to test the effects of specified environmental conditions on biological items, industrial products, materials, and electronic devices and components.

Such a chamber can be used:

1. as a stand-alone test for environmental effects on test specimens
2. as preparation of test specimens for further physical tests or chemical tests
3. as environmental conditions for conducting testing of specimens

Figure 6. Typical standard test sheet for permissible defectives for qualification approval test IEC publication 60384-4 ed.4.0 (Copyright © 1998 of publication IEC Geneva, Switzerland. www.iec.ch)

Group No.		Test	Sub-clause of this publication	Number of specimens (*n*) and number of permissible defectives (*pd*)						
				Per value [1]	For four or less values to be tested [1]			For six values to be tested [1]		
				n	4 *n*	*pd*	*pd* total	6 *n*	*pd*	*pd* total
0		High surge current [2]	4.21							
		Visual examination	4.2							
		Dimensions	4.2							
		Leakage current	4.3.1							
		Capacitance	4.3.2	30	120	1		180	2 [3]	
		Tangent of loss angle	4.3.3							
		Impedance [4]	4.3.4							
		Spare specimens		2	8			12		
	1A	Robustness of terminations	4.4	3	12	1		18	1	
		Resistance to soldering heat [5]	4.5							
	1B	Solderability [5]	4.6	6	24	1		36	2 [3]	
		Rapid change of temperature	4.7							
		Vibration	4.8							
		Bump or shock [4]	4.9 or 4.10							
1		Climatic sequence	4.11	9	36	2 [3]	4	54	2 [3]	6
2		Damp heat, steady state	4.12	5	20	1		30	2 [3]	
3		Endurance	4.13	5	20	1		30	2 [3]	
4A		Surge voltage	4.14	2	8	1		12	1	
4B		Reverse voltage [4]	4.15							
		Pressure relief [4]	4.16	2	8			12	1	
5A		Storage at high temperature	4.17	2	8	1		12	1	
5B		Storage at low temperature [4]	4.18	2	8			12	1	
6		Characteristics at high and low temperature	4.19	3	12	1		18	2 [3]	
		Charge and discharge [4]	4.20							

1) Case size/voltage combinations, see 3.4.1.

2) For solid electrolyte capacitors only and if prescribed by the detail specification.

3) Not more than one defective is permitted from any one value.

4) If prescribed in the detail specification.

5) Not applicable to capacitors with screw terminations or other terminations not designed to be soldered, as stated in the detail specification.

Thermal Tests

Temperature cycling (or temperature cycle) is the process of cycling through two temperature extremes, typically at relatively high rates of change. It is an environmental stress test used in evaluating product reliability as well as in manufacturing to catch early-term, latent defects by inducing failure through thermal fatigue.

Figures 7 and 8 shows a typical test chamber for thermal cycling test of electronic boards.

There are two types of temperature measurement: contact and contactless methods. Figures 9 to 11 show temperature measurement of a transformer winding with a temperature sensor which is mount on

Figure 7. A thermal cycling chamber, (Source: Garouk Co. with permission)

Figure 8. An environmental chamber, (Source: Garouk Co. with permission)

Figure 9. Mounting of a temperature sensor for test

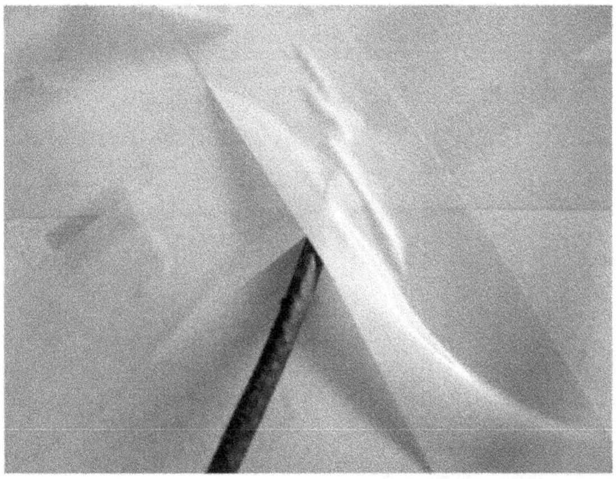

Figure 10. A temperature sensor

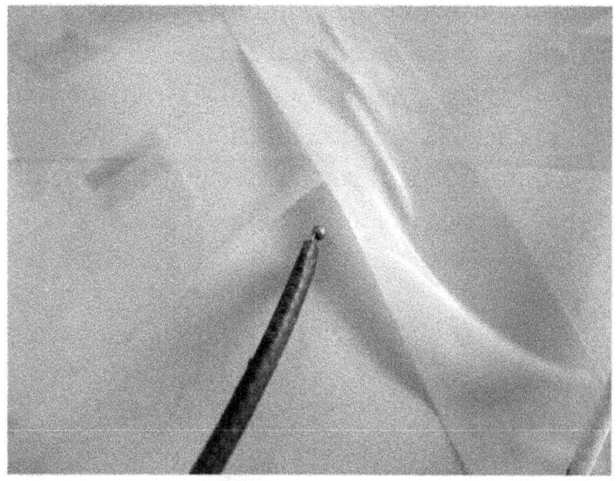

Figure 11. Temperature measurement of a transformer

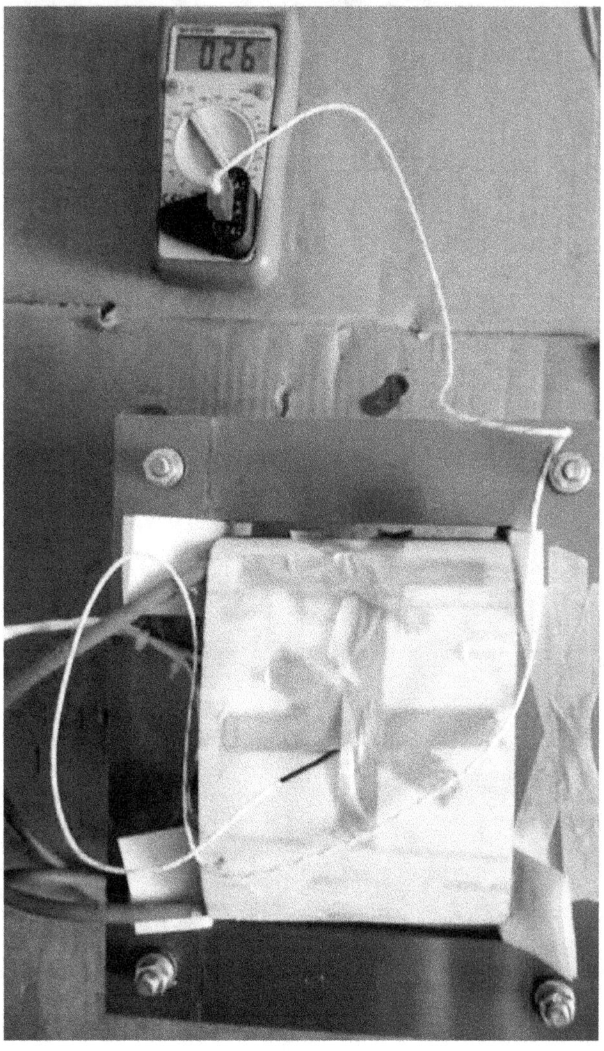

the winding. Figure 12 shows the same transformer under temperature rise test with contactless temperature measuring device. Figure 13 shows the result of temperature rise test on this transformer. Figures 14 and 15 show typical test scenario for thermal cycling and thermal shock tests.

If the temperature rise resulted from thermal tests is more than allowable value, we can continue the test to check the precise location of the fault. Figures 16, 17, 18, and 19 show the result of thermal test on a transformer. It can be seen that one side of the transformer winding is damaged. It may be because of lack of proper cooling in this side.

Figure 12. Remote temperature sensing

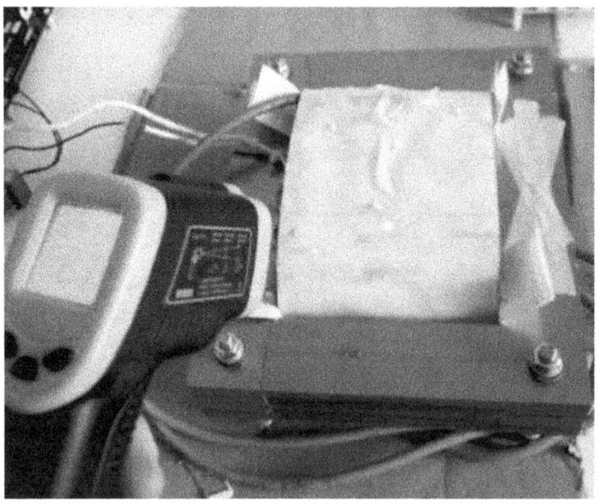

Figure 13. A typical temperature rise test result

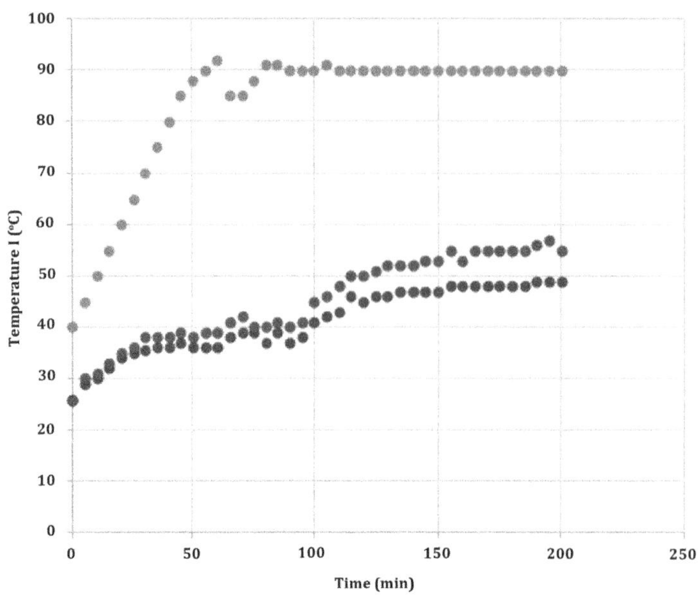

Figure 14. A scenario for thermal cycling

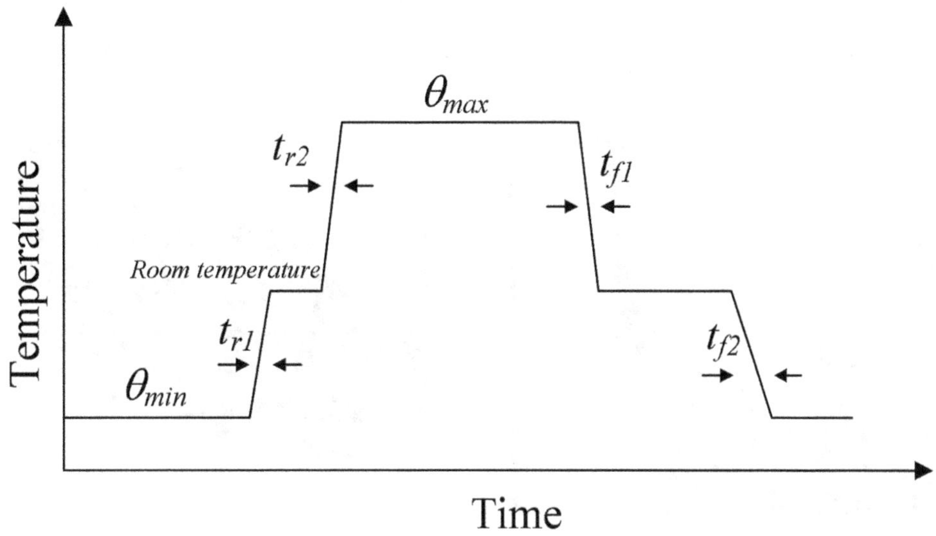

Figure 15. A scenario for thermal shock

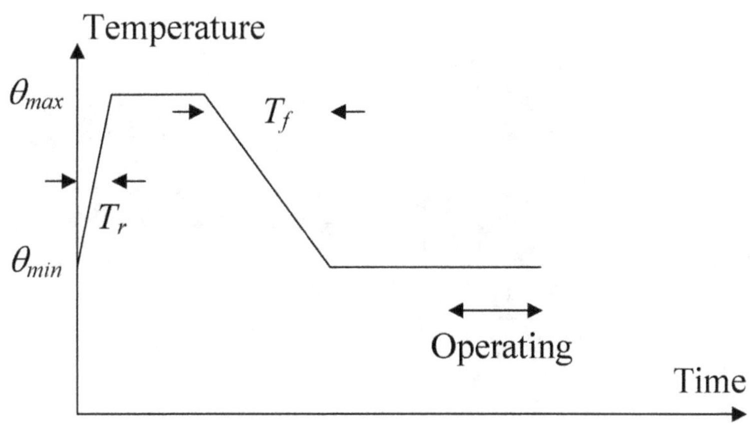

Electrical Tests

Electrical field is an important failure factors in power converters. Under normal conditions, any electrical device will produce a minimal amount of leakage current due to the voltages and internal capacitance present within the product. Yet due to design flaws or other factors, the insulation in a product can break down, resulting in excessive leakage current flow (Wen, Baosheng, Wei, Xuetao, Kai, Okamoto, & Suzuoki, 2014). This failure condition can cause shock or death to anyone that comes into contact with the faulty product (Thomson, & Fenger, 2001).

Figure 16. Transformer damage in thermal cycling test

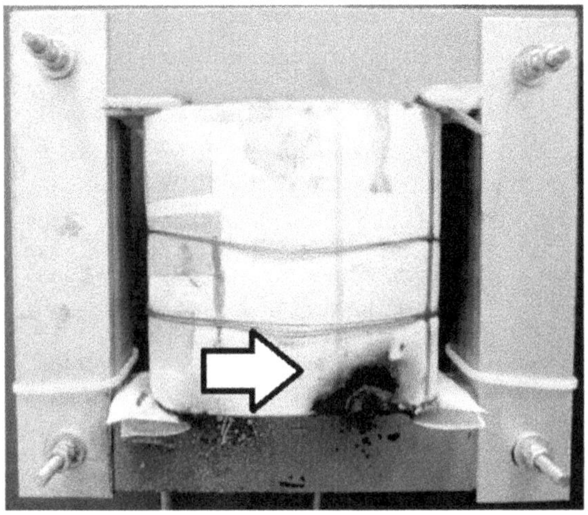

Figure 17. Transformer damage in high temperature test

Figure 18. Damage of winding in thermal cycling test of a transformer

Figure 19. Transformer insulation damage in thermal cycling test

Sample Test Equipment: Hipot

Traditionally, hipot is a term given to a class of electrical safety testing instruments used to verify electrical insulation in finished appliances, cables or other wired assemblies, printed circuit boards, electric motors, and transformers. The test is done as differential mode or common mode (Figure 20). The test is usually done with higher voltage (for example: twice the nominal voltage). To prevent core saturation in electrical machines, a double frequency source is used as shown in Figure 21. Figures 22, 23, 24, and 25 shows electric withstand test for some devices. A hipot tester is an electronic device used to verify the electrical insulation in a device or other wired assembly that could subject someone to a shock if it failed. It generally consists of:

Figure 20. Hi-pot test

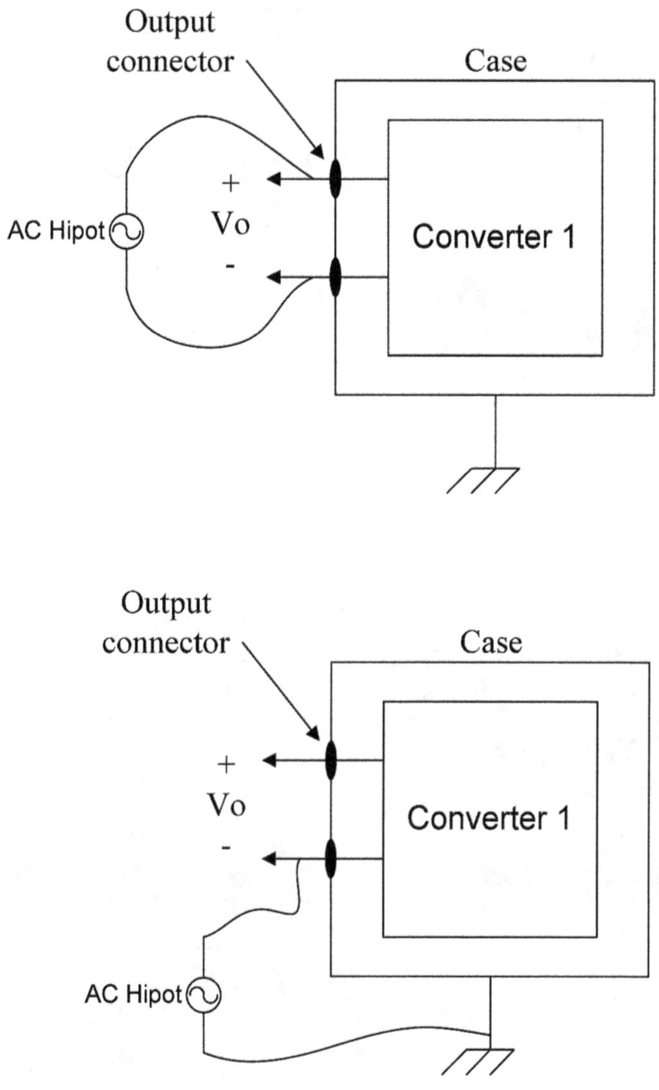

Figure 21. Double frequency hi-pot used for transformers

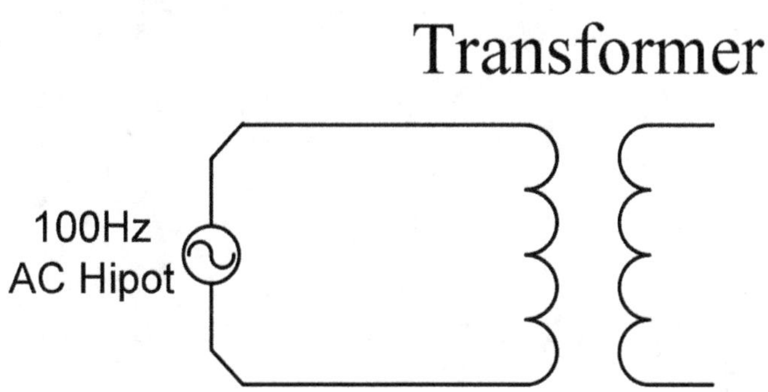

Figure 22. An onboard hi-pot

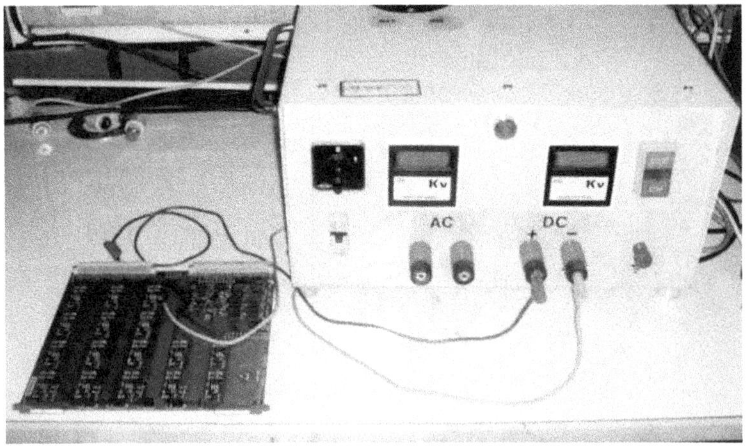

Figure 23. A high voltage test set up, (up): high voltage source, (left): high voltage capacitor, (right): cable under test

Figure 24. Typical test of a connector

Figure 25. Typical test of a connector, (up): high voltage electrode, (down): earth connection

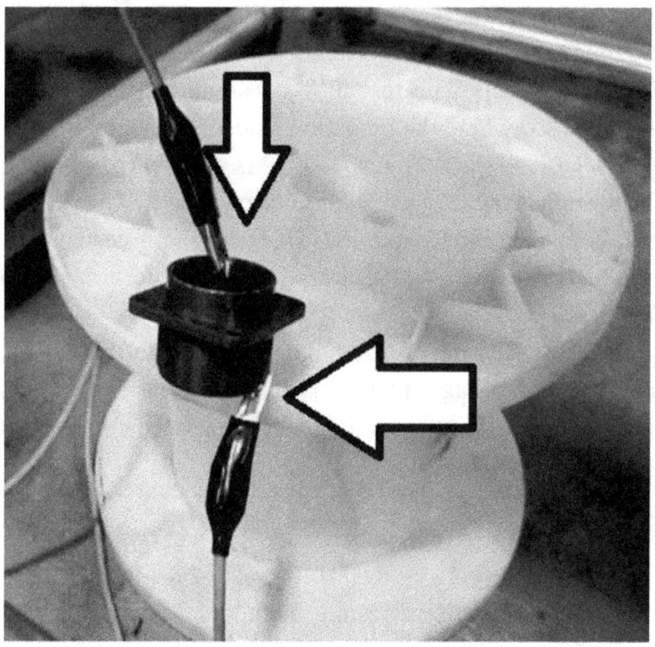

1. A source of high voltage,
2. A current meter,
3. A switching matrix used to connect the high voltage source and the current meter to all of the contact points in a cable.

Hipot testers may also have a microcontroller and a display to automate the testing process and display the testing results.

In a typical wired assembly a hipot test should connect all circuits in common to ground. Then, one by one the tester will disconnect a given circuit from ground and connect that circuit to high voltage. The current that flows is monitored to verify that it is low enough.

A hipot test or dielectric withstanding voltage test verifies that the insulation of a product or component is sufficient to protect the operator from electrical shock. In a typical hipot test, high voltage is applied between a product's current-carrying conductors and its metallic shielding. The resulting current that flows through the insulation, known as leakage current, is monitored by the hipot tester. The theory behind the test is that if a deliberate over-application of test voltage does not cause the insulation to break down, the product will be safe to use under normal operating conditions for a long time.

In addition to over-stressing the insulation, the test can also be performed to detect material and workmanship defects, most importantly small gap spacings between current-carrying conductors and earth ground. When a product is operated under normal conditions, environmental factors such as humidity, dirt, vibration, shock and contaminants can close these small gaps and allow current to flow. This condition can create a shock hazard if the defects are not corrected at the factory. No other test can uncover this type of defect as well as the Dielectric Withstand test 15.

Three types of hipot tests are commonly used. These three tests differ in the amount of voltage applied and the amount (or nature) of acceptable current flow:

- **Dielectric Breakdown Test:** The test voltage is increased until the dielectric fails, or breaks down, allowing too much current to flow. The dielectric is often destroyed by this test so this test is used on a random sample basis. This test allows designers to estimate the breakdown voltage of a product's design and to see where the breakdown occurred.
- **Dielectric Withstand Test:** A standard test voltage is applied (below the established Breakdown Voltage) and the resulting leakage current is monitored. The leakage current must be below a preset limit or the test is considered to have failed. This test is non-destructive and is usually required by safety agencies to be performed as a 100% production line test on all products before they leave the factory.
- **Insulation Resistance Test:** This test is used to provide a quantifiable resistance value for all of a product's insulation. The test voltage is applied in the same fashion as a standard hipot test, but is specified to be Direct Current (DC). The voltage and measured current value are used to calculate the resistance of the insulation.

Types of High Voltage Test

There are mainly four types of high voltage testing methods applied on high voltage equipment and these are

1. Sustained low frequency tests.
2. Constant DC test.
3. High frequency test.
4. Surge or impulse test.

All of these tests are done based on a defined scenario. Figure 26 shows the scenario for amplitude of applied voltage in low and high frequency tests and DC test. To perform the impulsive voltage test, we need an impulse generator as shown in Figure 27. The amplitude of applied voltage depends on test time duration as shown in Figure 28. Figure 29 shows a typical impulse voltage.

Low Frequency Test

This test is generally done at power frequency. This is most commonly used high voltage test, carried on H.V. equipment. This test i.e. sustained low frequency test are carried out on a specimen of insulating material to determine and ensure, dielectric strength, dielectric losses of the insulating material. This

Figure 26. Voltage profile in a high voltage test

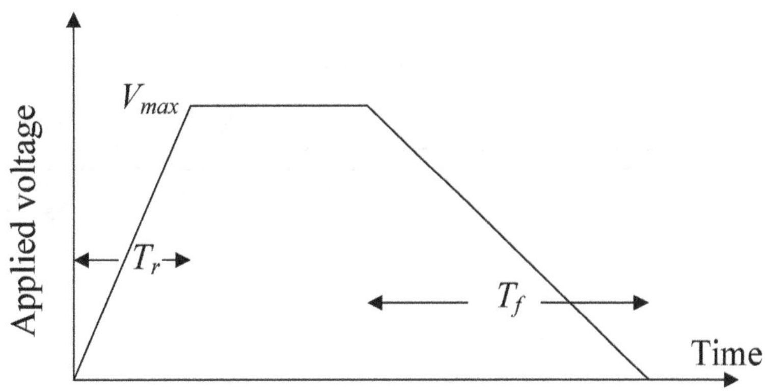

Figure 27. A surge voltage test setup (Source: Seraj Co. with permission)

Figure 28. Maximum expected overvoltage versus duration of transient IEC publication 60146-2 ed.2.0 (Copyright © 1999 of publication IEC Geneva, Switzerland. www.iec.ch)

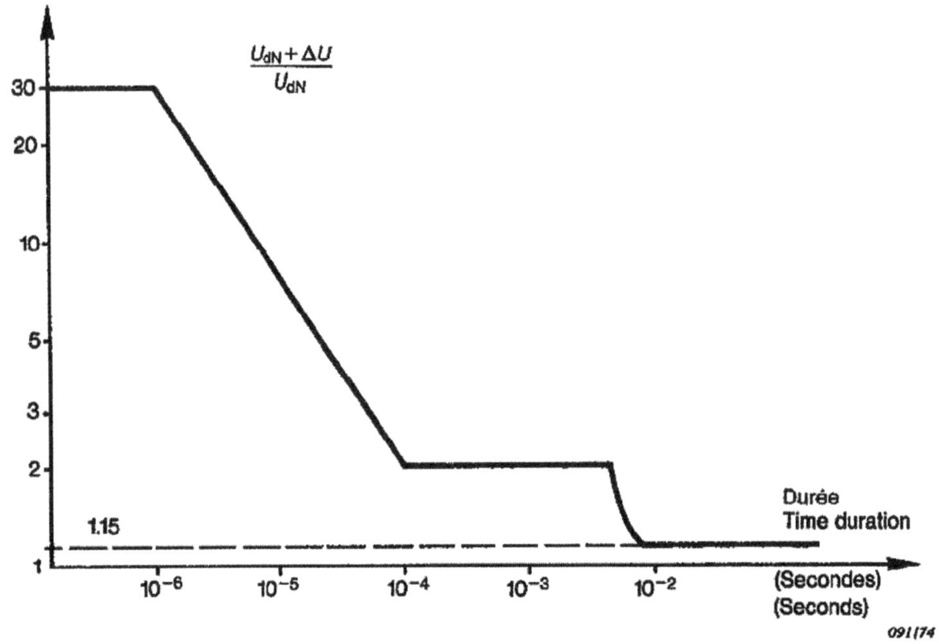

Figure 29. A typical surge voltage test

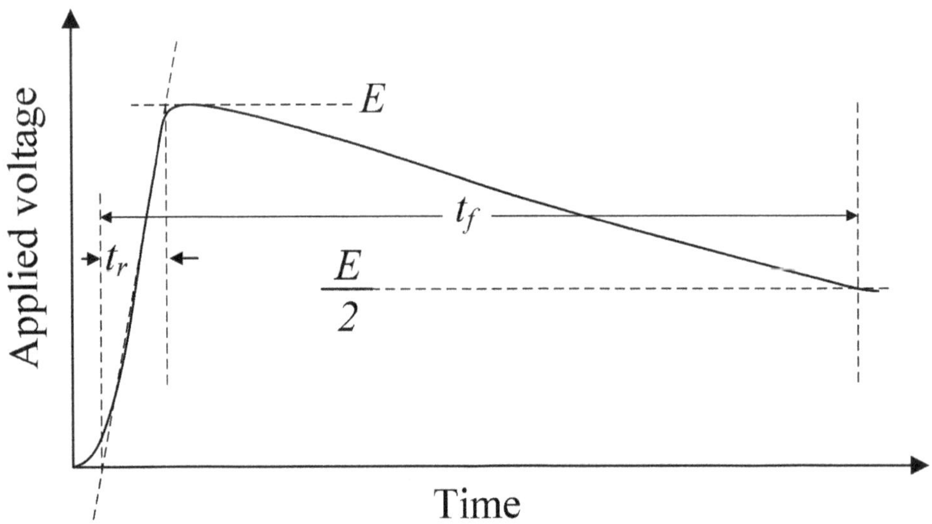

test is also carried out on high voltage equipment and high voltage electrical insulators to ensure the dielectric strength and losses of these equipment and insulators.

High voltage is applied across a specimen of insulation or equipment under test by means of a high voltage transformer. A resistor is connected with series with the transformer to limit the short circuit electric current in the event of breakdown occurred in the device under test. For this test the power frequency high voltage is applied to the specimen or equipment under test for a long specific period to ensure the continuous high voltage withstanding capability of the device.

High Voltage DC Test

High voltage DC test is normally applicable to those equipments which are used in high voltage DC transmission system. But this test is also applicable for high voltage AC equipments, when high voltage AC testing is not possible due to unavoidable condition. For example mainly at site, after installation of equipments it is quite difficult to arrange for high voltage alternating power as high voltage transformer may not be available at site. Hence, high voltage test with alternating power is not possible at site after installation of equipment. At that situation high voltage DC test is most suitable.

In high voltage direct electric current test of AC equipment, direct voltage about two times of the normal rated voltage is applied across the equipment under test for 15 minutes to 1.5 hours. Although high voltage DC test is not complete substitute of high voltage AC test but still it is applicable where HVAC test is not at all possible.

Figures 30 and 31 show low frequency test of a PCB for detecting the possible internal and external problems.

High Frequency Test

The insulators used at high voltage transmission system, may be subjected to breakdown or flash-over during high frequency disturbances. The high frequency disturbances occurs in the HV system due to switching operations or any other external causes. High frequency in power may cause failure of insulators even at comparatively low voltage due to high dielectric loss and heating. So the insulation of all high voltage equipment must ensure the high frequency voltage withstanding capacity during its normal life

Figure 30. High voltage external test of a PCB

Figure 31. High voltage internal test of a PCB

span. Mainly sudden interruption of line electric current during switching and open circuit fault, gives rise to the frequency of voltage wave form in the system.

It is found the dielectric loss for every cycle of the power is nearly constant. So at high frequency the dielectric loss per second becomes much higher than that of normal power frequency. This fast and large dielectric loss causes excessive heating of the insulator. Excessive heating ultimately results to insulation failure may be by blasting of insulators. So to ensure this high frequency voltage withstanding capacity, high frequency test is carried out on high voltage equipments 18.

Surge Test or Impulse Test

There may be great influence of surge or lighting on the transmission lines. These phenomena can breakdown transmission line insulator and it may also attack, the electrical power transformer connected at the end of the transmission lines. Surge test or impulse tests are very high or extra high voltage tests, carried out for investing the influences of surges or lightning on the transmission equipment. Normally direct lightening strokes on transmission line is very rare. But when a charged cloud comes closer to the transmission line, the line is oppositely charged due to the electrical charge inside the cloud. When this charged cloud is suddenly discharged due to lightning stroke nearby, the induced charge of the line no longer bound but travel through the line with velocity of light. So it is understood that even when the lightnings do not strike the transmission conductor, directly, still there will be a transient over voltage disturbance.

MECHANICAL TESTS

Mechanical forces are reason of crack generation.

Sample Test: Vibration

Vibration testing is accomplished by introducing a forcing function into a structure, usually with some type of shaker. A shaker is a device used in vibration testing to excite the structure either for endurance testing or modal testing. Figure 32 shows a typical block diagram of a magnetic shaker. Alternately, a DUT (device under test) is attached to the "table" of a shaker as shown in Figure 33. Vibration testing is performed to examine the response of a device under test (DUT) to a defined vibration environment. The measured response may be fatigue life, resonant frequencies or squeak and rattle sound output (NVH).

For relatively low frequency forcing, servohydraulic (electrohydraulic) shakers are used. For higher frequencies, electrodynamic shakers are used. Generally, one or more "input" or "control" points located on the DUT-side of a fixture is kept at a specified acceleration. Other "response" points experience maximum vibration level (resonance) or minimum vibration level (anti-resonance). It is often desirable to achieve anti-resonance in order to keep a system from becoming too noisy, or to reduce strain on certain parts of a system due to vibration modes caused by specific frequencies of vibration.

The most common types of vibration testing services conducted by vibration test labs are Sinusoidal and Random. Sine (one-frequency-at-a-time) tests are performed to survey the structural response of the device under test (DUT). A random (all frequencies at once) test is generally considered to more closely replicate a real world environment, such as road inputs to a moving automobile.

Figure 34 shows a typical frequency dependent vibration characteristic.

Most vibration testing is conducted in a 'single DUT axis' at a time, even though most real-world vibration occurs in various axes simultaneously. MIL-STD-810G, released in late 2008, Test Method 527, calls for multiple exciter testing. The vibration test fixture which is used to attach the DUT to the shaker table must be designed for the frequency range of the vibration test spectrum. Generally for smaller

Figure 32. A simplified diagram of a shaker

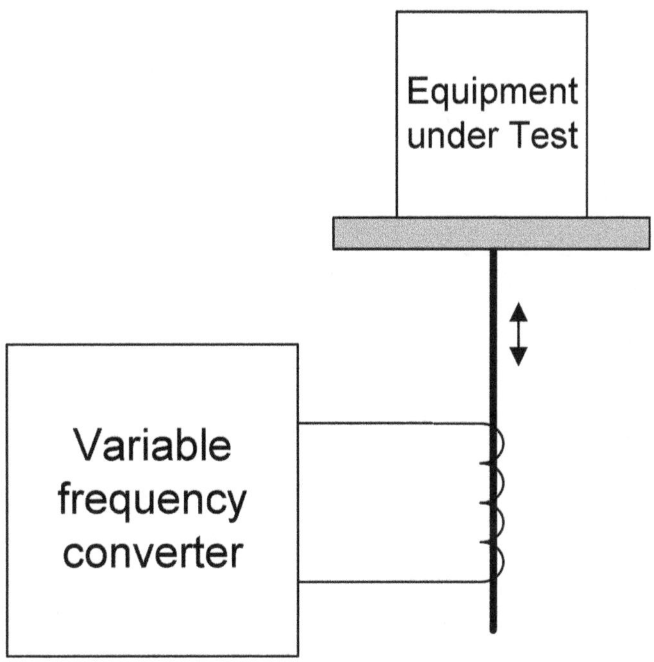

Figure 33. A practical shaker

Figure 34. Typical vibration test scenario

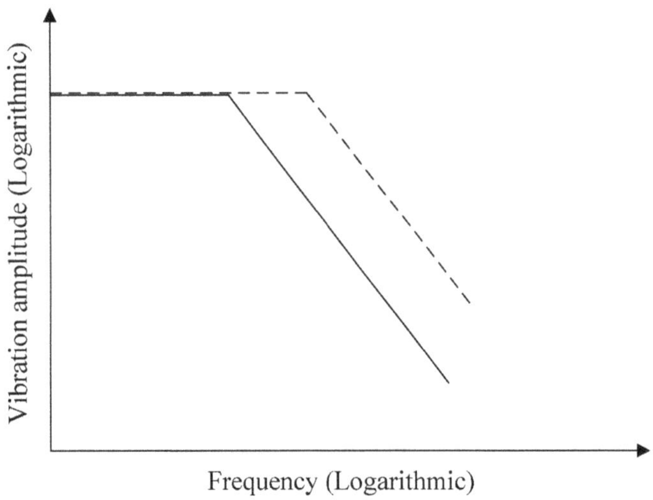

Figure 35. Mounting of vibration sensors on a motor IEC publication 60034-14 ed.3.0 (Copyright © 2003 of publication IEC Geneva, Switzerland. www.iec.ch)

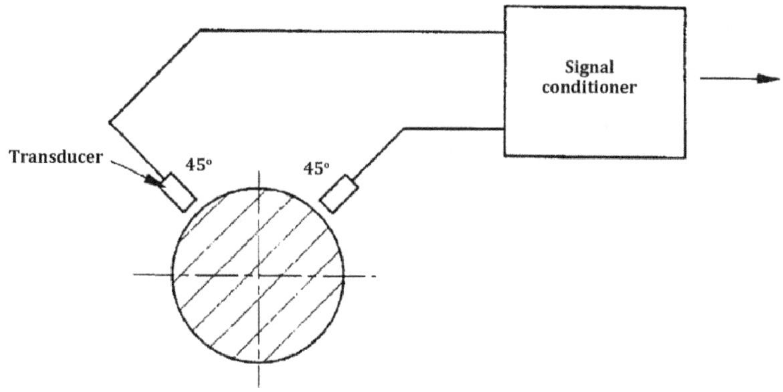

Figure 36. A vibration sensor

fixtures and lower frequency ranges, the designer targets a fixture design which is free of resonances in the test frequency range. This becomes more difficult as the DUT gets larger and as the test frequency increases, and in these cases multi-point control strategies can be employed to mitigate some of the resonances which may be present in the future. Figure 35 shows proper mounting position of vibration sensors on a rotating machine according to IEC standard. A typical vibration sensor is shown in Figure 36. Devices specifically designed to trace or record vibrations are called vibroscopes

Figures 37, 38, and 39 show various vibration test scenario for railway application. Figure 40 shows the allowable displacement of a motor.

Sample Test: Shock

Figure 41 shows a half-wave sinusoidal mechanical force for shock testing. Figure 42 shows allowable frequency response of shock test with the force shown in Figure 41.

Figure 37. Typical vibration standard test sheet for body mounted railway equipment IEC publication IEC 61373 ed.2.0 (Copyright © 2010 of publication IEC Geneva, Switzerland. www.iec.ch)

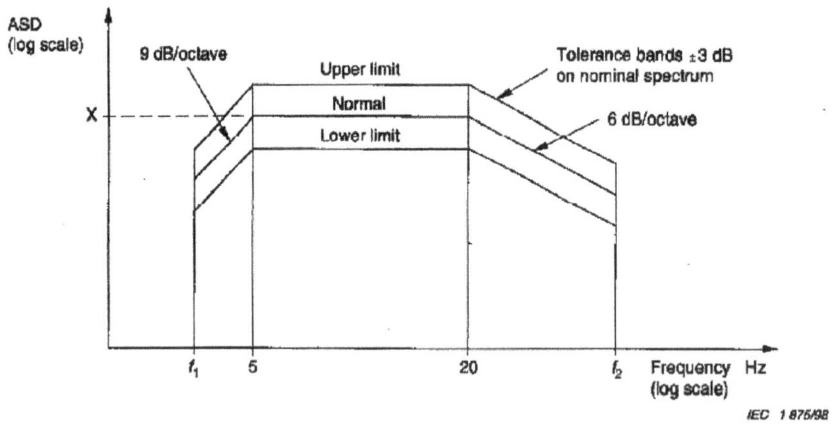

(a)

when mass <500 kg: $f_1 = 5$ Hz $f_2 = 150$ Hz

when mass >500 kg<1 250 kg: $f_1 = \frac{1\,250}{\text{mass}} \times 2$ Hz $f_2 = \frac{1\,250}{\text{mass}} \times 60$ Hz

when mass >1 250 kg: $f_1 = 2$ Hz $f_2 = 60$ Hz

	Vertical	Transverse	Longitudinal
Functional test ASD level $(m/s^2)^2/Hz$	0,0164	0,0041	0,0073
RMS value m/s^2 5 Hz to 150 Hz	0,75	0,37	0,50
Long life test ASD level $(m/s^2)^2/Hz$	1,034	0,250	0,452
RMS value m/s^2 5 Hz to 150 Hz	5,9	2,9.	3,9

NOTE 1 – For items with test frequencies less than 5 Hz the r.m.s. levels will be higher than those quoted above.

NOTE 2 – For items with test frequencies less than 150 Hz the r.m.s. levels will be lower than those quoted above.

NOTE 3 – If frequencies above f_2 are known to exist they may be included, the amplitude being established by extending the 6 dB/octave decay line until it intersects the maximum frequency required. In such cases the r.m.s. levels will be increased.

(b)

Figure 38. Typical vibration standard test sheet for bogie mounted railway equipment IEC publication IEC 61373 ed.2.0 (Copyright © 2010 of publication IEC Geneva, Switzerland. www.iec.ch)

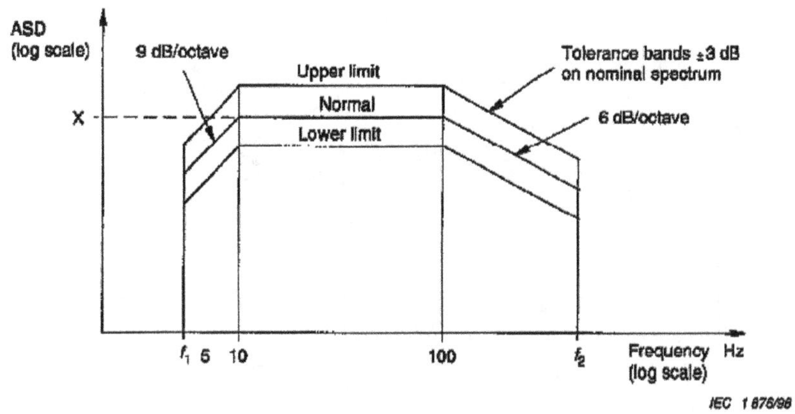

(a)

when mass <100 kg: $f_1 = 5$ Hz $f_2 = 250$ Hz

when mass >100 kg<250 kg: $f_1 = \frac{250}{\text{mass}} \times 2$ Hz $f_2 = \frac{250}{\text{mass}} \times 100$ Hz

when mass >250 kg: $f_1 = 2$ Hz $f_2 = 100$ Hz

	Vertical	Transverse	Longitudinal
Functional test ASD level $(m/s^2)^2/Hz$	0,190	0,144	0,0414
RMS value m/s^2 5 Hz to 250 Hz	5,4	4,7	2,5
Long life test ASD level $(m/s^2)^2/Hz$	11,83	8,96	2,62
RMS value m/s^2 5 Hz to 250 Hz	42,5	37,0	20,00

NOTE 1 – For items with test frequencies less than 5 Hz the r.m.s. levels will be higher than those quoted above.

NOTE 2 – For items with test frequencies less than 250 Hz the r.m.s. levels will be lower than those quoted above.

NOTE 3 – If frequencies above f_2 are known to exist they may be included, the amplitude being established by extending the 6 dB/octave decay line until it intersects the maximum frequency required. In such cases the r.m.s. levels will be increased.

(b)

Figure 39. Typical vibration standard test sheet for axle mounted railway equipment IEC publication IEC 61373 ed.2.0 (Copyright © 2010 of publication IEC Geneva, Switzerland. www.iec.ch)

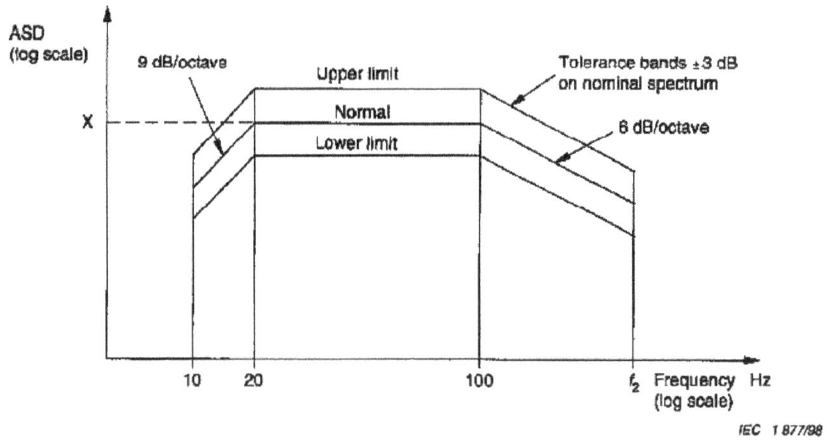

(a)

when mass <50 kg: f_2 = 500 Hz

when mass >50 kg<125 kg: $f_2 = \frac{125}{\text{mass}} \times 200$ Hz

when mass >125 kg: f_2 = 200 Hz

	Vertical	Transverse	Longitudinal
Functional test ASD level $(m/s^2)^2/Hz$	8,74	7,0	1,751
RMS value m/s^2 10 Hz to 500 Hz	38	34	17
Long life test ASD level $(m/s^2)^2/Hz$	545,2	441,2	110,3
RMS value m/s^2 10 Hz to 500 Hz	300	270	135

NOTE 1 – For items with test frequencies less than 500 Hz the r.m.s. levels will be lower than those quoted above.

NOTE 2 – If frequencies above f_2 are known to exist they may be included, the amplitude being established by extending the 6 dB/octave decay line until it intersects the maximum frequency required. In such cases the r.m.s. levels will be increased.

(b)

Figure 40. Elastic displacement of a motor IEC publication 60034-14 ed.3.0 (Copyright © 2003 of publication IEC Geneva, Switzerland. www.iec.ch)

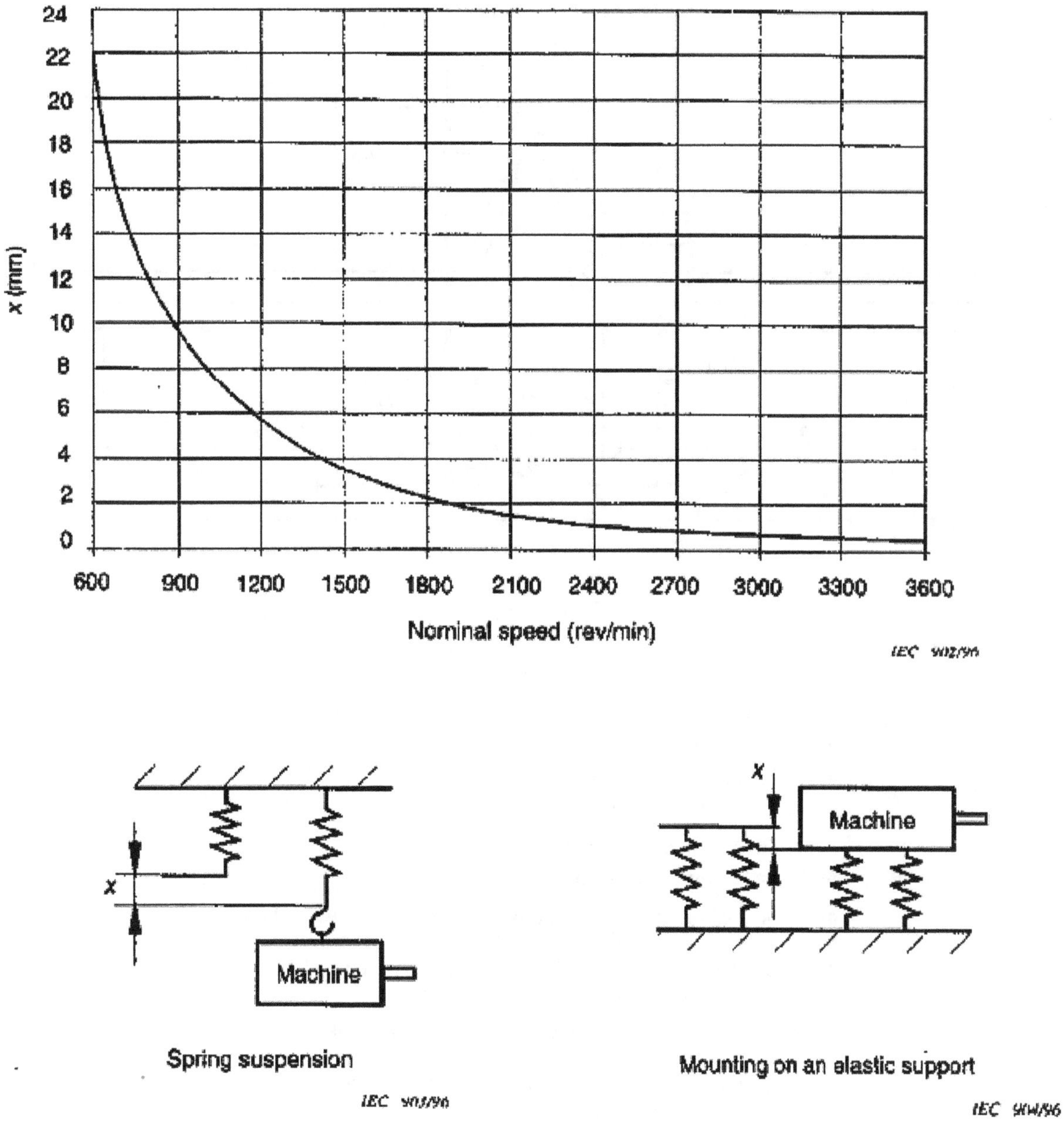

Sample Test: Dynamometer

A dynamometer can also be used to determine the torque and power required to operate a driven machine such as a pump. In that case, a motoring or driving dynamometer is used. A dynamometer that is designed to be driven is called an absorption or passive dynamometer. A dynamometer that can either drive or absorb is called a universal or active dynamometer.

Figure 41. Typical shock standard test sheet. IEC publication 60068-2-27 ed.4.0 (Copyright © 2008 of publication IEC Geneva, Switzerland. www.iec.ch)

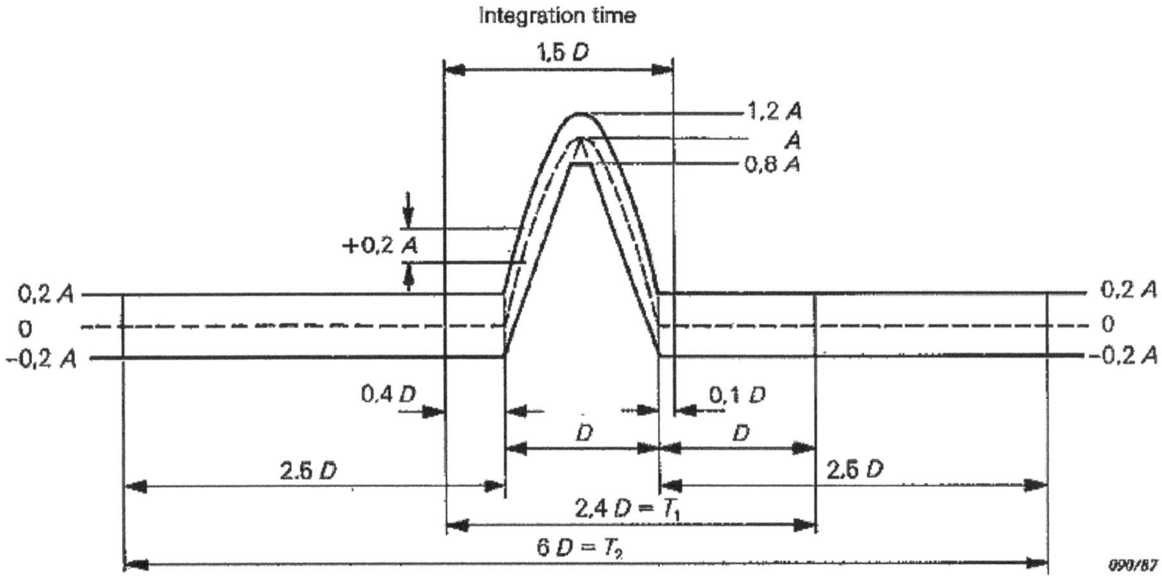

------ impulsion nominale
nominal pulse

—— limites des tolérances
limits of tolerance

D = durée de l'impulsion nominale
duration of nominal pulse

A = accélération de la crête de l'impulsion nominale
peak acceleration of nominal pulse

T_1 = durée minimale pendant laquelle il faut surveiller l'impulsion, dans le cas des chocs réalisés avec machines d'essai de chocs normales
minimum time during which the pulse shall be monitored for shocks produced using a conventional shock testing machine

T_2 = durée minimale pendant laquelle il faut surveiller l'impulsion, dans le cas des chocs réalisés avec générateur de vibrations
minimum time during which the pulse shall be monitored for shocks produced using a vibration generator

A dynamometer consists of an absorption (or absorber/driver) unit, and usually includes a means for measuring torque and rotational speed. An absorption unit consists of some type of rotor in a housing. The rotor is coupled to the engine or other equipment under test and is free to rotate at whatever speed is required for the test. Some means is provided to develop a braking torque between the rotor and housing of the dynamometer. The means for developing torque can be frictional, hydraulic, electromagnetic, or otherwise, according to the type of absorption/driver unit.

Figure 42. Typical shock standard test sheet. IEC publication 60068-2-27 ed.4.0 (Copyright © 2008 of publication IEC Geneva, Switzerland. www.iec.ch)

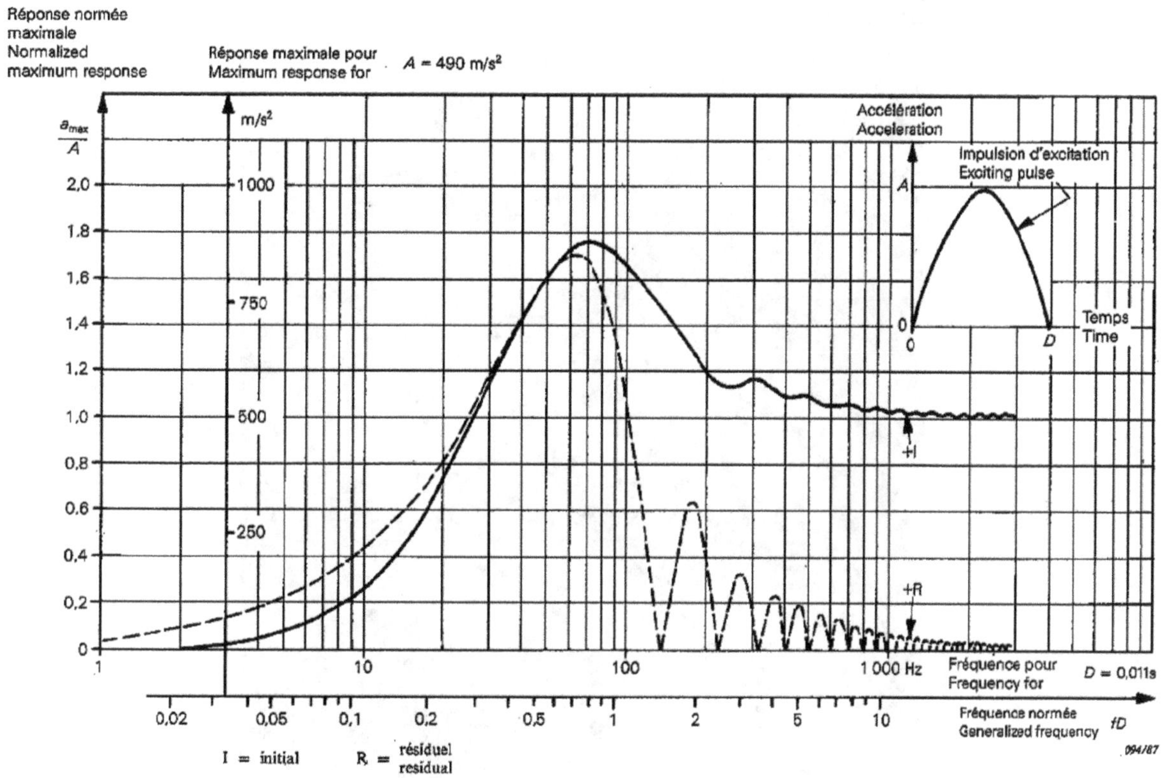

One means for measuring torque is to mount the dynamometer housing so that it is free to turn except as restrained by a torque arm. The housing can be made free to rotate by using trunnions connected to each end of the housing to support it in pedestal-mounted trunnion bearings. The torque arm is connected to the dyno housing and a weighing scale is positioned so that it measures the force exerted by the dyno housing in attempting to rotate. The torque is the force indicated by the scales multiplied by the length of the torque arm measured from the center of the dynamometer. A load cell transducer can be substituted for the scales in order to provide an electrical signal that is proportional to torque.

Another means to measure torque is to connect the engine to the dynamometer through a torque sensing coupling or torque transducer. A torque transducer provides an electrical signal that is proportional to the torque.

With electrical absorption units, it is possible to determine torque by measuring the current drawn (or generated) by the absorber/driver. This is generally a less accurate method and not much practiced in modern times, but it may be adequate for some purposes.

When torque and speed signals are available, test data can be transmitted to a data acquisition system rather than being recorded manually. Speed and torque signals can also be recorded by a chart recorder.

Figure 43 shows a typical dynamometer. T consists of a stand for fixing the device under test and a programmable rotary equipment. Speed and torque is programmed in the rotary equipment and then they applied to the device under test.

Figure 43. A practical dynamometer

Figure 44. Relation between vapor density and relative humidity and temperature

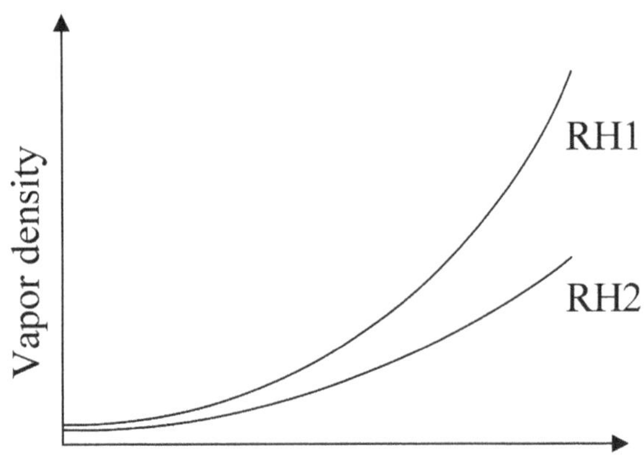

ENVIRONMENTAL TESTS

Sample Test: Corrosion

From general experience of life, we probably have some idea about what corrosion is and have experienced the higher levels of corrosion that occur in the presence of moisture and the hostile gas species that are often present in the atmosphere. Figure 44 shows increasing the water vapor density in the air with increasing the ambient temperature.

The phenomenon of corrosion involves reactions which lead to the creation of ionic species, by either loss or gain of electrons. Take the case of the rusting of iron, where metallic iron is converted into various oxides or hydroxides when exposed to moist air.

Figures 45 and 46 show the typical result of humidity test applied to transformers.

Figure 45. Effect of humidity on the core of a transformer

Figure 46. Effect of humidity on the core of a transformer

Sample Test: Electromigration

Electromigration is an electrochemical process where metal on an insulating material, in a humid environment and under an applied electric field, leaves its initial location in ionic form and redeposits somewhere else. Such migration may reduce isolation gaps and ultimately lead to an electrical short circuit. The process begins when a thin continuous film of water has been formed and a potential is applied between oppositely charged electrodes. Positive metal ions are formed at the positively biased electrode (the anode), and migrate toward the negatively charged cathode. Over time, these ions accumulate as metallic dendrites, reducing the spacing between the electrodes, and eventually creating a metal bridge. Electromigration is closely related to corrosion, with the anode being attacked, but which circuit element is the anode is determined by the applied field rather than the oxidation potential of the metal used.

The rate of electromigration increases with temperature and has four prerequisites – a mobile metal; a voltage gradient; a continuous film of moisture; soluble ions:

- Silver is the metal most susceptible to migration, since it is anodically very soluble and requires a low activation energy to initiate the migration process. Copper, zinc, and lead will also migrate, although only under much more severe conditions. Most other common electronic materials are not susceptible to migration: iron, nickel, and tin because of their low solubility in water; gold, platinum, and palladium because they are anodically stable.
- The severity of electromigration increases with applied potential gradient, the time to grow dendrites decreasing both with reducing electrode spacing and increasing voltage. The time to failure is a log-linear function of voltage at low voltages, but reaches a minimum as the effect becomes transport limited by the number of ions available.
- How much moisture is needed has been hotly debated. Whether the moisture needed represents just few monolayers, or several hundred monolayers, depends on the nature, structure, and porosity of the surface, and on its affinity for water. For example, phenolic resin laminates are more hygroscopic than ceramic substrates and are thus more susceptible to migration.
- The type of surface has another effect on electromigration, because nucleation sites are required for dendrites to initiate and grow, and these are provided by the surface roughness of the substrate and the presence of sharp corners and kinks.
- The rate of electromigration increases with relative humidity, but will become significantly worse if changing conditions encourage the formation of water droplets. This has been reported as a main factor in MLC failure due to tin and silver migration.
- A soluble ionic species is essential to provide the conductive medium for the migration to occur, and the nature of the ionic contaminants has a major impact on performance. The severity of electromigration depends on the particular ionic species involved, and factors such as its mobility, ionic radius, electronegativity, electron affinity, and charge to size. Ionic contaminants come from a wide range of sources such as activators in the flux, reaction products of the soldering process, the breakdown of cleaning solvents, and fingerprints.

The move to finer pitch components, and consequently reduced spacings, makes electromigration more likely, because the voltage gradients between conductors increase and it becomes more difficult to eliminate the minute amounts of ionic contaminants which are sufficient to cause problems.

Three main methods have been used to reduce the electromigration problem:

- Alloying any silver with an anodically stable metal such as palladium or platinum. This is one reason (as well as the improvement in solderability) why palladium-containing silver alloys are preferred to pure silver for MLC terminations
- Using a coating of solder resist to shield the PCB surface from humidity and ionic contaminants. Exposed epoxy glass is much more hydrophilic than most solder mask materials, and the improvement in Surface Insulation Resistance is typically several orders of magnitude
- Plating any silver with metals such as tin, nickel, or gold. In MLCs, the nickel barrier coating improves resistance to electromigration as well as enhancing performance on exposure to molten solder.

Figure 47. IP test of a converter

Figure 48. IP test setup to prove protection against dust IEC publication 60034-5 ed.4.0 (Copyright © 2000 of publication IEC Geneva, Switzerland. www.iec.ch)

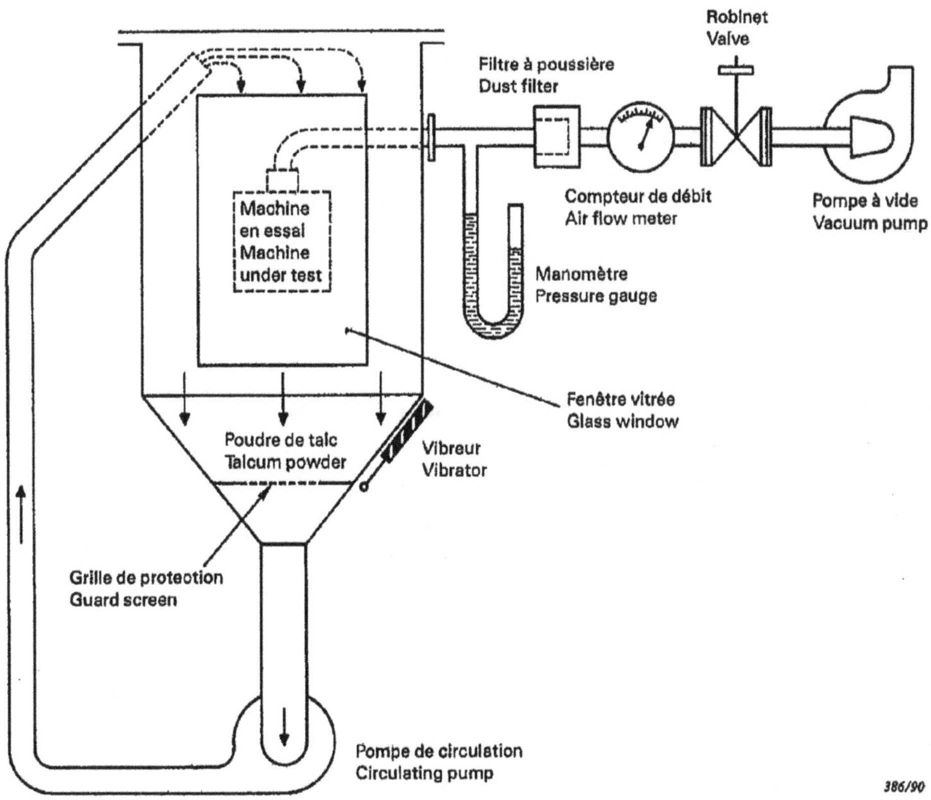

Figure 49. IP test setup to prove protection against splashing water IEC publication 60034-5 ed.4.0 (Copyright © 2000 of publication IEC Geneva, Switzerland. www.iec.ch)

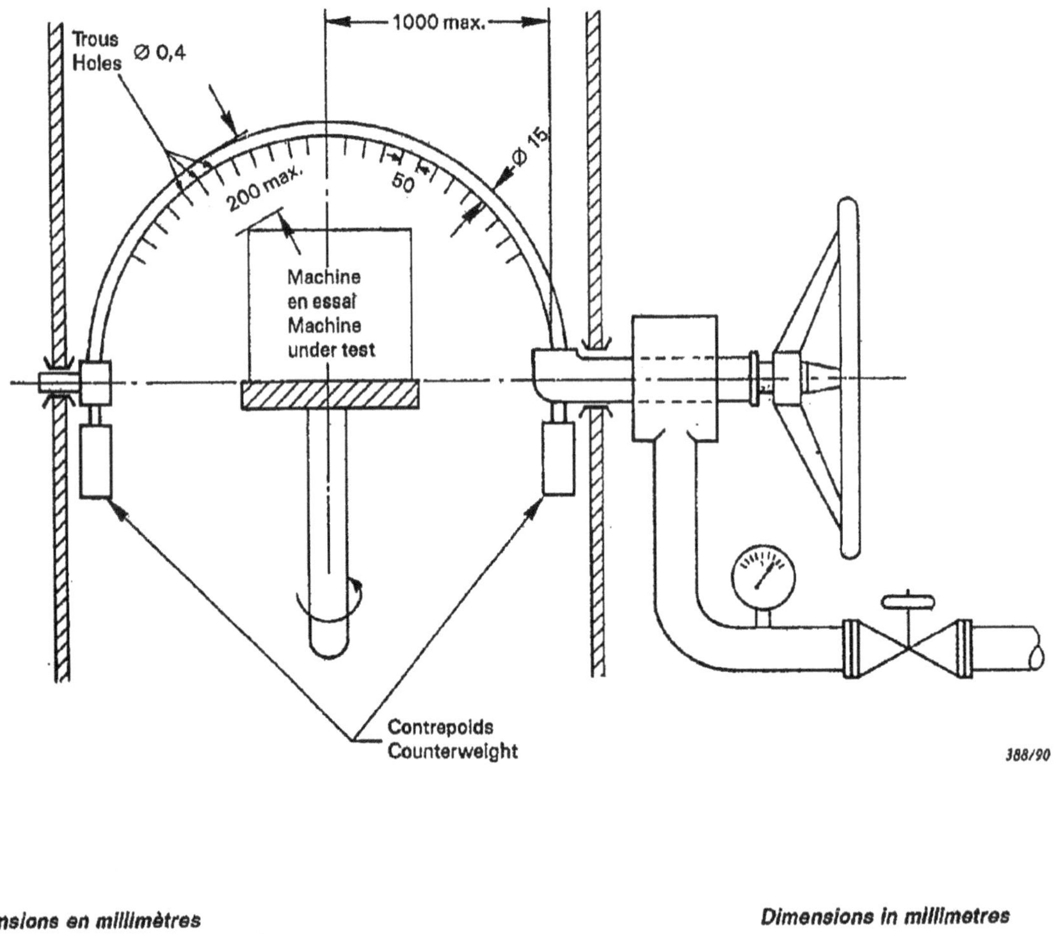

IP Test

The IP Code, International Protection Marking, IEC standard 60529, sometimes interpreted as Ingress Protection Marking, classifies and rates the degree of protection provided against intrusion (body parts such as hands and fingers), dust, accidental contact, and water by mechanical casings and electrical enclosures. It is published by the International Electrotechnical Commission (IEC). Figure 47 shows a converter under IP test. Figures 48 and 49 show standard test setups for IP test in IEC standard.

SUMMARY AND CONCLUSION

In this chapter, reliability measurement in electric power converter was described. The main approach for this goal is accelerated aging tests. The advantage of predicting from field and test data is that the reliability results can be accurately determined including the associated uncertainty of the estimate. The disadvantage is the difficulty of obtaining and assessing accurate field and test data.The main topics of this chapter are summarized as follows:

1. The purpose of accelerated life testing (ALT) is to induce field failure in the laboratory at a much faster rate by providing a harsher, but nonetheless representative environment. In this method, at each test one or some parameters of environmental conditions are stressed more than typical state to reduce test time to less than real state. Then, there are some determined relations between this accelerated test results and typical condition results that are used in finding failure rates and reliability evaluation. The failure rate of the product can be determined.
2. Limitation of tests is determined by whether the test is destructive or not. Non-destructive testing (NDT) is a wide group of analysis techniques used in science and industry to evaluate the properties of a material, component or system without causing damage. Instantaneous data logging of tests can be used as a tool for preventing system damage during tests.
3. Arrhenius' equation is a simple but remarkably accurate formula for the temperature dependence of reaction rates. Additional stress can reduce the time to failure. A programmed time dependent stress can be used to test real useful life of the converter.
4. Industry standards in the electrical industry guide the qualification testing process and usually include aging tests as well as comparisons to similar products. Manufacturers follow specific industry standards when manufacturing electrical products. To ensure their products meet those standards, the products must undergo automated or manual testing processes to ensure the specifications are met. Testing also is a way to uncover glitches and problems with a design before it's integrated into a larger project. Testing allows engineers to identify the difference between what they expect their products to do and what they actually accomplish

The data obtained from accelerated aging tests as well as reliability calculations shows an evaluation of the converter condition from reliability point of view. It is usually necessary to make an improvement on the converter to increase its useful life time. In the next chapters, we present some commonly used as well as some novel methods for reliability improvement in electric power converters.

REFERENCES

Lu, M. W., & Rudy, R. J. (2001). Laboratory reliability demonstration test considerations. *IEEE Transactions on Reliability*, *50*(1), 12–16. doi:10.1109/24.935011

Saha, T. K. (2003). Review of modern diagnostic techniques for assessing insulation condition in aged transformers. *IEEE Transactions on Dielectrics and Electrical Insulation*, *10*(5), 903–917. doi:10.1109/TDEI.2003.1237337

Thomson, W. T., & Fenger, M. (2001). Current signature analysis to detect induction motor faults. *IEEE Industry Applications Magazine*, *7*(4), 26–34. doi:10.1109/2943.930988

Wen, C., Baosheng, H., Wei, S., Xuetao, S., Kai, W., Okamoto, T., & Suzuoki, Y. (2014). An approach for economic assessment on oil-paper insulation diagnosis through accelerated aging experiments. *IEEE Transactions on Dielectrics and Electrical Insulation*, *21*(4), 1842–1850. doi:10.1109/TDEI.2014.004370

Wong, P., Albrecht, P., Allan, R., Billinton, R., Chen, Q., & Fong, C. et al. (1999). The IEEE Reliability Test System-1996. A report prepared by the Reliability Test System Task Force of the Application of Probability Methods Subcommittee. *IEEE Transactions on Power Systems*, *14*(3), 1010–1020. doi:10.1109/59.780914

ADDITIONAL READING

Benbouzid, M. E. H., & Kliman, G. B. (2003). What stator current processing-based technique to use for induction motor rotor faults diagnosis. *IEEE Transactions on Energy Conversion*, *18*(2), 238–244. doi:10.1109/TEC.2003.811741

Dissado, L. A., Mazzanti, G., & Montanari, G. C. (1997). The role of trapped space charges in the electrical aging of insulating materials. *IEEE Transactions on Dielectrics and Electrical Insulation*, *4*(5), 496–506. doi:10.1109/94.625642

Gorur, R. S., Cherney, E. A., Hackam, R., & Orbeck, T. (1988). The electrical performance of polymeric insulating materials under accelerated aging in a fog chamber. *IEEE Transactions on Power Delivery*, *3*(3), 1157–1164. doi:10.1109/61.193898

Perisse, F., Werynski, P., & Roger, D. (2007). A New Method for AC Machine Turn Insulation Diagnostic Based on High Frequency Resonances. *IEEE Transactions on Dielectrics and Electrical Insulation*, *14*(5), 1308–1315. doi:10.1109/TDEI.2007.4339494

Schneider, H. M., Guidi, W. W., Burnham, J. T., Gorur, R. S., & Hall, J. F. (1993). Accelerated aging and flashover tests on 138 kV nonceramic line post insulators. *IEEE Transactions on Power Delivery*, *8*(1), 32336. doi:10.1109/61.180353

Suehle, J. S., & Chaparala, P. (1997). Low electric field breakdown of thin SiO_2 films under static and dynamic stress. *IEEE Transactions on Electron Devices*, *44*(5), 801–808. doi:10.1109/16.568042

Trevisanello, L., Meneghini, M., Mura, G., Vanzi, M., Pavesi, M., Meneghesso, G., & Zanoni, E. (2008). Accelerated Life Test of High Brightness Light Emitting Diodes. *IEEE Transactions on Device and Materials Reliability*, *8*(2), 304–311. doi:10.1109/TDMR.2008.919596

Wang, M., Vandermaar, A. J., & Srivastava, K. D. (2002). Review of condition assessment of power transformers in service. *IEEE Electrical Insulation Magazine*, *18*(6), 12–25. doi:10.1109/MEI.2002.1161455

Xiaojun, L., Jin, Q., & Bernstein, J. B. (2008). Compact Modeling of MOSFET Wearout Mechanisms for Circuit-Reliability Simulation. *IEEE Transactions on Device and Materials Reliability*, *8*(1), 98–121. doi:10.1109/TDMR.2008.915629

KEY TERMS AND DEFINITIONS

Accelerated Aging Test: A test with more stress for shortening the time to failure.
Destructive Test: a test with damaging effect on a (some) part(s) of a system.

Standard Test: A programmed test used for acceptance of a product.
Stress: The forces on a system that leads to failure.
Test Chamber: Equipment for performing programmed tests.

Section 3

Methods for Preventing Faults

Chapter 6
Reliability as a Figure of Merit

ABSTRACT

An electric power converter is specified by its system performance indices. Many system performance indices like efficiency are important in design process of system. However, that doesn't mean the converter performance is necessarily adequate for a practical application. At the end of design process of a converter, some desired specifications may not be achieved. In this chapter, reliability as a figure of merit in design of a system is presented and compared with other indices. We want to highlight the effect of reliability considerations on the design methodology of a power converter. The most important specification of a power supply or power converter is its robustness. Because any failure in power supply leads to failure of the whole of the system. A power converter may have poor performance but operate reliably and vice versa. In fact, this is a reliability based design approach to achieve a long useful life. It is shown that in many systems, high efficiency is not a good choice for selection of system operating point. A system can be inefficient but very reliable. Two complex examples are presented to show undesired results of neglecting reliability in design process. Methods for more reliable operation of electric power converters than high performance operation are proposed. A discussion about correct and intelligent optimization of a system parameters and operating set point is presented.

INTRODUCTION

Quality of any system is expressed by some indices which are named as figure of merits. These indices help users to have a smart selection among some systems with a common application (Rajaei, Kaboli, & Emadi, 2008). For example, both electric and engine motors can be used for driving a vehicle. Selection between these two choices can be done by difference of their efficiency as a figure of merit. An electric power converter as a system also has various figure of merits (Kaboli, Haddadi, & Khaligh, 2008). Some of the most important figure of merits in a power electronic converter are listed as follows:

- Efficiency
- Power factor
- Total harmonic distortion of voltage and current (input or output)

DOI: 10.4018/978-1-4666-9429-3.ch006

- Duty cycle

There are similar indices for electric machines too (Nasirian, Kaboli, & Davoudi, 2012). However, some of these indices are more important in electric machines or in power electronics. For example, regarding to switching nature of power electronic converters, harmonic distortion indices are more important than electric machines.

It is a common scenes to use the above mentioned indices for characterizing the converters in many literatures. In this chapter, we want to open another window for reader and it is reliability as a figure of merit. Figure 1 shows the state of this chapter in flowchart of the book. Importance of this concept is due to its conflict with some of well stablished indices like efficiency.

Figure 1. State of chapter 6 in the flowchart of the book

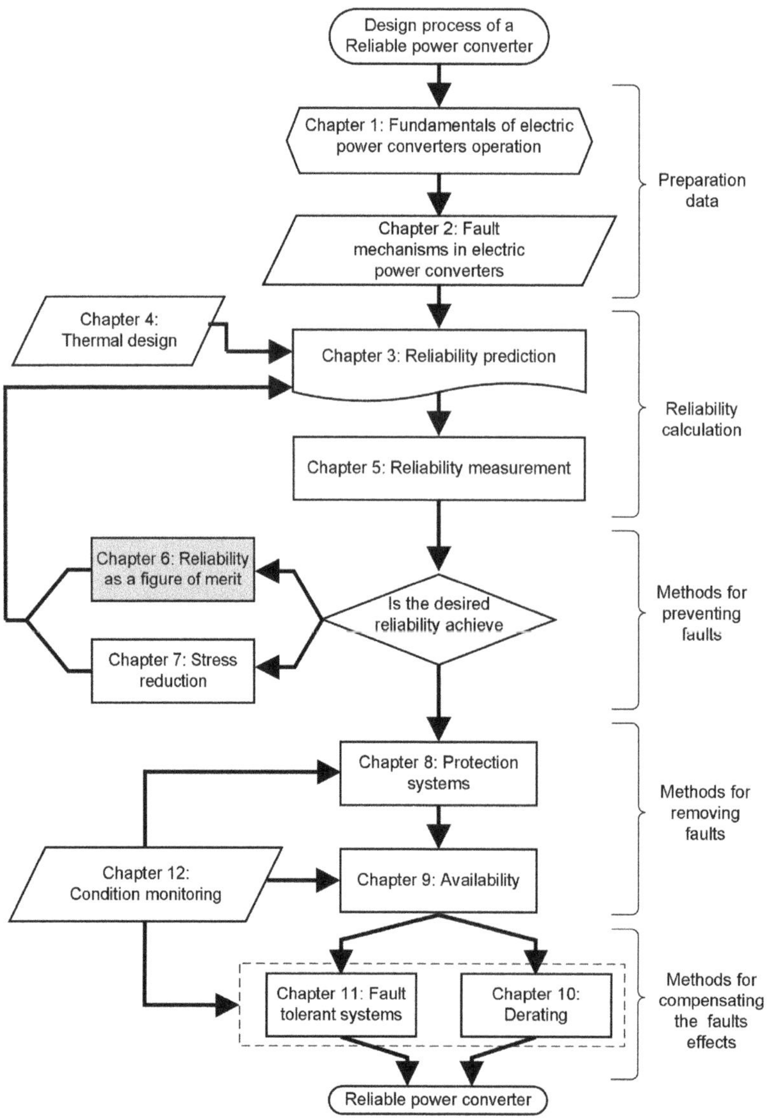

Reliability as a Forgotten Quality Index

The performance of power electronic converters, especially in terms of efficiency and power density, has been continuously improved by the intensive research and advancement in circuit topologies, control schemes (Hasanzadeh, Zolghadri, Kaboli, & Homaifar, 2003), semiconductors, passive components, and system integration technologies. In recent years, the automotive and aerospace industries have brought stringent reliability constraints on power electronic systems because of safety requirements. The industrial and energy sectors are also following the same trend to more reliable power electronic systems with cost-effective and sustainable solutions.

Of course, high efficiency is an important goal; however, that doesn't mean the converter's performance is necessarily adequate for a practical application. The published efficiency is usually a typical value measured at 25°C ambient, at the nominal input voltage. As the temperature rises, losses for semiconductor (usually MOSFETs) and copper traces can rise dramatically. At low line, there's increased input current loss. At high-line switching, losses increase; thus, efficiency goes down. With this in mind, worst-case efficiency is more important than the quoted typical efficiency. Whereas efficiency values can assist in comparing similar-sized converters with equal voltage and current ratings, they aren't worth much if the size, output voltage, or the output currents are different.

Design for Reliability (DFR) is not a new concept, but it has begun to receive a great deal of attention in recent years. In this chapter, we will try to answer these questions and, at the same time, we will propose a general DFR process that can be adopted and deployed with a few modifications across different industries in a way that will fit well into the overall Product Development Process

RELIABILITY ORIENTED APPROACH

In the first part of this chapter, it will be discussed the challenges and exciting new opportunities in the research on reliability of power electronics.

A figure of merit is a quantity used to characterize the performance of a device, system or method, relative to its alternatives. In engineering, figures of merit are often defined for particular materials or devices in order to determine their relative utility for an application. In commerce, such figures are often used as a marketing tool to convince consumers to choose a particular brand

It is usual that some of figure of merits such as efficiency are widely accepted by many users. However, reliability is less applicable than these labels. In this section, it is shown that reliability consideration leads to different results in design process.

Figure 2 shows a typical comparison between allowable surge current in various solid state switches. It can be seen that diode as an uncontrolled switch is higher margin than semi-controlled and fully-controlled switches. Figures 3, 4, and 5 shows real values for surge current in three various switches.

Example: Reliability in Motor Drive System

First, let us start with some basic clarifications. Traditional quality control assures that the product will work after assembly and as designed. Whereas reliability provides the probability that an item will perform its intended function for a designated period of time without failure under specified conditions. In other words, reliability looks at how long the product will work as designed, which is a very different

Figure 2. Surge current capacity of various switches

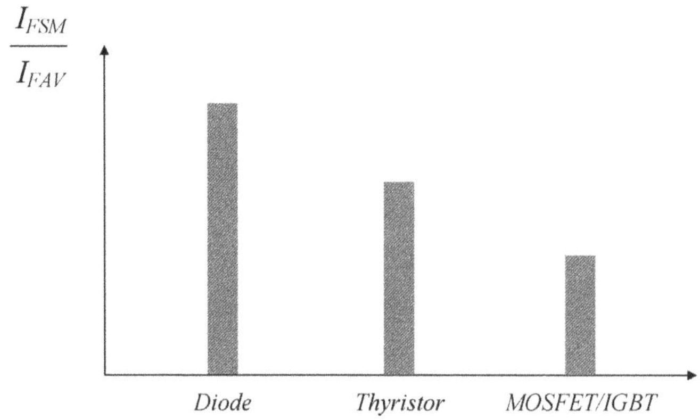

Figure 3. Absolute maximum ratings of a Diode, (Source: ON Semiconductor Co. with permission)

MAXIMUM RATINGS

Rating	Symbol	Value	Unit
Peak Repetitive Reverse Voltage Working Peak Reverse Voltage DC Blocking Voltage	V_{RRM} V_{RWM} V_R	200	V
Average Rectified Forward Current, T_C = 156°C Per Leg Total Device	$I_{F(AV)}$	 8.0 16	A
Peak Rectified Forward Current (Square Wave, 20 kHz), T_C = 154°C – Per Diode Leg	I_{FM}	16	A
Nonrepetitive Peak Surge Current (Surge applied at rated load conditions halfwave, single phase, 60 Hz)	I_{FSM}	100	A
Operating Junction Temperature and Storage Temperature	T_J, T_{stg}	−65 to +175	°C

Figure 4. Absolute maximum ratings of a MOSFET, (Published with permission of International Rectifier)

Absolute Maximum Ratings

	Parameter	Max.	Units
I_D @ T_C = 25°C	Continuous Drain Current, V_{GS} @ 10V	30	A
I_D @ T_C = 100°C	Continuous Drain Current, V_{GS} @ 10V	21	
I_{DM}	Pulsed Drain Current ①	120	
P_D @T_C = 25°C	Power Dissipation	214	W
	Linear Derating Factor	1.4	W/°C
V_{GS}	Gate-to-Source Voltage	± 20	V
E_{AS}	Single Pulse Avalanche Energy②	315	mJ
I_{AR}	Avalanche Current①	30	A
E_{AR}	Repetitive Avalanche Energy①	21	mJ
dv/dt	Peak Diode Recovery dv/dt ③	8.6	V/ns
T_J T_{STG}	Operating Junction and Storage Temperature Range	-55 to +175	°C
	Soldering Temperature, for 10 seconds	300 (1.6mm from case)	
	Mounting torque, 6-32 or M3 srew	10 lbf•in (1.1N•m)	

Figure 5. Absolute maximum ratings of an IGBT, (Published with permission of International Rectifier)

Absolute Maximum Ratings

	Parameter	Max.	Units
V_{CES}	Collector-to-Emitter Voltage	1000	V
I_C @ T_C = 25°C	Continuous Collector Current (Silicon Limited)	55	A
I_C @ T_C = 100°C	Continuous Collector Current (Silicon Limited)	35	
I_{CM}	Pulse Collector Current, V_{GE} = 15V ②	60	
I_{LM}	Clamped Inductive Load Current, V_{GE} = 20V ①	80	
V_{GE}	Continuous Gate-to-Emitter Voltage	±30	V
P_D @ T_C = 25°C	Maximum Power Dissipation	210	W
P_D @ T_C = 100°C	Maximum Power Dissipation	105	
T_J T_{STG}	Operating Junction and Storage Temperature Range	-55 to +175	°C
	Soldering Temperature, for 10 sec.	300 (0.063 in.(1.6mm) from case)	
	Mounting Torque, 6-32 or M3 Screw	10 lbf·in (1.1 N·m)	

objective than that of traditional quality control (Kaboli, Zolghadri, & Khaligh, 2008). Therefore, different tools and models apply to reliability that do not necessarily apply to quality and vice versa.

There are many various algorithms to implement a loss minimization control system for an ASD system. In previous years, when ASD loss was considerable, the motor and its ASD are considered as a unique part and loss minimization algorithm operates on general inputs and outputs of this set. In the recent years, the ASD loss can be neglected comparison with motor loss and new loss minimization techniques only operates on motor. In all of these methods, it is a common practice that ASD has not been considered as a separate part and its loss is not controlled. Therefore, the ASD loss may reaches to its limitation and it leads to low reliability in drive system.

In this section, a loss minimization study is performed on a motor drive system while ASD loss is considered as a separate part. Thus, there are at least two variables in this problem and it converts the loss minimization problem to a multi objective optimization problem. This work is done on a direct torque controlled (DTC) induction motor drive because it is used in many ac drives (Kaboli, Zolghadri, & Homaifar, 2003). Free parameters which are used in this optimization problem are the values of hysteresis band of flux controller (HBFC) and hysteresis band of torque controller (HBTC). Other free parameters of a DTC scheme like amplitude of flux reference and sampling frequency have an evident solution for optimal working of drive. The optimization problem is constructed using inverter loss from ASD section and motor loss from motor section. As the torque ripple is a major problem of DTC and it directly affects on the performance of drive system, the torque ripple is also participated in the optimization process. The preliminary study is presented based on analytical survey and is verified with computer simulation and experimental results. In this section, some of open questions about performance of DTC are answered too. Then, a cost function consisting motor losses, torque ripple and inverter losses is defined. Minimization of this cost function leads to the optimum values for the flux and torque hysteresis bands 5.

The basic idea of the DTC concept, which its block diagram is shown in Figure 6, is to choose the best vector of the voltage, which makes the flux rotate and produce the desired torque. During this rotation, the amplitude of the flux remains inside a pre-defined band. Circular trajectory of the stator flux is divided into six symmetrical sectors referred to the inverter voltage vectors. For each section, based on the torque and flux errors, a proper vector set is proposed. The certain vectors are applied to the motor so that the amplitude of the flux and torque remain constant.

Figure 6. Block diagram of the DTC method

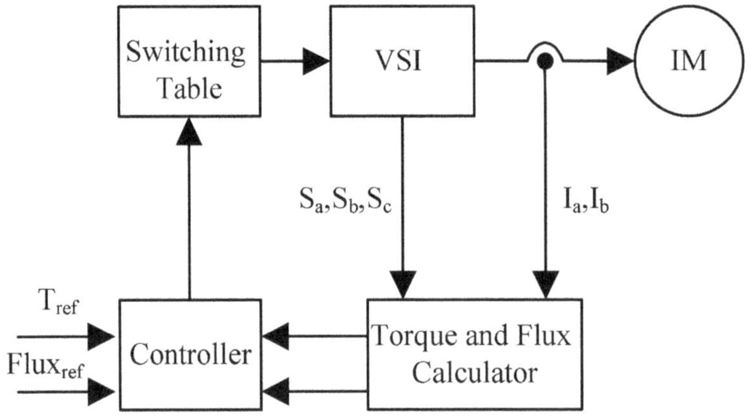

The most commonly used procedure to optimize several criteria simultaneously is to combine them linearly and introduce a weighting factor for each of them. It maight be possible to exist some limitations for variables. Thus, optimization problem for a DTC based drive can be presented by:

$$C(\bar{x}) = \frac{k_1}{P_{loss-con-\max}} P_{loss-con}(\bar{x}) + \frac{k_2}{P_{loss-mot-\max}} P_{loss-mot}(\bar{x}) + \frac{k_3}{TR_{\max}} TR(\bar{x})$$

$$\bar{x} = \begin{bmatrix} \Delta\psi \\ \Delta T \end{bmatrix}$$

$$\begin{bmatrix} 0 \\ 0 \end{bmatrix} \leq \bar{x} \leq \begin{bmatrix} \Delta\psi_{\max} \\ \Delta T_{\max} \end{bmatrix} \tag{1}$$

In this section, variation of DTC objective function components presented in (1) with hysteresis band value of flux and torque controller is investigated. To evaluate the validity of theoretical analysis, a DTC setup has been simulated with SIMULINK and implemented in laboratory. The specifications of induction motor used in this study are presented in Table 1. The experimental setup, shown in Figure 7 consists of an induction motor, insulated gate bipolar transistor (IGBT) based inverter, and digital signal processor (DSP) (TMS320C) based controller.

Motor Loss

Motor loss can be divided into two different categories: electrical loss and mechanical loss. In this study, the motor output power is kept constant because its output torque is constant. Therefore, the mechanical loss can be considered constant. The effect of torque ripple variation on the mechanical loss is neglected. Electrical loss includes core loss and copper loss (Kaboli, Zolghadri, 2003). Core loss consists of hysteresis loss and eddy current loss which are presented by (2).

Table 1. Characteristics of induction motor

Parameter	Value
Rated Power (KW)	5.5
Number of Poles	4
Stator Resistance (Ω)	0.18
Stator Inductance (mH)	56
Magnetizing Inductance (mH)	53
Rotor Resistance (Ω)	0.50
Rotor Inductance (mH)	56
Nominal Torque (N.m)	35
Nominal stator flux (Wb)	0.65

Figure 7. Block diagram of experimental setup

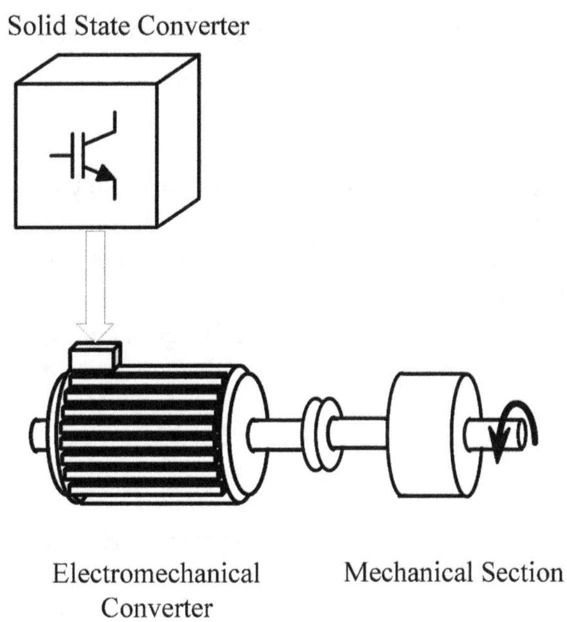

$$P_{hys} = K_h B^n f$$
$$P_{eddy} = K_e B^2 f^2 \qquad (2)$$

For each voltage harmonic a flux harmonic is generated in core. The relation between amplitude of voltage harmonics and its order with its respective flux component density is expressed by (3).

$$V_h = 4.44 N f B_h \qquad (3)$$

Therefore, Hysteresis and eddy current loss of core for each harmonics can be calculated by (2).

Figure 8. Variation of voltage harmonics in DTC versus torque and flux hysteresis band

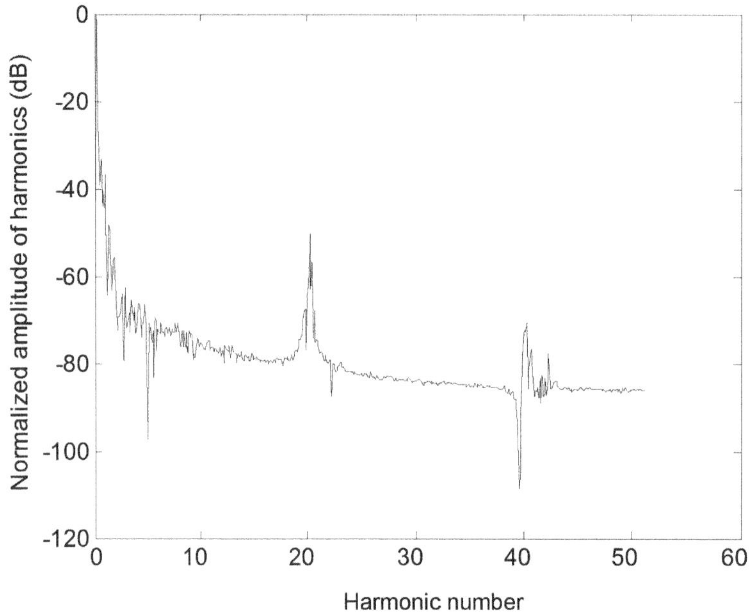

To calculate the core loss, the frequency spectrum of stator voltage is calculated. This spectrum has two major harmonics that vary versus parameters of flux and torque controller as shown in Figure 8. Then the Hysteresis and eddy current loss of each voltage harmonics are calculate by (2) and (3). Summation of these harmonic losses results motor core loss 9.

The motor copper loss is calculated in similar way. For each current harmonic, the copper loss is obtained by (4).

$$P_{cu} = R_{ac} I_{rms}^2 \tag{4}$$

Figure 9 shows variation of motor loss for different values of HBFC and HBTC (Kaboli, Zolghadri, & Emadi, 2003).

Referring to Figure 8 can explain these behaviors. This figure shows that harmonics of stator voltage in DTC which is produced by flux controller locate in high frequency for small values of HBFC. Equations (2) and (3) show that core hysteresis loss is decreased with increasing the harmonic frequency. Therefore, Hysteresis loss is a small part of motor loss for small value HBFC. For great values of HBFC, the generated harmonics shifted to low frequency and their amplitudes is increased according to (3). Thus, Hysteresis loss of motor core grows up. On the other hand, the harmonics produced by torque controller has such a high frequency that has not distinct effect on motor loss.

Figure 9. Variation of motor loss versus torque and flux hysteresis band

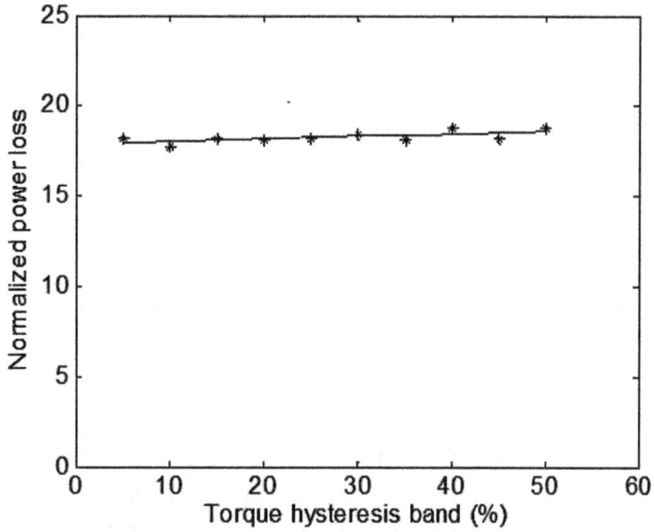

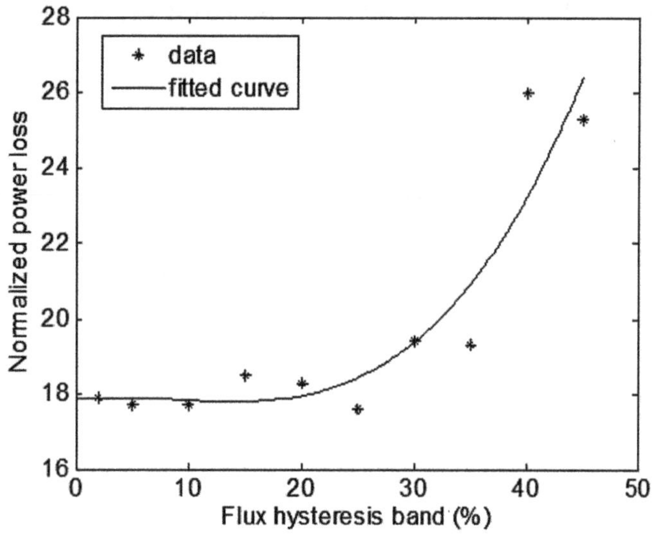

Inverter Loss

Each inverter switch is a combination of an insulated gate bipolar transistor (IGBT) with anti parallel connection of a diode. The inverter switch loss is divided into conduction loss and switching loss which are expressed in (5) and (6).

$$P_{con} = V_{sat} I_{av} + r_{on} I_{rms}^2 \tag{5}$$

$$E_{on}(i) = k_1 + k_2 i_c$$
$$E_{off}(i) = k_3 + k_4 i_c \tag{6}$$

The inverter diode conduction loss is also presented in (7). The diode switching loss is neglected in this study.

$$P_D = V_D I_{av} + r_D I_{rms}^2 \tag{7}$$

Figure 10 shows variation of inverter loss for different values of HBFC and HBTC.

Torque Ripple

The most important characteristic of DTC is its fast torque response. When a proper vector is applied to motor, the stator flux vector rotates very fast and the angle between the stator flux and rotor flux is increased. Therefore, the amplitude of the motor torque is increased. If sampling period Ts is not low enough, then torque waveform usually jumps to the out of torque hysteresis band. Therefore, motor which is derived with DTC usually suffers from large torque ripple. This large torque ripple has undesired effect on motor useful life and its load. The torque ripple ΔT is divided to two parts as shown in the following 11.

$$\Delta T = \Delta T_1 + \Delta T_2 \tag{8}$$

Where

$$\Delta T_1 = -T(\frac{1}{\tau_s} + \frac{1}{\tau_r})\frac{T_s}{\sigma} \tag{9}$$

$$\Delta T_2 = P\frac{L_m}{\sigma L_s L_r}[(V_s - j\omega_m \psi_s).j\psi_r]T_s \tag{10}$$

The first component is due to the stator and rotor resistances and acts in order to reduce the absolute value of the torque. This component is proportional to the torque value and is independent of the voltage vector and motor speed. The second component represents the effect of the applied voltage vector on the torque variation and is dependent on motor speed. As it can be seen in (9) and (10), the value of the motor flux has no effect on the first contribution of the torque ripple; while, it influences mainly the second contribution. The torque ripple is usually measured by (11).

$$\delta = \sqrt{\frac{1}{T_f}\int_0^{T_f}(\frac{T_e}{T_{e-av}} - 1)^2 dt} \tag{11}$$

Figure 10. Variation of inverter loss versus torque and flux hysteresis bands

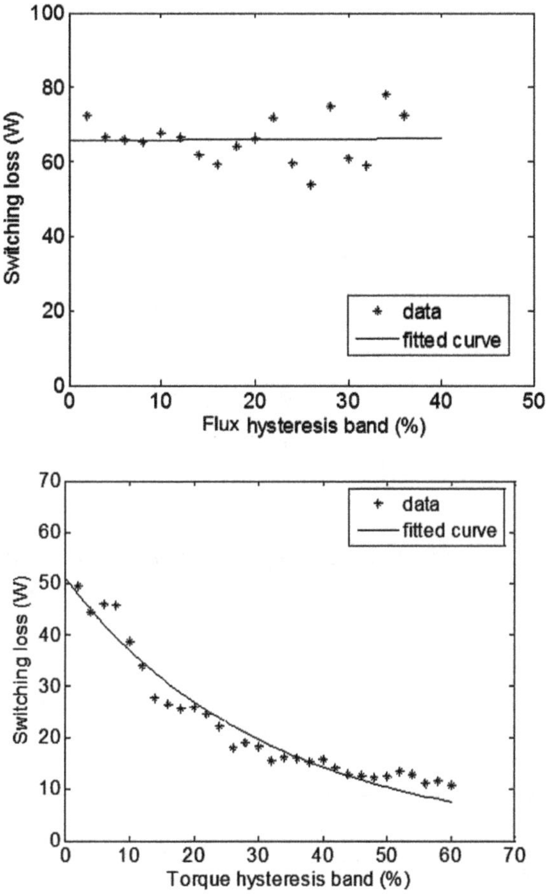

Figure 11. Variation of torque ripple versus torque and flux hysteresis bands

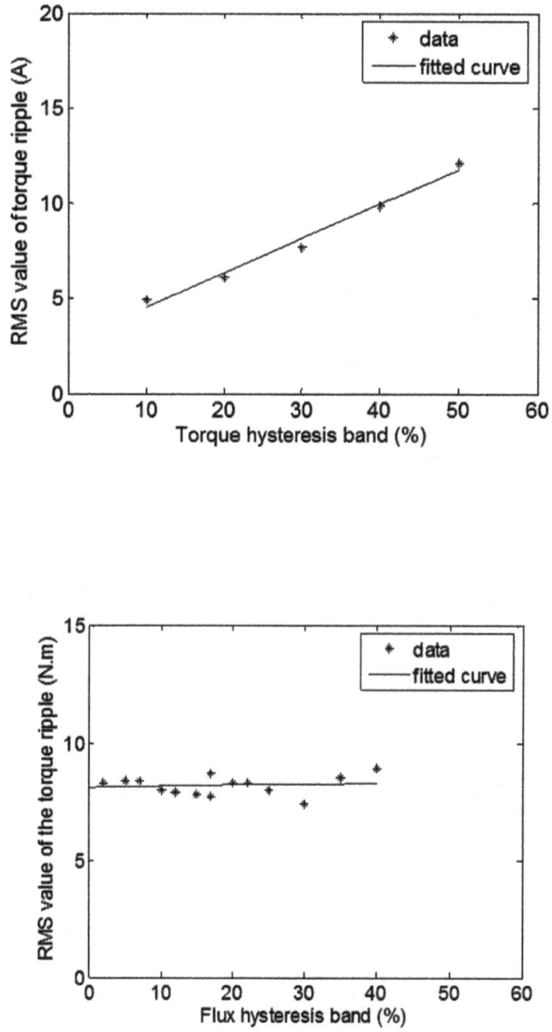

Figure 11 shows variation of torque ripple for different values of HBFC and HBTC.

Combination of results obtained in the previous section is used for optimization of DTC according to (1).

Figure 12 shows the variation of objective function versus HBFC and HBTC while motor and inverter loss are considered as a unique part and torque ripple is neglected. It can be seen that the optimization result leads to HBFC=HBTC=0. For this point, the normalized inverter loss is maximum and inverter reaches to its thermal limit. This result is achieved while torque ripple is considered as shown in Figure 13. Therefore, combination of inverter loss with motor loss leads to over temperature in inverter switches.

Figure 14 shows the variation of objective function versus HBFC and HBTC while motor and torque ripple are considered in objective function and inverter loss is neglected. The result of optimization is similar to previous one and inverter operates with undesired loss in optimum point.

Figure 12. Variation of objective function versus torque and flux hysteresis bands with motor and inverter loss as a unique part and without torque ripple

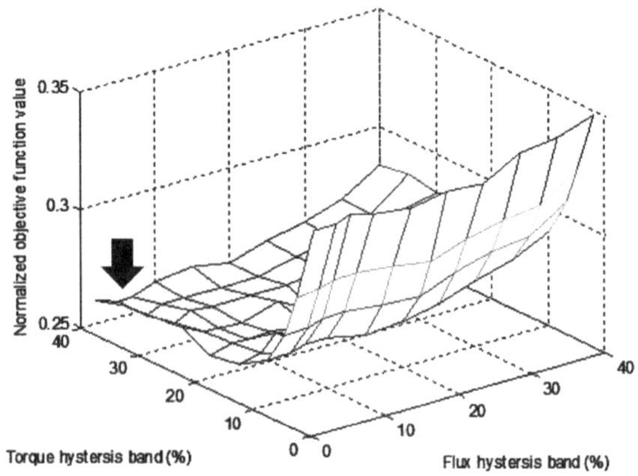

Figure 13. Variation of objective function versus torque and flux hysteresis bands with motor and inverter loss as a unique part and with torque ripple

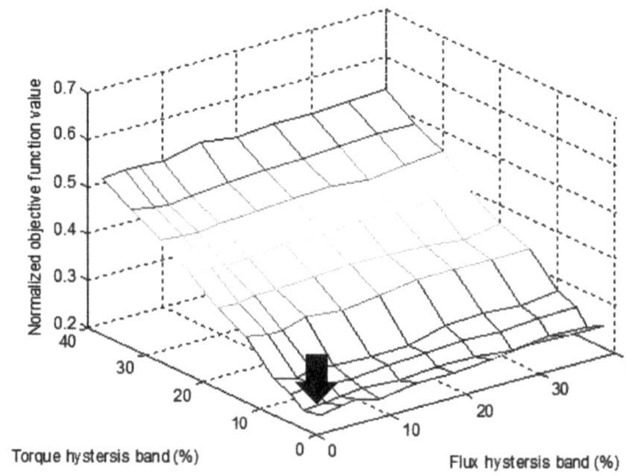

Figure 14. Variation of objective function versus torque and flux hysteresis bands without inverter loss

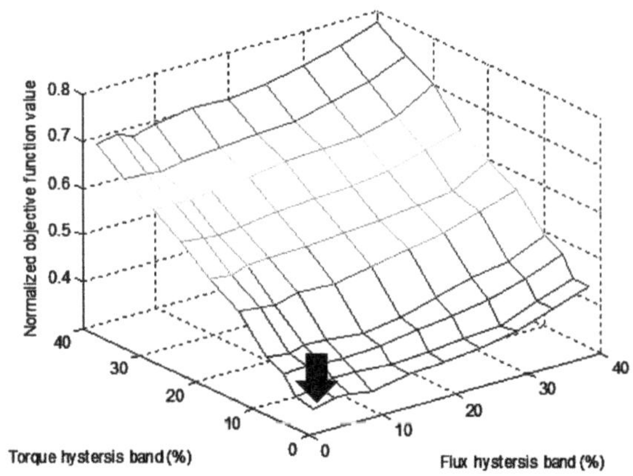

Figure 15. Variation of objective function including normalized inverter loss, motor loss and torque ripple

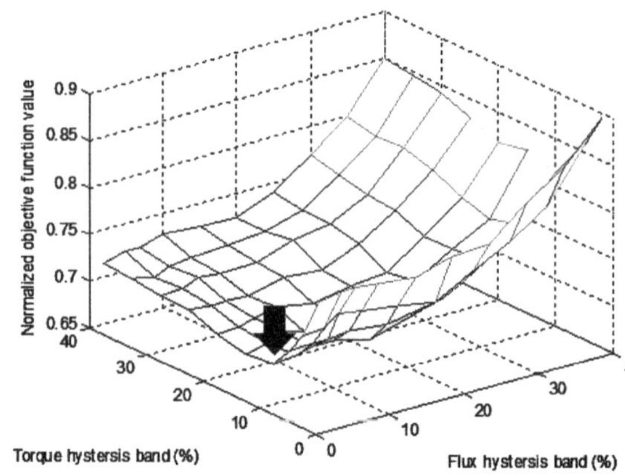

Figure 15 shows the variation of objective function versus HBFC and HBTC while three component of objective function are normalized to their maximum values and then optimization is performed. It can be seen that in the resulted optimum point all of these three components have suitable value.

As the HBFC has similar effect on all of objective function components, it is set to minimum value and optimization is done on HBTC.

Example: Reliability in Power Electronic Converter Design

In a turn off RCD snubber, capacitor value is generally optimized using the minimum switching loss method. In this section, capacitor value was optimized based on reliability considerations. The optimum capacitor value was calculated and compared with that found in the minimum switching loss method. It is shown that reliability of switch and its snubber increase with new capacitor value while their efficiency remains almost unchanged.

In switching power converters, presence of a capacitor in parallel with the switch prevents the switch voltage break down due to overvoltage during turn off time. In addition, it reduces the switch turn off loss. Combining a capacitor with a resistor as well as a diode forms the well-known Resistor-Capacitor-Diode (RCD) snubber in which a resistor is used to reduce the level of capacitor discharge current and a diode is used to bypass the resistor during the switch turn off (Figure 16). In spite of the development of various types of snubbers, RCD snubbers are more widely used due to their simple configuration and passive elements. In this type of snubbers, as the capacitor value increases, the switch turn off loss decreases and the capacitor stored energy increases which is dissipated in the resistor. In general, a capacitor value is selected which results in minimum power loss. Although this value gives maximum efficiency, it may not be the best choice from the reliability point of view. For obtaining the maximum efficiency, capacitor value is determined based on the turn off power loss in both the switch and its snubber. However, in terms of reliability, the amount of power dissipated in each element (i.e. (i) switch and (ii) its snubber) must be taken into account individually. Reliability is considerably affected by the amount of turn off power loss in switch and its snubber because the switching loss is a considerable portion of power loss in the converters with high switching frequency.

In this section, the design of RCD snubber was described based on reliability considerations. The power loss and reliability of switch and snubber was calculated for different capacitor values. Then, the optimum capacitor value for maximum reliability was compared with the one resulted from the maximum efficiency design method. Furthermore, the effect of variation in converter specifications such as its load and switching frequency on the proposed design was studied.

- **Loss Calculation:** Losses in semiconductor components are divided into two categories: conduction loss and switching loss. The voltage drop in the on state resistance and the diode forward voltage drop cause conduction losses. Switching losses appear in the power switches because the

Figure 16. RCD snubber to provide proper switching of an inductive load

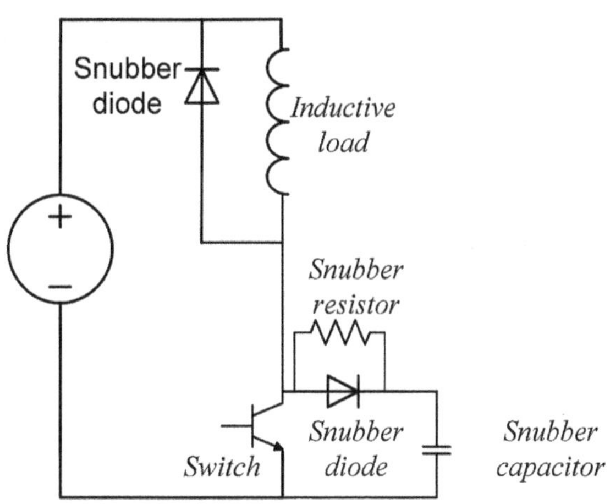

Figure 17. Voltage and current waveforms for MOSFET in switching

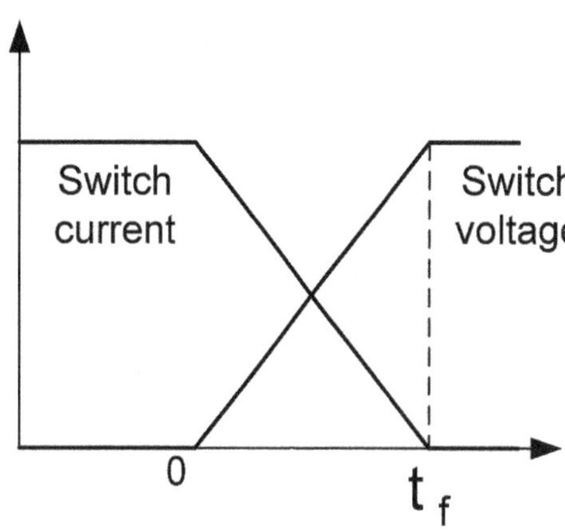

instantaneous power is non-zero during transition from ON state to OFF state and back. Switching characteristics and switching losses are presented in which is reviewed in the next sections. Table 1 shows the different parts of power loss which is dissipated by the elements of circuit.

- **Conduction Loss:** When the switch is ON, load current flows through it for DT and when it's OFF, load current flows through freewheeling diode for $(1 - D)T$.
- **Switching Loss:** Waveforms of a transistor voltage and current for switching the inductive load are shown in Figure 17. The case with no snubber is shown in Figure 17(a). The change of load current is negligible in the turn off interval and it can be modeled by a current source (Figure 18 and Figure 19).

Switching waveforms for small, normal and large snubber capacitor are shown in Figure 17(b), (c) and (d) respectively. In the case with snubber, when the switch is turned off its current decreases and reaches zero. The difference between the load and switch currents flows in the diode D and charges the capacitor C. A "normal" snubber capacitor is defined to be that which allows its voltage to reach the level V at the same time as the transistor current reaches zero, as indicated in Figure 17(c). A "small" and "large" series snubber capacitor allows the voltage to rise and reach full value V before or after the switch current falls to zero, respectively.

Turn Off Switching Losses

During turn off, the switch can be modeled as a current source which its current decreases linearly and reaches zero. The difference between the load and switch currents flows in the diode D and charges the capacitor C up to source voltage V.

In order to simplify the equations, a per unit capacitor value x based on "normal" capacitor value was used.

In order to obtain the formula of turn off switching losses as a function of x in presence of the snubber, two cases are considered; "small" capacitor and "large" capacitor. In both cases it is assumed that the switch current decreases linearly during the off time. When the capacitor is smaller than "normal" value, its voltage reaches full value V before the switch is completely turned off. So, at first, its current increases linearly and then becomes zero. At the same time the capacitor voltage increases in parabolic manner and then it is fixed in voltage V. When the capacitor is greater than the "normal" value, it is charged slowly so does not charge up to full value V during turn off time. Therefore the capacitor current is equal to difference between the load and switch currents in the time interval 0 to t_{off} and is equal to load current after turn off time. So the capacitor voltage increases in parabolic manner during turn off time and then increases linearly.

Turn off switching energy loss is obtained by integration of switch instantaneous power over the turn off time. Multiplying switching frequency by the switching energy results in turn off switching power loss.

Snubber diode energy losses is due to ON state resistance and forward voltage drop. Snubber diode current is equal to capacitor current at the turn off time so the snubber diode energy losses.

Solving the above integral results in snubber diode power loss.

Figure 18. Variation of OFF switching loss with different snubber capacitor, (up): C=1nF, (center): C=2nF, (down): C=10nF

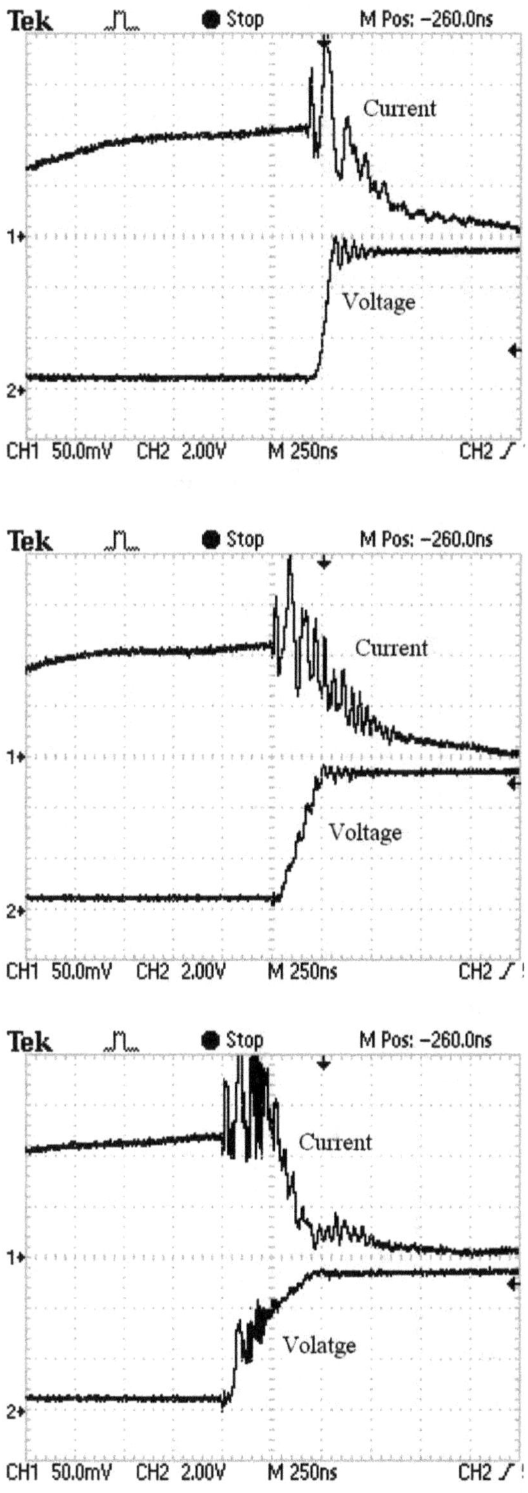

Figure 19. Variation of ON switching loss with different snubber capacitor

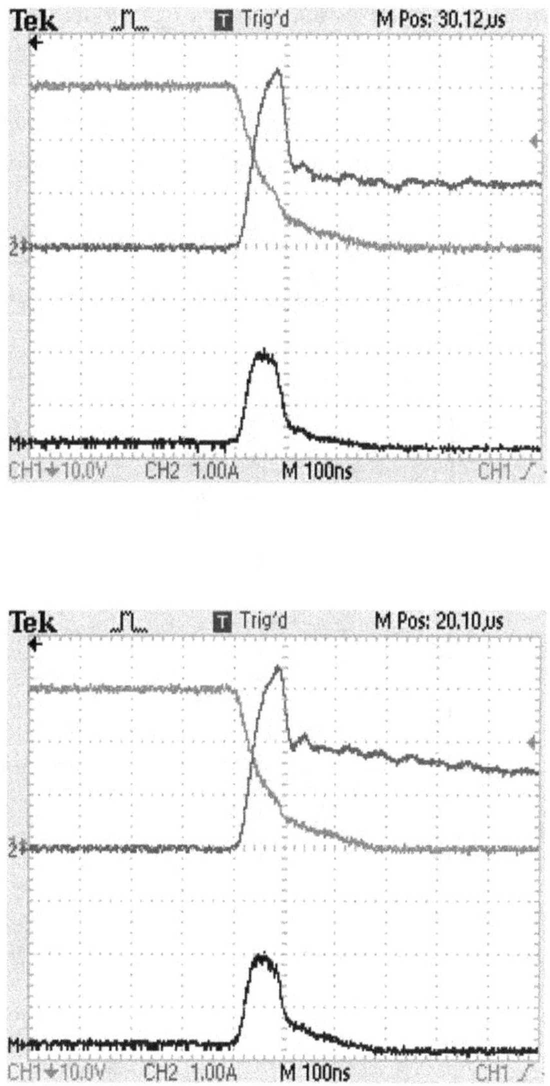

Turn on Switching Losses

When the gate to source voltage of MOSFET reaches threshold value, it starts to be turned on. At first drain current rises and reaches full value I for the time t_{ri} then the drain to source voltage falls and reaches zero for time t_{fv}. At the interval 0 to t_{ri} the freewheeling diode is still on and the drain to source voltage is equal to *V*.

When the voltage fall time starts, the freewheeling diode is reverse biased and turned off. Circuit model of converter at this interval is shown in Figure 20. According to this model the capacitor is dis-

Figure 20. Model of converter in the voltage fall time interval

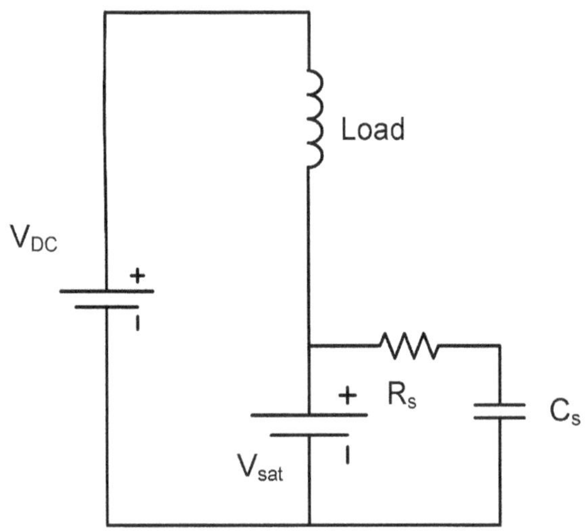

Figure 21. Model of converter after the voltage fall time interval

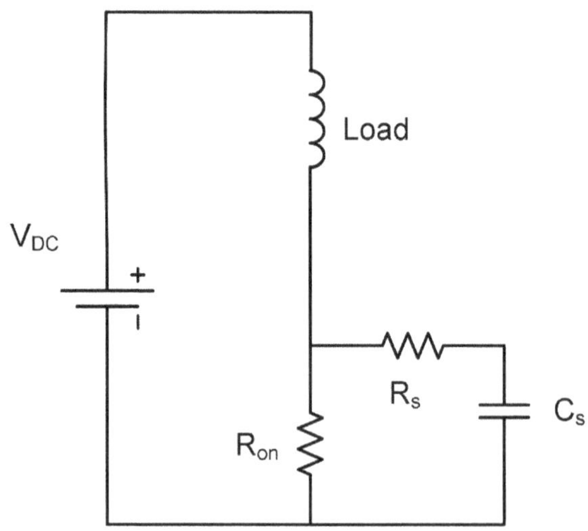

Figure 22. State space diagram of switch and its snubber with those failure rates

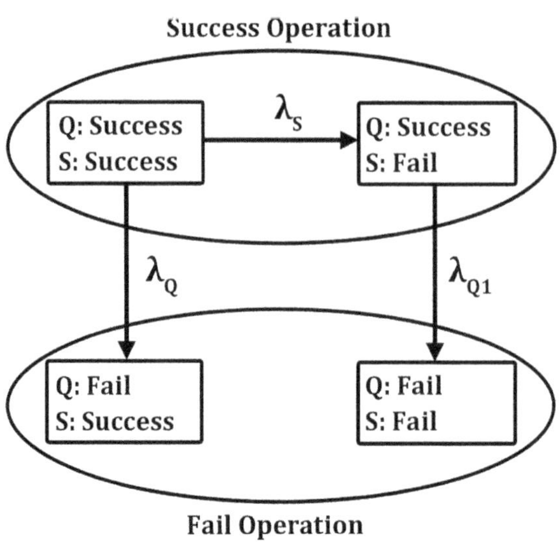

charged on the snubber resistor R and the voltage source V_Q. Loop current i is expressed by the following differential equation where it is assumed that the V_Q is linear with slope $-V/t_{fv}$.

Solution of this differential equation results in the loop current which is reaches I_f at the end of interval 0 to t_{fv} I_f.

After the voltage fall, switch can be modeled as a resistor which is called ON state resistor. Figure 21 shows the model of converter after time t_{fv}. In this case, the loop current is equal to discharge current of capacitor on snubber and switch ON state resistor and therefore decreases exponentially. (Figure 22)

Plot of MTTF versus per unit capacitor is shown in Figure 23, 24, 25, 26, 27, 28, 29, 30, 31, 32, 33, 34, 35, 36, & 37. As can be seen, by increasing the capacitor value, MTTF increase at first and then decreases. Optimum capacitor value which is obtained from this method is a function of different parameters such as voltage, load current, switching frequency, rise and fall time and etc. which are required to solve complex nonlinear equation.

RELIABLE OR HIGH PERFORMANCE

Reliability does not means high quality! This is the main statement of this chapter.

Example: Passive Current Limiting

A regulated power supply is an embedded circuit; it converts unregulated AC into a constant DC. With the help of a rectifier it converts AC supply into DC. Its function is to supply a stable voltage (or less often current), to a circuit or device that must be operated within certain power supply limits. The output from the regulated power supply may be alternating or unidirectional.

Figure 23. Power loss of snubber parts

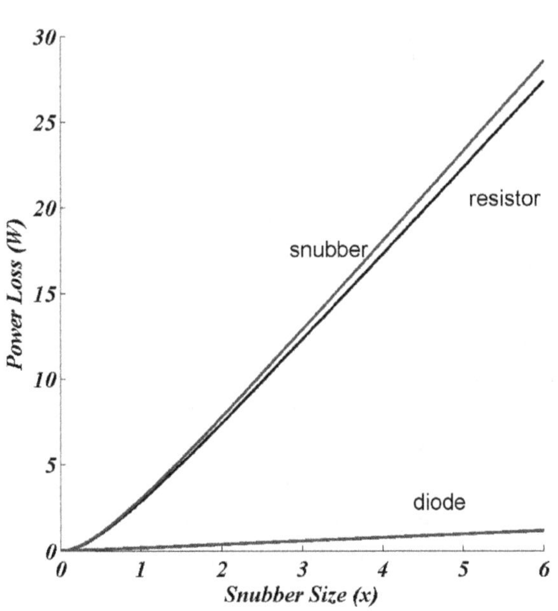

Figure 24. Various terms of power loss of snubber

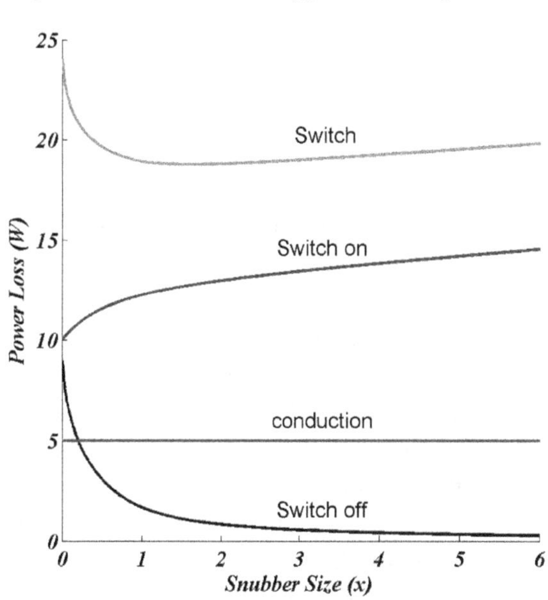

Figure 25. Variation of efficiency with snubber values

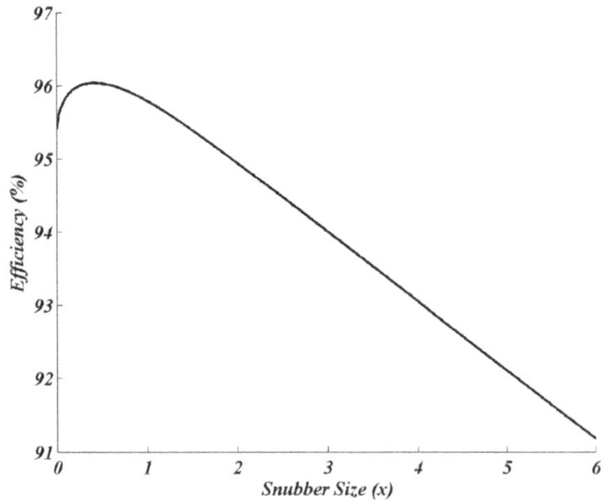

Figure 26. Variation of snubber diode power losses with snubber values

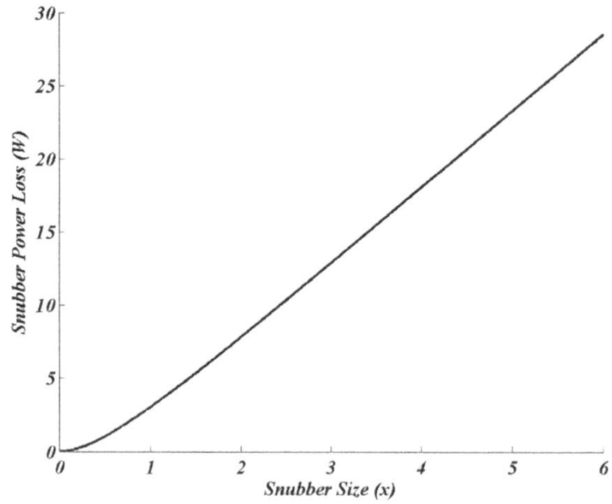

Figure 27. Variation of snubber power losses with snubber values

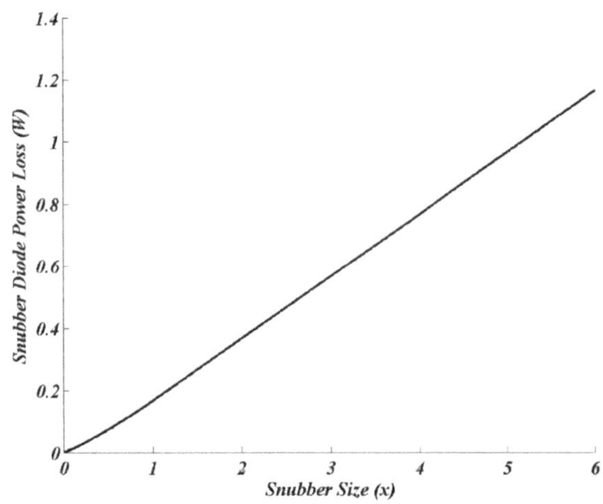

Figure 28. Variation of snubber resistor power losses with snubber values

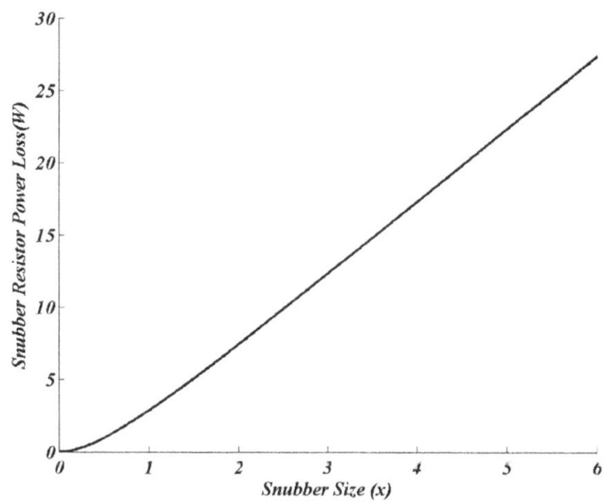

The type of stabilization used may be restricted to ensuring that the output remains within certain limits under various load conditions, or it may also include compensation for variations in its own supply source. The latter is much more common today.

Power supplies often have protection from short circuit or overload that could damage the supply or cause a fire. Fuses and circuit breakers are two commonly used mechanisms for overload protection.

A fuse contains a short piece of wire which melts if too much current flows. This effectively disconnects the power supply from its load, and the equipment stops working until the problem that caused the overload is identified and the fuse is replaced. Some power supplies use a very thin wire link soldered in place as a fuse. A circuit breaker contains an element that heats, bends and triggers a spring which shuts the circuit down. Once the element cools, and the problem is identified the breaker can be reset and the power restored.

Figure 29. Variation of switch power losses with snubber values

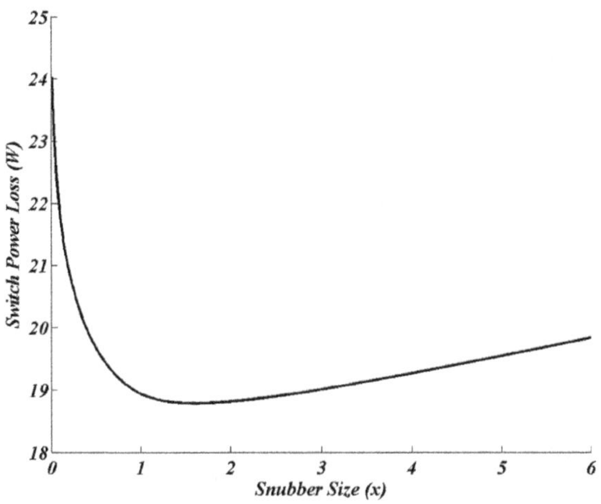

Figure 30. Variation of switch temperature with snubber values

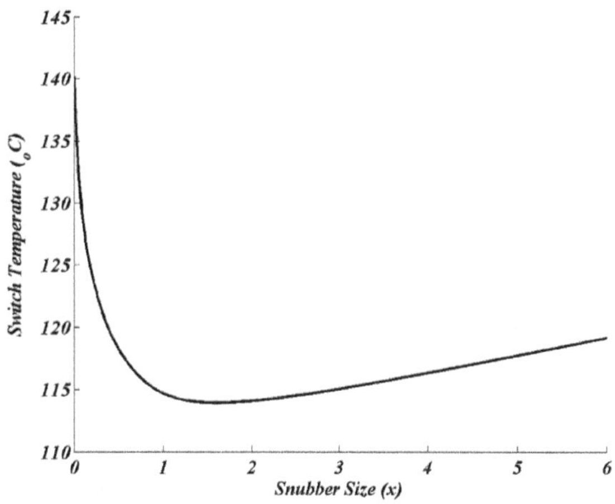

Figure 31. Variation of switching power losses with snubber values

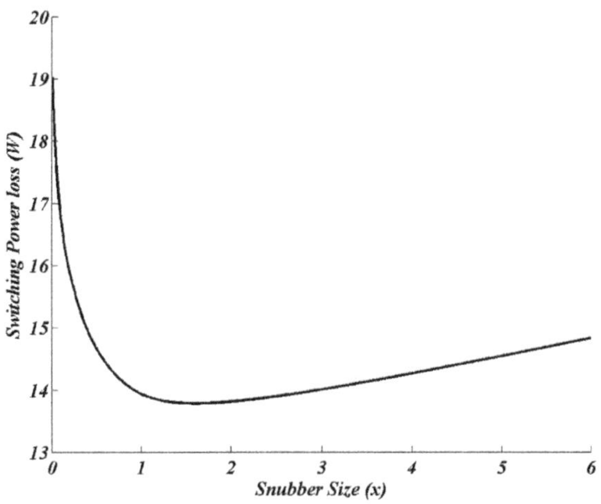

Figure 32. Variation of total power losses with snubber values

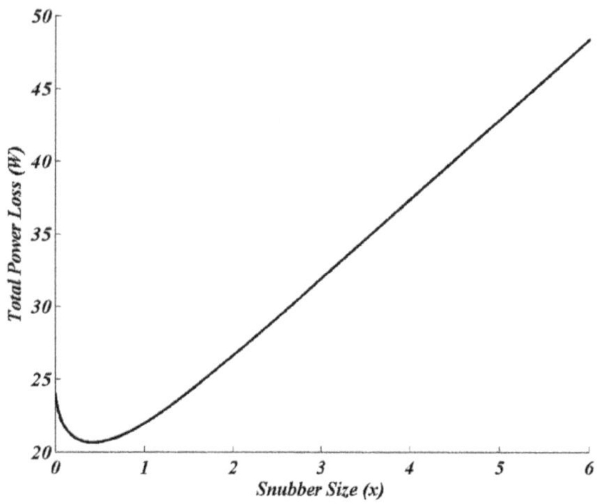

Some PSUs use a thermal cutout buried in the transformer rather than a fuse. The advantage is it allows greater current to be drawn for limited time than the unit can supply continuously. Some supplies use current limiting instead of cutting off power if overloaded. The two types of current limiting used are electronic limiting and impedance limiting. The former is common on lab bench PSUs, the latter is common on supplies of less than 3 watts output.

"Good" power supplies have a very low output impedance. This applies epecially to well designed regulated supplies. You can imagine it as an internal resistor in series with the output. Since the output varies very little with load, then this resistor must be very small.

Series limiting resistors in the multiplier assembly limit short circuit discharge currents to safe and predictable levels. Knowing what these levels are the trip point for the arc detection circuitry can be set

Figure 33. Variation of OFF switching power losses with snubber values

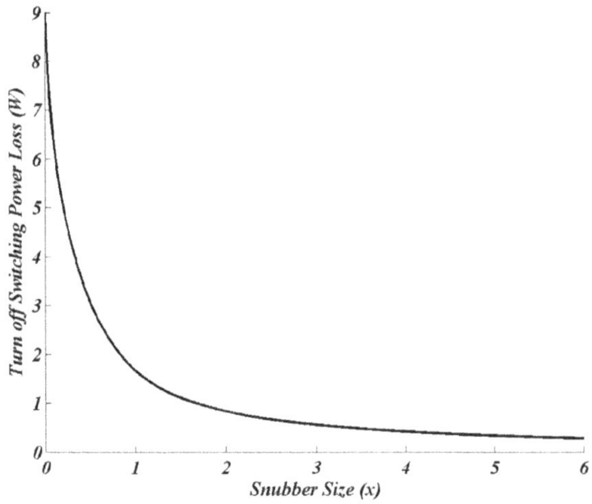

Figure 34. Variation of ON switching power losses with snubber values

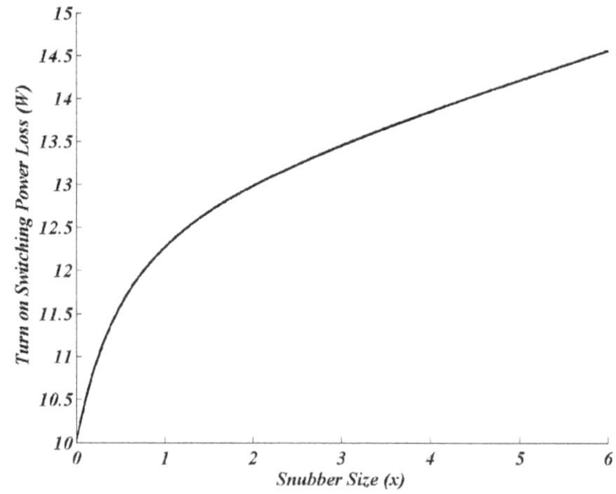

Figure 35. MTTF of switch and its snubber as a function of per unit capacitor

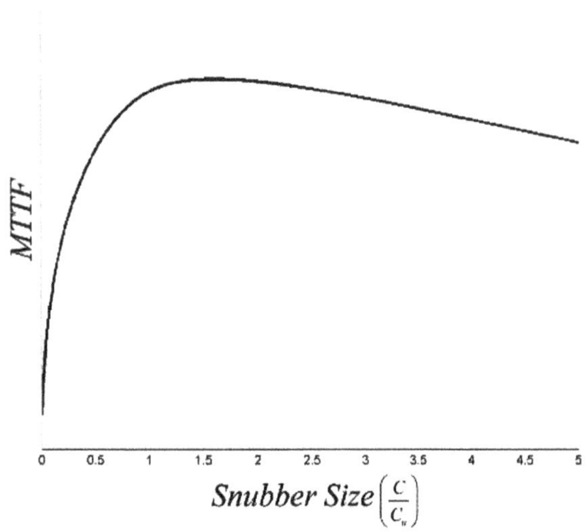

Figure 36. Experimental results for switch temperature

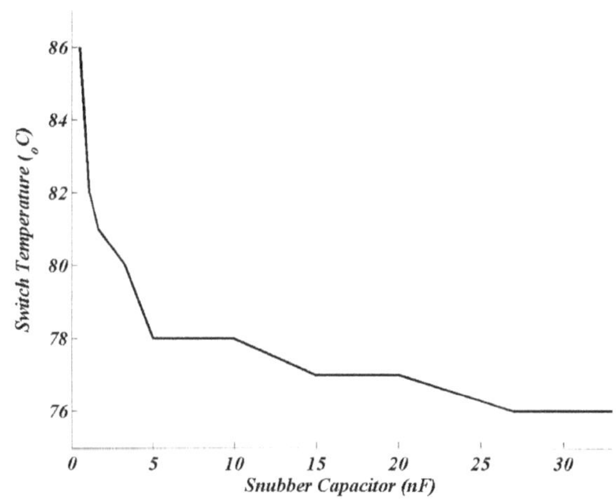

by Spellman that will protect the power supply from excessive arcing, while allowing normal power supply functionality. Figure 38 shows a converter with protected output by a series resistor.

The surge current may be generated by saturated machine cores as shown in Figure 39. In this case, series resistor is also used in series to input of the converter as shown in Figure 40. Figure 41 shows a high power resistor used for passive protection.

Figure 37. Experimental results for switch MTTF

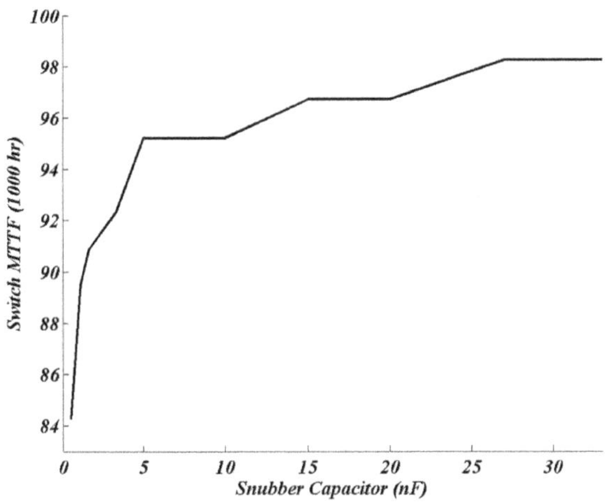

Figure 38. A power supply with series output resistor

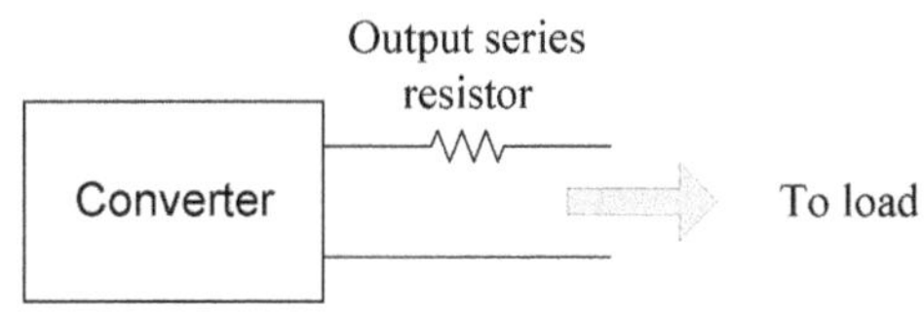

Example: Current Sharing

Passive current sharing is a method of paralleling the outputs of two or more power supplies or dc-dc converters so that they share the load near equally. This approach has strong appeal, because it is simple and inexpensive to implement, and it can be used with the latest low-cost, miniaturized power modules.

Although passive current sharing cannot be used to obtain double the current output you would get from a single converter (because one of the converters will always try to output more than half the total load current, and therefore exceed its maximum rating), it provides a highly scalable means of accommodating demands for more power, which typically result from increases in system size or functionality over time. Passive current sharing also can improve the reliability of N+1 power module configurations by reducing the stress on each converter in the system without the need for any additional active circuitry.

Unfortunately, the simplicity of this method of paralleling is not without tradeoffs — the biggest being loss of system efficiency and load regulation. Whether these tradeoffs are acceptable obviously is a design decision, and to a large degree depends upon the application. In the example presented in this article, load regulation is less of an issue because the paralleled converters are feeding an on-board intermediate bus to supply multiple point-of-load (POL) converters, which provide further down-conversion and regulation for their various silicon loads

Example: Ideal or Reliable Power Network?

By their nature and function, power distribution networks are often subject to incidental flaws. It is therefore important to install protection systems. This protection depends on the treatment of neutral. They are two methods of neutral earthing:

- Direct grounding of the neutral– The neutral system is directly connected to the earth. In such case, the short circuit current can reach 20 to 30 times the rated current.

Figure 39. Surge current in a saturated core, (up): Vi=0.5Vn, (center): Vi=Vn, (down): Vi=1.2Vn

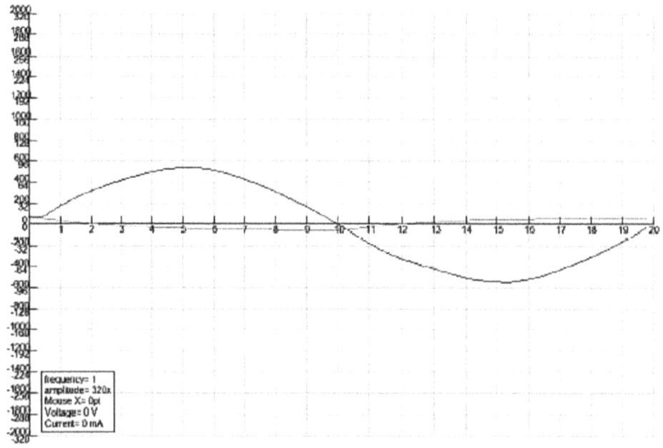

(a)

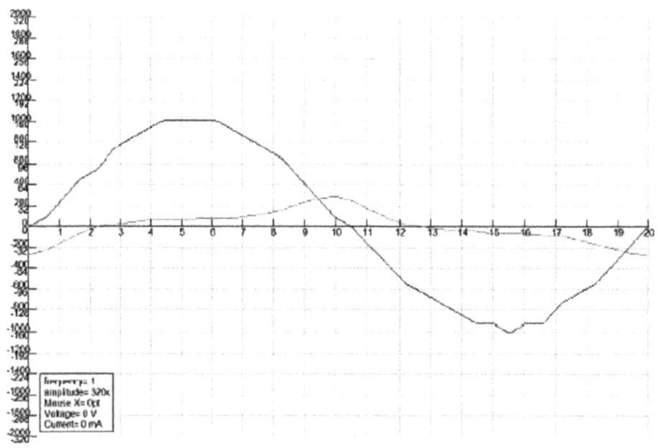

(b)

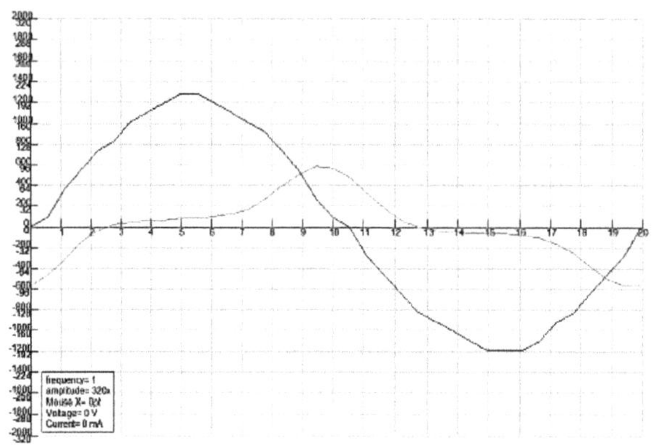

(c)

Figure 40. A power supply with series input resistor

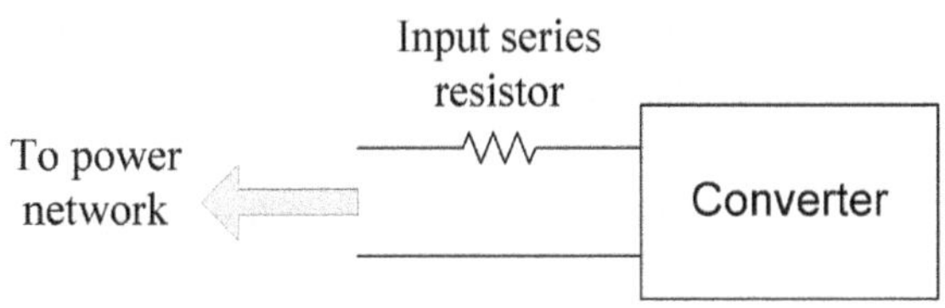

Figure 41. A high power series resistor

- Neutral grounding with limitation of fault current – This technique consists in inserting an impedance between neutral and earth. There are 3 types of impedances earthing:
  1. Earthing through resistor
  2. Earthing through reactance
  3. Suppression coil (Petersen coil)

The neutral earthing resistor is the most common, economical and sufficient solution. It is used when the neutral of the supply transformer is available and its own impedance is not enough to limit the fault current. Our Neutral grounding resistors, supplied all over the world, are known for their solidity, reliability, performance and low maintenance requirement

Distribution power systems may be solidly grounded, with one circuit conductor directly connected to an earth grounding electrode system. Alternatively, some amount of electrical impedance may be connected between the distribution system and ground, to limit the current that can flow to earth. The impedance may be a resistor, or an inductor (coil). In a high-impedance grounded system, the fault current is limited to a few amperes (exact values depend on the voltage class of the system); a low-impedance grounded system will permit several hundred amperes to flow on a fault. A large solidly-grounded distribution system may have thousands of amperes of ground fault current.

Example: Distribution Transformer

All the flux in transformer will not be able to link with both the primary and secondary windings. A small portion of flux will link either winding but not both. This portion of flux is called leakage flux. Due to this leakage flux in transformer, there will be a self - reactance in the concerned winding. This self-reactance of transformer is alternatively known as leakage reactance of transformer. This self - reactance associated with resistance of transformer is impedance. Due to this impedance of transformer, there will be voltage drops in both primary and secondary transformer windings. Leakage inductance depends on the geometry of the core and the windings. Voltage drop across the leakage reactance results in often undesirable supply regulation with varying transformer load. But it can also be useful for harmonic isolation (attenuating higher frequencies) of some loads.

Leakage inductance can be an undesirable property, as it causes the voltage to change with loading. In many cases it is useful. Leakage inductance has the useful effect of limiting the current flows in a transformer (and load) without itself dissipating power (excepting the usual non-ideal transformer losses). Transformers are generally designed to have a specific value of leakage inductance such that the leakage reactance created by this inductance is a specific value at the desired frequency of operation.

Commercial transformers are usually designed with a short-circuit leakage reactance impedance of between 3% and 10%. If the load is resistive and the leakage reactance is small (<10%) the output voltage will not drop by more than 0.5% at full load, ignoring other resistances and losses.

Figure 42 shows transformers with separated primary and secondary windings. This separation leads to increasing the leakage inductance of the transformer but it reduces the fault short circuit current of the transformer.

Figure 42. Separation of transformer windings, (up), separation in length, (down): separation in width

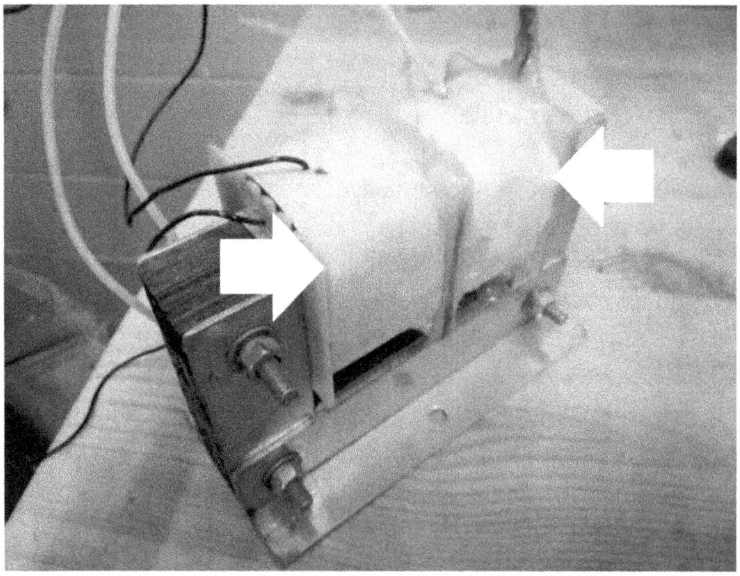

(a)

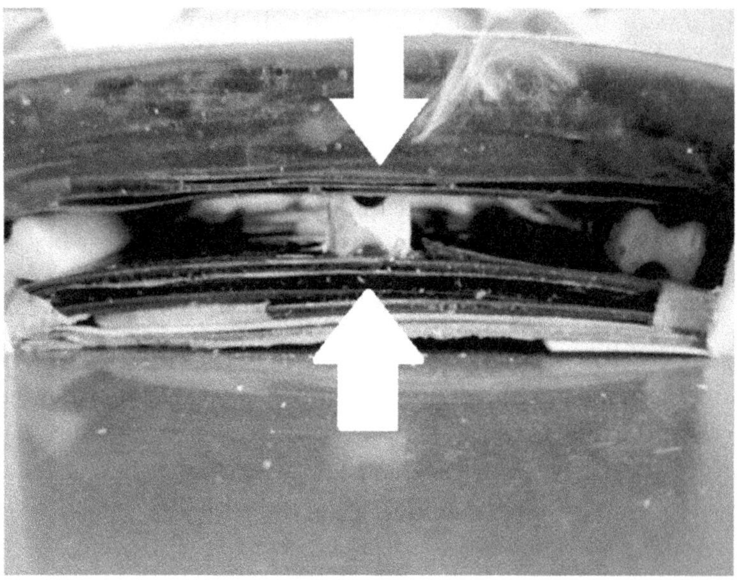

(b)

Figure 43. Short circuit characteristic of an AC generator

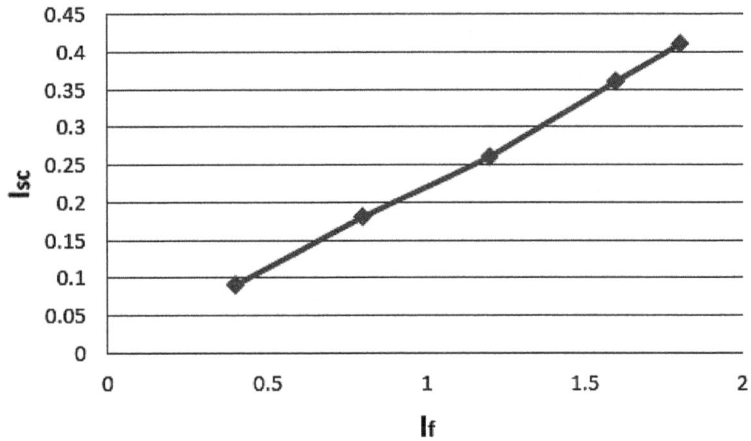

Example: Generator Series Impedance

Synchronous Reactance is an equivalent series per-phase impedance with current limiting effects. Figure 43 shows the short circuit current of a small synchronous generator.

Example: Resistor in Rotor of Induction

Induction motor inrush current is a common problem in starting of this types of motors as shown in Figure 44. Many faults of induction motor occur in the starting. Rotor external series resistor reduces the efficiency of the motor but it controls the startup inrush current as shown in Figure 45.

Example: Switching vs. Linear Power Supply

Advantages and Disadvantages

- **Linear:** One advantage of linear power supplies may be familiarty, because they have been available for many years. They are known to be relatively noise-free and reasonably reliable. They are generally easy to design and fairly inexpensive to manufacture.
  Because of the large transformers required, linear power supplies are generally heavy, which may be either an advantage or a disadvantage, depending on the need to balance weight distribution in a given application. As a general rule of thumb, a 16V-output linear power supply weighs about one pound per ampere. A possible disadvantage of linear power supplies relates to the power transistor used to regulate the load. Because the power transistor operates in its linear region, and all the output current must pass through it, it requires large heat sinks to dissipate energy loss. (Recall that the power transistor is in series with the load and acts as a variable resistor.) Except in rare instances where heat is wanted to warm interior space, the inefficiency of linear power supplies — 50% — has to be considered a disadvantage.
- **Switching power supplies:** Although switching power supplies have been available for a number of years, higher production costs, compared to linear power supplies, have limited their use

Figure 44. Starting transient of an induction motor

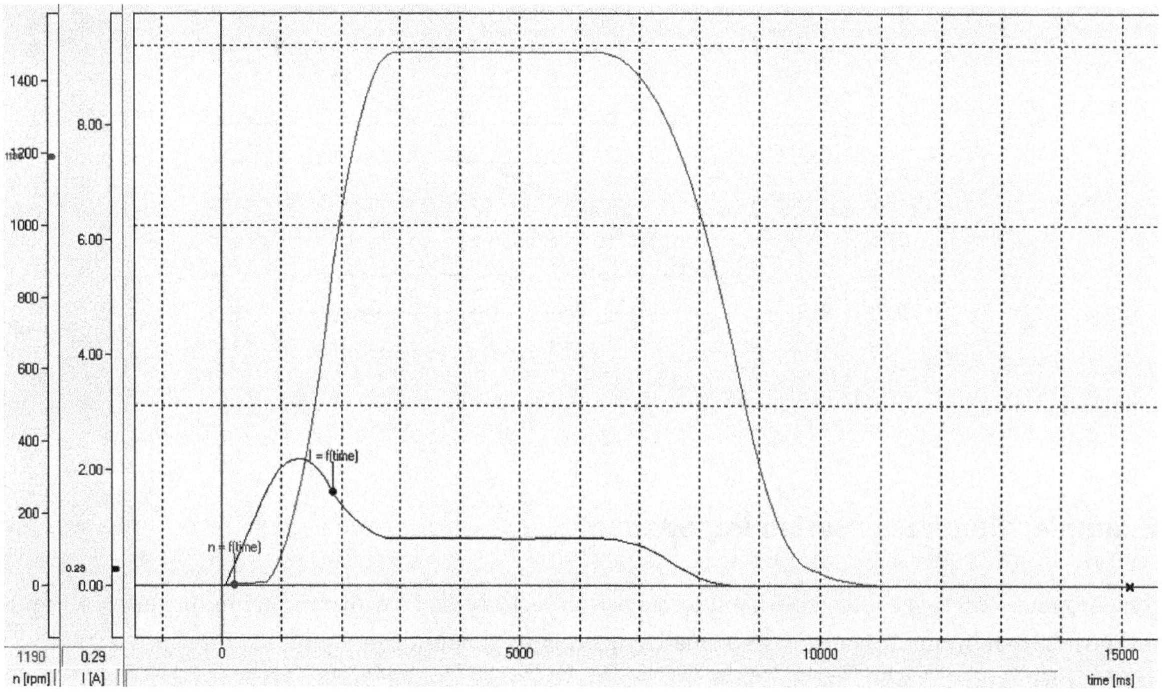

in some applications. Early switching power supplies used discrete components to control pulse width, and transistors instead of MOSFETs as main switch components. As a result, the disadvantages of switching power supplies once included uneven reliability and radiated EMI (electrical noise). Although they were known to be noisy, unreliable and difficult to mass produce, switching power supplies had the advantage of being lighter and smaller than their linear counterparts. In the last few years, big improvements in PWM and MOSFET design have been made. Today, when all design considerations have been taken into account, switching power supplies are highly reliable and virtually noise-free. Production costs have come down because application-specific components are being designed for use in switching power supplies.

Switching power supplies are about 80%-90% efficient. Higher efficiency usually is an advantage, because heat normally is considered to be wasted energy (at the least) and potentially damaging to nearby electronic components

DESIGN FOR RELIABILITY

Reliability design begins with the development of a (system) model. Reliability and availability models use block diagrams and fault trees to provide a graphical means of evaluating the relationships between different parts of the system. These models may incorporate predictions based on failure rates taken from historical data. While the (input data) predictions are often not accurate in an absolute sense, they are valuable to assess relative differences in design alternatives. Maintainability parameters, for example MTTR, are other inputs for these models.

Figure 45. Effect of rotor resistance on starting torque of an induction motor, (up): without rotor external resistance, (center): small rotor external resistance, (down): large rotor external resistance

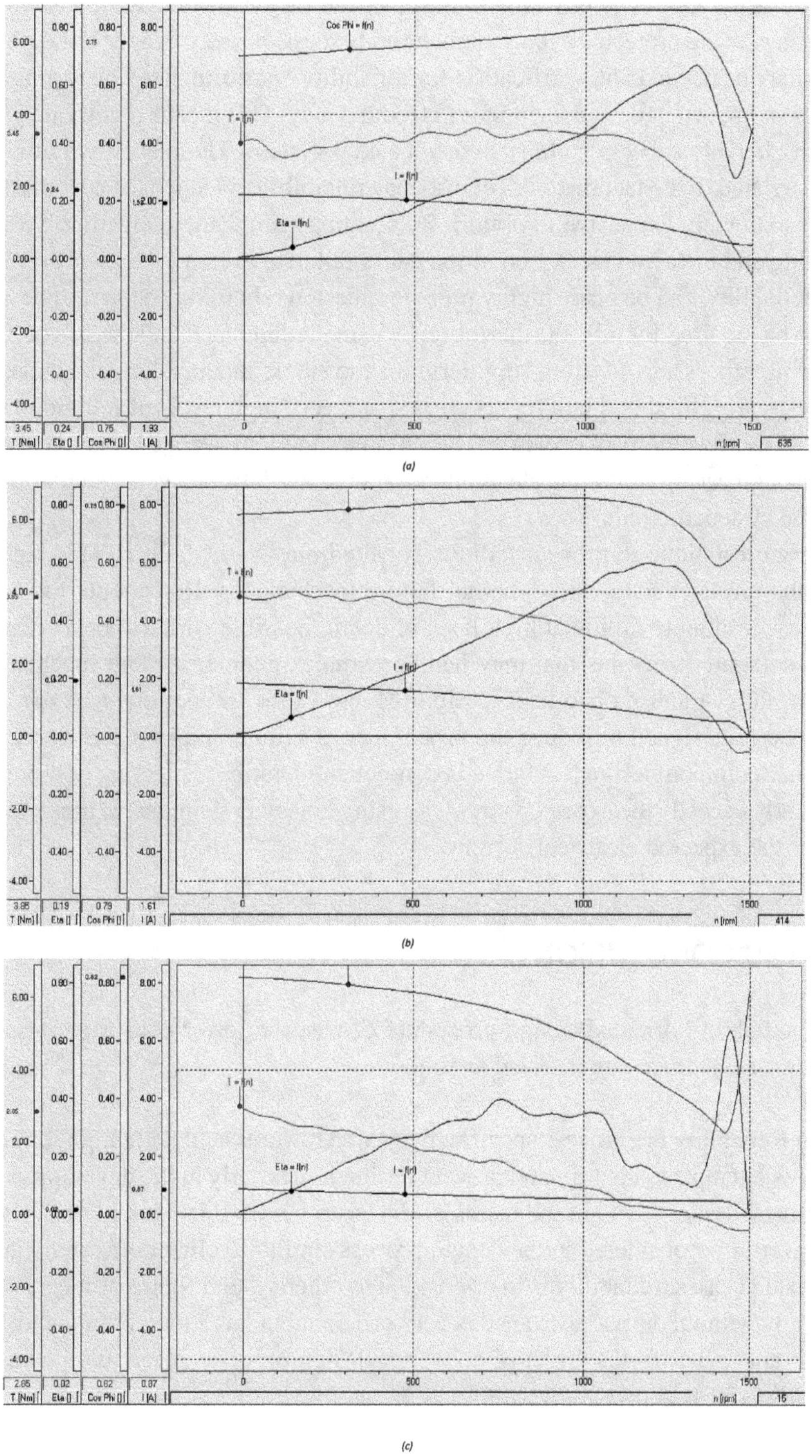

The most important fundamental initiating causes and failure mechanisms are to be identified and analyzed with engineering tools. A diverse set of practical guidance and practical performance and reliability requirements should be provided to designers so they can generate low-stressed designs and products that protect or are protected against damage and excessive wear. Proper Validation of input loads (requirements) may be needed and verification for reliability "performance" by testing may be needed

One of the most important design techniques is redundancy. This means that if one part of the system fails, there is an alternate success path, such as a backup system. The reason why this is the ultimate design choice is related to the fact that high confidence reliability evidence for new parts / items is often not available or extremely expensive to obtain. By creating redundancy, together with a high level of failure monitoring and the avoidance of common cause failures, even a system with relative bad single channel (part) reliability, can be made highly reliable (mission reliability) on system level. No testing of reliability has to be required for this. Furthermore, by using redundancy and the use of dissimilar design and manufacturing processes (different suppliers) for the single independent channels, less sensitivity for quality issues (early childhood failures) is created and very high levels of reliability can be achieved at all moments of the development cycles (early life times and long term). Redundancy can also be applied in systems engineering by double checking requirements, data, designs, calculations, software and tests to overcome systematic failures.

Another design technique to prevent failures is called physics of failure. This technique relies on understanding the physical static and dynamic failure mechanisms. It accounts for variation in load, strength and stress leading to failure at high level of detail, possible with use of modern finite element method (FEM) software programs that may handle complex geometries and mechanisms like creep, stress relaxation, fatigue and probabilistic design (Monte Carlo simulations / DOE). The material or component can be re-designed to reduce the probability of failure and to make it more robust against variation. Another common design technique is component derating: Selecting components whose tolerance significantly exceeds the expected stress, as using a heavier gauge wire that exceeds the normal specification for the expected electrical current

SUMMARY AND CONCLUSION

In this chapter, reliability oriented design procedure of electric power converters was described. The main topics of this chapter are summarized as follows:

1. Design for Reliability is a process specifically geared towards achieving high long-term reliability. This process attempts to identify and prevent design issues early in the development phase, instead of having these issues found in the hands of the customer. In this method of design, reliability of power converter is considered in the design process similar to other converter figures of merit.
2. In some special cases, reliability is in opposition to other system figure of merit. As an example, a power supply without output resistance is very efficient but has a high output short circuit current. It is important to clarify the target of converter design process. A reliable power converter may have a poor performance.

Considering reliability as a figure of merit, the designer is forced to use some tools for achieving the desired reliability. In the next chapters, we describe the tools for reliability improvement in both design and operating process.

REFERENCES

Hasanzadeh, A., Zolghadri, M. R., Kaboli, S. H., & Homaifar, A. (2003). A genetic algorithm based programmed PWM optimum switching pattern calculation. In *Proceedings of The Fifth International Conference on Power Electronics and Drive Systems* (pp.1081-1085). IEEE. doi:10.1109/PEDS.2003.1283122

Kaboli, S., Haddadi, A., & Khaligh, A. (2008). Axiomatic design of step down DC/DC converter. In *Proceedings of IEEE International Conference on Industrial Technology* (pp.1-5). IEEE. doi:10.1109/ICIT.2008.4608313

Kaboli, S., & Zolghadri, M. R. (2003). Effect of motor speed on the optimum operating point of direct torque controlled induction motor. In *Proceedings of Sixth International Conference on Electrical Machines and Systems* (pp.619-622). IEEE.

Kaboli, S., Zolghadri, M. R., & Emadi, A. (2003). Hysteresis band determination of direct torque controlled induction motor drives with torque ripple and motor-inverter loss considerations. In *Proceedings of IEEE 34th Annual Power Electronics Specialist Conference* (pp. 1107-1111). IEEE. doi:10.1109/PESC.2003.1216604

Kaboli, S., Zolghadri, M. R., & Khaligh, A. (2008). Reliability Improvement of Industrial Drives Using Multi-Objective Optimization. In *Proceedings of IEEE Industry Applications Society Annual Meeting* (pp.1-6). IEEE. doi:10.1109/08IAS.2008.198

Kaboli, S. H., Zolghadri, M. R., & Homaifar, A. (2003). Effects of sampling time on the performance of direct torque controlled induction motor drive. In *Proceedings of IEEE International Symposium on Industrial Electronics* (pp.1049-1052). IEEE. doi:10.1109/ISIE.2003.1267968

Nasirian, V., Kaboli, S., & Davoudi, A. (2012). Output power maximization and optimal symmetric freewheeling excitation for Switched Reluctance Generators. In *Proceedings of IEEE/IAS 48th Industrial & Commercial Power Systems Technical Conference* (pp.1-10). IEEE. doi:10.1109/ICPS.2012.6229595

Rajaei, A. H., Kaboli, S., & Emadi, A. (2008). Sliding-mode control of z-source inverter. In *Proceedings of 34th Annual Conference of IEEE Industrial Electronics* (pp. 947-952). IEEE.

ADDITIONAL READING

Abdelsalam, A. K., Massoud, A. M., Ahmed, S., & Enjeti, P. (2011). High-Performance Adaptive Perturb and Observe MPPT Technique for Photovoltaic-Based Microgrids. *IEEE Transactions on Power Electronics*, *26*(4), 1010–1021. doi:10.1109/TPEL.2011.2106221

Araujo, S. V., Zacharias, P., & Mallwitz, R. (2010). Highly Efficient Single-Phase Transformerless Inverters for Grid-Connected Photovoltaic Systems. *IEEE Transactions on Industrial Electronics*, *57*(9), 3118–3128. doi:10.1109/TIE.2009.2037654

Băjenescu, T. I., & Bâzu, M. I. (1999). *Reliability of Electronic Components*. Berlin, Germany: Springer. doi:10.1007/978-3-642-58505-0

Benbouzid, M. E. H., Vieira, M., & Theys, C. (1999). Induction motors' faults detection and localization using stator current advanced signal processing techniques. *IEEE Transactions on Power Electronics*, *14*(1), 14–22. doi:10.1109/63.737588

Borup, U., Blaabjerg, F., & Enjeti, P. N. (2001). Sharing of nonlinear load in parallel-connected three-phase converters. *IEEE Transactions on Industry Applications*, *37*(6), 1817–1823. doi:10.1109/28.968196

Crowe, D. (2001). *Design for reliability*. NY, USA: John Wiley & Sons. doi:10.1201/9781420040845

Jahns, T. M. (1980). Improved Reliability in Solid-State AC Drives by Means of Multiple Independent Phase Drive Units. *IEEE Transactions on Industry Applications*, *IA-16*(3), 321–331. doi:10.1109/TIA.1980.4503793

Kastha, D. K., & Bose, B. K. (1994). Investigation of fault modes of voltage-fed inverter system for induction motor drive. *IEEE Transactions on Industry Applications*, *30*(4), 1028–1038. doi:10.1109/28.297920

Kermarrec, A.-M., Massoulie, L., & Ganesh, A. J. (2003). Probabilistic reliable dissemination in large-scale systems. *IEEE Transactions on Parallel and Distributed Systems*, *14*(3), 248–258. doi:10.1109/TPDS.2003.1189583

Matsui, N. (1996). Sensorless PM brushless DC motor drives. *IEEE Transactions on Industrial Electronics*, *43*(2), 300–308. doi:10.1109/41.491354

McPherson, J. W. (2010). *Reliability Physics and Engineering*. Berlin, Germany: Springer. doi:10.1007/978-1-4419-6348-2

Mecrow, B. C., Jack, A. G., Haylock, J. A., & Coles, J. (1996). Fault-tolerant permanent magnet machine drives. *IEE Proceedings. Electric Power Applications*, *143*(6), 437–442. doi:10.1049/ip-epa:19960796

Peng, F. Z., Li, H., Su, G.-J., & Lawler, J. S. (2004). A new ZVS bidirectional DC-DC converter for fuel cell and battery application. *IEEE Transactions on Power Electronics*, *19*(1), 54–65. doi:10.1109/TPEL.2003.820550

Tuladhar, A., Hua Jin, , Unger, T., & Mauch, K. (2000). Control of parallel inverters in distributed AC power systems with consideration of line impedance effect. *IEEE Transactions on Industry Applications*, *36*(1), 131–138. doi:10.1109/28.821807

Wai, R. J., Wang, W. H., & Lin, Ch. Y. (2008). High-Performance Stand-Alone Photovoltaic Generation System. *IEEE Transactions on Industrial Electronics*, *55*(1), 240–250. doi:10.1109/TIE.2007.896049

Way Kuo, & Prasad, V.R. (2000). An annotated overview of system-reliability optimization. *IEEE Transactions on Reliability*, *49*(2), 17187.

Yilmaz, M., & Krein, P. T. (2013). Review of Battery Charger Topologies, Charging Power Levels, and Infrastructure for Plug-In Electric and Hybrid Vehicles. *IEEE Transactions on Power Electronics*, *28*(5), 2151–2169. doi:10.1109/TPEL.2012.2212917

KEY TERMS AND DEFINITIONS

Efficiency: A system index about useful output power of a system.

Reliability: An index about rate of failure in a system.

Optimization: A trade-off between set-point of system parameters and variables to reach desired operating condition for all parts of system.

System Performance Index: Quantities for characterizing a system performance such as efficiency and power density.

Figure of Merit: An index for system performance classification.

Chapter 7
Stress Reduction

ABSTRACT

After evaluation of reliability in the previous chapters and its consideration as a converter figure of merit, in this and the next chapters, guidelines for improvement of reliability are presented. These methods are used in both design and operation process of the converter. The focus of this chapter is on the component stress reduction in the design process. Based on background of chapter two, reliability of a converter increases if it operates at a set point with low stress. It is assumed that the converter is under design process or operates without fault. The methods for reliability improvement in faulty converters are discussed in the next chapters. In this chapter, methods for reducing electric field are described at both system and printed circuit board level. Low temperature operating conditions for an electric power converter are described and tools for this goal are presented. Series connection for voltage sharing and parallel connection for current sharing is explained. Novel control methods of power converters for reducing the complexity and reliable operation are presented. Control of inrush current as a typical transient problem in electric power converters is presented. Methods for preventing the over stress condition on the components in faulty cases are described. Techniques for reducing mechanical and environmental stress are expressed. Mechanical dampers for preventing the high amplitude vibration and insulating colors against humidity are presented. Industrial and real samples are presented to demonstrate application of the proposed methods.

INTRODUCTION: STRESS ON THE COMPONENTS

This chapter is starting chapter of the second part of this book. By now, we studied reliability calculation and methods of reliability testing. In the next five chapters of this book, we present the methods for reliability improvement.

Following the materials presented in Chapter 2 about mechanisms of fault in power converters, any method that reduces those failure factors can be considered as a technique for reliability improving.

The failure mechanism of a converter is started immediately after starting. In the beginning, the converter operates normally but under stress of failure factors:

DOI: 10.4018/978-1-4666-9429-3.ch007

- Power losses in the converter causes to temperature rise in various parts of the converter (Kaboli, Vahdati-Khajeh, Zolghadri, & Homaifar, 2005),
- Applied voltage causes to apply an electric field to insulators,
- Mechanical forces leads to vibrations
- Environmental factors

These factors acts from the beginning of converter application. In long term, they cause to age the converter and its failure. The time interval for changing a stress factor to a failure factor is directly related to value of stress. Higher stress leads to shorter time to failure and vice versa. Thus, the first method of reliability improvement, is reducing the stress factors on the converter. This is the subject of the current chapter. Figure 1 shows the state of this chapter in the flowchart of the book.

Figure 1. State of chapter 7 in the flowchart of the book

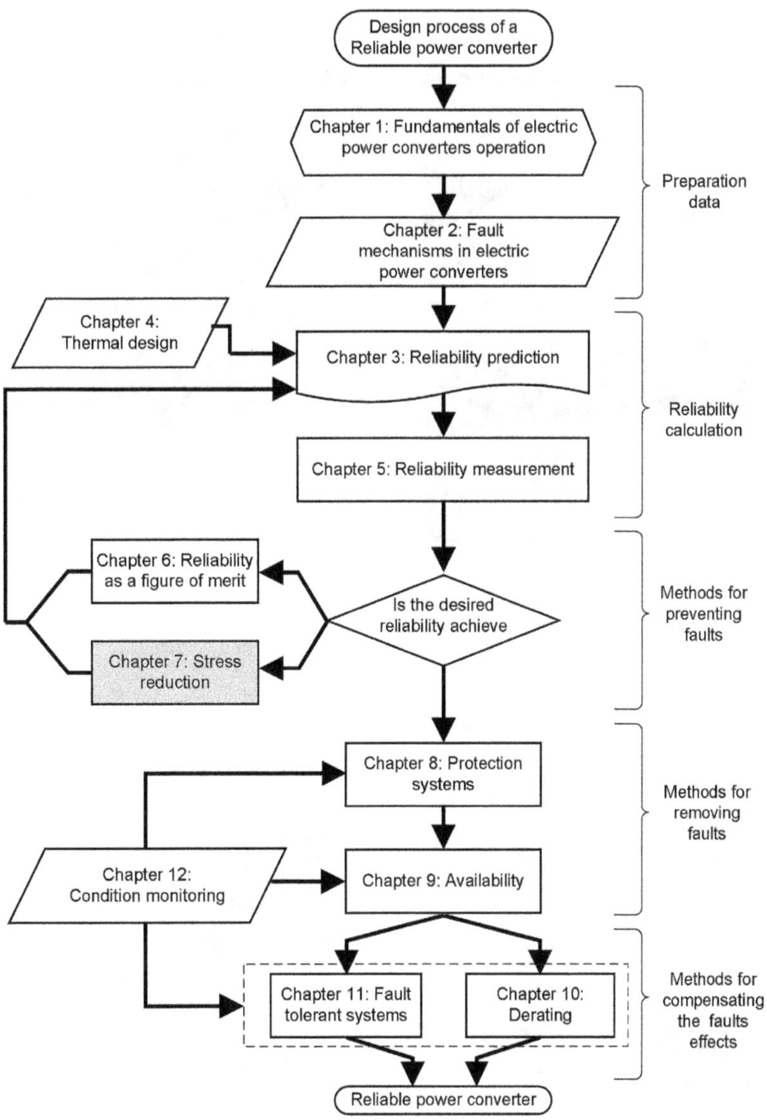

A cascading failure is a failure in a system of interconnected parts in which the failure of a part can trigger the failure of successive parts. Such a failure may happen in many types of systems, including power transmission, computer networking, finance and bridges.

Some of the methods for reliability improvement act in hardware level. Aim of these methods is usually reduction of hot spot temperature or reduction of electric field applied to devices.

Figures 2 to 4 show some methods for reduction of electrical field stress on equipments. Figure 2 shows application of insulator spacer between primary and secondary of a transformer. Figure 3 shows application of isolator base for inductors in high voltage application. Figure 4 shows a three phase transformer with spacing between high voltage windings.

Figure 2. Spacing between primary and secondary windings of a transformer

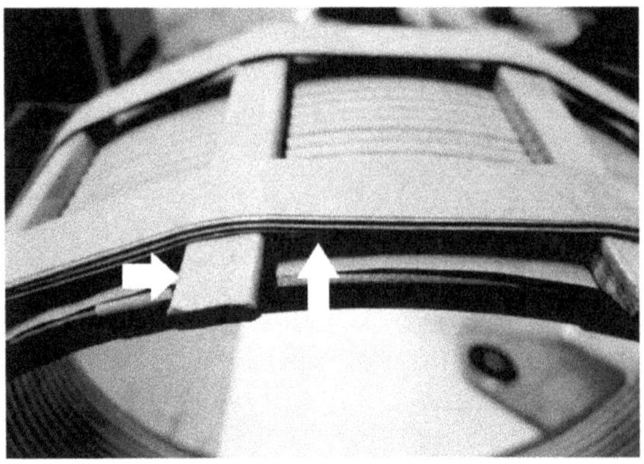

Figure 3. Isolator base for high voltage inductors

Figure 4. Spacing between different phases of a 3-phase transformer

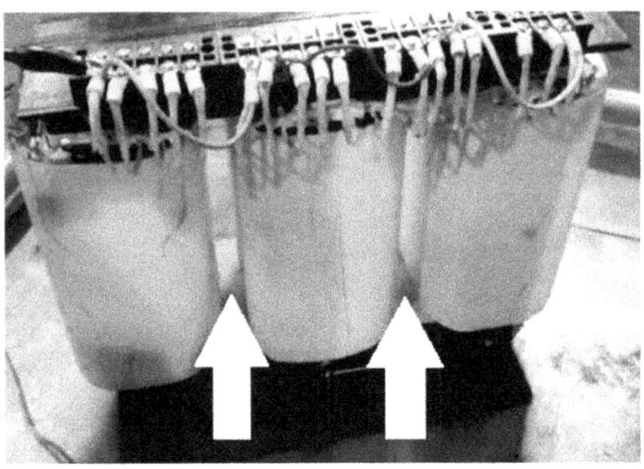

Figure 5. A heat sink for double side cooling of a thyristor

THERMAL STRESS FACTORS

Thermal Management

In electronic systems, a heat sink is a passive heat exchanger that cools a device by dissipating heat into the surrounding medium. In computers, heat sinks are used to cool central processing units or graphics processors. Heat sinks are used with high-power semiconductor devices such as power transistors where the heat dissipation ability of the basic device is insufficient to moderate its temperature (Kaboli, Zolghadri, Vahdati-Khajeh, & Homaifar, 2005).

A heat sink is designed to maximize its surface area in contact with the cooling medium surrounding it, such as the air. Air velocity, choice of material, protrusion design and surface treatment are factors that affect the performance of a heat sink. Heat sink attachment methods and thermal interface materials also affect the die temperature of the integrated circuit. Thermal adhesive or thermal grease improve the heat sink's performance by filling air gaps between the heat sink and the device.

Figure 5 shows a thyristor which is mounted on between two heat sink for double side cooling.

A heat sink transfers thermal energy from a higher temperature device to a lower temperature fluid medium. The fluid medium is frequently air, but can also be water, refrigerants or oil. If the fluid medium is water, the heat sink is frequently called a cold plate. In thermodynamics a heat sink is a heat reservoir that can absorb an arbitrary amount of heat without significantly changing temperature. Practical heat sinks for electronic devices must have a temperature higher than the surroundings to transfer heat by convection, radiation, and conduction 2.

Figure 6 shows transient thermal impedance of a thyristor. It can be seen that the thermal impedance is very low in short time. This means that the device can dissipate more power loss than its nominal value in steady state. Figure 7 shows correct installation method of fans for an equipment. In this figure, the fans work in opposite direction to provide a proper ventilation for the equipment. Figure 8 shows a thermal switch which is implemented on an interface metal base. As the switch has a flat surface, it can not be mounted on the round surfaces and it leads to poor thermal connection. To solve this problem, a metal interface is used with round surface in one side and flat surface in another side. Figures 9 and 10 shows a flexible thermal pad for electrical isolation of heat sink. Figure 11 shows application of internal

Figure 6. Transient thermal impedance of a thyristor, (Source: On-semiconductor Co. with permission)

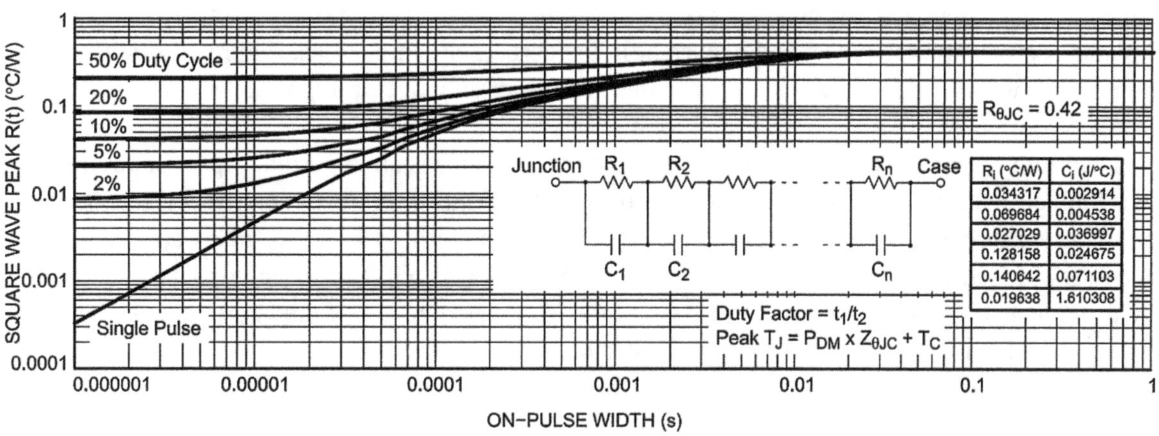

fans for proper air ventilation inside the equipment. Figures 12 and 13 show various types of heat sink. Figure 14 shows dependence of thermal resistance of a heat sink to its length. It can be seen that the minimum thermal resistance of a heat sink is saturated for long length. Figure 15 shows inside an IGBT with thermally conductive grease for proper heat transfer. Figure 16 shows an optimum thickness for the materials used for heat transferring. Figure 17 shows a sample of this materials used for an IGBT. Figure 18, 19, 20, and 21 shows a heat sink without electrical isolation to switch.

Thermally Conductive Tape

Thermally conductive tape is one of the most cost-effective heat sink attachment materials. It is suitable for low-mass heat sinks and for components with low power dissipation. It consists of a thermally conductive carrier material with a pressure-sensitive adhesive on each side.

Thermal Interface Materials

Thermal contact resistance occurs due to the voids created by surface roughness effects, defects and misalignment of the interface. The voids present in the interface are filled with air. Heat transfer is therefore due to conduction across the actual contact area and to conduction (or natural convection) and radiation across the gaps. If the contact area is small, as it is for rough surfaces, the major contribution to the resistance is made by the gaps. To decrease the thermal contact resistance, the surface roughness can be decreased while the interface pressure is increased. However, these improving methods are not always practical or possible for electronic equipment. Thermal interface materials are a common way to overcome these limitations.

Properly applied thermal interface materials displace the air that is present in the gaps between the two objects with a material that has a much-higher thermal conductivity. Air has a thermal conductivity of 0.022 W/m•K while TIMs have conductivities of 0.3 W/m•K and higher.

Figure 22 shows internal and external fans which are mounted in a cabinet for proper air ventilation. Figures 23 and 24 shows a stack of thyristors with forced air cooling. Figure 25 shows two alternative for air path direction in an equipment case. The air path shown in Figure 25(a) does not affect all parts in the case and it is by passed into a short path. However, the air path shown in Figure 25(b) passes through all parts in the case. Figure 26 shows a group of high power resistors and an IGBT which are mounted on a heat sink. The heat sink is cooled with a fan. Figures 27 and 28 shows the method for eliminating the eddy current loss in screw connections of transformer core laminations. In this method, an electrical isolated screw is used to prevent eddy current in the corners of laminations. Figure 29 shows application of Litz wire to reduce the skin effect in the winding of an inductor. Figure 30 shows a high power rectifier consists of two main transformer for current sharing.

Determining appropriate trace sizes for current requirements is an important aspect of circuit board development. Since copper is not a perfect conductor, it presents a certain amount of impedance to current flowing through it, and some of the energy is lost in the form of heat. For many applications it is necessary to predict the temperature rise caused by this loss, which has been accomplished traditionally by using a chart created over fifty years ago by the National Bureau of Standards, or by using one of several calculators based upon it. The chart shows the relationship between current, conductor temperature rise and conductor cross-sectional area. If any two of these are known, the third can be approximated.

Figure 7. Correct direction of fans in an equipment

Figure 8. A thermal switch and its metal medium

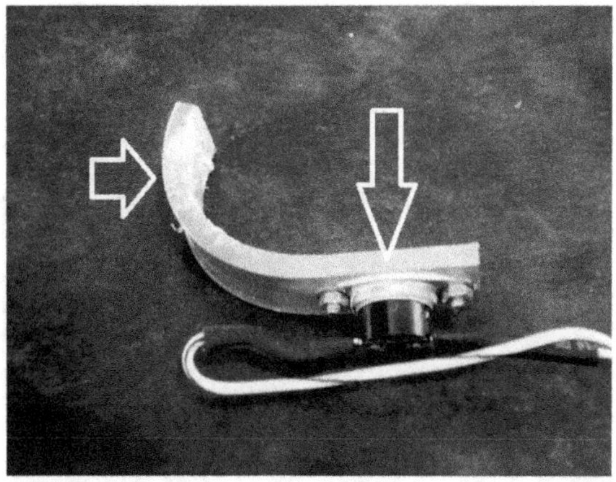

Figure 9. Flexible silicon pad

Figure 10. Flexibility of a silicon pad

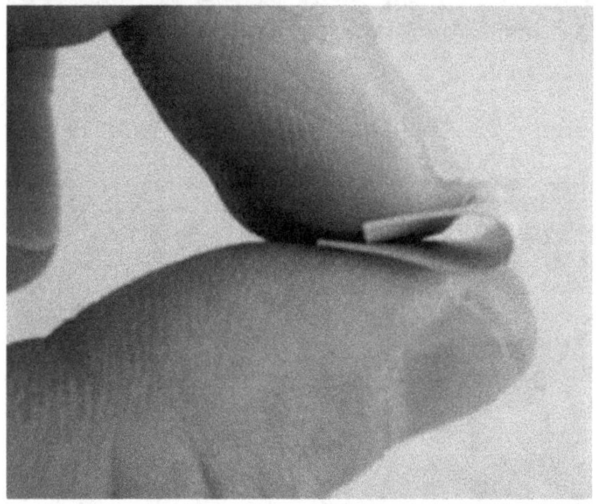

Figure 11. Internal fans for heatsinks inside a power supply

Figure 12. Various types of heatsink

Figure 13. A high performance heatsink

Figure 14. Variation of thermal resistance of heatsink with length

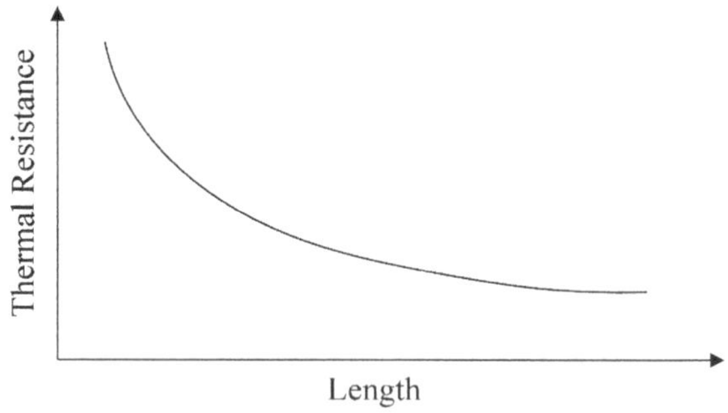

Figure 15. Application of jell as thermal medium

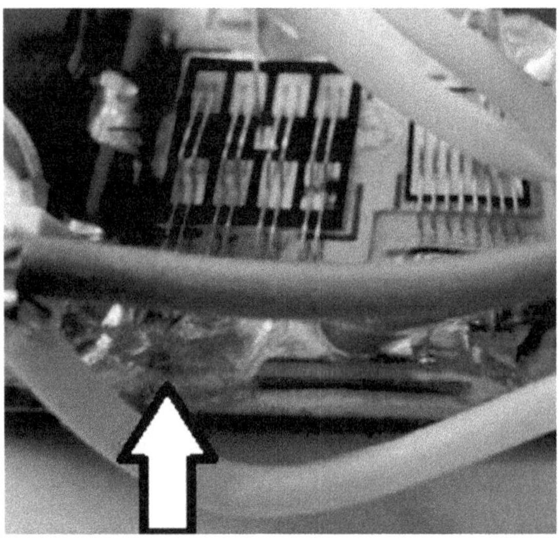

Figure 16. Variation of thermal resistance of medium grease with thickness

Figure 31 shows permissible current in traces of a PCB. The trace current increases if higher temperature rise is acceptable.

Example: Force Cooling in Transformers

The main source of heat generation in transformer is its copper loss. Although there are other factors contribute heat in transformer such as hysteresis & eddy current losses but contribution of IR loss dominate

Figure 17. Application of thermal grease on an IGBT

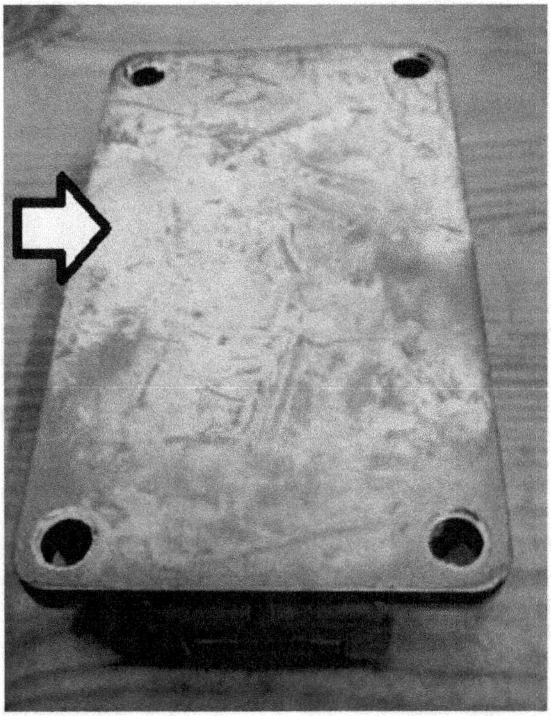

Figure 18. A rectifier with non-isolated heat-sinks

Figure 19. Forced air cooling in a rectifier with non-isolated heatsinks

them. If this heat is not dissipated properly, the temperature of the transformer will rise continually which may cause damages in paper insulation and liquid insulation medium of transformer. So it is essential to control the temperature with in permissible limit to ensure the long life of transformer by reducing thermal degradation of its insulation system 15.

Figure 20. Air duct in a rectifier with non-isolated heatsinks

Figure 21. One of the heat sinks without galvanic isolation with high voltage switch in a rectifier with non-isolated heatsinks

Figure 22. Direction of Fans in a cabinet

Figure 23. Application of fan on a heat sink

Some of transformers are cooled with oil. Here natural convectional flow of hot oil is utilized for cooling. In convectional circulation of oil, the hot oil flows to the upper portion of the transformer tank and the vacant place is occupied by cold oil. This hot oil which comes to upper side, will dissipate heat in the atmosphere by natural conduction, convection & radiation in air and will become cold. In this way the oil in the transformer tank continually circulate when the transformer put into load. As the rate of

Figure 24. Application of fan on a heat sink

Figure 25. Two various patches for air flow in a power supply box, (up) incorrect air path, (down): correct air path

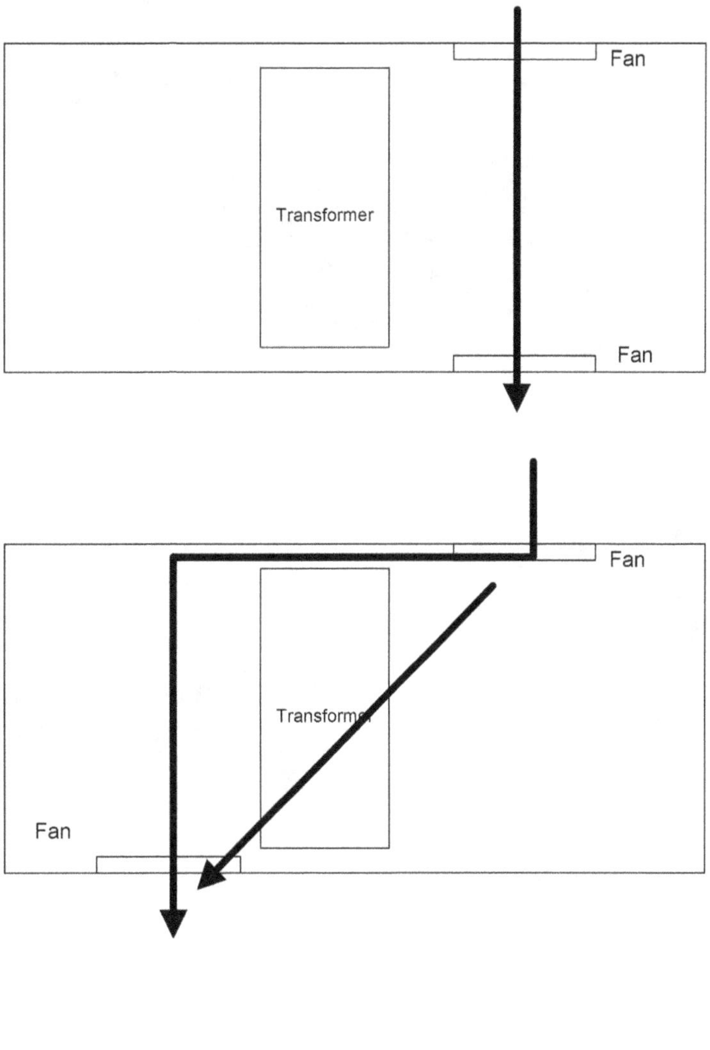

Figure 26. A heat sink for an IGBT with thermostat protection

Figure 27. Isolation of laminated sheets in the core of a transformer

Figure 28. Isolation of screw in a laminated core

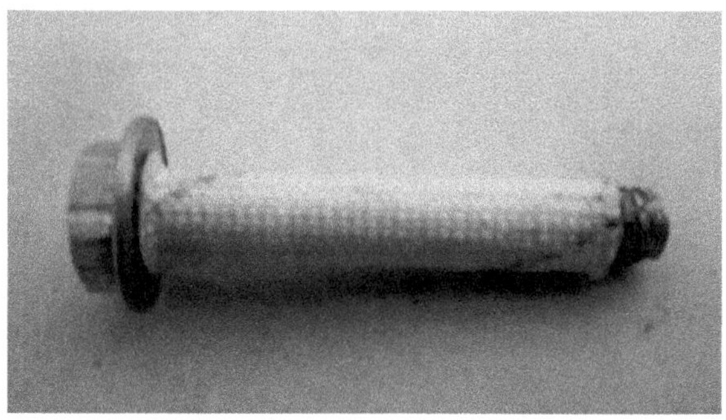

Figure 29. Litz wire for reduction of high frequency conduction lossesFigure 30. Current sharing between two transformers

Figure 30. Current sharing between two transformers

Figure 31. PCB trace width for different temperature rise

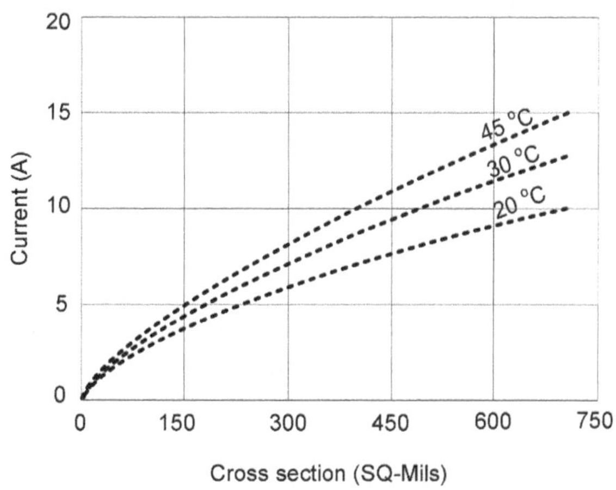

Figure 32. Cooling fan of a transformer

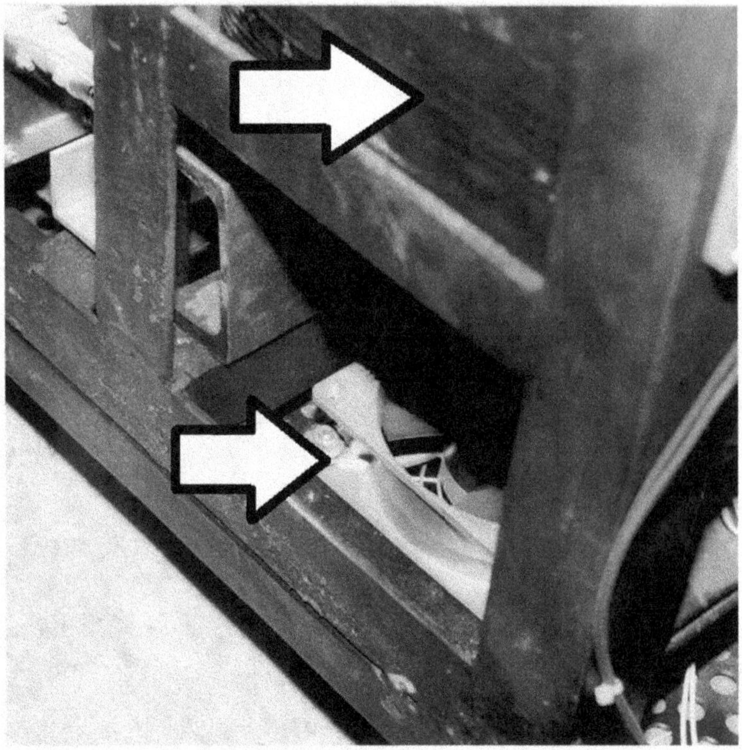

dissipation of heat in air depends upon dissipating surface of the oil tank, it is essential to increase the effective surface area of the tank. So additional dissipating surface in the form of tubes or radiators connected to the transformer tank. This is known as radiator of transformer or radiator bank of transformer.

Figure 32 shows a transformer which is cooled with a fan mounted behind the transformer core. Figures 33, 34, and 35 show cooling ducts in dry transformers with natural air cooling. Figures 36 and 37 show the cooling channel for a dry type forced cooled transformer. Figures 38, 39, 40, and 41 show the cooling channel for an oil cooled transformer.

Transient Thermal Control

Start Up Considerations

An inrush current limiter is a device or group of devices used to limit inrush current. Negative temperature coefficient (NTC) thermistors and resistors are two of the simplest options, with cool-down time and power dissipation being their main drawbacks, respectively. More complex solutions can be used when design constraints make simpler options infeasible 17. Figure 42 shows transient inrush current t of a switch mode power electronic converter. Figure 43 shows the transient inrush current of an induction motor during start up. The amplitude of this inrush current is reduced by applying a series resistor with stator of induction motor as shown in Figure 43(b). Figure 44.shows a typical characteristic of a NTC resistor. Figure 45 shows effect of soft starter to control the inrush current of a power electronic converter.

Figure 33. Cooling ducts in a three phase transformer

Figure 34. Cooling ducts in a three phase transformer

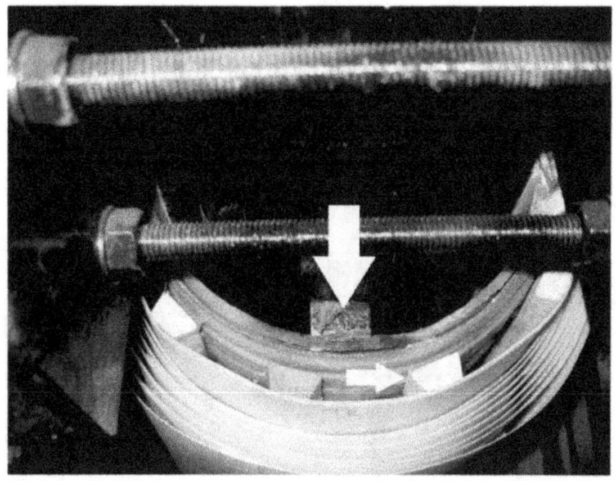

Figure 35. Cooling ducts in a three phase transformer

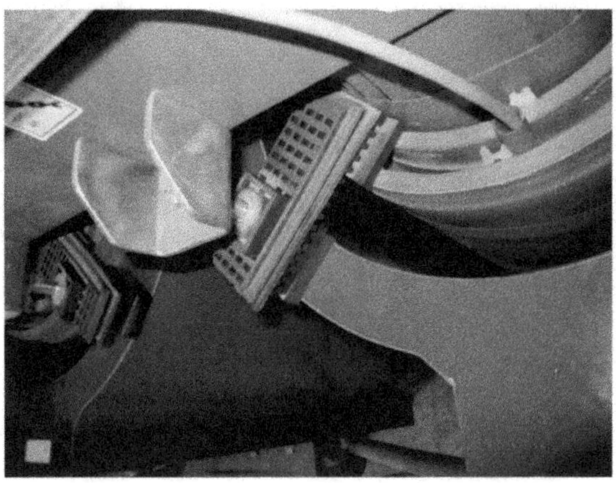

Figure 36. Cooling ducts in a single phase transformer

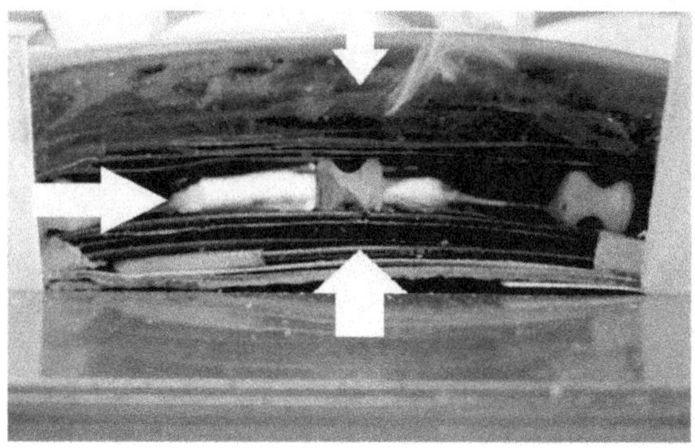

Figure 46 shows effect of rotor resistance to control the inrush current of an induction motor with wound type rotor. Figure 47 shows a rectifier with large output capacitor. In this converter, the inrush current is controlled by an inductor in series with capacitors. Figure 48 shows a commonly used method for control the inrush current. In this method, some series resistor are in series with input of the converter for a short time interval. They are by passed when the inrush current decreases to a acceptable value.

Figure 37. Cooling ducts in a single phase transformer with forced cooling

Figure 38. Cooling ducts in a high voltage transformer with liquid cooling

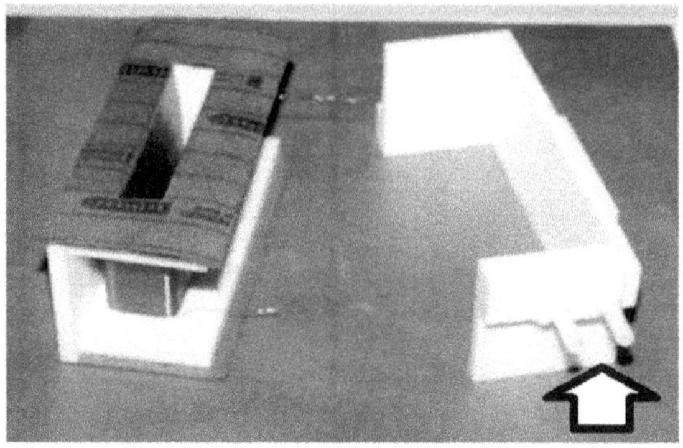

Figure 39. High voltage winding of the transformer with high voltage output

Figure 40. Mounting of cooling ducts for a liquid cooled transformer

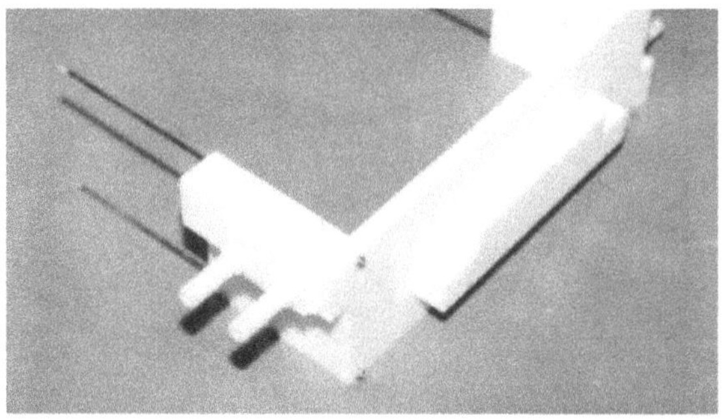

Transient Control

A thermistor is a type of resistor whose resistance varies significantly with temperature, more so than in standard resistors. The word is a portmanteau of thermal and resistor. Thermistors are widely used as inrush current limiters, temperature sensors, self-resetting overcurrent protectors, and self-regulating heating elements. Figure 49 shows typical characteristic of a PTC.

Figure 41. Cooling ducts in a single phase transformer

Figure 42. Inrush current of a converter

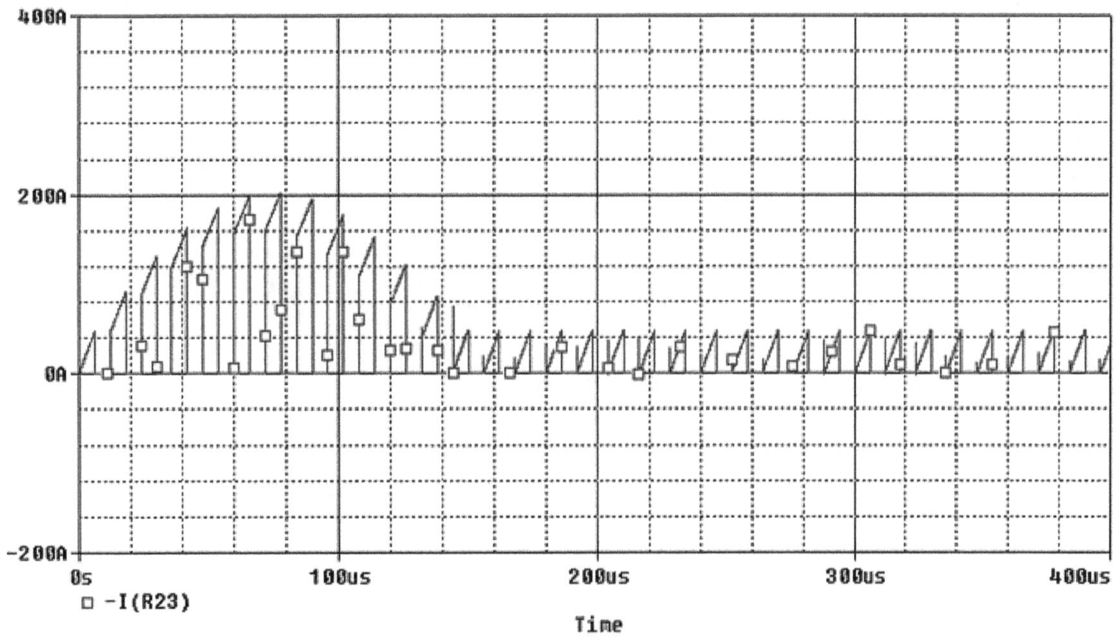

ELECTRICAL STRESS FACTORS

Steady State Electric Field

Corona discharge usually forms at highly curved regions on electrodes, such as sharp corners, projecting points, edges of metal surfaces, or small diameter wires. The high curvature causes a high potential gradient at these locations, so that the air breaks down and forms plasma there first. In order to sup-

Figure 43. Effect of series resistor on inrush current of an induction motor, (a): without series resistor, (b): with series resistor

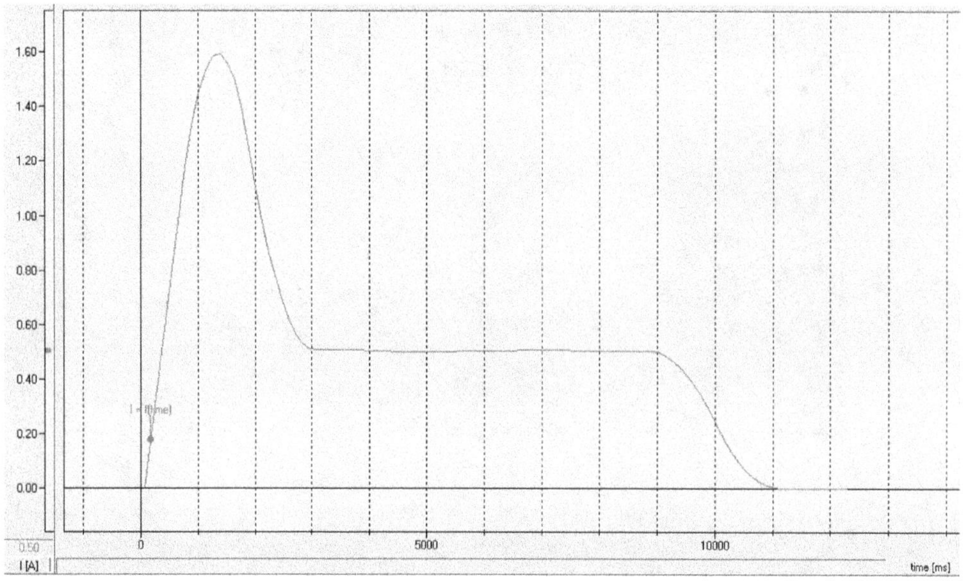

(a)

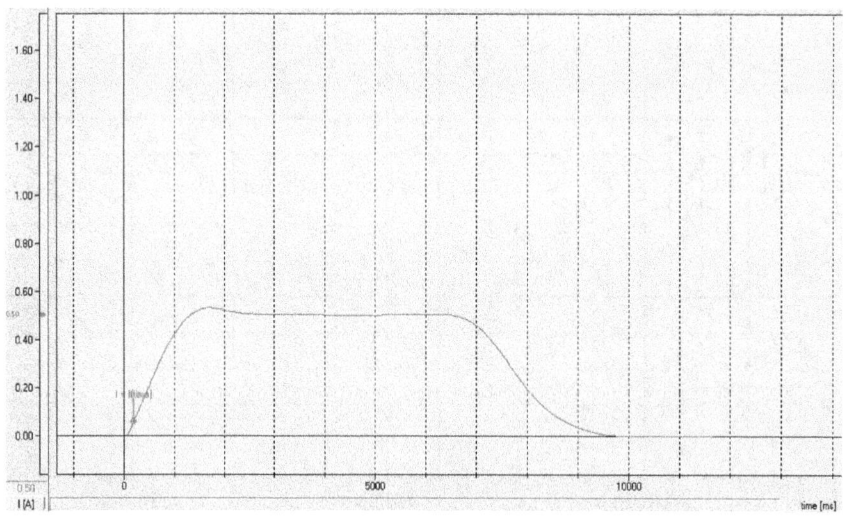

(b)

press corona formation, terminals on high voltage equipment are frequently designed with smooth large diameter rounded shapes like balls or toruses, and corona rings are often added to insulators of high voltage transmission lines.

Figure 50 shows a high voltage transformer with round head of high voltage outputs. Figures 51 and 52 show application of this technique for a PCB.

Figure 44. A typical characteristic of a NTC

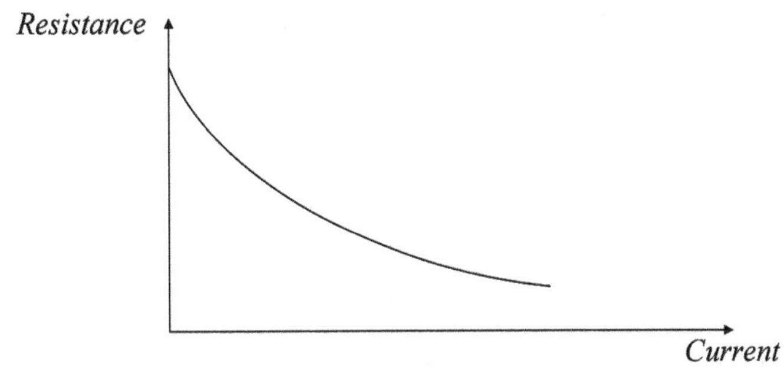

Figure 45. Effect of soft start on inrush current of a converter, (a): without soft start, (b):with soft start

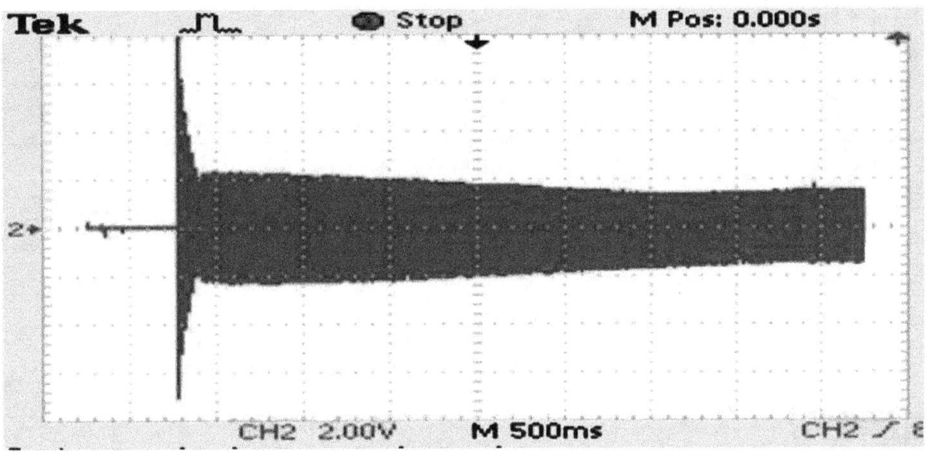

(a)

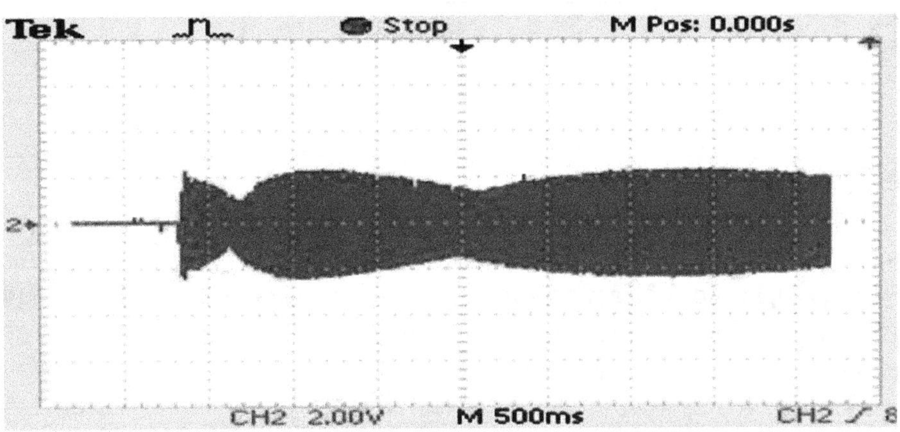

(b)

Figure 46. Inrush current of an induction motor for various values of rotor resistor

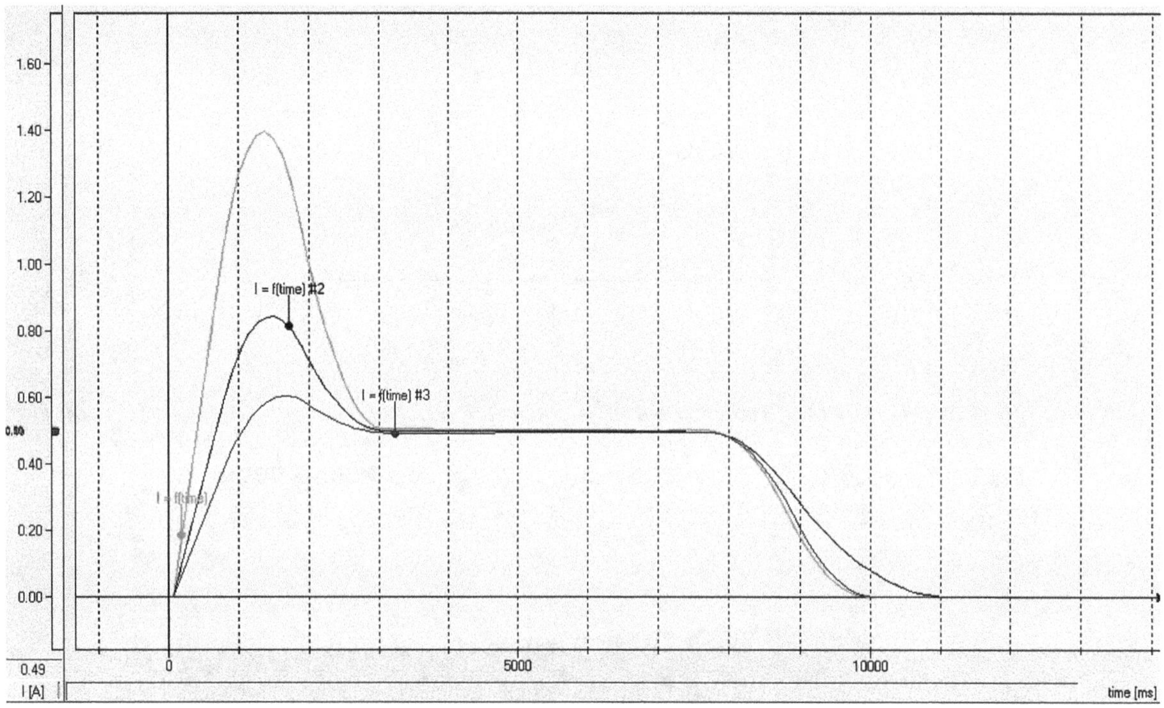

Figure 47. A three phase rectifier with inductive filter

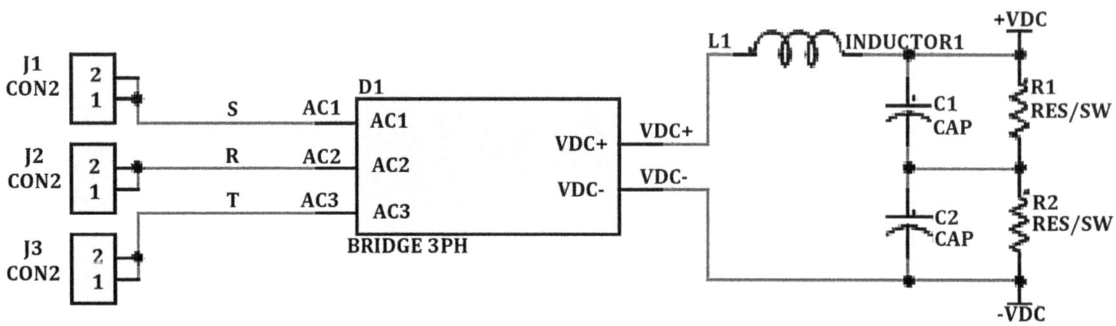

PCB Trace Spacing

Track width and clearances for most designs is primarily a limitation of the PCB manufacturing process.

Figure 53 shows application of air gap for increasing the insulator resistance between two parts of a PCB. Figure 54 shows a PCB containing sections with different voltage levels. The traces of low voltage sections are designed to have maximum distance from high voltage section.

Figure 48. Series resistors for inrush current limiting and its bypass switches

Figure 49. Typical characteristic of a PTC

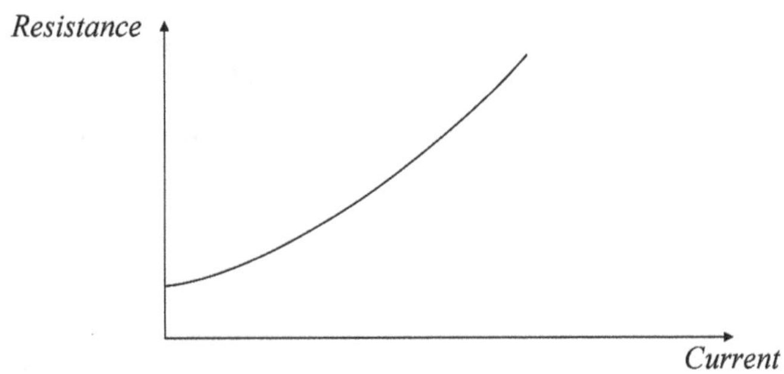

Figure 50. Round head of a high voltage transformer

Figure 51. Round soldering on a high voltage PCB

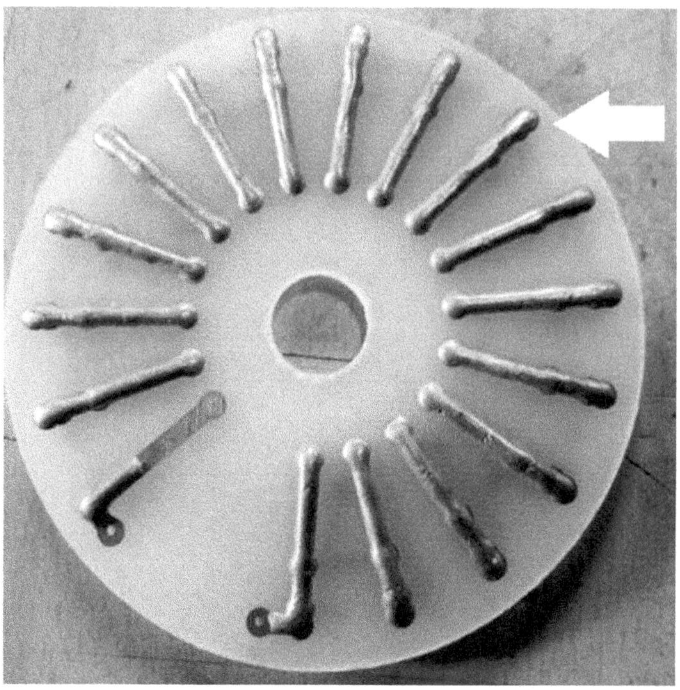

Figure 52. Round pad soldering on a high voltage PCB

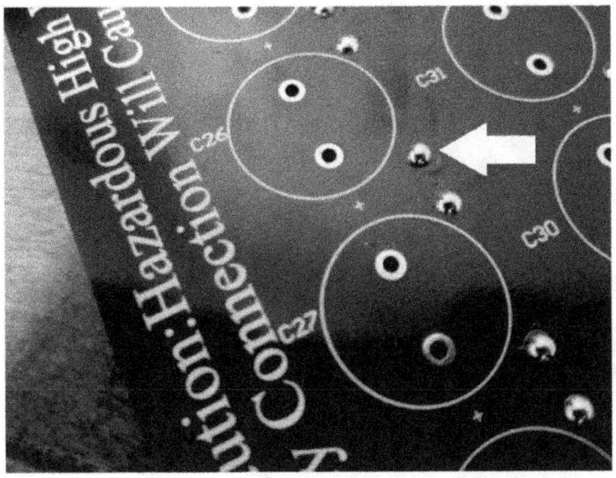

Figure 53. Application of air gap for high isolation

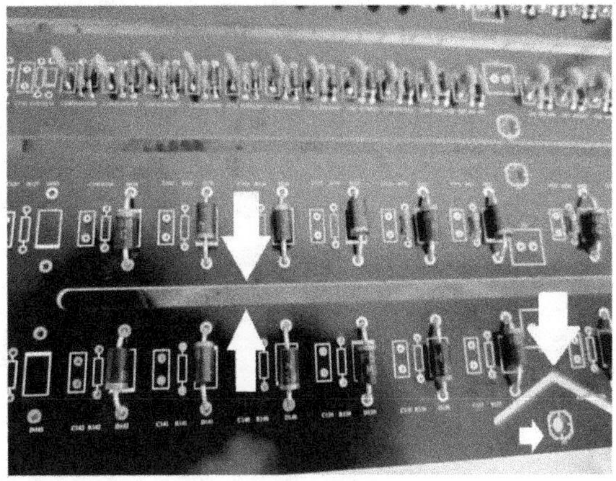

Figure 54. Separation of low and high voltage zones on a PCB

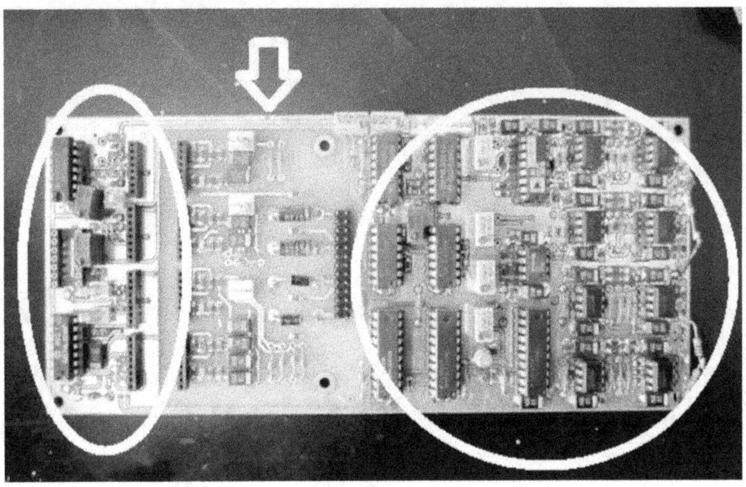

Voltage Sharing

A voltage sharing circuit which allows use of low voltage semiconductor devices to obtain higher output voltages is made up of a series stack of the low voltage devices with diode limiters provided to bias the low voltage devices such that the maximum voltage across any one of the low voltage devices is limited to a predetermined value.

If required the modules can be connected in series for higher voltage ratings or in parallel for higher current ratings, although when connected in parallel the combined current ratings should be reduced by 20%.

Figure 55 shows voltage sharing between series connection of diode in a high voltage diode stack. Figure 56 shows the voltage sharing between series capacitors.

Figure 55. Series diodes for voltage sharing

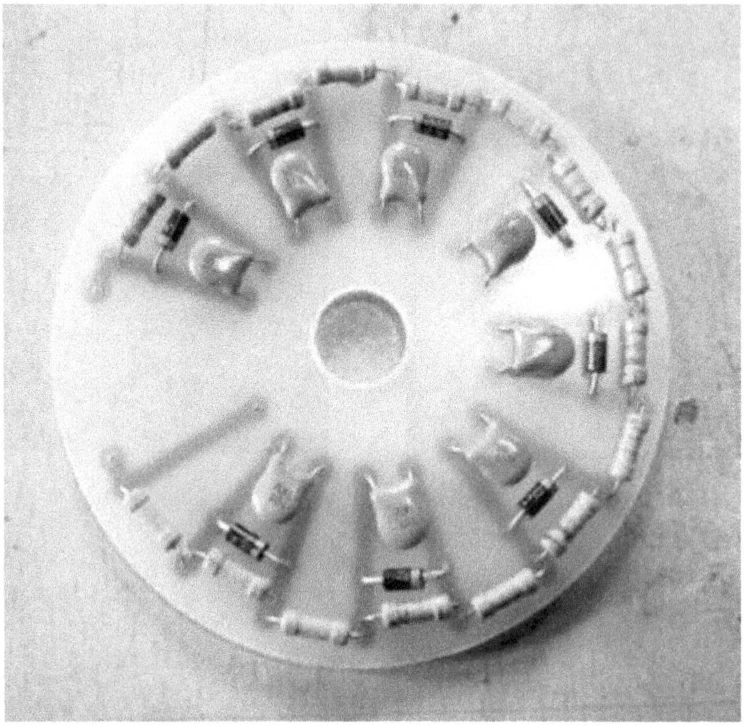

Figure 56. Series capacitors for voltage sharing

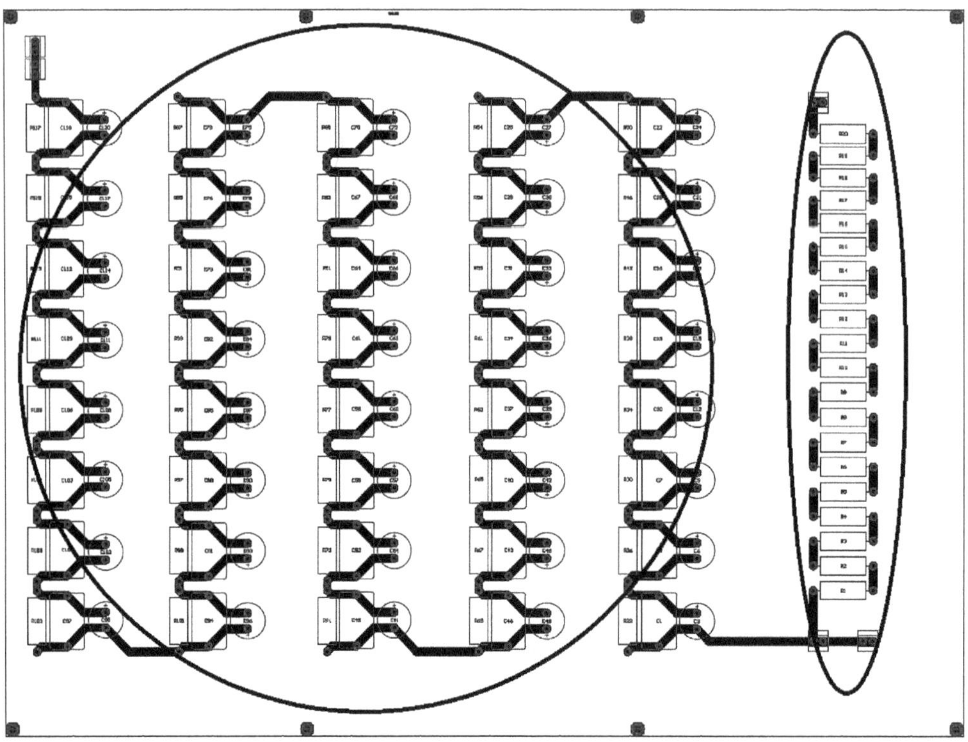

Printed Circuit Boards

As the voltage and current in power electronic converters are high values, the printed circuit used for these converters should meet standards for minimum spacing between traces and also the width of traces. Figure 57 shows the minimum spacing between traces versus their voltage difference in various standards. Figures 58, 59, 60, 61, 62, 63, 64, 65, and 66 shows the effect of various techniques for reducing the electrical field between traces in a PCB.

Figure 57. PCB minimum spacing in different standards

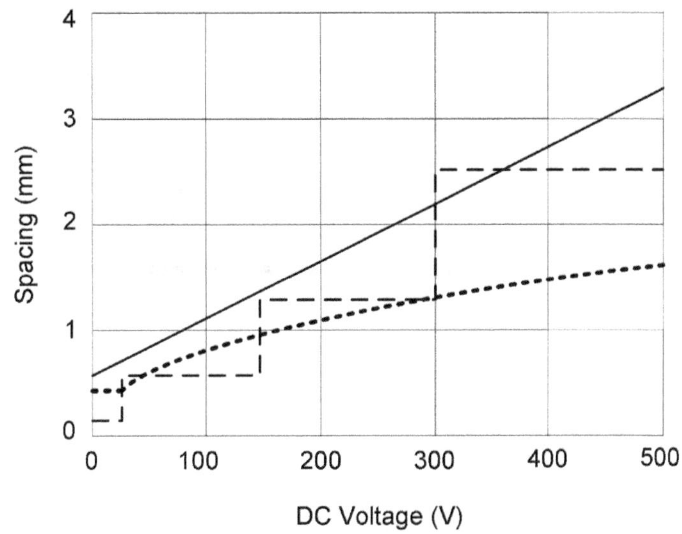

Figure 58. Electric field between two parallel traces on PCB

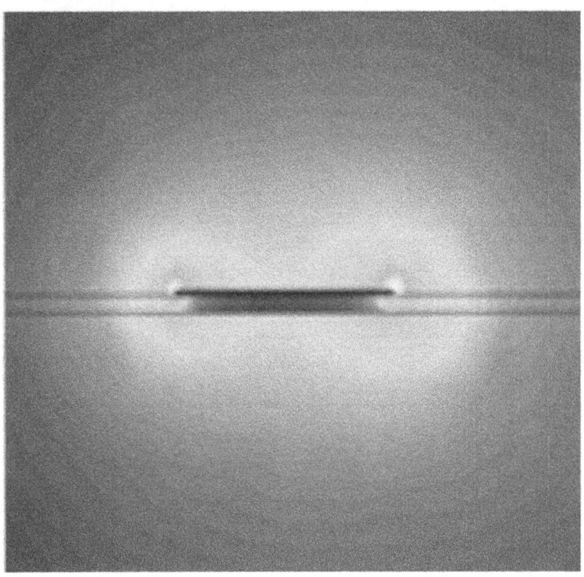

Figure 59. Electric field (down) between two parallel traces on PCB (up)

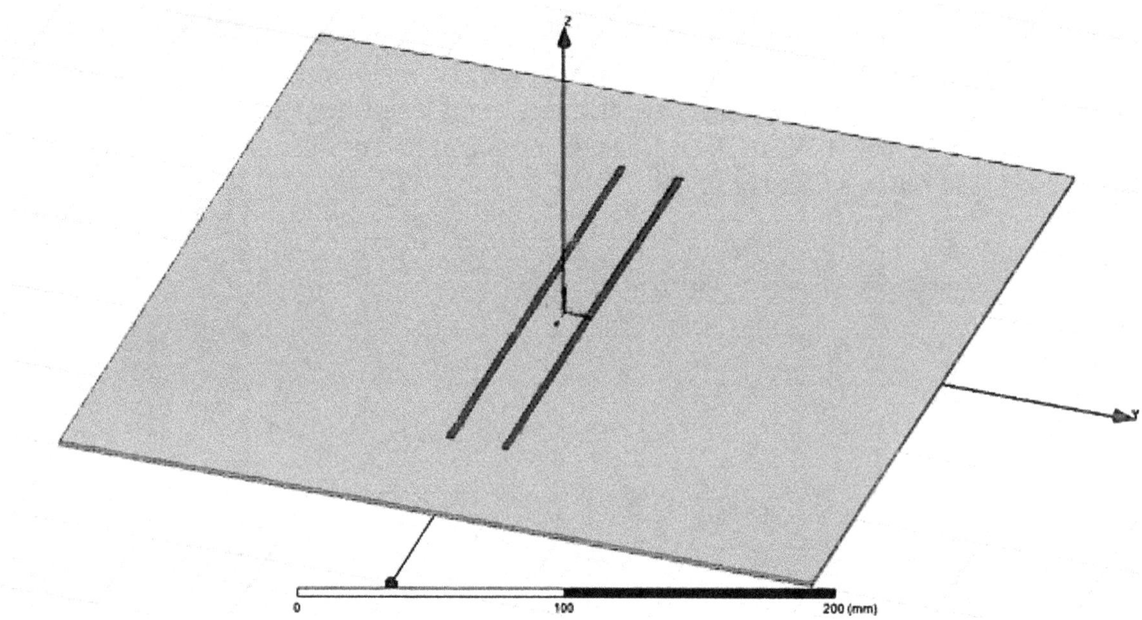

(a)

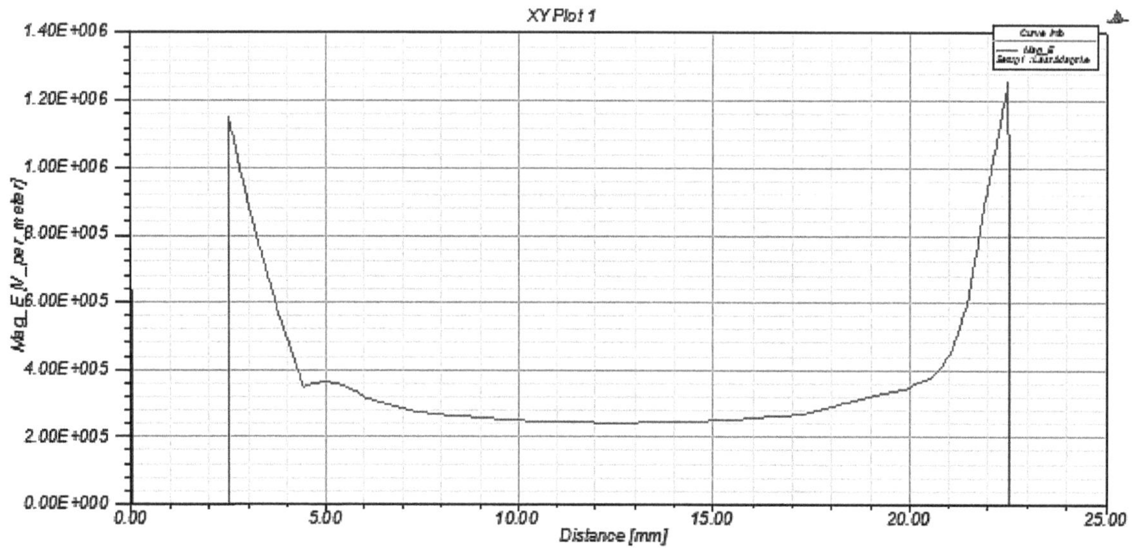

(b)

Figure 60. Electric field (down) between two parallel traces on PCB with symmetric air gap (up)

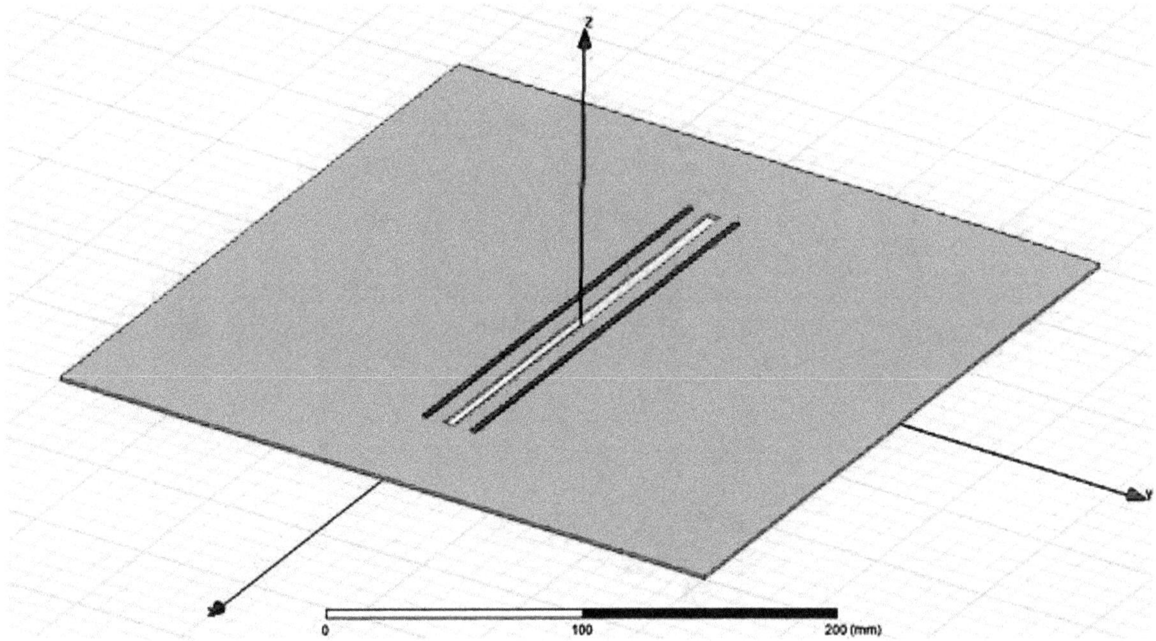

(a)

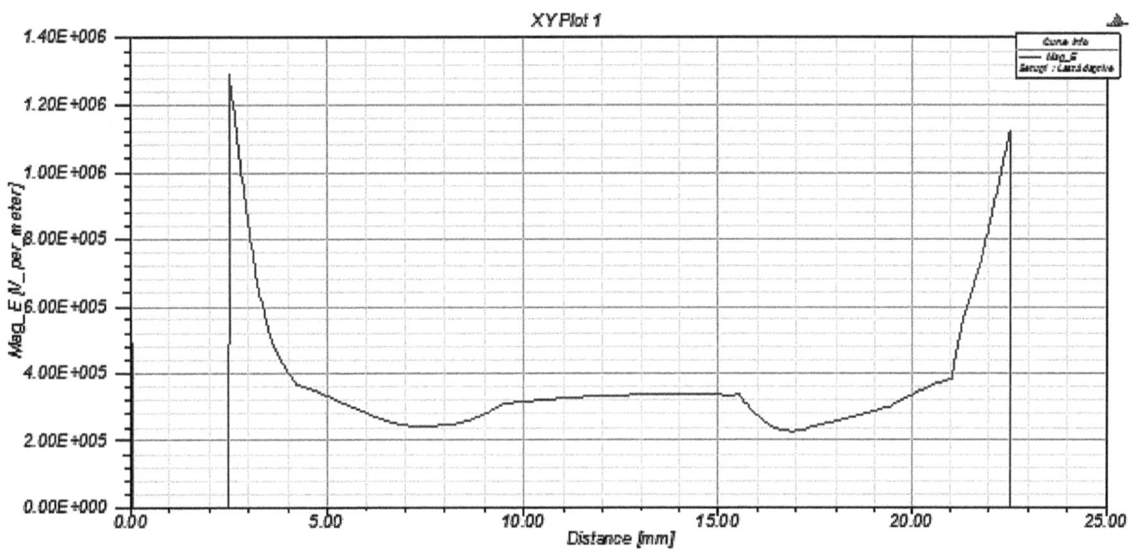

(b)

Figure 61. Electric field (down) between two parallel traces with asymmetric air gap on PCB (up)

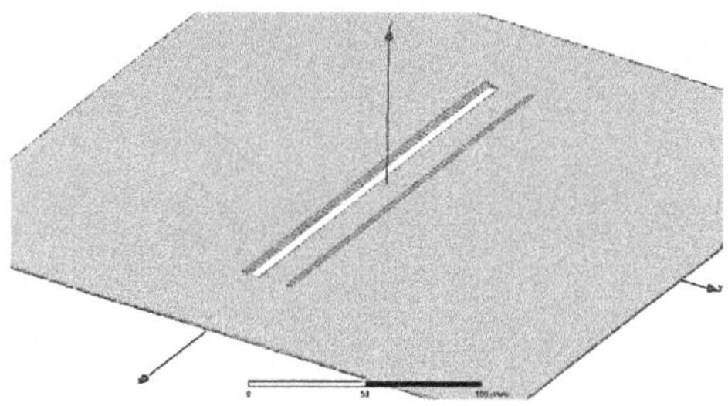

(a)

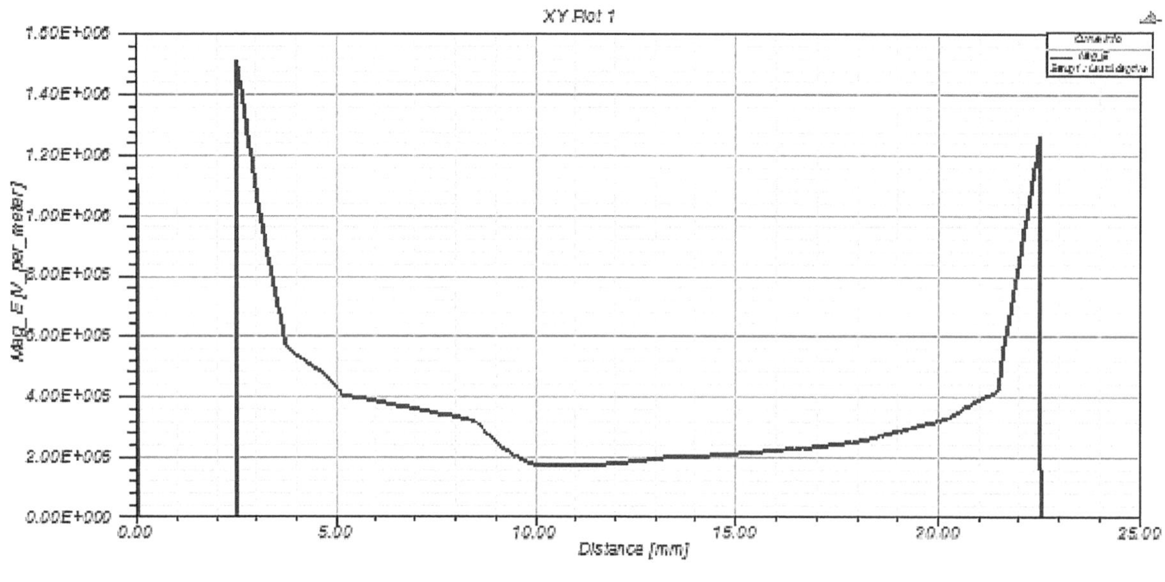

(b)

Fault Diverting

A crowbar circuit is an electrical circuit used to prevent an overvoltage condition of a power supply unit from damaging the circuits attached to the power supply. It operates by putting a short circuit or low resistance path across the voltage source, much as if one dropped a tool of the same name across the output terminals of the power supply. Crowbar circuits are frequently implemented using a thyristor, TRIAC, trisil or thyratron as the shorting device. Once triggered, they depend on the current-limiting circuitry of the power supply or, if that fails, the blowing of the line fuse or tripping the circuit breaker.

Figure 62. Electric field between two parallel rounded traces on PCB

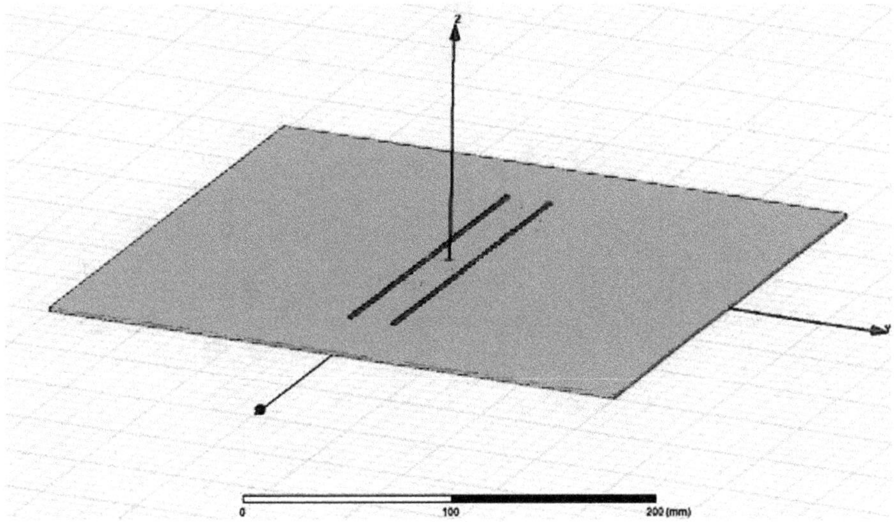

(a)

(b)

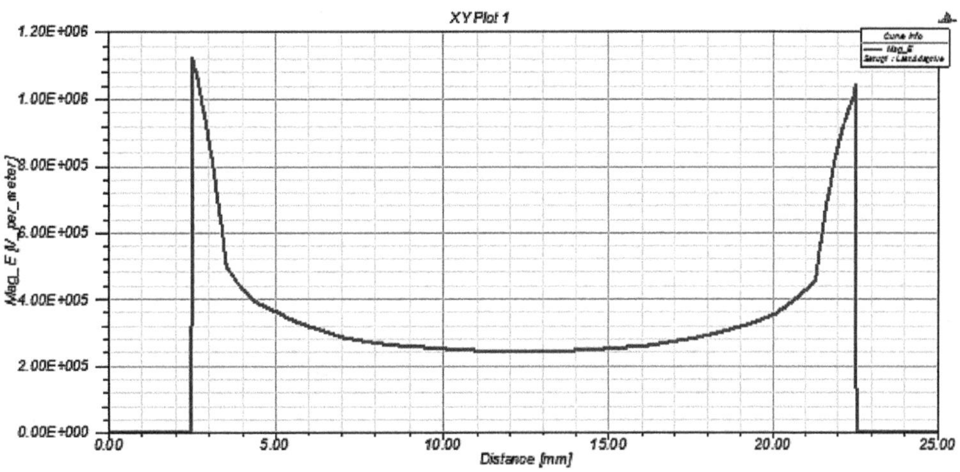

(c)

Figure 63. Electric field (down) between two parallel traces on PCB including a 90 degree corner (up)

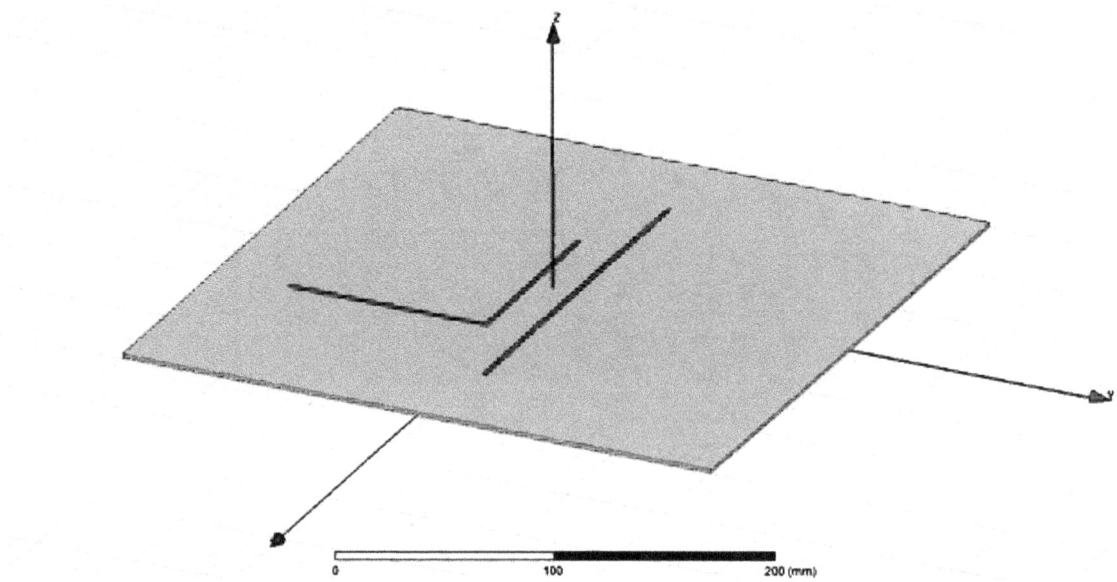

(a)

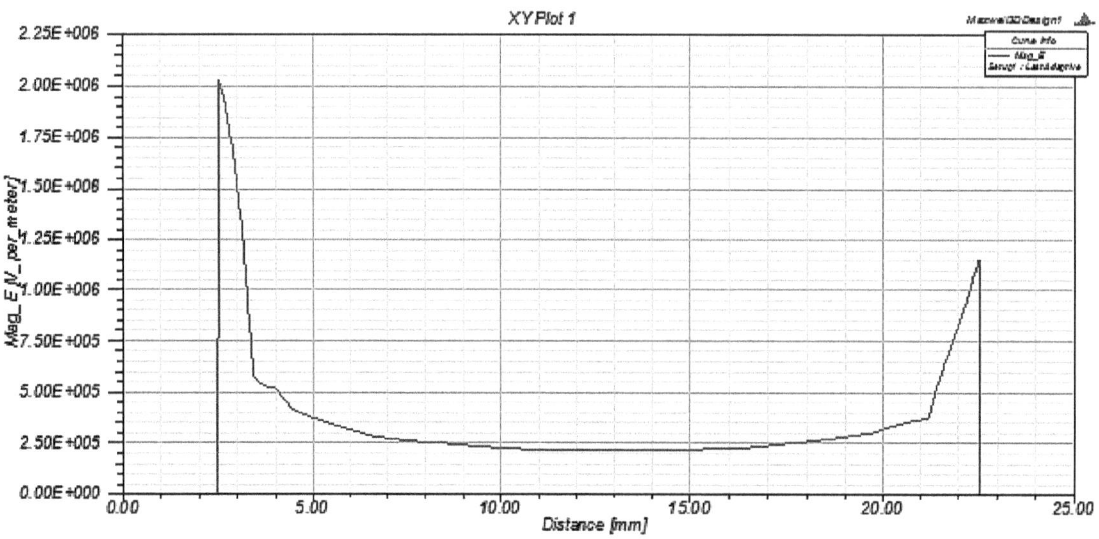

(b)

Figure 64. Electric field (down) between two parallel traces on PCB including a 90 degreecorner and air gap (up)

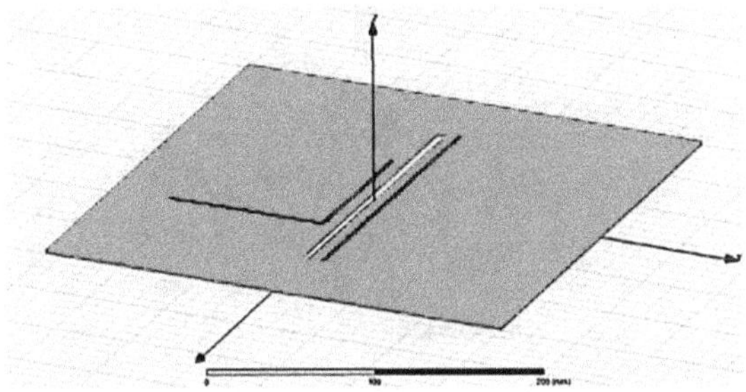

(a)

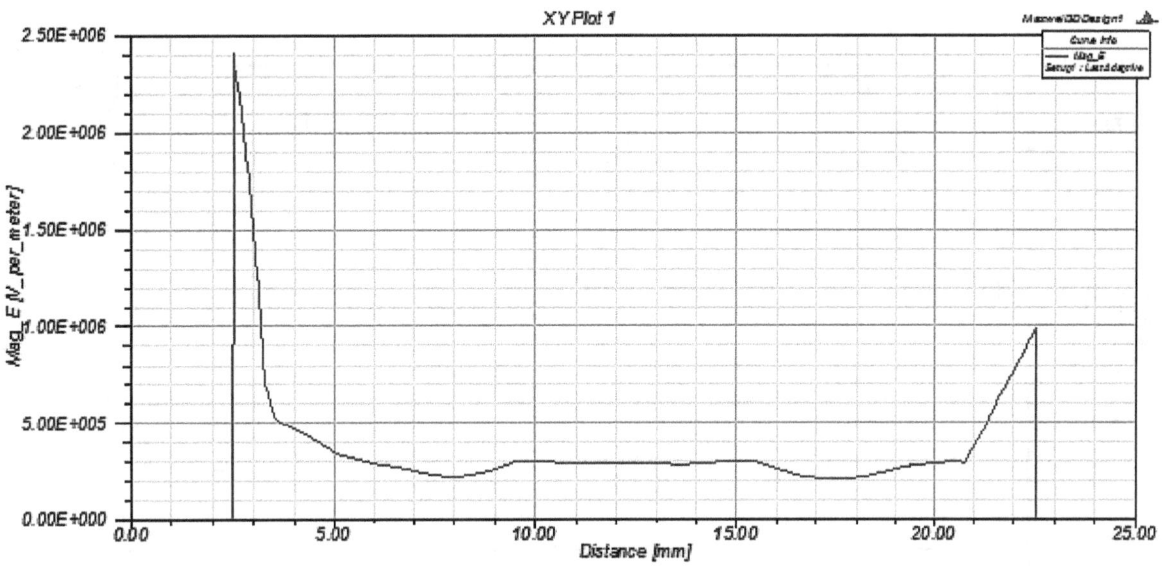

(b)

High voltage crowbars are used for HV tube (Klystron and IOT) protection. A crowbar circuit is distinct from a clamp in that, once triggered, it pulls the voltage below the trigger level, usually close to ground. A clamp prevents the voltage from exceeding a preset level. Thus, a crowbar will not automatically return to normal operation when the overvoltage condition is removed; power must be removed entirely to stop its conduction.

An active crowbar is a crowbar that can remove the short circuit when the transient is over thus allowing the device to resume normal operation. Active crowbars use a transistor, gate turn off (GTO) thyristor or forced commutated thyristor instead of a thyristor to short the circuit. Active crowbars are commonly used to protect the frequency converter in the rotor circuit of doubly fed generators against

Figure 65. Electric field (down) between two parallel traces on PCB including a rounded 90 degree corner (up)

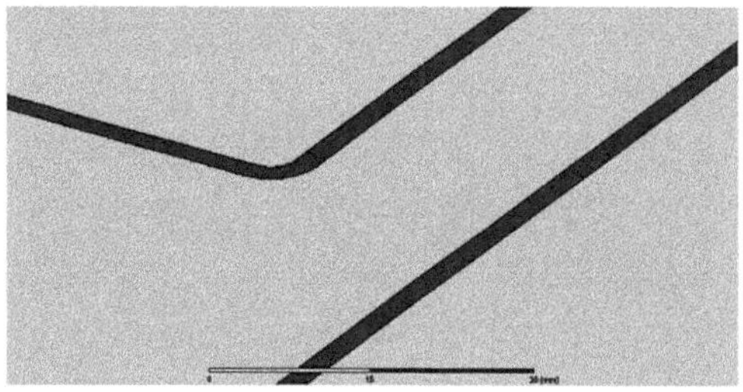

(a)

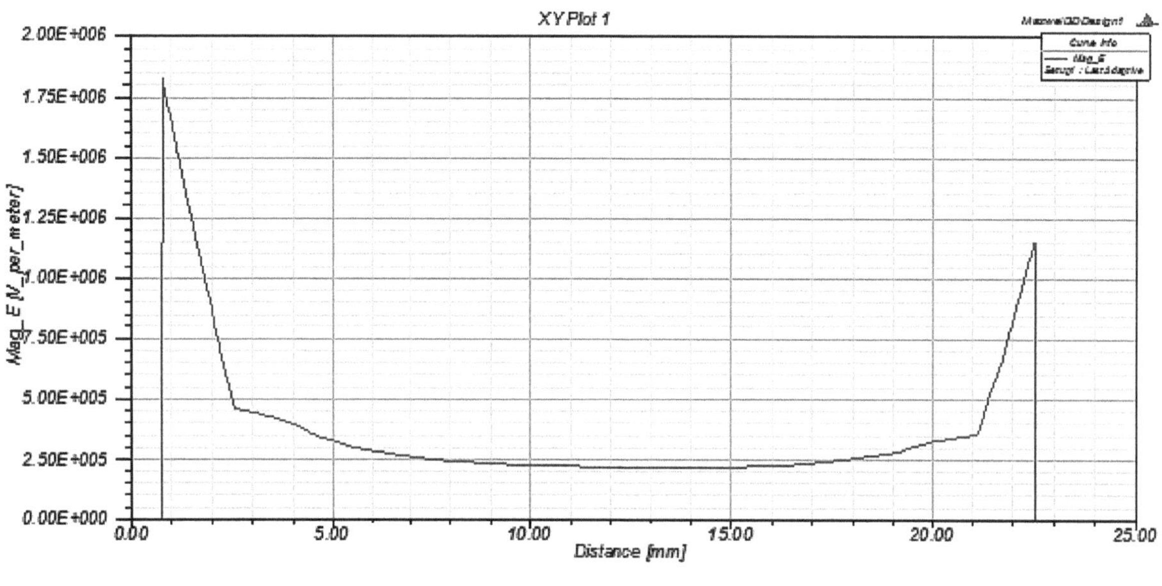

(b)

high voltage and current transients caused by the voltage dips in the power network. Thus the generator can ride through the fault and quickly continue the operation even during the voltage dip.

The advantage of a crowbar over a clamp is that the low holding voltage of the crowbar lets it carry higher fault current without dissipating much power (which could otherwise cause overheating). Also, a crowbar is more likely than a clamp to deactivate a device (by blowing a fuse or tripping a breaker), bringing attention to the faulty equipment. An example crowbar circuit is shown in Figure 67.

Figure 66. Electric field (down) between two parallel traces on PCB including a 135 degree corner (up)

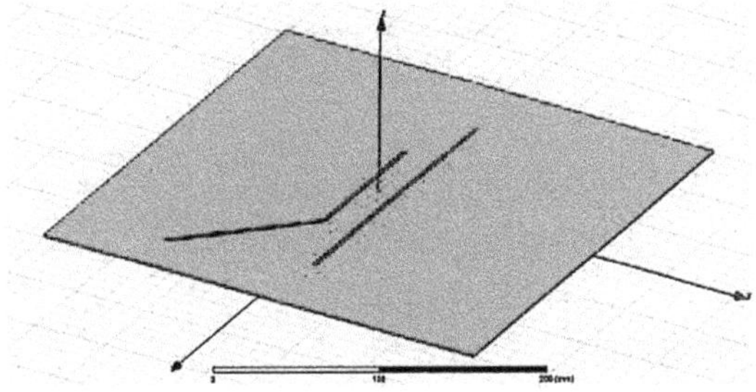

(a)

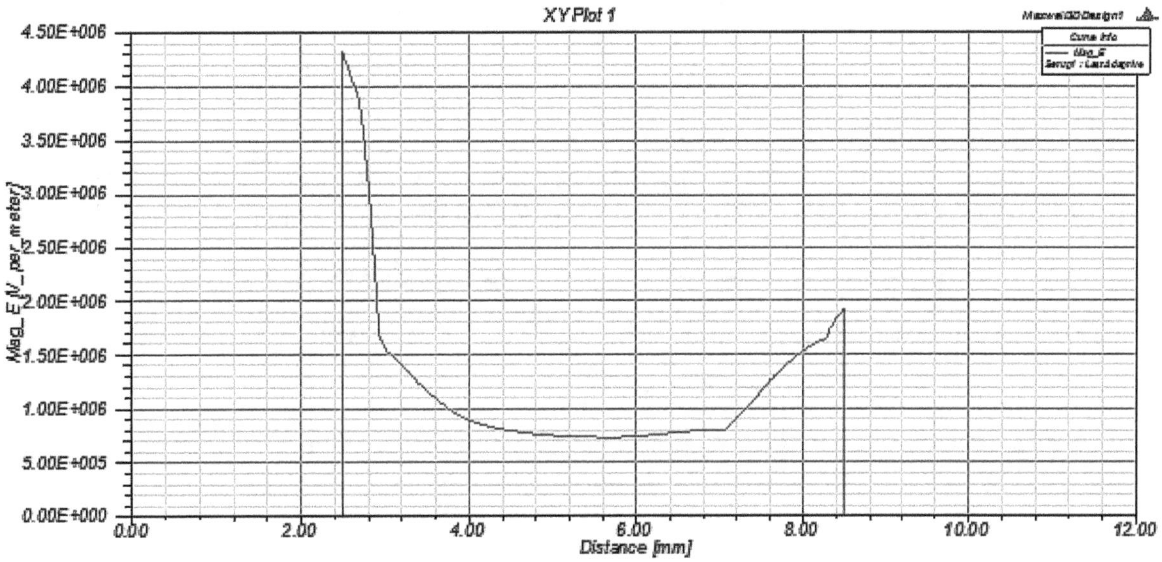

(b)

Transient Voltage Spike

In industrial applications, high voltage power supply spikes with durations ranging from a few micro-seconds to hundreds of milliseconds are commonly encountered. The electronics within these systems must not only survive transient voltage spikes, but in many cases also operate reliably throughout the event. In systems where power is distributed over long wires severe transients are generated by load steps. Corroded connections between a power source and load can lead to an abrupt interruption of current flow, and a high value of dI/dt. The best example of this condition is automotive load dump, where there is a sudden break in the connection to the battery caused by vibration and corroded terminals. These transients pose a difficult challenge for engineers trying to protect sensitive electronics Figure 68 shows various over voltage in IEC standard.

Figure 67. A typical crowbar circuit

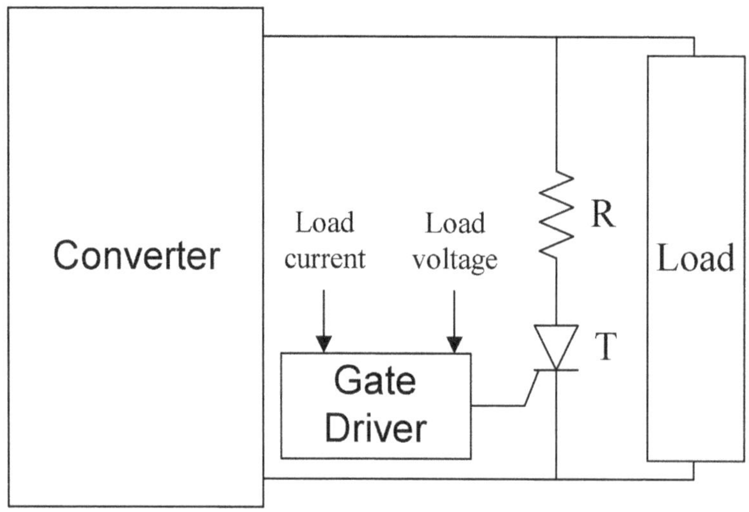

Figure 68. Various types of over voltages applied to an equipment via AC electric network (IEC publication 60146-1-2 ed.4.0 Copyright ©2001 of publication IEC Geneva, Switzerland. www.iec.ch)

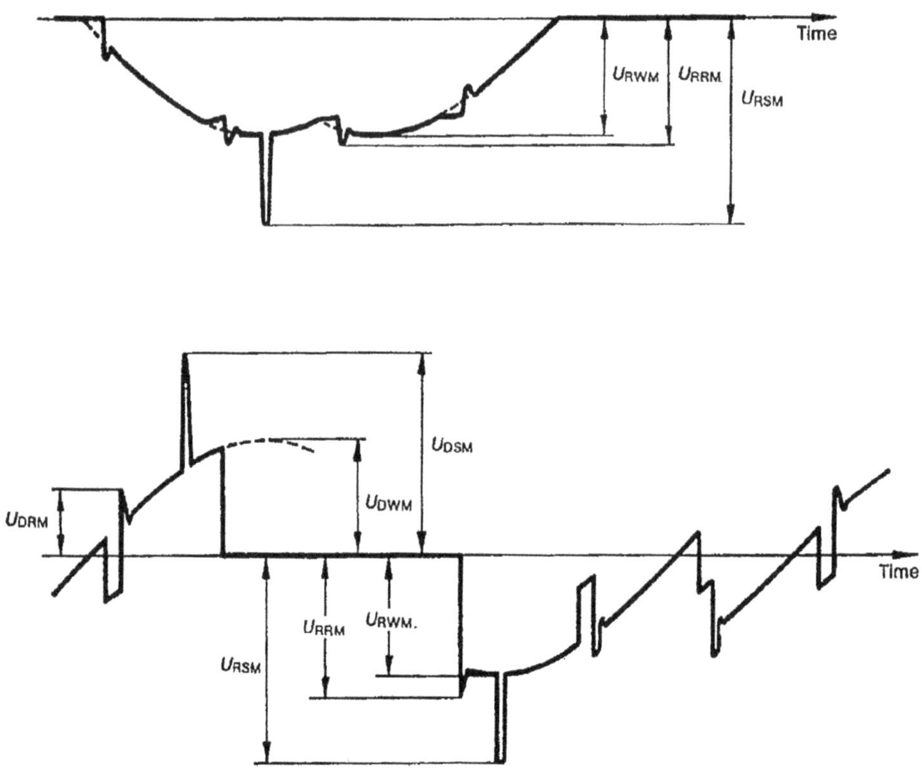

Figure 69. Effect of Drain-Gate capacitor on a MOSFET driver

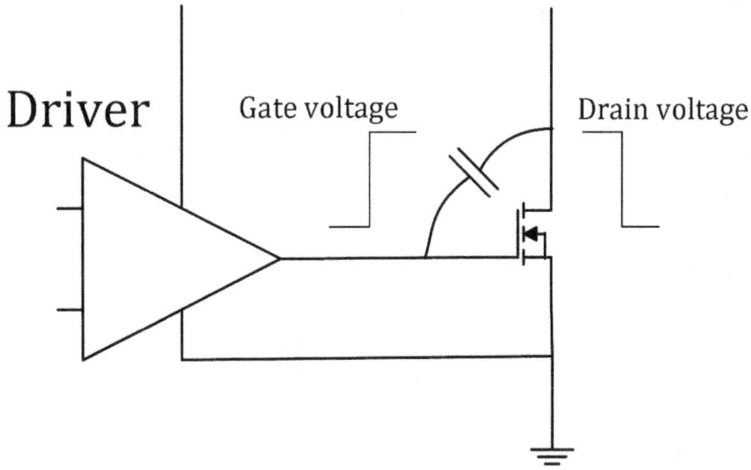

The voltage spike may have a low amplitude but with high rise time rate. This high rise time rate leads to increasing the capacitor coupling between electronic parts. Figure 69 shows an example of this problem. In this figure, rapid fall of drain in a MOSFET causes to a capacitor coupling between drain and gate of the MOSFET. The injected current leads to decrease the gate charge and unwanted ringing in the gate.

Isolation

An optoisolator, also known as an optical coupler or optocoupler, is a semiconductor device that allows signals to be transferred between circuits or systems, while keeping those circuits or systems electrically isolated from each other. Optoisolators are used in a wide variety of communications, control, and monitoring systems.

In its simplest form, an optoisolator consists of an light-emitting diode (LED), IRED (infrared-emitting diode), or laser diode for signal transmission, and a photosensor for signal reception. The “transmitter” takes the electrical signal and converts it into a beam of modulated visible light or infrared (IR). This beam travels across a transparent gap and is picked up by the “receiver,” which converts the modulated light or IR back into an electrical signal. The electrical output waveform is identical to the electrical input waveform, although the input and output amplitudes (signal strengths) often differ. The optoisolator is enclosed in a single package, and has the appearance of an integrated circuit (IC) or a transistor with extra leads.

MECHANICAL STRESS FACTORS

Modal Analysis

The goal of modal analysis in structural mechanics is to determine the natural mode shapes and frequencies of an object or structure during free vibration (Kaboli, Zolghadri, Haghbin, & Homaifar, 2004). It is common to use the finite element method (FEM) to perform this analysis because, like other calculations using the FEM, the object being analyzed can have arbitrary shape and the results of the calculations are acceptable. The types of equations which arise from modal analysis are those seen in eigensystems. The physical interpretation of the eigenvalues and eigenvectors which come from solving the system are that they represent the frequencies and corresponding mode shapes. Sometimes, the only desired modes are the lowest frequencies because they can be the most prominent modes at which the object will vibrate, dominating all the higher frequency modes.

Mechanical Damper

A shock absorber is a mechanical device designed to smooth out or damp shock impulse, and convert kinetic energy to another form of energy (usually thermal energy, which can be easily dissipated). It is a type of dashpot.

In a vehicle, shock absorbers reduce the effect of traveling over rough ground, leading to improved ride quality and vehicle handling. While shock absorbers serve the purpose of limiting excessive suspension movement, their intended sole purpose is to damp spring oscillations (Sheikhi, Oraee, Kaboli, & Dorkhah, 2009). Shock absorbers use valving of oil and gasses to absorb excess energy from the springs. Spring rates are chosen by the manufacturer based on the weight of the vehicle, loaded and unloaded. Some people use shocks to modify spring rates but this is not the correct use. Along with hysteresis in the tire itself, they damp the energy stored in the motion of the unsprung weight up and down (Nasirian, Davoudi, Kaboli, & Edrington, 2013). Effective wheel bounce damping may require tuning shocks to an optimal resistance.

Spring-based shock absorbers commonly use coil springs or leaf springs, though torsion bars are used in torsional shocks as well. Ideal springs alone, however, are not shock absorbers, as springs only store and do not dissipate or absorb energy. Vehicles typically employ both hydraulic shock absorbers and springs or torsion bars. In this combination, “shock absorber” refers specifically to the hydraulic piston that absorbs and dissipates vibration (Kaboli, Zolghadri, Haghbin, & Emadi, 2003).

There are dampers that are designed to dampen the vibrations in a mechanical system. A shock absorber on a car in a good example of this. There are dampers that remove or lessen pressure surges in piping systems. Water hammer is an example of a pressure surge. Finally some types of valves are referred to as dampers. They are common in heating and air conditioning duct work. Figure 70 show application of RTV adhesive for mechanical stability of electronic parts. Figure 71 shows a typical mechanical damper. Figure 72 shows a rotary inverter in railway application which is mounted via mechanical damper behind the bogie of a train.

ENVIRONMENTAL STRESS FACTORS

Cooling process water filtration is becoming more common as companies discover the relative low cost of the filtration systems that help:

- **Increase cooling tower efficiencies:** Solid contaminants in cooling towers reduce cooling tower efficiencies. Deposits on heat transfer surfaces increase the fouling factor which significantly increases energy consumption. These same solid contaminants can clog small spray nozzle orifices

Figure 70. Application of RTV grease for vibration reduction, (a): for height capacitors, (b): for transformers in an airborne power supply

(a)

(b)

Figure 71. A typical mechanical damper

Figure 72. Application of damper for a rotary power supply in railway

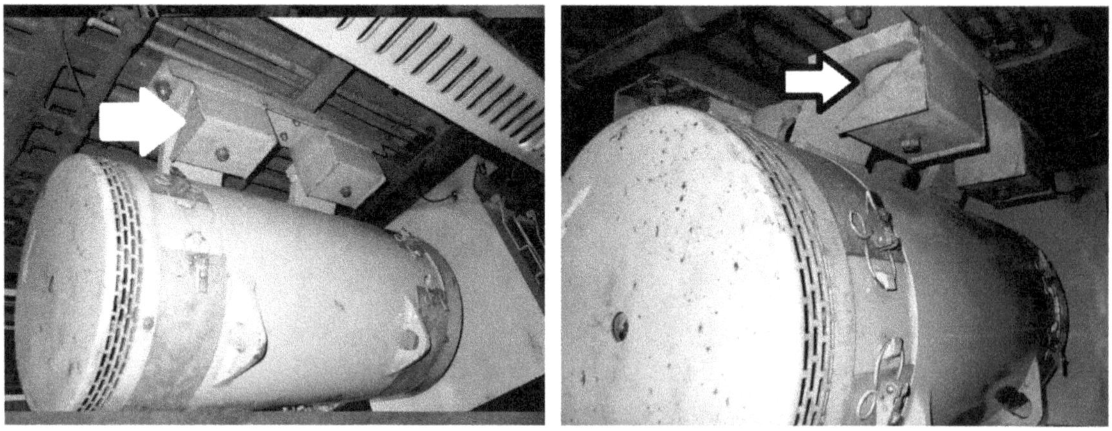

(a) (b)

Figure 73. Cooling water filters

(a) (b)

causing poor distribution through the fill. Filtration minimizes deposits on heat transfer surfaces and prevents nozzle clogging

- **Increase cooling tower life:** The life of the cold water basin may determine the life of the cooling tower. A thick layer of solid contaminants in the cold water basin reduces the benefit of treatment chemicals such as corrosion inhibitors or oxidizers to reach the basin. Filtered cooling tower water prevents solid contaminants from building up in the basin.
- **Reduce treatment chemical costs:** A layer of solid contaminants in the cold water basin provides an excellent environment for bacteria and algae growth. Also, suspended solid contaminants may buffer the beneficial effects of treatment chemicals, hence requiring additional chemical treatment. Filtering cooling tower water prevents solid contaminant contribution to bacteria and algae growth, and minimizes the need for additional chemicals.
- **Reduce maintenance:** A layer of solid contaminants will eventually require removal from the cold water basin. Filtration of the cooling tower water will minimize the need for manually cleaning the basin.

- Reduce shut-down time for the manually cleaning of cooling towers

Figure 73 shows water filters used in a cooling system.
Solid contaminants enter cooling towers from three primary sources:

- **Ambient air dirt load:** Cooling towers make excellent air scrubbers. Relatively high volumes of air pass through cooling towers and most of the contaminants in the air end up in the cooling tower basin.
- **Circulation water build-up:** Calcium carbonate scale that forms in the tower can flake off. Treatment chemical residue and algae will also build up in the circulation water.
- **Make-up water:** Depending on the source, the make-up water may also contribute to contaminant build-up

SUMMARY AND CONCLUSION

In this chapter, requirements for operation of an electric power converter in low stress conditions were described. Based on the background of chapters 2 and 3, stress reduction leads to long useful life of the systems. The main topics of this chapter are summarized as follows:

1. Any method that reduces the stress of factors affected reliability can be used for better system design. Some of the methods for reliability improvement act at hardware level. Aim of these methods is usually reduction of hot spot temperature or reduction of electric field applied to devices. In rotary power converters, they also reduce the mechanical tension on the structure of the machines. Other methods act as control strategies for low stress operation of the converter.
2. Application of insulators with high electric strength and increasing of space between points with high potential difference help reduce the risk of electric breakdown.
3. Using series and parallel connection of devices helps reduction of voltage and thermal stress, respectively. This method is a commonly used tool for designing a reliable system.
4. Application of mechanical dampers prevent damages due to mechanical vibration and shock. A shock absorber is a mechanical device designed to smooth out or damp shock impulse, and convert kinetic energy to another form of energy.
5. Isolator coatings are useful tools for reduction of undesired influence of environmental parameters on devices.

In this chapter, we discussed the method of reliability improvement for a converter under design process or operating without fault. The next chapter presents the methods for reliability improvement in a faulty converter that is not damaged. The tool for this goal is protection system.

REFERENCES

Kaboli, S., Vahdati-Khajeh, E., Zolghadri, M. R., & Homaifar, A. (2005). A fast optimal flux search controller with improved steady state behavior for DTC based induction motor drives. In *Proceedings of IEEE International Conference on Electric Machines and Drives* (pp.1732-1736). IEEE. doi:10.1109/IEMDC.2005.195954

Kaboli, S., Zolghadri, M. R., Haghbin, S., & Emadi, A. (2003). Torque ripple minimization in DTC of induction motor based on optimized flux value determination. In *Proceedings of The 29th Annual Conference of the IEEE Industrial Electronics Society* (pp.431-435). IEEE. doi:10.1109/IECON.2003.1280019

Kaboli, S., Zolghadri, M. R., Haghbin, S., & Homaifar, A. (2004). A control strategy for reducing the torque ripple in low speed operation of direct torque controlled induction motor. In *Proceedings of IEEE International Symposium on Industrial Electronics* (pp.1203-1208). IEEE. doi:10.1109/ISIE.2004.1571984

Kaboli, S., Zolghadri, M. R., Vahdati-Khajeh, E., & Homaifar, A. (2005). On the Performance of Optimal Flux Search Controller for DTC Based Induction Motor Drives. In *Proceedings of IEEE International Conference on Electric Machines and Drives* (pp.1752-1756). IEEE. doi:10.1109/IEMDC.2005.195957

Nasirian, V., Davoudi, A., Kaboli, S., & Edrington, C. S. (2013). Excitation Shifting: A General Low-Cost Solution for Eliminating Ultra-Low-Frequency Torque Ripple in Switched Reluctance Machines. *IEEE Transactions on Magnetics*, *49*(9), 5135–5149. doi:10.1109/TMAG.2013.2251651

Park, K. B., Choi, S. W., Kim, Ch. E., Moon, G. W., & Youn, M. J. (2009). An AC-PDP Single Sustaining Driver Employing the Voltage Stress Reduction Technique. *IEEE Transactions on Power Electronics*, *24*(4), 1124–1128. doi:10.1109/TPEL.2008.2008993

Sheikhi, A., Oraee, H., Kaboli, S., & Dorkhah, M. (2009). A new configuration of switched reluctance motor for reducing the torque ripple. In *Proceedings of International Conference on Electric Power and Energy Conversion Systems* (pp.1-5). IEEE.

ADDITIONAL READING

Asiminoaei, L., Aeloiza, E., Enjeti, P. N., & Blaabjerg, F. (2008). Shunt Active-Power-Filter Topology Based on Parallel Interleaved Inverters. *IEEE Transactions on Industrial Electronics*, *55*(3), 1175–1189. doi:10.1109/TIE.2007.907671

Blume, S. W. (2011). *High Voltage Protection for Telecommunications*. NY, USA: IEEE. doi:10.1002/9781118127018

Chu, R. C., Simons, R. E., Ellsworth, M. J., Schmidt, R. R., & Cozzolino, V. (2004). Review of cooling technologies for computer products. *IEEE Transactions on Device and Materials Reliability*, *4*(4), 568–585. doi:10.1109/TDMR.2004.840855

Gillot, C., Schaeffer, C., Massit, C., & Meysenc, L. (2001). Double-sided cooling for high power IGBT modules using flip chip technology. *IEEE Transactions on Components and Packaging Technologies*, *24*(4), 698–704. doi:10.1109/6144.974963

Gupta, R. K., Mohapatra, K. K., Somani, A., & Mohan, N. (2010). Direct-Matrix-Converter-Based Drive for a Three-Phase Open-End-Winding AC Machine with Advanced Features. *IEEE Transactions on Industrial Electronics*, *57*(12), 4032–4042. doi:10.1109/TIE.2010.2043045

Gurevich, V. (2003). *Protection Devices and Systems for High-Voltage Applications*. IL, USA: Marcel Dekker, Inc. doi:10.1201/9780203911648

IEEE. (2007). *IEEE Recommended Practice for the Design of Reliable Industrial and Commercial Power Systems*. NY, USA: IEEE.

Jung, J.-W., Lee, B.-H., Kim, D.-J., Hong, J.-P., Kim, J.-Y., Jeon, S.-M., & Song, D.-H. (2012). Mechanical Stress Reduction of Rotor Core of Interior Permanent Magnet Synchronous Motor. *IEEE Transactions on Magnetics*, *48*(2), 911–914. doi:10.1109/TMAG.2011.2172582

Karppinen, J. S., Li, J., & Paulasto-Kröckel, M. (2013). The Effects of Concurrent Power and Vibration Loads on the Reliability of Board-Level Interconnections in Power Electronic Assemblies. *IEEE Transactions on Device and Materials Reliability*, *13*(1), 16176. doi:10.1109/TDMR.2012.2226462

Kim, E. S., Joe, K. Y., Kye, M. H., Kim, Y. H., & Yoon, B. D. (1999). An improved soft-switching PWM FB DC/DC converter for reducing conduction losses. *IEEE Transactions on Power Electronics*, *14*(2), 258–264. doi:10.1109/63.750178

Kim, S.-H., Cherney, E. A., Hackam, R., & Rutherford, K. G. (1994). Chemical changes at the surface of RTV silicone rubber coatings on insulators during dry-band arcing. *IEEE Transactions on Dielectrics and Electrical Insulation*, *1*(1), 106–123. doi:10.1109/94.300238

Mudawar, I. (2001). Assessment of high-heat-flux thermal management schemes. *IEEE Transactions on Components and Packaging Technologies*, *24*(2), 122–141. doi:10.1109/6144.926375

Nussbaumer, T., Raggl, K., & Kolar, J. W. (2009). Design Guidelines for Interleaved Single-Phase Boost PFC Circuits. *IEEE Transactions on Industrial Electronics*, *56*(7), 2559–2573. doi:10.1109/TIE.2009.2020073

Pan, Ch.-T., & Lai, Ch.-M. (2010). A High-Efficiency High Step-Up Converter With Low Switch Voltage Stress for Fuel-Cell System Applications. *IEEE Transactions on Industrial Electronics*, *57*(6), 1998–2006. doi:10.1109/TIE.2009.2024100

Qun, Z., Lee, F. C., & Tsai, F.-S. (2002). Voltage and current stress reduction in single-stage power factor correction AC/DC converters with bulk capacitor voltage feedback. *IEEE Transactions on Power Electronics*, *17*(4), 47484.

Qun, Zh., Lee, F. C., & Tsai, F.-S. (2002). Voltage and current stress reduction in single-stage power factor correction AC/DC converters with bulk capacitor voltage feedback. *IEEE Transactions on Power Electronics*, *17*(4), 47484.

Shemshadi, A., Akbari, A., & Bathaee, T. (2013). A novel approach for reduction of electric field stress in vacuum interrupter chamber using advanced soft computing algorithms. *IEEE Transactions on Dielectrics and Electrical Insulation*, *20*(5), 1951–1958. doi:10.1109/TDEI.2013.6633729

Sheng, W. W., & Colino, R. P. (2005). *Power Electronic Modules*. Florida, USA: CRC.

Sippola, M., & Sepponen, R. E. (2002). Accurate prediction of high-frequency power-transformer losses and temperature rise. *IEEE Transactions on Power Electronics*, *17*(5), 835–847. doi:10.1109/TPEL.2002.802193

Tang, Y., Xie, Sh., Zhang, Ch., & Xu, Z. (2009). Improved Z-Source Inverter With Reduced Z-Source Capacitor Voltage Stress and Soft-Start Capability. *IEEE Transactions on Power Electronics*, *24*(2), 409–415. doi:10.1109/TPEL.2008.2006173

Vargas, R., Ammann, U., & Rodriguez, J. (2009). Predictive Approach to Increase Efficiency and Reduce Switching Losses on Matrix Converters. *IEEE Transactions on Power Electronics*, *24*(4), 894–902. doi:10.1109/TPEL.2008.2011907

Vargas, R., Ammann, U., Rodriguez, J., & Pontt, J. (2008). Predictive Strategy to Control Common-Mode Voltage in Loads Fed by Matrix Converters. *IEEE Transactions on Industrial Electronics*, *55*(12), 4372–4380. doi:10.1109/TIE.2008.2007016

Wang, H., & Blaabjerg, F. (2014). Reliability of Capacitors for DC-Link Applications in Power Electronic Converters-An Overview. *IEEE Transactions on Industry Applications*, *50*(5), 3569–3579. doi:10.1109/TIA.2014.2308357

Wang, P., Montanari, G. C., & Cavallini, A. (2014). Partial Discharge Phenomenology and Induced Aging Behavior in Rotating Machines Controlled by Power Electronics. *IEEE Transactions on Industrial Electronics*, *61*(12), 7105–7112. doi:10.1109/TIE.2014.2320226

Xu, D., Zhao, Ch., & Fan, H. (2004). A PWM plus phase-shift control bidirectional DC-DC converter. *IEEE Transactions on Power Electronics*, *19*(3), 666–675. doi:10.1109/TPEL.2004.826485

Yamazaki, K., & Ishigami, H. (2010). Rotor-Shape Optimization of Interior-Permanent-Magnet Motors to Reduce Harmonic Iron Losses. *IEEE Transactions on Industrial Electronics*, *57*(1), 61–69. doi:10.1109/TIE.2009.2025285

Yang, J., & Lee, F. C. (2015). New Modulation Scheme for Three-Level Active Neutral-Point-Clamped Converter with Loss and Stress Reduction. *IEEE Transactions on Industrial Electronics*, *62*(9), 5468–5479. doi:10.1109/TIE.2015.2405505

KEY TERMS AND DEFINITIONS

Coating: A thin layer used for hiding a surface against external environment.
Current Sharing: Achieved by parallel connection of devices and leads to low temperature operation.
Damper: A device for reducing the effect of any oscillation (mechanical) on system.
Inrush Current: The input current of a system or a part of a system with high value for a short period.
Voltage Sharing: A tool for reducing applied voltage on devices via series connection.

Section 4

Methods for Removing Faults

Chapter 8
Protection Systems

ABSTRACT

In the previous chapter, we assumed that there is no fault in the converter. To achieve a converter without failure, we presented the methods for stress reduction. However, a fault may occur in a system even operating with low stress. In the current chapter, we take one further step and assume that a fault occurs in the converter but there is a short time interval between fault occurrence and catastrophic damage to the converter. Therefore, the topic of this chapter is the methods for saving the converter in this condition. In this chapter, protection methods for saving the system against damaging faults are presented. Based on background of chapter two, protection systems should be able to bypass the effect of failure factors on electric power converter. Methods for current limiting and voltage clamping as the usual factors of failure in converters are described. Circuit diagram of a snubber is presented and its operation is described based on safe operating area of solid state power switches. Operating diagrams of fuses as emergency circuit breakers are presented. Measurement methods and devices used in protection systems are explained. Experimental samples and standard diagrams are presented to clarify the theoretical notes in all cases.

INTRODUCTION: PROTECTION FOR RAPID ISOLATION

In previous chapter, we tried to postpone the failure time with reducing the stress on converter. However, the converter must operate and it is logically affected by stress. It is true that we reduce the stress but the stress still exists and cause to failure in long term. Now, we are faced with a fault in the converter. What should we do? This chapter deals with one of the methods for preventing catastrophic damage in faulty converters: protection systems. Protection systems acts when a fault occurs in converter. Their performance is very important; isolation of the converter is not always the best choice because this strategy has a bad effect on availability of the converter. We talk about this manner in the next chapter. In the current chapter, we are not sensitive to this concern. However, we consider this concern from another view; any failure factor needs to a time interval for damaging the converter. If the protection system is very fast, it protects the converter but conflict with noise is possible. Therefore, protection systems should be fast enough to protect the converter but not so fast to incorrect operation. Figure 1 shows the

DOI: 10.4018/978-1-4666-9429-3.ch008

Figure 1. State of chapter 8 in the flowchart of the book

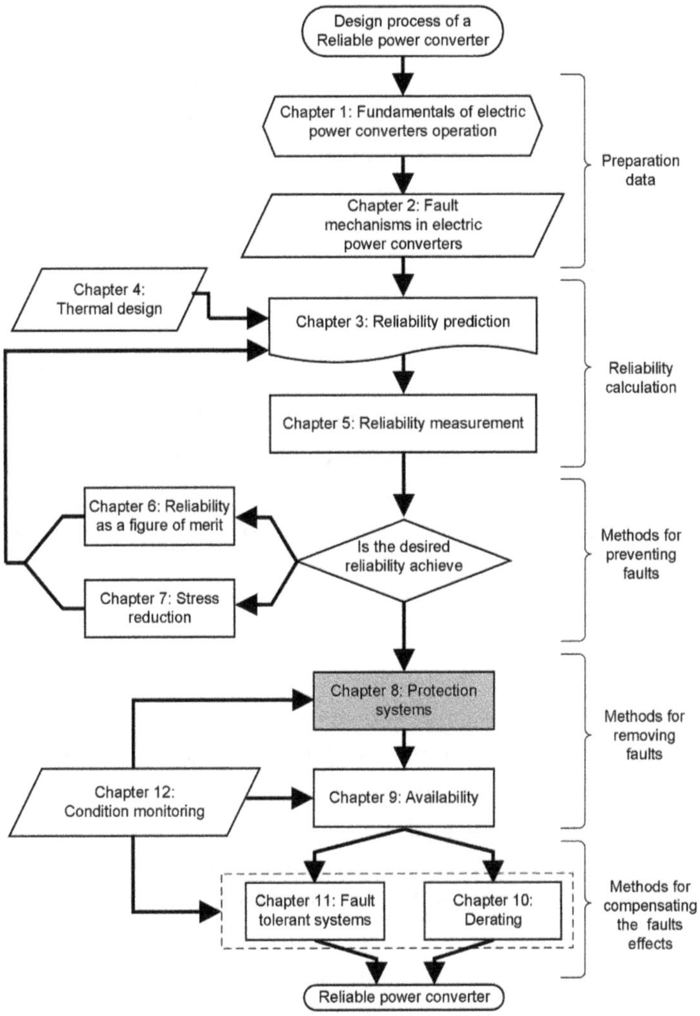

state of this chapter in the flowchart of the book. In an electric power system, a fault is any abnormal electric current. For example, a short circuit is a fault in which current bypasses the normal load. An open-circuit fault occurs if a circuit is interrupted by some failure. In three-phase systems, a fault may involve one or more phases and ground, or may occur only between phases. In a "ground fault" or "earth fault", charge flows into the earth. The prospective short circuit current of a fault can be calculated for power systems. In power systems, protective devices detect fault conditions and operate circuit breakers and other devices to limit the loss of service due to a failure.

In a polyphase system, a fault may affect all phases equally which is a "symmetrical fault". If only some phases are affected, the resulting "asymmetrical fault" becomes more complicated to analyze due to the simplifying assumption of equal current magnitude in all phases being no longer applicable. The analysis of this type of fault is often simplified by using methods such as symmetrical components.

Design of systems to detect and interrupt power system faults is the main objective of power system protection.

Transient Fault

A transient fault is a fault that is no longer present if power is disconnected for a short time and then restored. Many faults in overhead power lines are transient in nature. When a fault occurs, equipment used for power system protection operate to isolate the area of the fault. A transient fault will then clear and the power-line can be returned to service. Typical examples of transient faults include:

- Momentary tree contact
- Bird or other animal contact
- Lightning strike
- Conductor clashing

Transmission and distribution systems use an automatic re-close function which is commonly used on overhead lines to attempt to restore power in the event of a transient fault. This functionality is not as common on underground systems as faults there are typically of a persistent nature. Transient faults may still cause damage both at the site of the original fault or elsewhere in the network as fault current is generated.

Fault Management

Cascading failures usually begin when one part of the system fails. When this happens, nearby nodes must then take up the slack for the failed component. This in turn overloads these nodes, causing them to fail as well, prompting additional nodes to fail one after another in what is also known as vicious circle

Cascading failure is common in power grids when one of the elements fails (completely or partially) and shifts its load to nearby elements in the system. Those nearby elements are then pushed beyond their capacity so they become overloaded and shift their load onto other elements. Cascading failure is a common effect seen in high voltage systems, where a single point of failure (SPF) on a fully loaded or slightly overloaded system results in a sudden spike across all nodes of the system. This surge current can induce the already overloaded nodes into failure, setting off more overloads and thereby taking down the entire system in a very short time.

This failure process cascades through the elements of the system like a ripple on a pond and continues until substantially all of the elements in the system are compromised and/or the system becomes functionally disconnected from the source of its load. For example, under certain conditions a large power grid can collapse after the failure of a single transformer.

Monitoring the operation of a system, in real-time, and judicious disconnection of parts can help stop a cascade. Another common technique is to calculate a safety margin for the system by computer simulation of possible failures, to establish safe operating levels below which none of the calculated scenarios is predicted to cause cascading failure, and to identify the parts of the network which are most likely to cause cascading failures.

One of the primary problems with preventing electrical grid failures is that the speed of the control signal is no faster than the speed of the propagating power overload, i.e. since both the control signal and the electrical power are moving at the same speed, it is not possible to isolate the outage by sending a warning ahead to isolate the element. To ameliorate this systemic defect, superconducting magnetic energy storage units at critical junctions can store or release power for a few seconds to allow control systems to catch up and actuate isolating procedures

A catastrophic failure is a sudden and total failure from which recovery is impossible. Catastrophic failures often lead to cascading systems failure. The term is most commonly used for structural failures, but has often been extended to many other disciplines where total and irrecoverable loss occurs. Such failures are investigated using the methods of forensic engineering, which aims to isolate the cause or causes of failure.

Catastrophic failure is also observed in the steam turbine rotor occurring due to peak stress acting on the rotor and also the stress concentration increases which ultimately leads to the failure of the disc.

In firearms, catastrophic failure usually refers to a rupture or disintegration of the barrel or receiver of the gun when firing it. Some possible causes for this is the gun being out of battery, having improper headspace, using incorrect ammunition for the firearm, using ammunition with an incorrect propellant charge, an obstructed or partially obstructed barrel, or weakened metal in the barrel or receiver 1.

Types of Protection

- **Generator Sets:** In a power plant, the protective relays are intended to prevent damage to alternators or to the transformers in case of abnormal conditions of operation, due to internal failures, as well as insulating failures or regulation malfunctions. Such failures are unusual, so the protective relays have to operate very rarely. If a protective relay fails to detect a fault, the resulting damage to the alternator or to the transformer might require costly equipment repairs or replacement, as well as income loss from the inability to produce and sell energy.
- **High-Voltage Transmission Network:** Protection on the transmission and distribution serves two functions: Protection of plant and protection of the public (including employees). At a basic level, protection looks to disconnect equipment which experience an overload or a short to earth. Some items in substations such as transformers might require additional protection based on temperature or gas pressure, among others.
- **Overload and Back-Up for Distance (Overcurrent):** Overload protection requires a current transformer which simply measures the current in a circuit. There are two types of overload protection: instantaneous overcurrent and time overcurrent (TOC). Instantaneous overcurrent requires that the current exceeds a predetermined level for the circuit breaker to operate. TOC protection operates based on a current vs time curve. Based on this curve if the measured current exceeds a given level for the preset amount of time, the circuit breaker or fuse will operate.
- **Earth Fault ("Ground Fault" in the United States):** Earth fault protection again requires current transformers and senses an imbalance in a three-phase circuit. Normally the three phase currents are in balance, i.e. roughly equal in magnitude. If one or two phases become connected to earth via a low impedance path, their magnitudes will increase dramatically, as will current imbalance. If this imbalance exceeds a pre-determined value, a circuit breaker should operate. Restricted earth fault protection is a type of earth fault protection which looks for earth fault between two sets current transformers (hence restricted to that zone).
- **Distance (Impedance Relay):** Distance protection detects both voltage and current. A fault on a circuit will generally create a sag in the voltage level. If the ratio of voltage to current measured at the relay terminals, which equates to an impedance, lands within a predetermined level the circuit breaker will operate. This is useful for reasonable length lines, lines longer than 10 miles, because

its operating characteristics are based on the line characteristics. This means that when a fault appears on the line the impedance setting in the relay is compared to the apparent impedance of the line from the relay terminals to the fault. If the relay setting is determined to be below the apparent impedance it is determined that the fault is within the zone of protection. When the transmission line length is too short, less than 10 miles, distance protection becomes more difficult to coordinate. In these instances the best choice of protection is current differential protection.

- **Back-Up:** The objective of protection is to remove only the affected portion of plant and nothing else. A circuit breaker or protection relay may fail to operate. In important systems, a failure of primary protection will usually result in the operation of back-up protection. Remote back-up protection will generally remove both the affected and unaffected items of plant to clear the fault. Local back-up protection will remove the affected items of the plant to clear the fault.
- **Low-Voltage Networks:** The low-voltage network generally relies upon fuses or low-voltage circuit breakers to remove both overload and earth faults.

Protective Device

A digital (numeric) multi-function protective relay for distribution networks. A single such device can replace many single-function electromechanical relays, and provides self-testing and communication functions.

- Protective relays control the tripping of the circuit breakers surrounding the faulted part of the network
- Automatic operation, such as auto-reclosing or system restart
- Monitoring equipment which collects data on the system for post event analysis

While the operating quality of these devices, and especially of protective relays, is always critical, different strategies are considered for protecting the different parts of the system. Very important equipment may have completely redundant and independent protective systems, while a minor branch distribution line may have very simple low-cost protection.

There are three parts of protective devices:

- Instrument transformer: current or potential (CT or VT)
- Relay
- Circuit breaker

Advantages of protected devices with these three basic components include safety, economy, and accuracy.

- **Safety:** Instrument transformers create electrical isolation from the power system, and thus establishing a safer environment for personnel working with the relays.
- **Economy:** Relays are able to be simpler, smaller, and cheaper given lower-level relay inputs.
- **Accuracy:** Power system voltages and currents are accurately reproduced by instrument transformers over large operating ranges.

Coordination

Protective device coordination is the process of determining the "best fit" timing of current interruption when abnormal electrical conditions occur. The goal is to minimize an outage to the greatest extent possible. Historically, protective device coordination was done on translucent log–log paper. Modern methods normally include detailed computer based analysis and reporting.

Protection coordination is also handled through dividing the power system into protective zones. If a fault were to occur in a given zone, necessary actions will be executed to isolate that zone from the entire system. Zone definitions account for generators, buses, transformers, transmission and distribution lines, and motors. Additionally, zones possess the following features: zones overlap, overlap regions denote circuit breakers, and all circuit breakers in a given zone with a fault will open in order to isolate the fault. Overlapped regions are created by two sets of instrument transformers and relays for each circuit breaker. They are designed for redundancy to eliminate unprotected areas; however, overlapped regions are devised to remain as small as possible such that when a fault occurs in an overlap region and the two zones which encompass the fault are isolated, the sector of the power system which is lost from service is still small despite two zones being isolated.

THERMAL PROTECTION

Protection Against Over Current

Power loss generated by over current in circuits is a reason of thermal failure. Protective sections in power converters prevent this fail (Aghaei, Karimi, & Kaboli, 2014). Figure 2 shows the simplest way to measure the current in a converter. In this method, a series resistor is used to converter the current to voltage. Resistance of this resistor is very low to prevent considerable voltage drop and conduction loss. The produced voltage may be need to be amplified as shown in Figure 3. Figure 4 shows a typical current limiter circuit used in output stage of a power supply. Figure 5 shows the performance of current limiter circuit during an overcurrent fault in the converter.

In an electric power system, overcurrent or excess current is a situation where a larger than intended electric current exists through a conductor, leading to excessive generation of heat, and the risk of fire or

Figure 2. A shunt resistor for current sensing

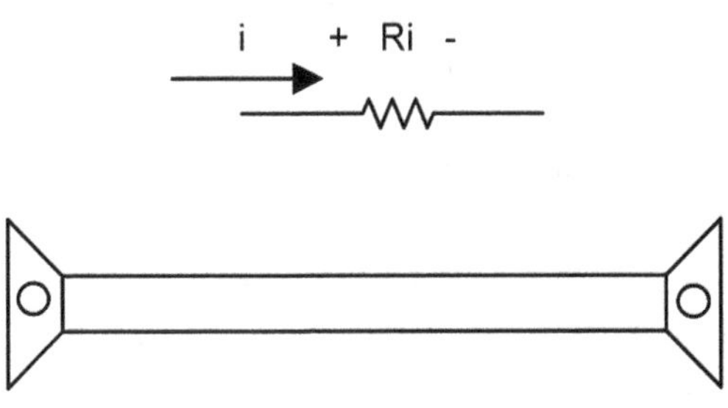

Figure 3. A typical circuit for current sensing

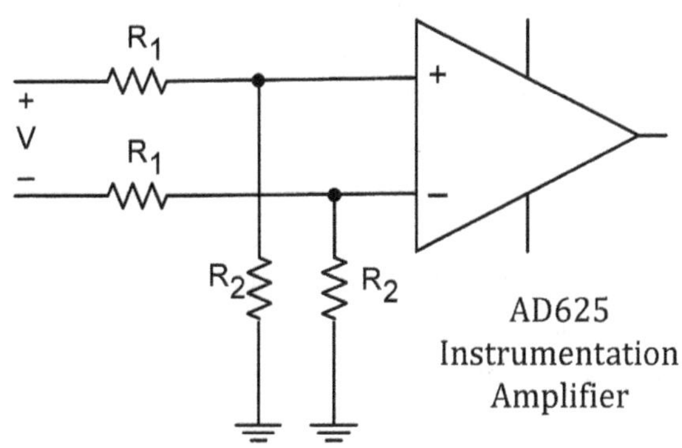

Figure 4. Current protection of a power MOSFET

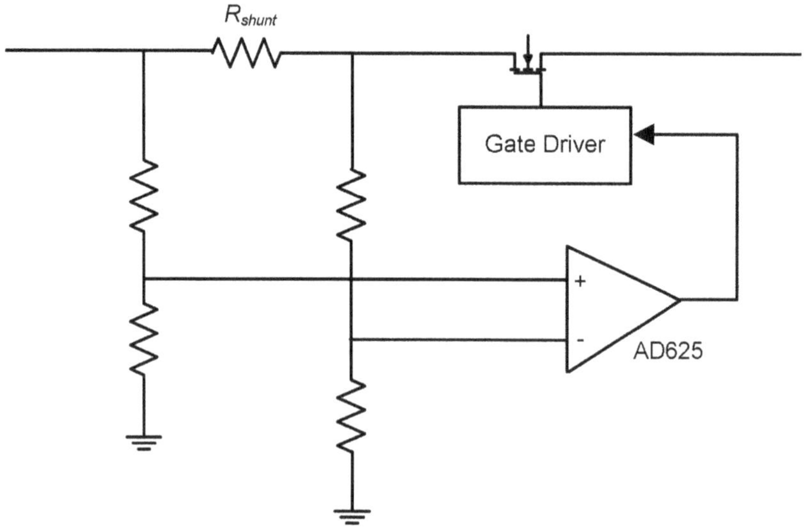

Figure 5. Typical operation of a current limiter

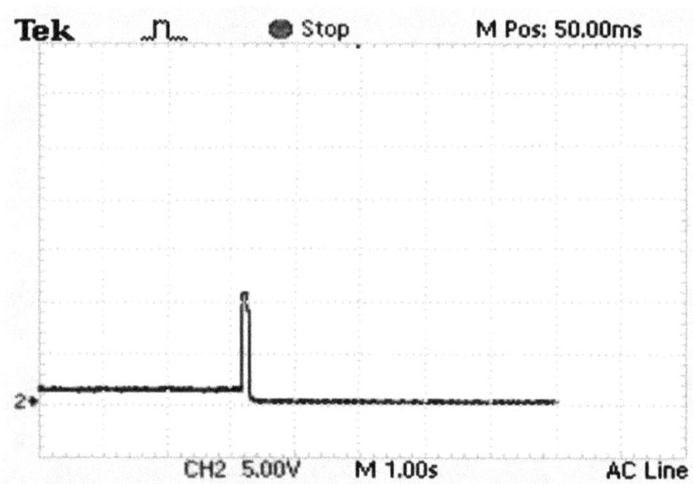

damage to equipment. Possible causes for overcurrent include short circuits, excessive load, and incorrect design. Fuses, circuit breakers, temperature sensors and current limiters are commonly used protection mechanisms to control the risks of overcurrent.

Current limiting is the practice in electrical or electronic circuits of imposing an upper limit on the current that may be delivered to a load with the purpose of protecting the circuit generating or transmitting the current from harmful effects due to a short-circuit or similar problem in the load.

The simplest form of current limiting for mains is a fuse. As the current exceeds the fuse's limits it blows thereby disconnecting the load from the source (Najmi, Ebrahimi, & Oraee, 2012). This method is most commonly used for protecting the house-hold mains. A circuit breaker is another device for mains current limiting.

Compared to circuit breakers, fuses attain faster current limitation by means of arc quenching. Since fuses are passive elements, they are inherently secure. Their drawback is that once blown, they need to be replaced.

A circuit breaker is an automatically operated electrical switch designed to protect an electrical circuit from damage caused by overload or short circuit. Its basic function is to detect a fault condition and interrupt current flow. Unlike a fuse, which operates once and then must be replaced, a circuit breaker can be reset (either manually or automatically) to resume normal operation. Circuit breakers are made in varying sizes, from small devices that protect an individual household appliance up to large switchgear designed to protect high voltage circuits feeding an entire city.

Once a fault is detected, contacts within the circuit breaker must open to interrupt the circuit; some mechanically-stored energy (using something such as springs or compressed air) contained within the breaker is used to separate the contacts, although some of the energy required may be obtained from the fault current itself. Small circuit breakers may be manually operated, larger units have solenoids to trip the mechanism, and electric motors to restore energy to the springs.

Thermal magnetic circuit breakers, which are the type found in most distribution boards, incorporate both techniques with the electromagnet responding instantaneously to large surges in current (short circuits) and the bimetallic strip responding to less extreme but longer-term over-current conditions. The thermal portion of the circuit breaker provides an "inverse time" response feature, which trips the circuit breaker sooner for larger overcurrents. Figure 6 shows application of various fuses in a PDU.

Fuse

A fuse interrupts excessive current (blows) so that further damage by overheating or fire is prevented. Wiring regulations often define a maximum fuse current rating for particular circuits. Overcurrent protection devices are essential in electrical systems to limit threats to human life and property damage. The time and current operating characteristics of fuses are chosen to provide adequate protection without needless interruption. Slow blow fuses are designed to allow harmless short term currents over their rating while still interrupting a sustained overload. Fuses are manufactured in a wide range of current and voltage ratings to protect wiring systems and electrical equipment. Self-resetting fuses automatically restore the circuit after the overload has cleared, and are useful in environments where a human replacing a blown fuse would be difficult or impossible, for example in aerospace or nuclear applications.

The speed at which a fuse blows depends on how much current flows through it and the material of which the fuse is made. The operating time is not a fixed interval, but decreases as the current increases. Fuses have different characteristics of operating time compared to current, characterized as fast-blow, slow-blow, or time-delay, according to time required to respond to an overcurrent condition. A standard fuse may require twice its rated current to open in one second, a fast-blow fuse may require twice its rated current to blow in 0.1 seconds, and a slow-blow fuse may require twice its rated current for tens of seconds to blow (Eric, & Lehn, 2006).

Fuse selection depends on the load's characteristics. Semiconductor devices may use a fast or ultrafast fuse as semiconductor devices heat rapidly when excess current flows. The fastest blowing fuses are designed for the most sensitive electrical equipment, where even a short exposure to an overload current could be very damaging. Normal fast-blow fuses are the most general purpose fuses. The time delay fuse (also known as anti-surge, or slow-blow) are designed to allow a current which is above the rated value of the fuse to flow for a short period of time without the fuse blowing. These types of fuse are used on

Figure 6. Application of fuse in various branches of a PDU with coordination

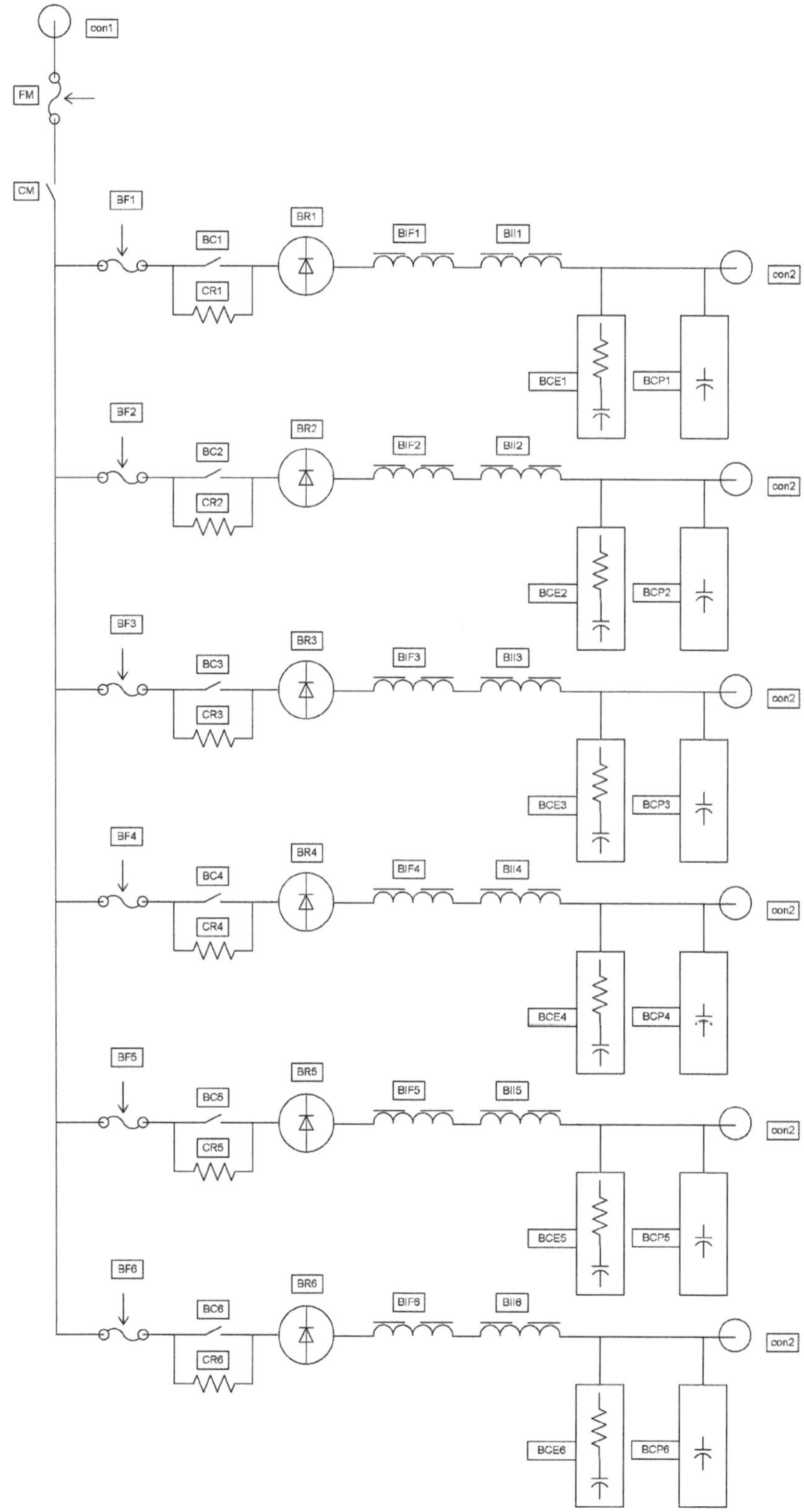

Figure 7. Conventional overload curve of a fuse (IEC publication 60146-6 ed1.0 Copyright ©1992 of publication IEC Geneva, Switzerland. www.iec.ch)

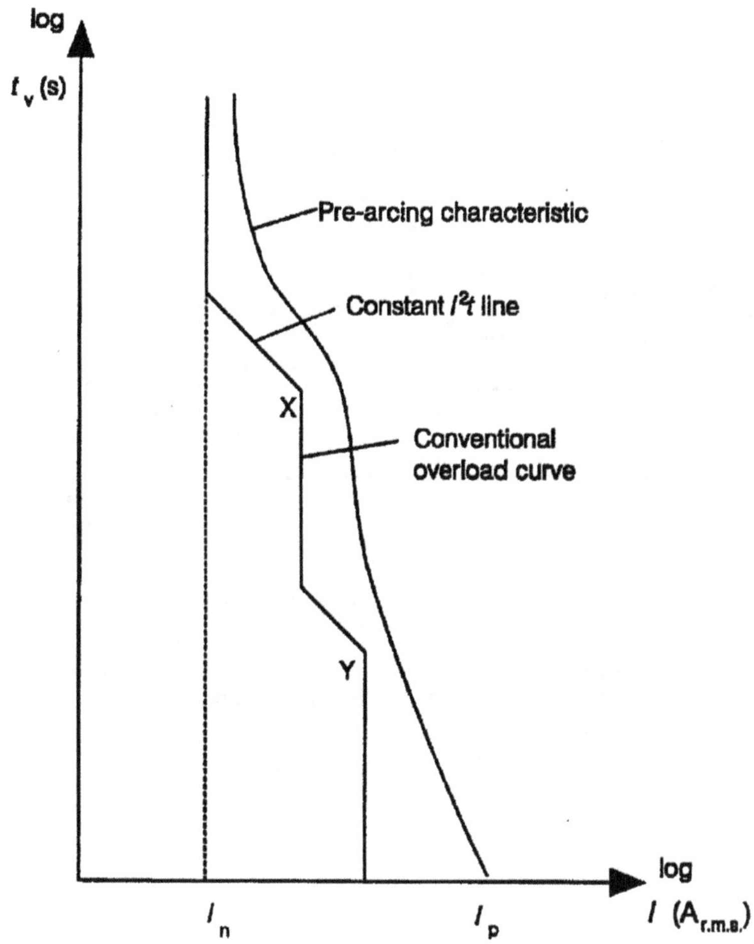

equipment such as motors, which can draw larger than normal currents for up to several seconds while coming up to speed. Figure 7 shows Time-Current characteristics of typical fuses. Figures 8 and 9 show induction motor initial current in two various starting method which are used for setting the fuse rating.

Shoot Through and Short Circuit

Shoot Through

Shoot-through failure is one of the most dangerous failure modes encountered in conventional inverter circuits. Shoot-through failures occur when two series semiconductor switches are inadvertently turned on simultaneously, causing the flow of extremely high currents that destroy the device (Wheeler, Rodriguez, Clare, Empringham, & Weinstein, 2002).

Figure 10 shows an inverter leg. This is a commonly used configuration in DC to AC converters which is constructed by series connection of two switches. Gate pulses of the high side and low side

Figure 8. Starting current and torque of an induction motor with autotransformer as a starter (IEC publication 60947-4-1 ed.3.1 Copyright ©2012 of publication IEC Geneva, Switzerland. www.iec.ch)

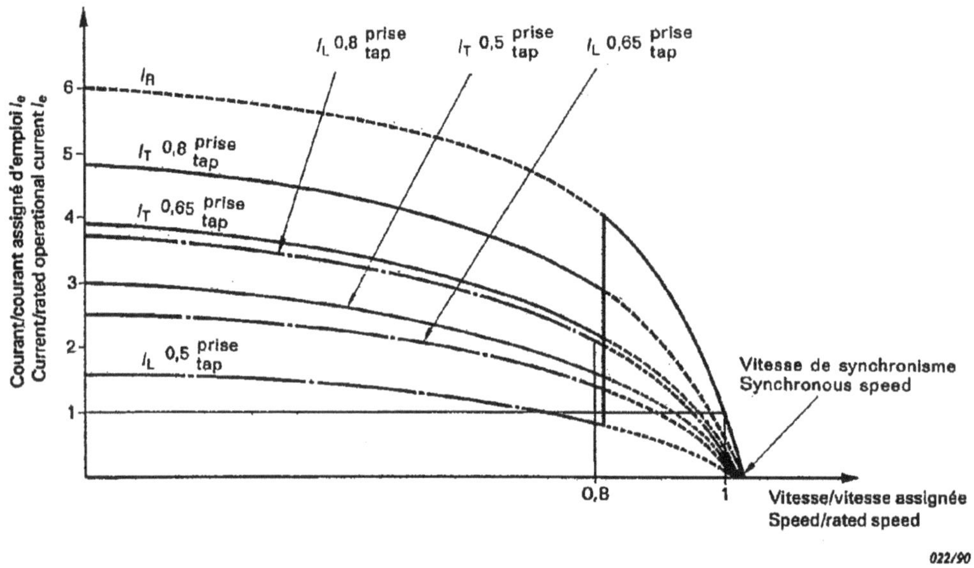

I_R = courant dans le moteur sous tension nominale
motor current at rated voltage

I_T = courant dans le moteur sous tension réduite
motor current at reduced voltage

I_L = courant en ligne sous tension réduite
line current at reduced voltage

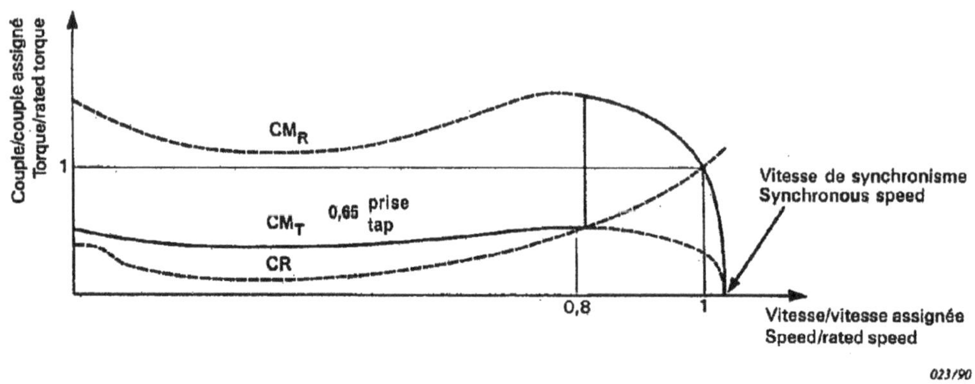

switches are shown in Figure 11. To prevent shoot through in the leg, each switch should be turn off before another switch is turn on as shown in Figure 12.

Example

Power transformers in service experience different types of transient overvoltages. One of them is effects of lightning. Lightning overvoltages are characterized by very steep initial rate of rise of voltage

Figure 9. Starting current and torque of an induction motor with star-delta start (IEC publication 60947-4-1 ed.3.1 Copyright © 2012 of publication IEC Geneva, Switzerland. www.iec.ch)

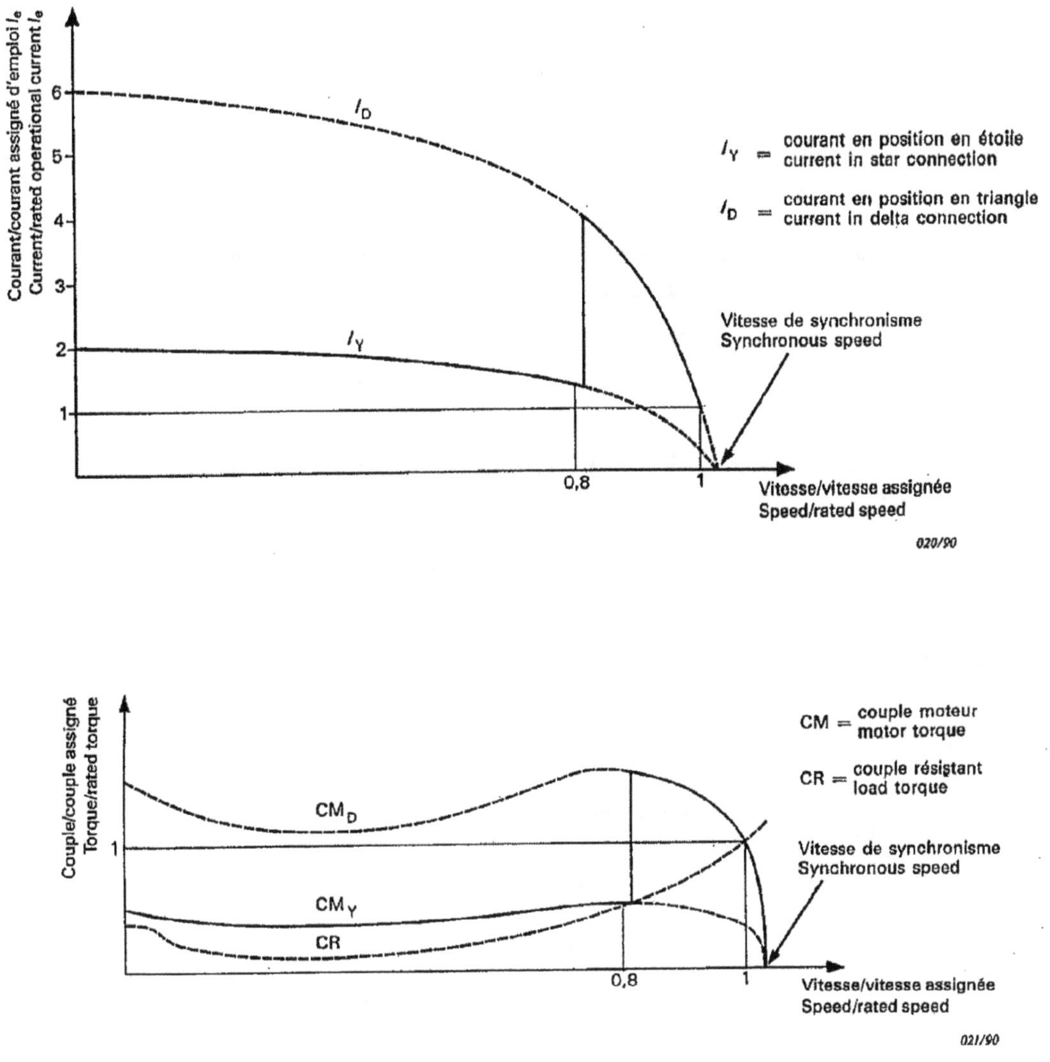

Figure 10. An inverter leg consists of two power switches

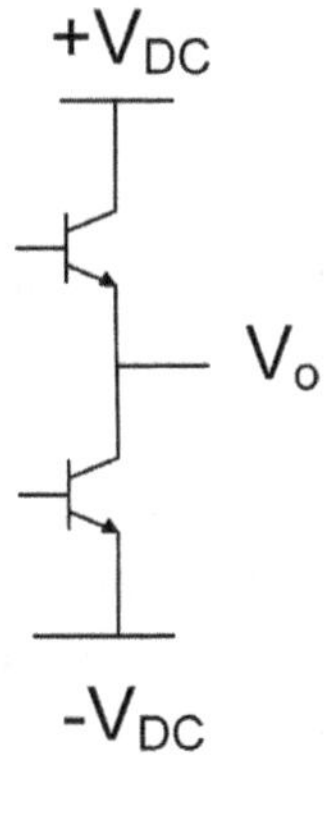

Figure 11. Gates pulses of a leg

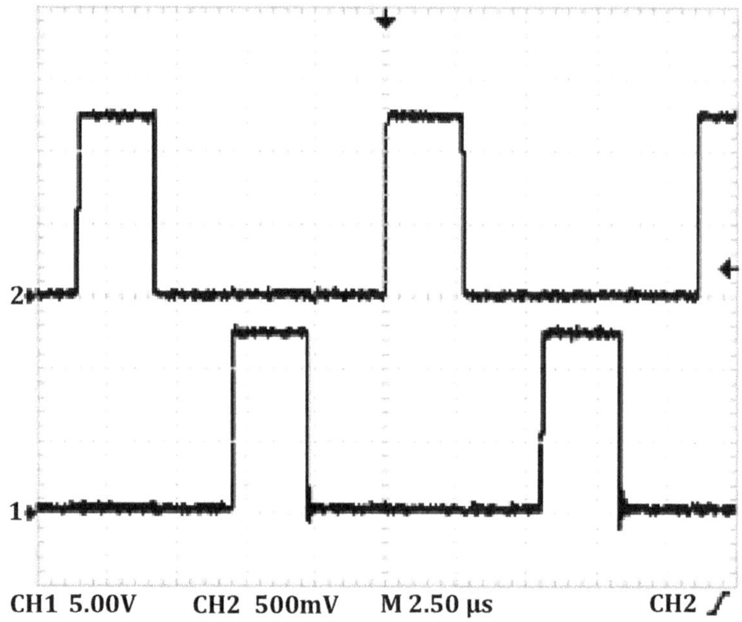

Figure 12. Dead zone between gate pulses of a leg

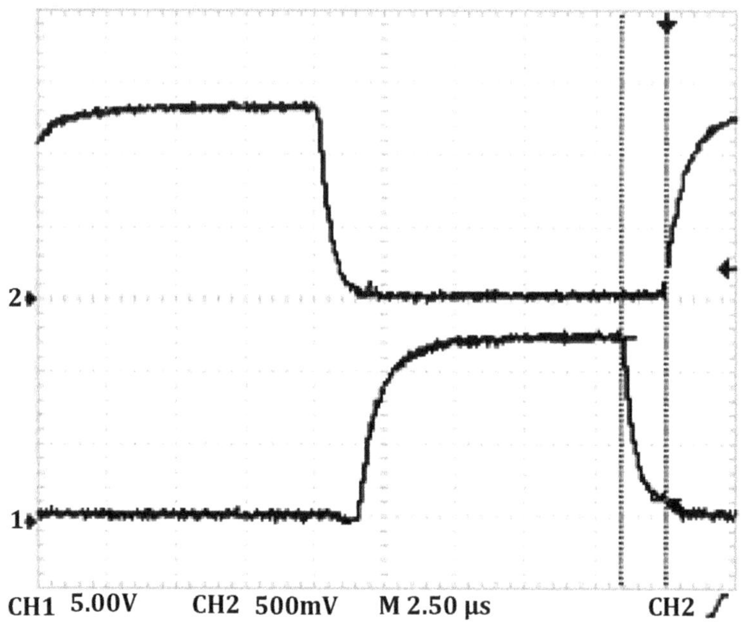

and relatively slower rate of fall of voltage with respect to time. Surge voltages with steep front time are most important type of transient overvoltages which can cause damage to insulation of transformer windings. The practice of design and construction of HV transformer windings to withstand effects of surge voltages with steep fronts appearing at its line terminal has been to provide a surge voltage distribution controlling ring at line end which is also called as static end ring. Other method of improving the steep front surge voltage withstand performance of power transformer HV windings is to increase the series capacitance between turns of sections or discs of winding known as interleaving 1-3.

MOA blocks have been known to be able to provide very good surge protection to equipments used in electrical power stations and substations for more than a decade. Their technology has advanced to highly satisfactory levels as to be able to provide surge absorbers having required type of characteristics to suit different applications 4, 6, 7.

Line Conditioners

In addition to high-voltage and current conditions, other problems can occur with incoming power. The voltage might dip below the level needed to run the system, resulting in a brownout. Forms of electrical noise other than simple voltage surges or spikes might travel through the power line, such as radio-frequency interference or electrical noise caused by motors or other inductive loads.

Remember two things when you wire together digital devices (such as computers and their peripherals):

- Any wire can act as an antenna and have voltage induced in it by nearby electromagnetic fields, which can come from other wires, telephones, CRTs, motors, fluorescent fixtures, static discharge, and, of course, radio transmitters.
- Digital circuitry responds with surprising efficiency to noise of even a volt or two, making those induced voltages particularly troublesome. The electrical wiring in your building can act as an antenna, picking up all kinds of noise and disturbances.

A line conditioner can handle many of these types of problems. It filters the power, bridges brownouts, suppresses high-voltage and current conditions, and generally acts as a buffer between the power line and the system. A line conditioner does the job of a surge suppressor, and much more (& 2007). It is more of an active device, functioning continuously, rather than a passive device that activates only when a surge is present. A line conditioner provides true power conditioning and can handle myriad problems. It contains transformers, capacitors, and other circuitry that can temporarily bridge a brownout or low-voltage situation

Types of Current Sensing

There are two types of current sensing: direct and indirect. Indirect current sensing is based on Ampere's and Faraday's laws. By placing a coil (for instance Rogowski coil) around a current-carrying conductor, a voltage is induced across the coil that is proportional to the current. This allows for a non-invasive measurement where the sensing circuitry is not electrically connected to the monitored system. Since there is no direct connection between the sensing circuitry and the system, the system is inherently isolated. Indirect current sensing typically is used for load currents in the 100A-1000A range. This type of sensing, however, requires relatively expensive sensors and is not conducive to sensing currents on a PCB. Figure 13 and 14 shows the equivalent circuit if a CT for measuring the output current of a converter and its simplified structure. Time delay in measuring the current is an important limitation of indirect current measurement methods. This time delay may lead to damage the converter because the fault current is removed with delay. Figure 15 and 16 shows a comparison between two methods of cur-

Figure 13. Application of CT for current sensing

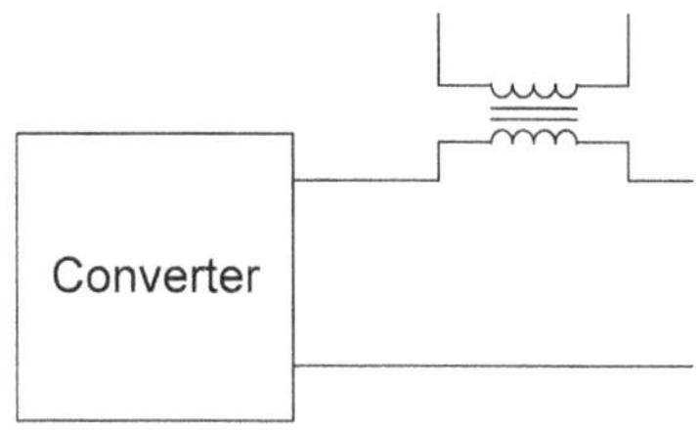

Figure 14. Simplified structure of a CT

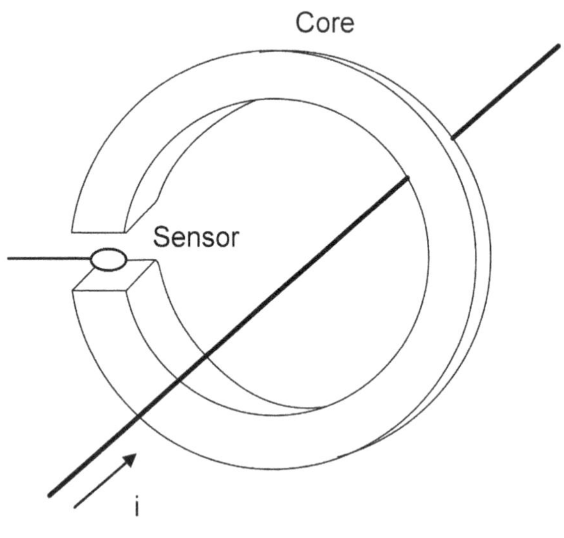

rent sensing. There is not considerable delay in in direct method and the over current fault is removed immediately as shown in Figure 15. However, there is a considerable time delay in indirect method as shown in Figure 16.

Direct current sensing is based on Ohm's law. By placing a shunt resistor in series with the system load, a voltage is generated across the shunt resistor that is proportional to the system load current. The voltage across the shunt can be measured by differential amplifiers such as current shunt monitors (CSMs), operational amplifiers (op amps), difference amplifiers (DAs), or instrumentation amplifiers (IAs). This method is an invasive measurement of the current since the shunt resistor and sensing circuitry are electrically connected to the monitored system. Therefore, direct sensing typically is used when galvanic isolation is not required. The shunt resistor also dissipates power, which may not be desirable. Direct current sensing typically is implemented for load currents <100A. The remainder of this particular article series will focus on direct current sensing 1.

Figure 15. Operation of a current limiter without delay between fault current and switch command

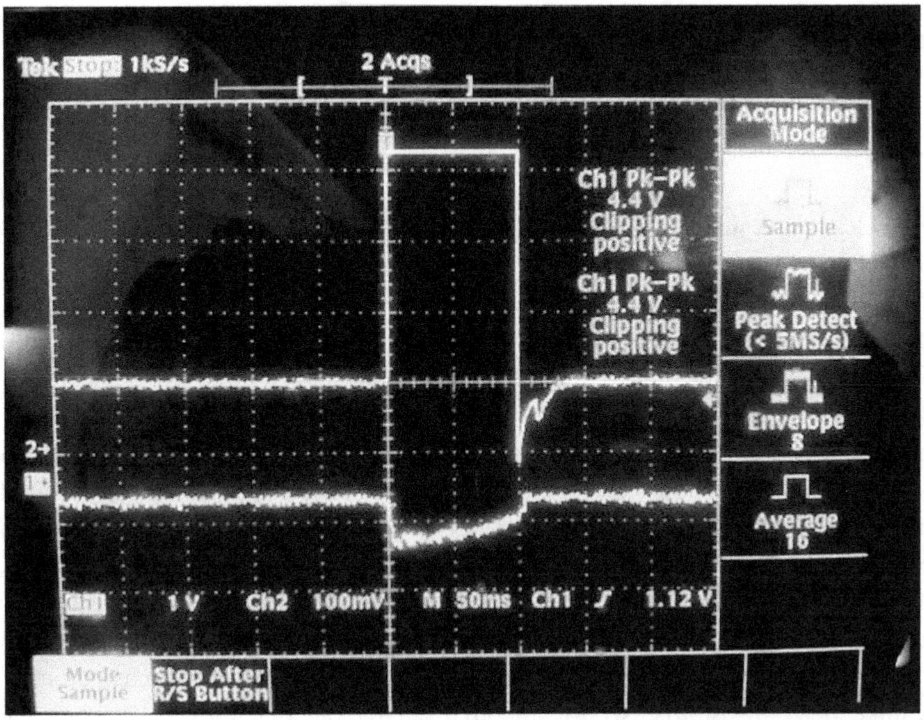

Figure 16. Operation of a current limiter with delay between fault current and switch command

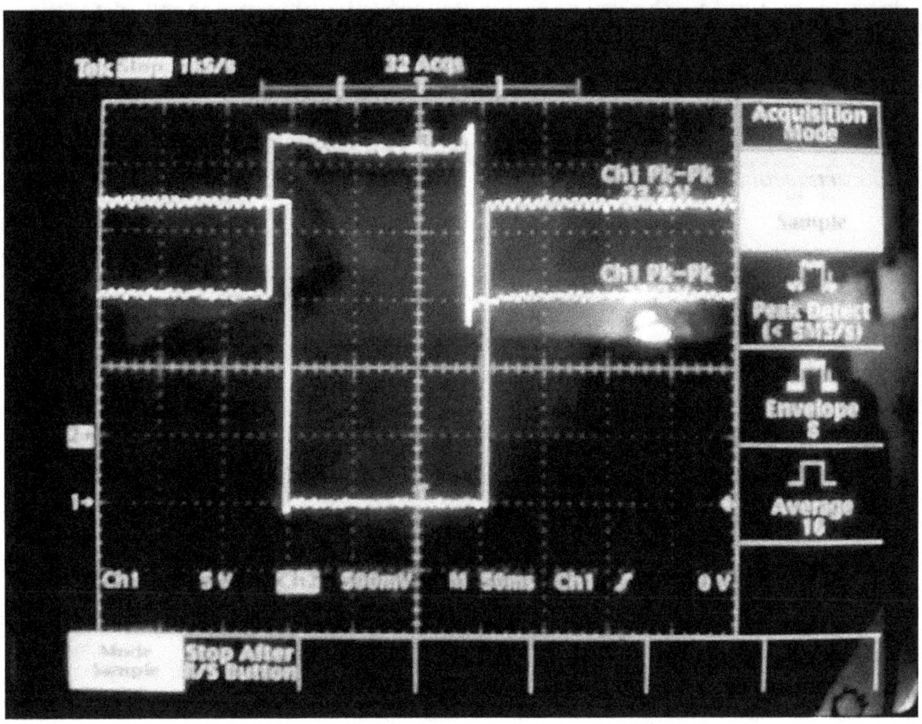

ELECTRICAL PROTECTION

Protection Against Over Voltage

Snubber

Snubbers are frequently used in electrical systems with an inductive load where the sudden interruption of current flow leads to a sharp rise in voltage across the current switching device, in accordance with Faraday's law. This transient can be a source of electromagnetic interference (EMI) in other circuits. Additionally, if the voltage generated across the device is beyond what the device is intended to tolerate, it may damage or destroy it. The snubber provides a short-term alternative current path around the current switching device so that the inductive element may be discharged more safely and quietly. Inductive elements are often unintentional, but arise from the current loops implied by physical circuitry. While current switching is everywhere, snubbers will generally only be required where a major current path is switched, such as in power supplies. Snubbers are also often used to prevent arcing across the contacts of relays and switches and the electrical interference and welding/sticking of the contacts that can occur. Figure 17 shows a switching circuit with inductive load. The series inductor can be generated with network cables or leakage inductance of transformers and motors. Figure 18 shows a turn of snubber which is implemented across the switch. RCD turn off snubber is one of the commonly used snubbers in power electronics as shown in Figure 19. The key element of this snubber is its capacitor for preventing the voltage spike. Series resistor is used for smooth discharge of capacitor in the next switch turn on as shown in Figure 20.

A simple snubber uses a small resistor (R) in series with a small capacitor (C). This combination can be used to suppress the rapid rise in voltage across a thyristor, preventing the erroneous turn-on of the thyristor; it does this by limiting the rate of rise in voltage (dV/dt) across the thyristor to a value which will not trigger it. An appropriately-designed RC snubber can be used with either DC or AC loads. This sort of snubber is commonly used with inductive loads such as electric motors. The voltage across a

Figure 17. An inductive load in a switching circuit

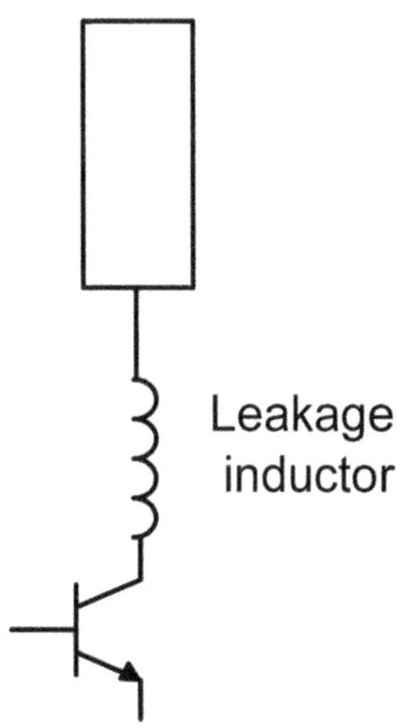

Figure 18. Placement of a parallel turn off snubber

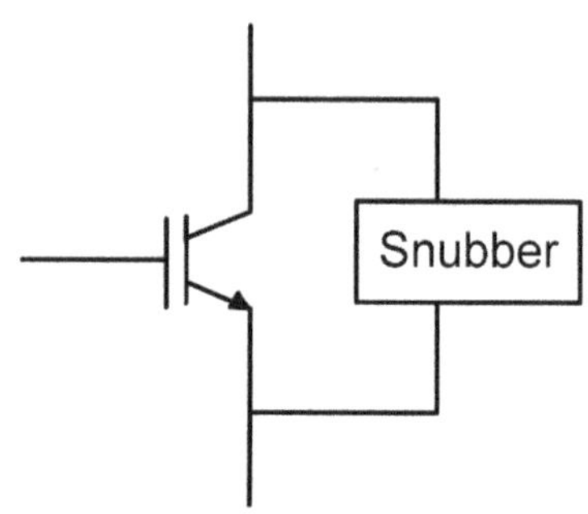

Figure 19. Circuit diagram of a turn off RCD snubber

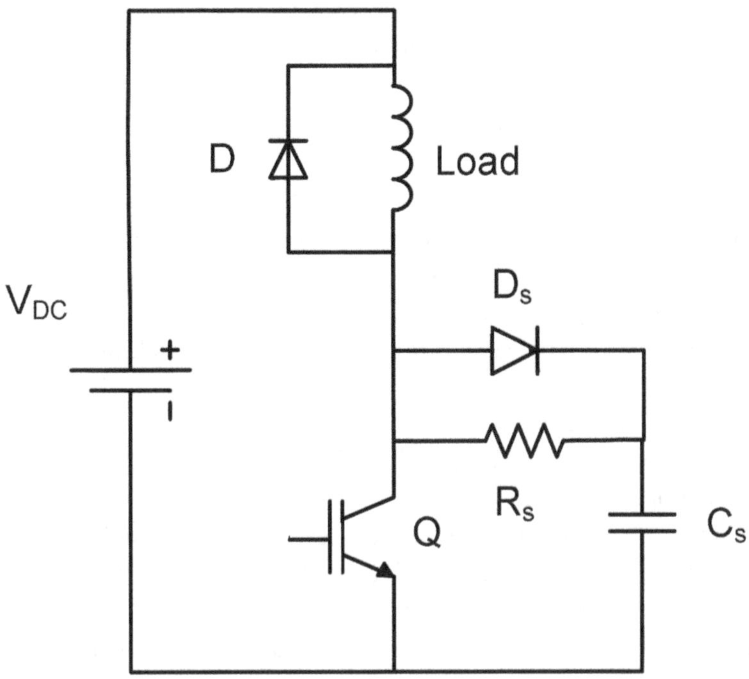

Figure 20. RC snubber of a power diode

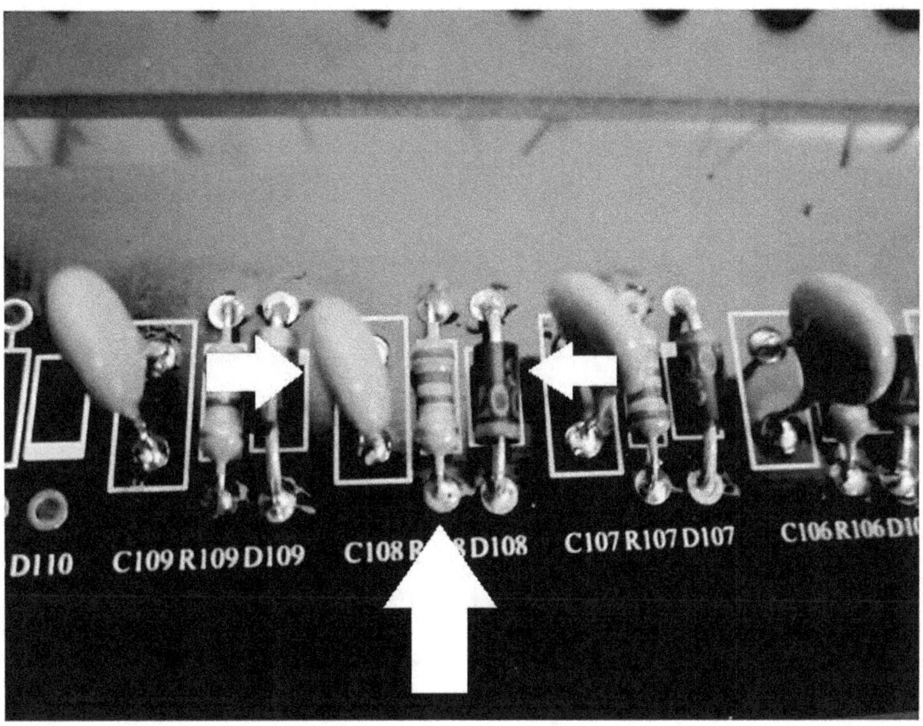

capacitor cannot change instantaneously, so a decreasing transient current will flow through it for a small fraction of a second, allowing the voltage across the switch to increase more slowly when the switch is opened. Determination of voltage rating can be difficult owing to the nature of transient waveforms, and may be defined simply by the power rating of the snubber components and the application. RC snubbers can be made discretely and are also built as a single component

A transient voltage suppressor or TVS is a general classification of an array of devices that are designed to react to sudden or momentary overvoltage conditions. One such common device used for this purpose is known as the transient voltage suppression diode that is simply a Zener diode designed to protect electronics device against overvoltages. Another design alternative applies a family of products that are known as metal-oxide varistor (MOV) that protect electronic circuits and electrical equipment.

The characteristic of a TVS requires that it respond to overvoltages faster than other common overvoltage protection components such as varistors or gas discharge tubes. This makes TVS devices or components useful for protection against very fast and often damaging voltage spikes. These fast overvoltage spikes are present on all distribution networks and can be caused by either internal or external events, such as lightning or motor arcing.

Applications of transient voltage suppression diodes are used for unidirectional or bidirectional electrostatic discharge protection of transmission or data lines in electronic circuits. MOV based TVSs are utilized to protect home electronics, distribution systems and may accommodate industrial level power distribution disturbances saving downtime and damage to equipment. The level of energy in a transient overvoltage can be equated to energy measured in joules or related to electric current when devices are rated for various applications. These bursts of overvoltage can be measured with specialized electronic meters that can show power disturbances of thousands of volts amplitude that last for a few microseconds or less

Line Capacitor

Figure 21 shows application of a capacitor as DC line suppressor. This capacitor should be as close as possible to inverter leg as shown in Figures 23 and 24. Any extra path leads to series inductor and prevent proper operation of this capacitor as shown in Figure 22.

Figure 21. Line capacitor as the snubber of an inverter leg

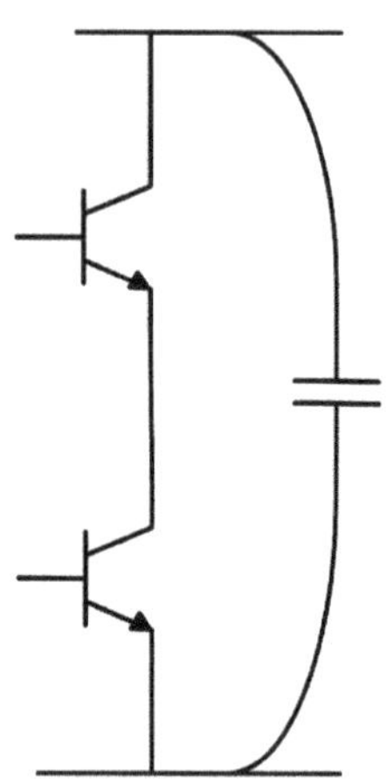

Figure 22. Incorrect connection of line snubber in an inverter

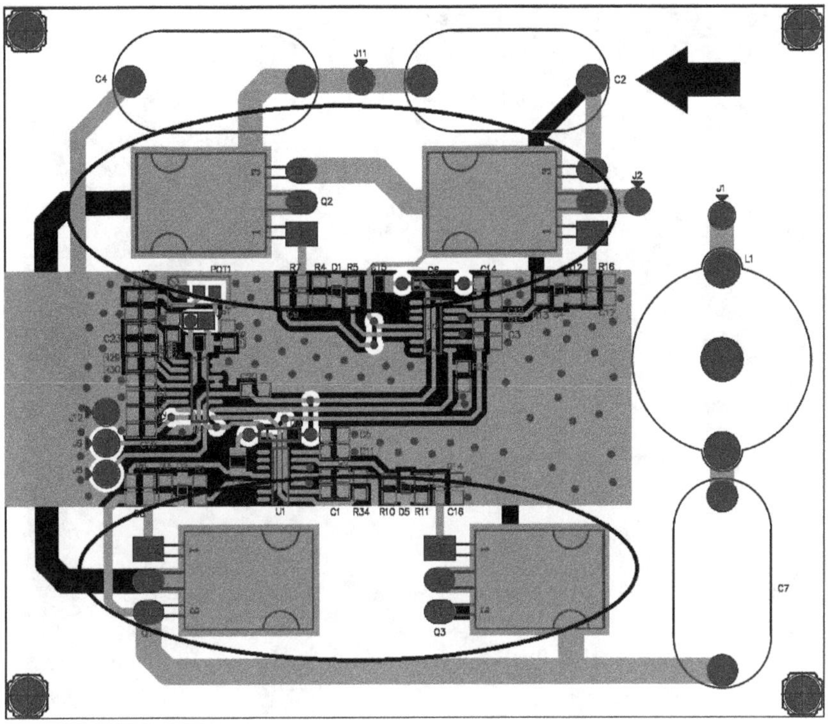

Figure 23. Line snubbers of a three phase inverter

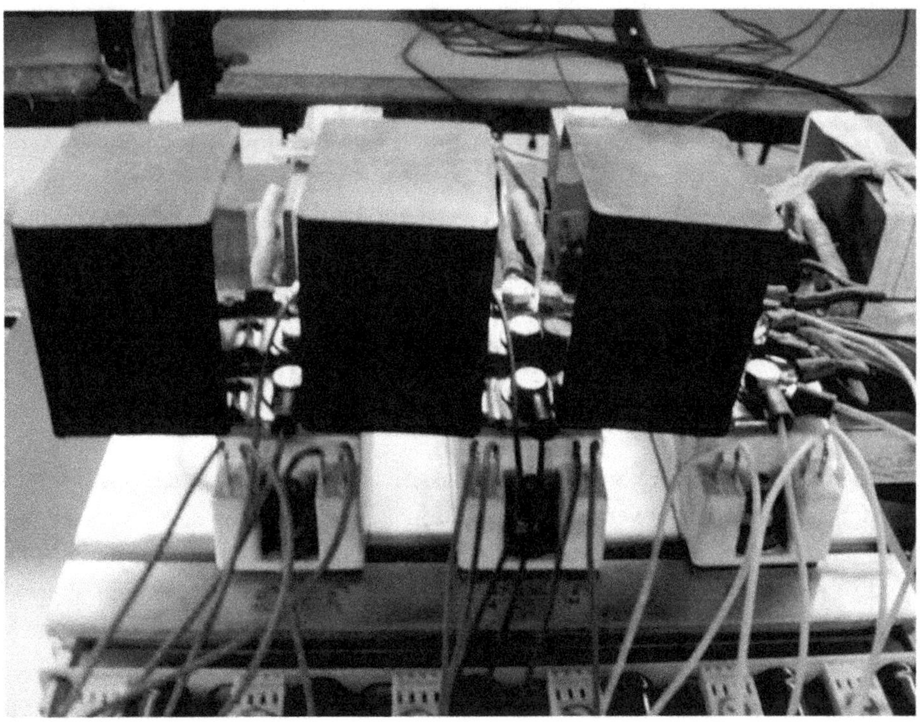

Figure 24. An IGBT of an inverter leg and its driver

Crow-Bar

A crowbar circuit is an electrical circuit used to prevent an overvoltage condition of a power supply unit from damaging the circuits attached to the power supply. It operates by putting a short circuit or low resistance path across the voltage source, much as if one dropped a tool of the same name across the output terminals of the power supply. Figures 25 and 26 show a comparison between fault current in load with and without application of a crowbar.

The advantage of a crowbar over a clamp is that the low holding voltage of the crowbar lets it carry higher fault current without dissipating much power (which could otherwise cause overheating). Also, a crowbar is more likely than a clamp to deactivate a device (by blowing a fuse or tripping a breaker), bringing attention to the faulty equipment.

Transient Over Voltage

A transient voltage suppressor or TVS is a general classification of an array of devices that are designed to react to sudden or momentary overvoltage conditions (Kastha, & Bose, 1994). One such common device used for this purpose is known as the transient voltage suppression diode that is simply a Zener diode designed to protect electronics device against overvoltages. Another design alternative applies a family of products that are known as metal-oxide varistor (MOV) that protect electronic circuits and electrical equipment.

Figure 25. Fault current in a power supply without crowbar

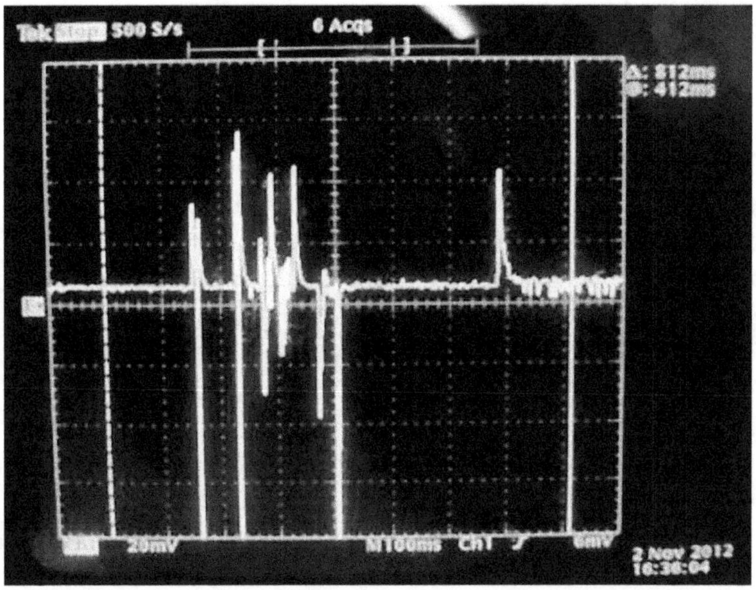

Figure 26. Fault current in a power supply with crowbar

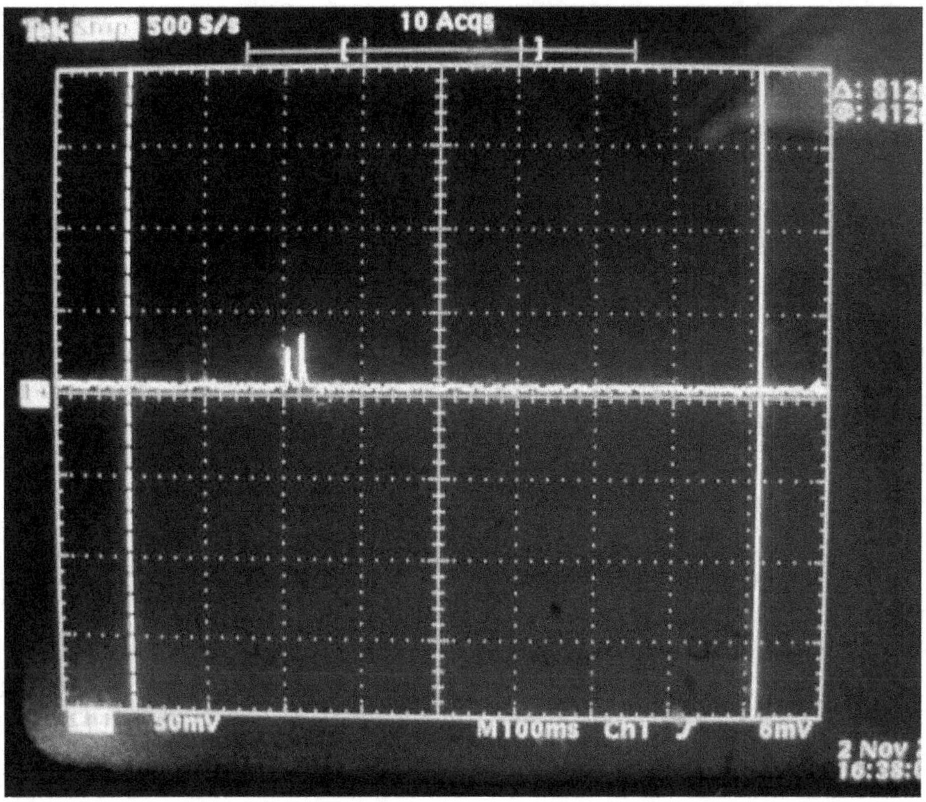

The characteristic of a TVS requires that it respond to overvoltages faster than other common overvoltage protection components such as varistors or gas discharge tubes. This makes TVS devices or components useful for protection against very fast and often damaging voltage spikes. These fast overvoltage spikes are present on all distribution networks and can be caused by either internal or external events, such as lightning or motor arcing.

Applications of transient voltage suppression diodes are used for unidirectional or bidirectional electrostatic discharge protection of transmission or data lines in electronic circuits. MOV based TVSs are utilized to protect home electronics, distribution systems and may accommodate industrial level power distribution disturbances saving downtime and damage to equipment. The level of energy in a transient overvoltage can be equated to energy measured in joules or related to electric current when devices are rated for various applications. These bursts of overvoltage can be measured with specialized electronic meters that can show power disturbances of thousands of volts amplitude that last for a few microseconds or less

A metal oxide varistor consists of a bulk semiconductor material (typically sintered granular zinc oxide) that can conduct large currents (effectively short-circuits) when presented with a voltage above its rated voltage. MOVs typically limit voltages to about 3 to 4 times the normal circuit voltage by diverting surge current elsewhere than the protected load. MOVs may be connected in parallel to increase current capability and life expectancy, providing they are matched sets (unmatched MOVs have a tolerance of approximately ±20% on voltage ratings, which is not sufficient). For more details on the effectiveness of parallel-connected MOVs. MOVs have finite life expectancy and "degrade" when exposed to a few large transients, or many smaller transients. As a MOV degrades, its triggering voltage falls lower and lower. If the MOV is being used to protect a low-power signal line, the ultimate failure mode typically is a partial or complete short circuit of the line, terminating normal circuit operation.

If used in a power filtering application, eventually the MOV behaves as a part-time effective short circuit on an AC (or DC) power line, which will cause it to heat up, starting a process called thermal runaway. As the MOV heats up, it may degrade further, causing a catastrophic failure that can result in a small explosion or fire, if the line current is not otherwise limited. An undersized MOV fails when "Absolute Maximum Ratings" in manufacturer's data-sheet are significantly exceeded.

MOVs are often connected in series with a thermal fuse, so that the fuse disconnects before catastrophic failure can happen. When this happens, only the MOV is disconnected. A failing MOV is a fire risk, which is a reason for the National Fire Protection Association's (NFPA) UL1449 in 1986 and subsequent revisions in 1998 and 2009. NFPA's primary concern is protection from fire.

When used in power applications, MOVs usually are thermal fused or otherwise protected to avoid persistent short circuits and other fire hazards. In a typical power strip, the visible circuit breaker is distinct from the internal thermal fuse, which is not normally visible to the end user. The circuit breaker has no function related to disconnecting an MOV. A thermal fuse or some equivalent solution protects from MOV generated hazards.

If a surge current is so excessively large as to exceed the MOV parameters and blow the thermal fuse, then a light found on some protectors would indicate unacceptable failure. Even adequately sized MOV protectors will eventually degrade beyond acceptable limits, with or without a failure light indication. Therefore, all MOV-based protectors intended for long-term use should have an indicator that the protective components have failed, and this indication must be checked on a regular basis to insure that protection is still functioning

A gas discharge tube (GDT) is a sealed glass-enclosed device containing a special gas mixture trapped between two electrodes, which conducts electric current after becoming ionized by a high voltage spike. GDTs can conduct more current for their size than other components. Like MOVs, GDTs have a finite life expectancy, and can handle a few very large transients or a greater number of smaller transients. The typical failure mode occurs when the triggering voltage rises so high that the device becomes ineffective, although lightning surges can occasionally cause a dead short.

GDTs take a relatively long time to trigger, permitting a higher voltage spike to pass through before the GDT conducts significant current. It is not uncommon for a GDT to let through pulses of 500 V or more of 100 ns in duration. In some cases, additional protective components are necessary to prevent damage to a protected load, caused by high-speed let-through voltage which occurs before the GDT begins to operate.

GDTs create an effective short circuit when triggered, so that if any electrical energy (spike, signal, or power) is present, the GDT will short this. Once triggered, a GDT will continue conducting (called follow-on current) until all electric current sufficiently diminishes, and the gas discharge quenches. Unlike other shunt protector devices, a GDT once triggered will continue to conduct at a voltage less than the high voltage that initially ionized the gas; this behavior is called negative resistance. Additional auxiliary circuitry may be needed in DC (and some AC) applications to suppress follow-on current, to prevent it from destroying the GDT after the initiating spike has dissipated. Some GDTs are designed to deliberately short out to a grounded terminal when overheated, thereby triggering an external fuse or circuit breaker.

Many GDTs are light-sensitive, in that exposure to light lowers their triggering voltage. Therefore, GDTs should be shielded from light exposure, or opaque versions that are insensitive to light should be used. The CG2 SN series of surge arrestors formerly produced by C P Clare, are advertised as being non-radioactive, and the datasheet for that series states that some members of the CG/CG2 series (75-470V) are radioactive.

Due to their exceptionally low capacitance, GDTs are commonly used on high frequency lines, such as are used in telecommunications equipment. Because of their high current handling capability, GDTs can also be used to protect power lines, but the follow-on current problem must be controlled. Figure 27 shows a typical varistor. Figure 29 shows application of a line suppressor in an inverter. Figure 28 shows a capacitor in the role of a snubber. Figure 30 shows a power MOSFET which is protected by back to back zener diodes in the gate terminal.

A varistor is an electronic component with a "diode-like" nonlinear current–voltage characteristic. The name is a portmanteau of variable resistor. Varistors are often used to protect circuits against excessive transient voltages by incorporating them into the circuit in such a way that, when triggered, they will shunt the current created by the high voltage away from the sensitive components. A varistor is also known as Voltage Dependent Resistor or VDR. A varistor's function is to conduct significantly increased current when voltage is excessive

The most common type of varistor is the Metal Oxide Varistor (MOV). This contains a ceramic mass of zinc oxide grains, in a matrix of other metal oxides (such as small amounts of bismuth, cobalt, manganese) sandwiched between two metal plates (the electrodes). The boundary between each grain and its neighbour forms a diode junction, which allows current to flow in only one direction. The mass of randomly oriented grains is electrically equivalent to a network of back-to-back diode pairs, each pair in parallel with many other pairs. When a small or moderate voltage is applied across the electrodes, only a tiny current flows, caused by reverse leakage through the diode junctions. When a large voltage

Figure 27. A typical varistor and its V-I characteristic, (Source: On-semiconductor Co. with permission)

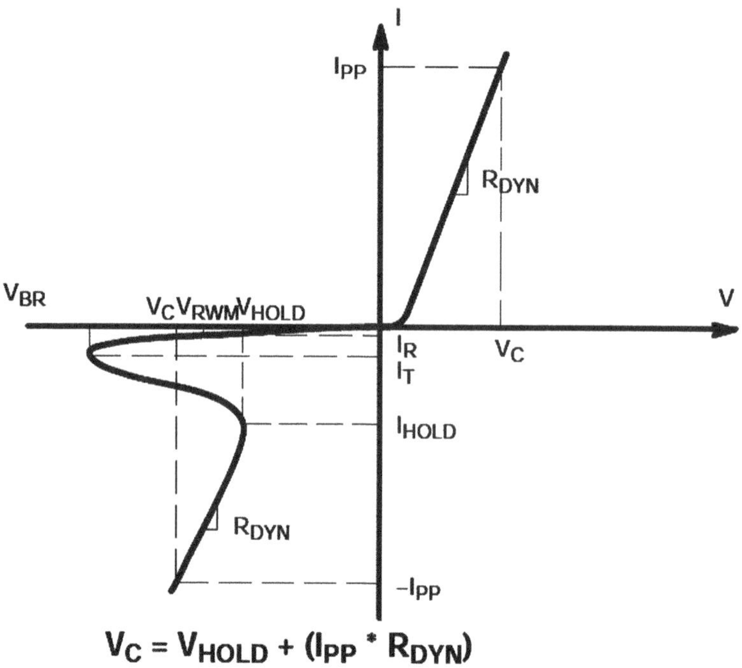

Figure 28. Line suppressor capacitors used as snubber for DC-link of a BUCK converter

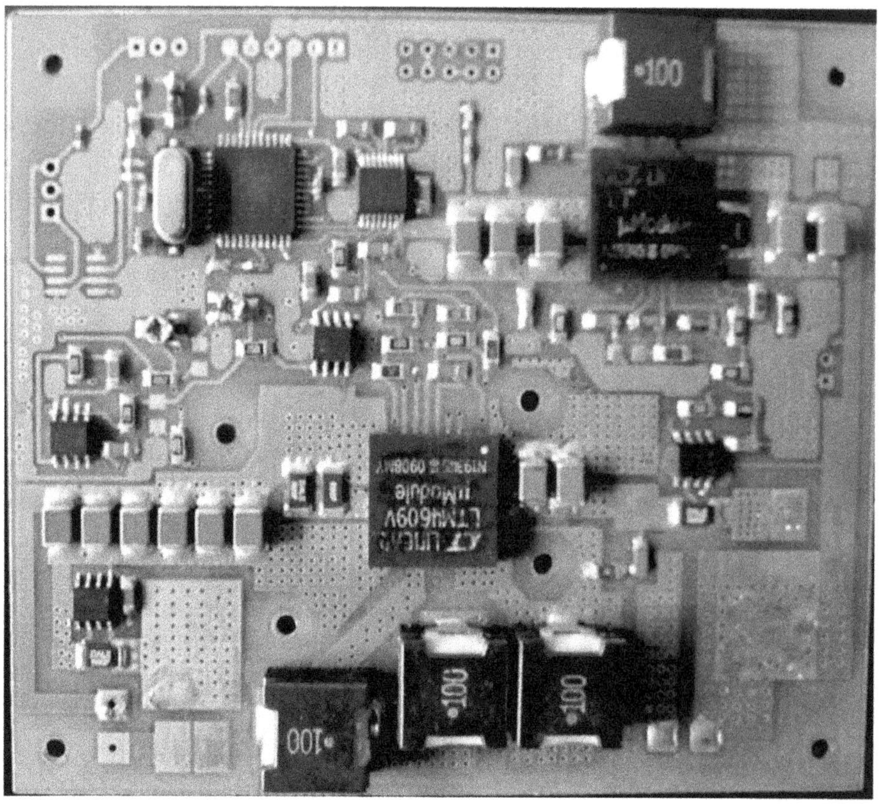

Figure 29. Application of film capacitor as a line suppressor snubber

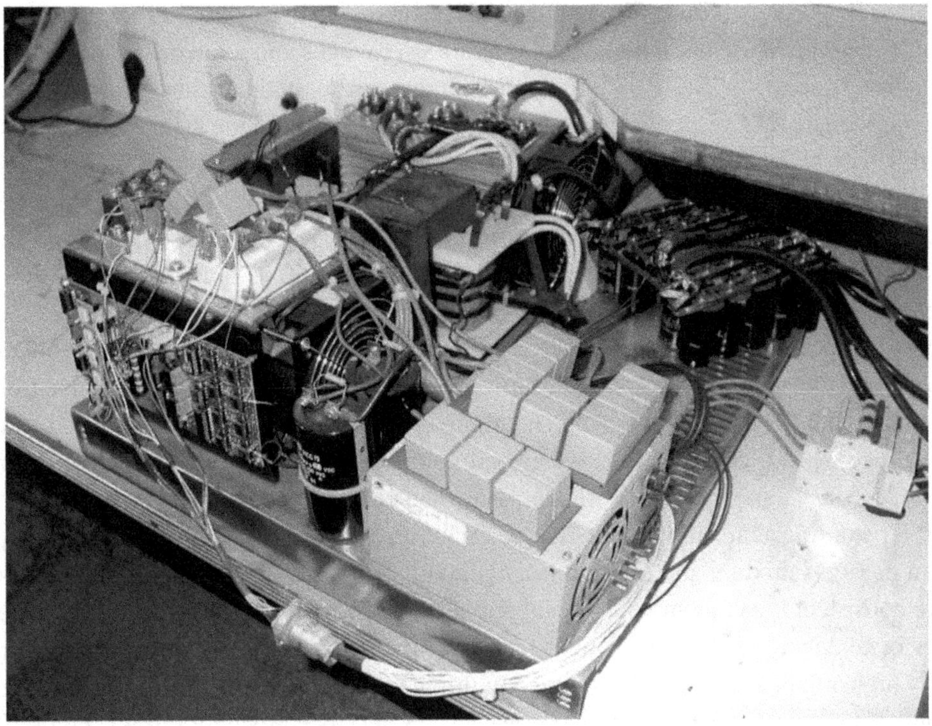

Figure 30. Internal protection in the gate terminal of a power MOSFET

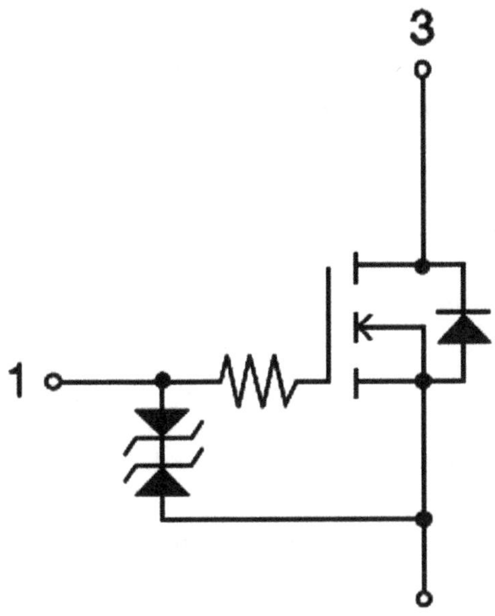

is applied, the diode junction breaks down due to a combination of thermionic emission and electron tunneling, and a large current flows. The result of this behavior is a highly nonlinear current-voltage characteristic, in which the MOV has a high resistance at low voltages and a low resistance at high voltages

The response time of the MOV is largely ambiguous, as no standard has been officially defined. The sub-nanosecond MOV response claim is based on the material's intrinsic response time, but will be slowed down by other factors such as the inductance of component leads and the mounting method. That response time is also qualified as insignificant when compared to a transient having an 8 μs rise-time, thereby allowing ample time for the device to slowly turn-on. When subjected to a very fast, <1 ns rise-time transient, response times for the MOV are in the 40–60 ns range

Another method for suppressing voltage spikes is the transient-voltage-suppression diode (TVS). Although diodes do not have as much capacity to conduct large surges as MOVs, diodes are not degraded by smaller surges and can be implemented with a lower "clamping voltage". MOVs degrade from repeated exposure to surges and generally have a higher "clamping voltage" so that leakage does not degrade the MOV. Both types are available over a wide range of voltages. MOVs tend to be more suitable for higher voltages, because they can conduct the higher associated energies at less cost.

Another type of transient suppressor is the gas-tube suppressor. This is a type of spark gap that may use air or an inert gas mixture and often, a small amount of radioactive material such as Ni-63, to provide a more consistent breakdown voltage and reduce response time. Unfortunately, these devices may have higher breakdown voltages and longer response times than varistors. However, they can handle significantly higher fault currents and withstand multiple high-voltage hits (for example, from lightning) without significant degradation. Figure 31 shows a TVS and its equivalent circuit. Figure 32 shows a typical V-I characteristic of a unidirectional TVS. Figures 33, 34, 35, and 36 show the results of application of a TVS in ESD protection. It can be seen that the time response of TVS is very Fast.

MECHANICAL PROTECTION SYSTEMS

Dynamic electric motor testing is often called on-line testing because it requires the motor to be running and generally assumes the motor is in its natural environment. Dynamic testing involves the connection of voltage probes and current transformers. Connecting dynamic test equipment is safe, quick and non-intrusive. Data is acquired and results are displayed in a summary format. The collected data is compared to the user-entered nameplate information and is presented in a pass/fail format with both current test data and trending logs displayed after each successive test.

Every reliability technician knows that costs associated with motor failures can be devastating to any business operation. Finding that a motor is operating with conditions that create excessive heat or stress is a guide to the technician to make changes in the motor's operation and to monitor its insulation. Knowing that a motor is in imminent danger of failing provides the technician with time to schedule repairs at his convenience rather than having the motor dictate to him due to a catastrophic failure. Reducing unscheduled downtime while increasing efficiency and profitability are common goals of all reliability technicians. Dynamic motor testing and monitoring is a relatively new concept aiding and advancing the capabilities of those responsible for the safe and continuous operation of electric motors and related equipment.

Figure 31. Equivalent circuit diagram of a TVS, (Source: On-semiconductor Co. with permission)

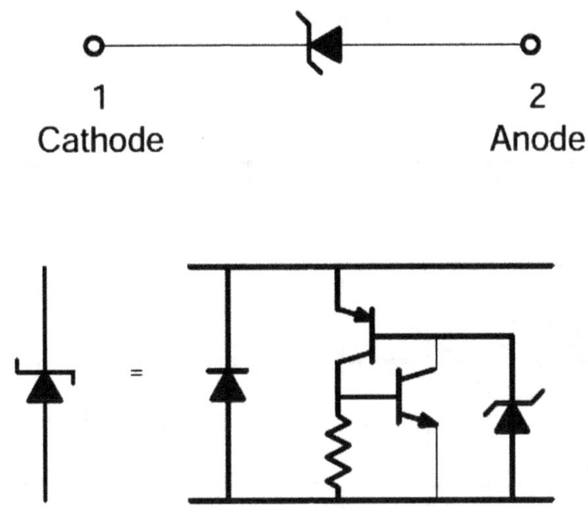

Figure 32. V-I characteristic of a TVS, (Source: On-semiconductor Co. with permission)

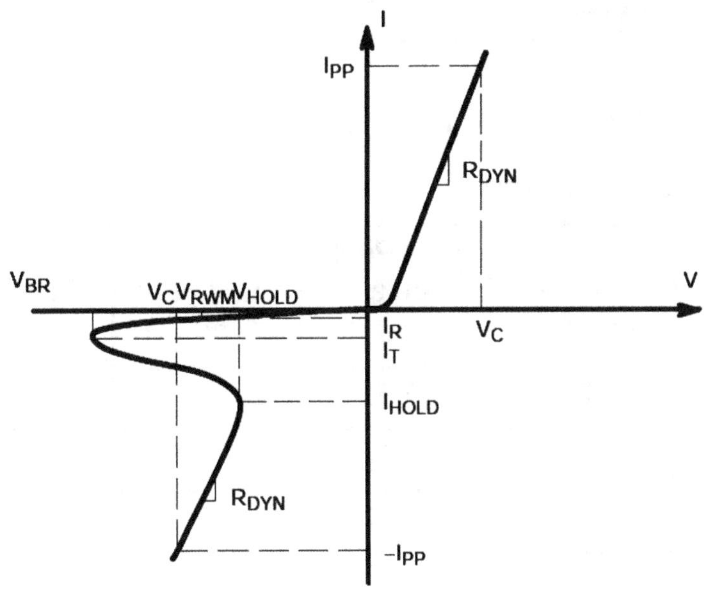

$$V_C = V_{HOLD} + (I_{PP} * R_{DYN})$$

A motor is one part of a complete system that includes incoming power quality, the motor and the driven load. Many motor problems are created by poor incoming power quality, and many more problems can be attributed to the load and load-related issues.

State-of-the-art dynamic motor test equipment is capable of separating electrical issues from mechanical issues as well as defining power-related problem areas. Good test equipment will provide an enormous amount of information regarding the incoming power, including voltage levels, imbalances and harmonic content. A small amount of voltage imbalance will result in a much larger amount of current

Figure 33. Test waveform of a TVS in IEC 61000-4-2 (Source: On-semiconductor Co. with permission)

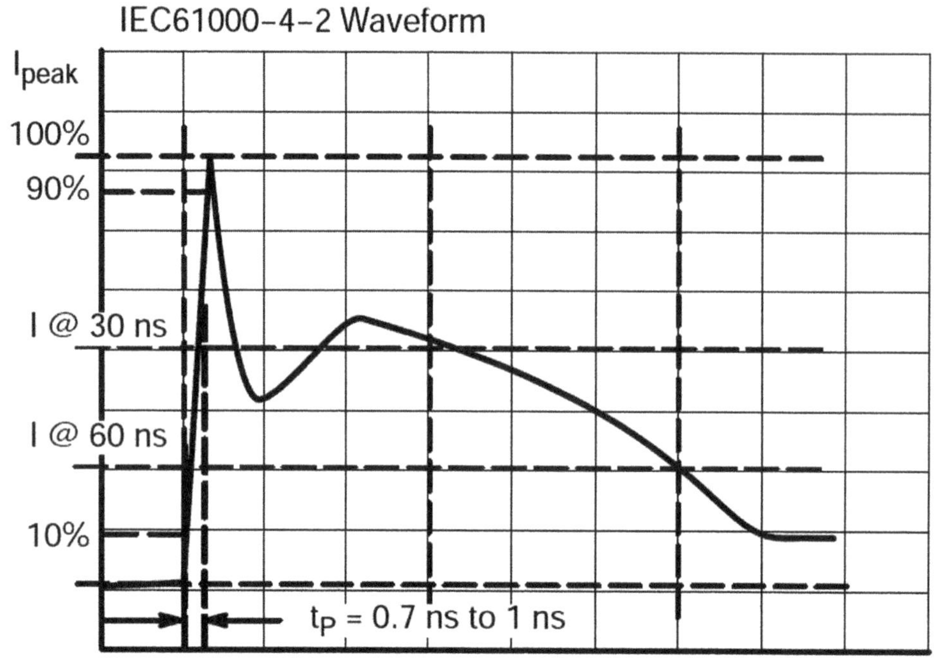

Figure 34. Sample points in test waveform of a TVS, (Source: On-semiconductor Co. with permission)

IEC 61000-4-2 Spec.

Level	Test Voltage (kV)	First Peak Current (A)	Current at 30 ns (A)	Current at 60 ns (A)
1	2	7.5	4	2
2	4	15	8	4
3	6	22.5	12	6
4	8	30	16	8

imbalance and increase losses within the motor. Harmonic distortion also results in wasted energy causing overheating due mainly to non-sinusoidal sine waves. These issues directly affect a motor's performance and its ability to handle its load. Overall, poor power quality manifests itself as higher heat within the stator and rotor, reducing efficiency and eventually resulting in premature motor failures. Monitoring power quality and making necessary adjustments are essential in maintaining motor longevity.

Besides power condition, dynamic testing provides extensive information about the motor's behavior and offers evidence of potential mechanical problem areas. The test equipment tracks current levels

Figure 35. Test setup for TVS test, (Source: On-semiconductor Co. with permission)

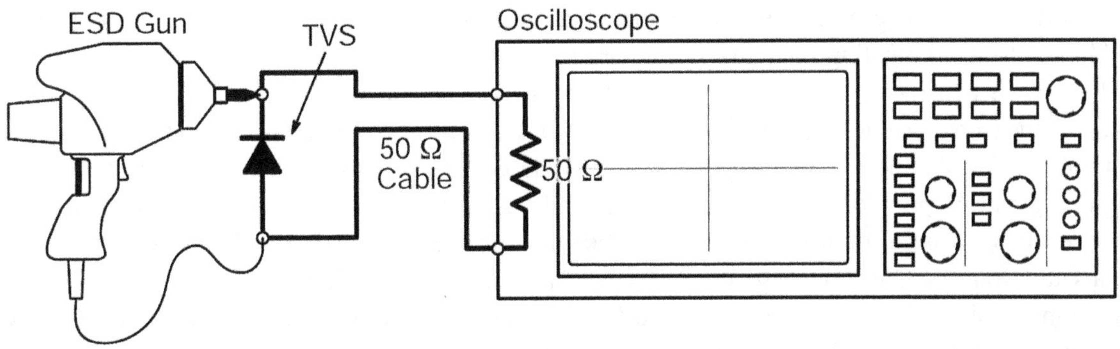

Figure 36. Test results of a TVS, (Source: On-semiconductor Co. with permission)

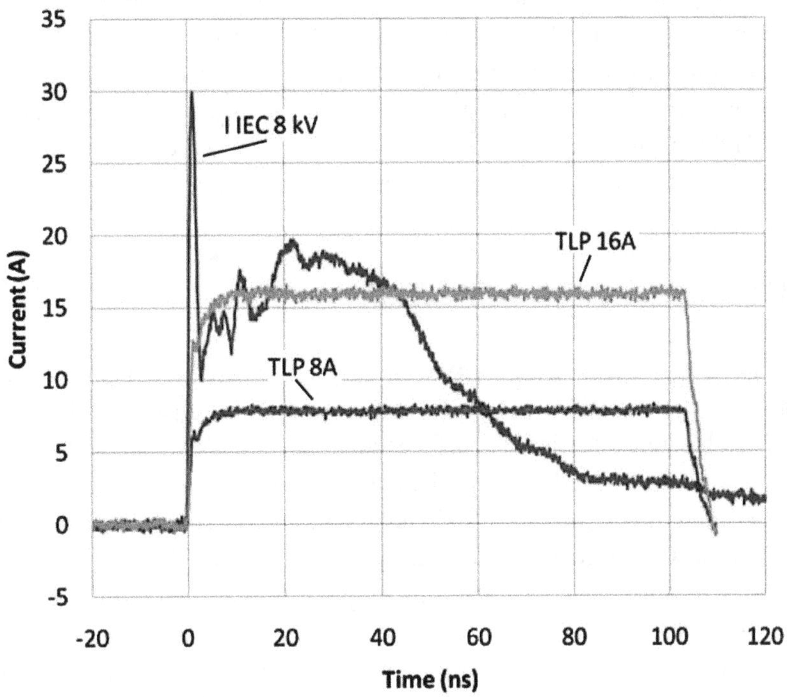

and unbalances, load levels and torque-related information. Combining this data with the power quality information, the equipment can predict de-rating factors that indicate potential problem areas.

Torque and torque ripple add another piece of the puzzle that is required for consistent and accurate diagnosis of the motor's health. Torque ripple is defined as the division of maximal torque divided by average torque during the acquisition period. The torque ripple itself is a measure of how small the torque band is that surrounds a steady state average torque. Torque ripple is independent of power condition and current level. It provides a visual look at how the driven load is performing and is an indicator of rotor stress.

Dynamic testing can identify rotor bar problems with a high degree of accuracy, and trending logs make tracking them over time easy and predictable. Rotor bar and cage defects result in lost efficiency and higher heat culminating in premature motor failure.

Measuring and tracking efficiency is a very difficult task. The operating efficiency of a motor cannot be easily measured in a field application. Many standards have a number of requirements that commonly can only be fulfilled in a laboratory environment. These standards also usually concentrate on ensuring a proper description of a motor's capabilities under good operating voltage conditions. In the field, however, there is little room for requirements like uncoupling a motor or regulating the voltage level for a saturation run. Questions regarding a particular motor's capabilities are found to be secondary in importance when compared to the operating efficiency under the given conditions in the field. The result of such an environment is that true efficiencies are unrealistic to obtain.

Operating efficiencies, however, are of crucial importance to an energy-conscious management. The requirements for a true measurement of operating efficiency in a field environment are ample and unrealistic (like installing torque transducers on the shaft of the motor and measuring the input power to the motor at the motor terminals, frequently at high voltage levels). Instead of true efficiency measurement, efficiency estimation becomes the only field-friendly approach for energy management. The difference between operating efficiency measurement and operating efficiency estimation is that the former attempts to find the true operating efficiency via direct measurement, while the latter accepts a small measure of inaccuracy for severely increased user friendliness.

ENVIRONMENTAL PROTECTION SYSTEMS

This is enough to increase the relative humidity from 60% to 100% at a temperature of 20 degrees. This humidity poses a considerable risk of mould formation and also subjects the building fabric of your property to risk. With Pluggit refresh fresh air systems, the humidity is systematically extracted and discharged by constant, controlled indoor air exchange. This protects the building fabric of your property from moisture damage and mould formation and your tenants from damage to their health. Furthermore, a convenience ventilation system reduces the heating costs by up to 70% and increases the living convenience and hence the quality of life. These factors increase the value of your property, create a basis for higher rental receipts and ensure a competitive advantage in the market

Silica gel is a granular, vitreous, porous form of silicon dioxide made synthetically from sodium silicate. Silica gel is tough and hard; it is more solid than common household gels like gelatin or agar. It is a naturally occurring mineral that is purified and processed into either granular or beaded form. As a desiccant, it has an average pore size of 2.4 nanometers and has a strong affinity for water molecules.

Silica gel is most commonly encountered in everyday life as beads in a small (typically 2 x 3 cm) paper packet. In this form, it is used as a desiccant to control local humidity to avoid spoilage or degradation of some goods. Because silica gel can have added chemical indicators (see below) and adsorbs moisture very well, silica gel packets usually bear warnings for the user not to eat the content

SUMMARY AND CONCLUSION

The subject of this chapter was about a faulty but not damaged converter. In this chapter, protection methods for a faulty electric power converter were described. The main topics of this chapter are summarized as follows:

1. Current and voltage protection are the main tools for saving the converters in faulty states. There are specific devices for protection of the converter against over voltage and over current in both steady state and transient states. A transient voltage suppressor or TVS is a general devices that are designed to react to sudden or overvoltage conditions.
2. Snubbers are passive or active devices for transferring power losses from power solid state switches into a more reliable resistor. A simple snubber uses a small resistor (R) in series with a small capacitor (C). This combination can be used to suppress the rapid rise in voltage across a switch,
3. Fuses are devices for interrupting the current of a faulty converter with very reliable operation. The speed at which a fuse blows depends on how much current flows through it and the material of which the fuse is made.
4. Delay in protection can disable the effect of protection system.

Protection systems usually prevent damage to the converter due to faults. However, a safe but out of service converter is not the goal of reliability. In the next chapter, we present the concept of availability for characterization of a safe converter..

REFERENCES

Aghaei, M., Karimi, Y., & Kaboli, S. (2014). Analysis of Phase-Shifted Full-Bridge Based dc-dc Converter considering transformer parasitic elements in discontinuous current mode. In *Proceedings of 5th Power Electronics, Drive Systems and Technologies Conference* (pp. 366-372). IEEE. doi:10.1109/PEDSTC.2014.6799402

Eric, W., & Lehn, P. W. (2006). Digital Current Control of a Voltage Source Converter with Active Damping of LCL Resonance. *IEEE Transactions on Power Electronics*, *21*(5), 1364–1373. doi:10.1109/TPEL.2006.880271

Kastha, D. K., & Bose, B. K. (1994). Investigation of fault modes of voltage-fed inverter system for induction motor drive. *IEEE Transactions on Industry Applications*, *30*(4), 1021038. doi:10.1109/28.297920

Morren, J., & de Haan, S. W. H. (2007). Short-Circuit Current of Wind Turbines With Doubly Fed Induction Generator. *IEEE Transactions on Energy Conversion*, *22*(1), 174–180. doi:10.1109/TEC.2006.889615

Najmi, V., Ebrahimi, S., & Oraee, H. (2012). Output voltage quality intensification of diode clamped multilevel inverters using FM PWM technique. In *Proceedings of 3rd Power Electronics and Drive Systems Technology* (p. 9102). IEEE. doi:10.1109/PEDSTC.2012.6183305

Wheeler, P. W., Rodriguez, J., Clare, J. C., Empringham, L., & Weinstein, A. (2002). Matrix converters: A technology review. *IEEE Transactions on Industrial Electronics*, *49*(2), 276–288. doi:10.1109/41.993260

ADDITIONAL READING

Abdi, B., Ranjbar, A. H., Gharehpetian, G. B., & Milimonfared, J. (2009). Reliability Considerations for Parallel Performance of Semiconductor Switches in High-Power Switching Power Supplies. *IEEE Transactions on Industrial Electronics*, *56*(6), 2133–2139. doi:10.1109/TIE.2009.2014306

Abramovitz, A., Liao, Ch. Sh., & Smedley, K. (2013). State-Plane Analysis of Regenerative Snubber for Flyback Converters. *IEEE Transactions on Power Electronics*, *28*(11), 5323–5332. doi:10.1109/TPEL.2013.2243845

Altintas, N., Bakan, A. F., & Aksoy, I. (2014). A Novel ZVT-ZCT-PWM Boost Converter. *IEEE Transactions on Power Electronics*, *29*(1), 256–265. doi:10.1109/TPEL.2013.2252197

Amini, M. R., & Farzanehfard, H. (2009). Novel Family of PWM Soft-Single-Switched DC–DC Converters with Coupled Inductors. *IEEE Transactions on Industrial Electronics*, *56*(6), 2102114. doi:10.1109/TIE.2009.2016509

Bauman, J., & Kazerani, M. (2011). A Novel Capacitor-Switched Regenerative Snubber for DC/DC Boost Converters. *IEEE Transactions on Industrial Electronics*, *58*(2), 514–523. doi:10.1109/TIE.2010.2046576

Behjati, H., & Davoudi, A. (2013). Reliability Analysis Framework for Structural Redundancy in Power Semiconductors. *IEEE Transactions on Industrial Electronics*, *60*(10), 4376–4386. doi:10.1109/TIE.2012.2216238

Busca, C., Teodorescu, R., Blaabjerg, F., Munk-Nielsen, S., Helle, L., Abeyasekera, T., & Rodriguez, P. (2011). An overview of the reliability prediction related aspects of high power IGBTs in wind power applications. *Microelectronics and Reliability*, *51*(9), 1903–1907. doi:10.1016/j.microrel.2011.06.053

Calkin, E. T., & Hamilton, B. H. (1976). Circuit Techniques for Improving the Switching Loci of Transistor Switches in Switching Regulators. *IEEE Transactions on Industry Applications*, *12*(4), 364–369. doi:10.1109/TIA.1976.349438

Choe, H. J., Chung, Y. C., Sung, C. H., Yun, J. J., & Kang, B. (2014). Passive Snubber for Reducing Switching-Power Losses of an IGBT in a DC–DC Boost Converter. *IEEE Transactions on Power Electronics*, *29*(12), 6332–6341. doi:10.1109/TPEL.2014.2304551

De León-Aldaco, S. E., Calleja, H., Chan, F., & Jiménez-Grajales, H. R. (2013). Effect of the Mission Profile on the Reliability of a Power Converter Aimed at Photovoltaic Applications-A Case Study. *IEEE Transactions on Power Electronics*, *28*(6), 2993007. doi:10.1109/TPEL.2012.2222673

Dhople, S. V., & Dominguez-Garcia, A. D. (2012). Estimation of Photovoltaic System Reliability and Performance Metrics. *IEEE Transactions on Power Systems*, *27*(1), 554–563. doi:10.1109/TPWRS.2011.2165088

Di Capua, G., & Femia, N. (2014). A Versatile Method for MOSFET Commutation Analysis in Switching Power Converter Design. *IEEE Transactions on Power Electronics*, *29*(2), 920–935. doi:10.1109/TPEL.2013.2256152

El-Refaie, A. M. (2011). Fault-tolerant permanent magnet machines: A review. *IET Electric Power Applications*, *5*(1), 59–74. doi:10.1049/iet-epa.2009.0117

Fernández, R. et al.. (2006). FinFET and MOSFET preliminary comparison of gate oxide reliability. *Microelectronics and Reliability*, *46*(9), 1601611.

Finney, S. J., Williams, B. W., & Green, T. C. (1996). RCD snubber revisited. *IEEE Transactions on Industry Applications*, *32*(1), 155–160. doi:10.1109/28.485827

Harb, S., & Balog, R. S. (2013). Reliability of Candidate Photovoltaic Module-Integrated-Inverter (PV-MII) Topologies-A Usage Model Approach. *IEEE Transactions on Power Electronics*, *28*(6), 3019–3027. doi:10.1109/TPEL.2012.2222447

Holdsworth, L., Wu, X. G., Ekanayake, J. B., & Jenkins, N. (2003). Comparison of fixed speed and doubly-fed induction wind turbines during power system disturbances. *IEE Proceedings. Generation, Transmission and Distribution*, *150*(3), 343–352. doi:10.1049/ip-gtd:20030251

Jin, Y., Fletcher, J. E., & O'Reilly, J. (2010). A Series-Dynamic-Resistor-Based Converter Protection Scheme for Doubly-Fed Induction Generator During Various Fault Conditions. *IEEE Transactions on Energy Conversion*, *25*(2), 422–432. doi:10.1109/TEC.2009.2037970

Kaur, H., Kabra, S., Bindra, S., Haldar, S., & Gupta, R. S. (2007). Impact of graded channel (GC) design in fully depleted cylindrical/surrounding gate MOSFET (FD CGT/SGT) for improved short channel immunity and hot carrier reliability. *Solid-State Electronics*, *51*(3), 39404. doi:10.1016/j.sse.2007.01.025

Khosroshahi, A., Abapour, M., & Sabahi, M. (2015). Reliability Evaluation of Conventional and Interleaved DC–DC Boost Converters. *IEEE Transactions on Power Electronics*, *30*(10), 5821–5828. doi:10.1109/TPEL.2014.2380829

Kostandyan, E. E., & Ma, K. (2012). Reliability estimation with uncertainties consideration for high power IGBTs in 2.3 MW wind turbine converter system. *Microelectronics and Reliability*, *52*(9), 2403–2408. doi:10.1016/j.microrel.2012.06.152

Laaksonen, H. J. (2010). Protection Principles for Future Microgrids. *IEEE Transactions on Power Electronics*, *25*(12), 2910–2918. doi:10.1109/TPEL.2010.2066990

Lu, D. D.-C., Cheng, D. K.-W., & Lee, Y.-S. (2003). A single-switch continuous-conduction-mode boost converter with reduced reverse-recovery and switching losses. *IEEE Transactions on Industrial Electronics*, *50*(4), 767–776. doi:10.1109/TIE.2003.814989

McMurray, W. (1980). Selection of Snubbers and Clamps to Optimize the Design of Transistor Switching Converters. *IEEE Transactions on Industry Applications*, *16*(4), 513–523. doi:10.1109/TIA.1980.4503823

Menzies, R. W., Steimer, P., & Steinke, J. K. (1994). Five-level GTO inverters for large induction motor drives. *IEEE Transactions on Industry Applications*, *30*(4), 93944. doi:10.1109/28.297910

Pahlevaninezhad, M., Das, P., Drobnik, J., Jain, P. K., & Bakhshai, A. (2012). A ZVS Interleaved Boost AC/DC Converter Used in Plug-in Electric Vehicles. *IEEE Transactions on Power Electronics*, *27*(8), 3513–3529. doi:10.1109/TPEL.2012.2186320

Park, K. B., Moon, G. W., & Youn, M. J. (2012). High Step-up Boost Converter Integrated With a Transformer-Assisted Auxiliary Circuit Employing Quasi-Resonant Operation. *IEEE Transactions on Power Electronics*, *27*(4), 1974–1984. doi:10.1109/TPEL.2011.2170223

Park, S., & Choi, S. (2010). Soft-Switched CCM Boost Converters with High Voltage Gain for High-Power Applications. *IEEE Transactions on Power Electronics*, *25*(5), 1211–1217. doi:10.1109/TPEL.2010.2040090

Park, S. H., Cha, G. R., Jung, Y. Ch., & Won, Ch. Y. (2010). Design and Application for PV Generation System Using a Soft-Switching Boost Converter With SARC. *IEEE Transactions on Industrial Electronics*, *57*(2), 515–522. doi:10.1109/TIE.2009.2036025

Park, Y., Jung, B., & Choi, S. (2012). Nonisolated ZVZCS Resonant PWM DC–DC Converter for High Step-Up and High-Power Applications. *IEEE Transactions on Power Electronics*, *27*(8), 3563575. doi:10.1109/TPEL.2012.2187342

Petit, C., Meinertzhagen, A., Zander, D., Simonetti, O., Fadlallah, M., & Maurel, T. (2005). Low voltage SILC and P-and N-MOSFET gate oxide reliability. *Microelectronics and Reliability*, *45*(3), 479–485. doi:10.1016/j.microrel.2004.08.002

Ren, Y., Xu, M., Zhou, J., & Lee, F. C. (2006). Analytical loss model of power MOSFET. *IEEE Transactions on Power Electronics*, *21*(2), 310–319. doi:10.1109/TPEL.2005.869743

Richardeau, F., & Pham, T. T. L. (2013). Reliability Calculation of Multilevel Converters: Theory and Applications. *IEEE Transactions on Industrial Electronics*, *60*(10), 4225–4233. doi:10.1109/TIE.2012.2211315

Salomonsson, D., Soder, L., & Sannino, A. (2009). Protection of Low-Voltage DC Microgrids. *IEEE Transactions on Power Delivery*, *24*(3), 1045–1053. doi:10.1109/TPWRD.2009.2016622

Seong, H. W., Kim, H. S., Park, K. B., Moon, G. W., & Youn, M. J. (2012). High Step-Up DC-DC Converters Using Zero-Voltage Switching Boost Integration Technique and Light-Load Frequency Modulation Control. *IEEE Transactions on Power Electronics*, *27*(3), 1383–1400. doi:10.1109/TPEL.2011.2162966

Song, Y., & Wang, B. (2013). Survey on Reliability of Power Electronic Systems. *IEEE Transactions on Power Electronics*, *28*(1), 591–604. doi:10.1109/TPEL.2012.2192503

Wai, R. J., Lin, Ch. Y., Duan, R. Y., & Chang, Y. R. (2007). High-Efficiency DC-DC Converter with High Voltage Gain and Reduced Switch Stress. *IEEE Transactions on Industrial Electronics*, *54*(1), 354–364. doi:10.1109/TIE.2006.888794

Wang, H., & Blaabjerg, F. (2014). Reliability of capacitors for DC-link applications - An overview. *IEEE Transactions on Industry Applications*, *50*(5), 3569–3578. doi:10.1109/TIA.2014.2308357

Wang, J., Chung, H. S.-H., & Li, R. T.-H. (2013). Characterization and Experimental Assessment of the Effects of Parasitic Elements on the MOSFET Switching Performance. *IEEE Transactions on Power Electronics*, *28*(1), 573–590. doi:10.1109/TPEL.2012.2195332

Xiaogao, X., Junming, Z., Chen, Z., Zhuo, Z., & Zhaoming, Q. (2007). Analysis and Optimization of LLC Resonant Converter with a Novel Over-Current Protection Circuit. *IEEE Transactions on Power Electronics*, *22*(2), 435–443. doi:10.1109/TPEL.2006.889919

Yang, Sh., Bryant, A., Mawby, P., Xiang, D., Ran, L., & Tavner, P. (2011). An Industry-Based Survey of Reliability in Power Electronic Converters. *IEEE Transactions on Industry Applications*, *47*(3), 1441–1451. doi:10.1109/TIA.2011.2124436

Yazdani, D., Mojiri, M., Bakhshai, A., & Joos, G. (2009). A Fast and Accurate Synchronization Technique for Extraction of Symmetrical Components. *IEEE Transactions on Power Electronics*, *24*(3), 674–684. doi:10.1109/TPEL.2008.2010321

Yun, J. J., Choe, H. J., Hwang, Y. H., Park, Y. K., & Bongkoo, K. (2012). Improvement of Power-Conversion Efficiency of a DC-DC Boost Converter Using a Passive Snubber Circuit. *IEEE Transactions on Industrial Electronics*, *59*(4), 1801814. doi:10.1109/TIE.2011.2141095

Zhang, W., Xu, D., Li, X., Xie, R., Li, H., Dong, D., & Chen, M. et al. (2013). Seamless Transfer Control Strategy for Fuel Cell Uninterruptible Power Supply System. *IEEE Transactions on Power Electronics*, *28*(2), 717–729. doi:10.1109/TPEL.2012.2204777

Zhao, Y., Li, W., & He, X. (2012). Single-Phase Improved Active Clamp Coupled-Inductor-Based Converter with Extended Voltage Doubler Cell. *IEEE Transactions on Power Electronics*, *27*(6), 2869–2878. doi:10.1109/TPEL.2011.2176752

KEY TERMS AND DEFINITIONS

Current Limiter: A circuit for preventing over loading of an electric power converter.
Current Transformer: A transformer used for current monitoring in an electric power converter.
Protection: A method for saving a faulty system.
Snubber: An electric network for operation of a solid state switch in low stress condition.
Voltage Clamp: A device or circuit for reducing the transient (usually) voltages applied to a converter.

Chapter 9
Availability

ABSTRACT

Protection methods, which were described in the previous chapter, save the converter against non-catastrophic faults. However, this method saves the converter but it also takes the converter out of the service. The subject of this chapter is converters that are not damaged but can not operate normally. In this chapter, availability of electric power converters as a most important but usually forgotten parameter is described. The concept of availability was originally developed for repairable systems that are required to operate continuously. It is explained that a system may be unavailable while none of its parts damaged. In fact, there is an important difference between reliability and availability. A converter may be highly reliable but unavailable and vice versa. One of the most important factors for this undesired state is influence of noise. In this chapter, electromagnetic interference and certain methods for reducing its undesired effects on electric power converters are presented. Electric power converters are usually the source of electromagnetic noise due to high operating voltage and/or current. Various techniques for safe operation of sensitive systems that operate close to these converters are described. In the last part of chapter, alarm management is presented based on availability concept. This method is used to prevent fast shutdown of important systems due to dispensable faults.

INTRODUCTION: AVAILABLE OR SAFE?

Figure 1 shows the state of this chapter in the flowchart of the book. The concept of availability was originally developed for repairable systems that are required to operate continuously, i.e., round-the-clock, and are at any random point in time either operating or "down" because of failure and are being worked upon so as to restore their operation in minimum time. In this original concept, a system is considered to be in only two possible states - - operating or in repair -- and availability is defined as the probability that a system is operating satisfactorily at any random point in time, t, when subject to a sequence of "up" and "down" cycles which constitute an alternating renewal process (Kwasinski, Krishnamurthy, Song, & Sharma, 2012).

A trade-off is a rational selection among alternatives in order to optimize some system parameter that is a function of two or more variables which are being compared (traded off). Examples of system

DOI: 10.4018/978-1-4666-9429-3.ch009

Figure 1. State of chapter 9 in the flowchart of the book

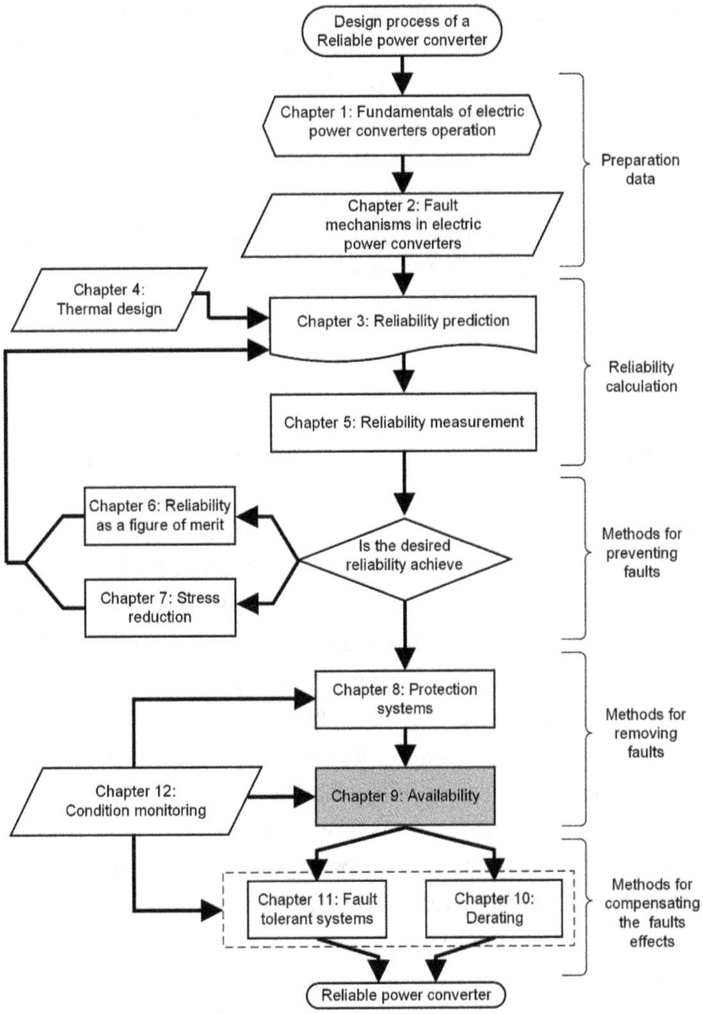

trade-offs involve performance, reliability, maintainability, cost, schedule, and risk. A trade-off may be quantitative or qualitative. Insofar as possible, it is desirable that trade-offs be based on quantifiable, analytic, or empirical relationships. Where this is not possible, then semiquantitative methods using ordinal rankings or weighting factors are often used.

The methodology for structuring and performing trade-off analyses is part of the system engineering process described in Section 4. The basic steps, summarized here are:

1. Define the trade-off problem and establish the trade-off criteria and constraints
2. Synthesize alternative design configurations
3. Analyze these alternative configurations
4. Evaluate the results of the analyses with respect to the criteria, eliminating those which violate constraint boundaries
5. Select the alternative which best meets criteria and constraint boundaries or iterate the design alternatives, repeating Steps 2 through 5 to obtain improved solutions.

System effectiveness and cost effectiveness models provide the best tools for performing tradeoff studies on the system level. Through the computerized models, any changes in any of the multitude of reliability, maintainability, performance, mission profile, logistic support, and other parameters can be immediately evaluated as to their effect on the effectiveness and total cost of a system. Thus, cost effectiveness modeling and evaluation, besides being used for selecting a specific system design approach from among several competing alternatives, is a very powerful

tool for performing parametric sensitivity studies and trade-offs down to component level when optimizing designs to provide the most effective system for a given budgetary and life cycle cost constraint or the least costly system for a desired effectiveness level.

AVAILABILITY

Availability means the probability that a system is operational at a given time, i.e. the amount of time a device is actually operating as the percentage of total time it should be operating. High-availability systems may report availability in terms of minutes or hours of downtime per year. Availability features allow the system to stay operational even when faults do occur. A highly available system would disable the malfunctioning portion and continue operating at a reduced capacity. In contrast, a less capable system might crash and become totally nonoperational. Availability is typically given as a percentage of the time a system is expected to be available, e.g., 99.999 percent.

Availability of a system is typically measured as a factor of its reliability - as reliability increases, so does availability. Availability of a system may also be increased by the strategy on focusing on increasing testability & maintainability and not on reliability. Improving maintainability is generally easier than reliability. Maintainability estimates (Repair rates) are also generally more accurate. However, because the uncertainties in the reliability estimates are in most cases very large, it is likely to dominate the availability (prediction uncertainty) problem, even while maintainability levels are very high.

When reliability is not under control more complicated issues may arise, like manpower (maintainers / customer service capability) shortage, spare part availability, logistic delays, lack of repair facilities, extensive retro-fit and complex configuration management costs and others. The problem of unreliability may be increased also due to the "domino effect" of maintenance induced failures after repairs. Only focusing on maintainability is therefore not enough. If failures are prevented, none of the others are of any importance and therefore reliability is generally regarded as the most important part of availability.

Reliability needs to be evaluated and improved related to both availability and the cost of ownership (due to cost of spare parts, maintenance man-hours, transport costs, storage cost, part obsolete risks etc.). Often a trade-off is needed between the two. There might be a maximum ratio between availability and cost of ownership. Testability of a system should also be addressed in the availability plan as this is the link between reliability and maintainability. The maintenance strategy can influence the reliability of a system (e.g. by preventive and/or predictive maintenance), although it can never bring it above the inherent reliability. So, Maintainability and Maintenance strategies influences the availability of a system. In theory this can be almost unlimited if one would be able to always repair any fault in an infinitely short time. This is in practice impossible. Repair-ability is always limited due to testability, manpower and logistic considerations.

An availability plan should clearly provide a strategy for availability control. Whether only Availability or also Cost of Ownership is more important depends on the use of the system. For example, a system that is a critical link in a production system - e.g. a big oil platform – is normally allowed to have a very high cost of ownership if this translates to even a minor increase in availability, as the unavailability of the platform results in a massive loss of revenue which can easily exceed the high cost of ownership. A proper reliability plan should always address RAMT analysis in its total context. RAMT stands in this case for Reliability, Availability, Maintainability/Maintenance and Testability in context to the customer needs (Kwasinski, 2011).

All reliability professionals are familiar with the terms Weibull Analysis and/or Life Data Analysis. In fact, for many, these analysis techniques have become almost synonymous with reliability and achieving high reliability. The reality, though, is that although life data analysis is an important piece of the pie, performing just this type of analysis is not enough to achieve reliable products. Rather, there are a variety of activities involved in an effective reliability program and in arriving at reliable products. Achieving the organization's reliability goals requires strategic vision, proper planning, sufficient organizational resource allocation and the integration and institutionalization of reliability practices into development projects.

Design for Reliability, however, is more specific than these general ideas. It is actually a process. Specifically, DFR describes the entire set of tools that support product and process design (typically from early in the concept stage all the way through to product obsolescence) to ensure that customer expectations for reliability are fully met throughout the life of the product with low overall life-cycle costs. In other words, DFR is a systematic, streamlined, concurrent engineering program in which reliability engineering is weaved into the total development cycle. It relies on an array of reliability engineering tools along with a proper understanding of when and how to use these tools throughout the design cycle. This process encompasses a variety of tools and practices and describes the overall order of deployment that an organization needs to follow in order to design reliability into its products

As the previous sections demonstrate, there is a clear distinction between the goals and tools employed to assure quality versus those employed to analyze and improve reliability. Of course, there are also many natural affinities between the two disciplines and it is understandable that many organizations have traditionally combined both quality and reliability under the same umbrella. In some cases, when the organization clearly understands the distinction between quality and reliability and applies the appropriate tools for both objectives, this combination can be appropriate and effective. However, when there is not a clear understanding of the essential differences in the tools involved, this can lead to very poor outcomes resulting from the improper use of tools and data. The rest of this article attempts to distinguish the specific processes and techniques that are necessary to ensure a product's reliability by presenting a high-level overview of a general DFR process.

On the other hand, Design for Reliability is a process specifically geared toward achieving high long-term reliability. This process attempts to identify and prevent design issues early in the development phase, instead of having these issues found in the hands of the customer. As mentioned previously, a variety of tools are used in order to accomplish this objective. These tools are different than those used in DFSS, even though there is some overlap. Figure 1 illustrates the different tools used in DFSS and DFR, as well as the overlap between the two. As you can see from this graphic, the types of tools used in DFR are based on modeling the life of the product, understanding the operating stresses and the physics of failure. The common area between DFSS and DFR includes tools such as Voice of the Customer (VOC), Design of Experiments (DOE) and Failure Modes and Effects Analysis (FMEA), which are essential elements in any kind of product improvement program.

At times, however, especially in the case of the more simple systems, trade-offs may be limited to achieving a required system availability while meeting the specified reliability and maintainability requirements. Comparatively simple trade-off techniques can then be used as shown in the following paragraphs.

The reliability-maintainability-availability relationship provides a measure of system effectiveness within which considerable trade-off potential usually exists, e.g., between reliability, maintainability, and logistic support factors. This potential should be re-evaluated at each successive stage of system development to optimize the balance between reliability, maintainability, and other system effectiveness parameters with respect to technical risks, life cycle cost, acquisition schedule, and operating and maintenance requirements. The latter become increasingly more important as complexity of system design increases, dictating the need for integration of system monitoring and checkout provisions in the basic design.

There are obvious practical limits which must be considered in trade-off optimization. These are called constraints, and all purposeful optimization must be bounded by constraints into feasible regions. For example, there are practical limits as to how high a value for MTBF can be achieved or how low MTTR can be made. In the one case, the reliability of system components or the required redundancy might be so high that the desired reliability could not be realistically achieved within the state-of-the-art or would be so expensive as to violate cost constraints.

Similarly, MTTRs close to zero would require extreme maintainability design features, such as completely built-in test features or automatic test and checkout to allow fault isolation to each individual replaceable module, with perhaps automatic switchover from a failed item to a standby item. This also could easily violate state-of-the-art or cost constraints.

It follows, then, that trade-offs not only involve relationships among system parameters and variables but also that they are bounded by both technical and economic constraints. In a sense, all trade-offs are economic ones, requiring cost-benefit analysis (not necessarily in terms of dollar costs but rather in terms of the availability and consumption of resources, of which dollars are often the most convenient measure). Resource constraints may also include manpower and skill levels, schedule or time availability, and the technical state-of-the-art capability. Later sections of this chapter deal with the cost problem.

There are two general classes of trade-offs. In the first, the contributing system variables are traded-off against one another without increasing the value of the higher level system parameter; for example, trading-off reliability and maintainability along an isoavailability contour (no change in availability). This might be done for reasons of standardization or safety or for operational reasons such as the level at which the system and its equipments will be maintained (Song, Krishnamurthy, Kwasinski, & Sharma, 2013).

The other class of trade-off is one in which the system variables are varied in order to obtain the highest value of the related system parameters within cost or other constraints. For example, reliability and maintainability might be traded-off in order to achieve a higher availability. This could result in moving from one isoavailability curve to another in Figure 3, perhaps along an isocline (a line connecting equal slopes).

As stated earlier in this section, reliability and maintainability jointly determine the inherent availability of a system. Thus, when an availability requirement is specified, there is a distinct possibility of trading-off between reliability and maintainability since, in the steady state, availability depends only on the ratio or ratios of MTTR/MTBF that is referred to as maintenance time ratio (MTR)

Example: UPS

An uninterruptible power supply, also uninterruptible power source, UPS or battery/flywheel backup, is an electrical apparatus that provides emergency power to a load when the input power source, typically mains power, fails. A UPS differs from an auxiliary or emergency power system or standby generator in that it will provide near-instantaneous protection from input power interruptions, by supplying energy stored in batteries, supercapacitors, or flywheels. The on-battery runtime of most uninterruptible power sources is relatively short (only a few minutes) but sufficient to start a standby power source or properly shut down the protected equipment.

A UPS is typically used to protect hardware such as computers, data centers, telecommunication equipment or other electrical equipment where an unexpected power disruption could cause injuries, fatalities, serious business disruption or data loss. UPS units range in size from units designed to protect a single computer without a video monitor (around 200 volt-ampere rating) to large units powering entire data centers or buildings. The world's largest UPS, the 46-megawatt Battery Electric Storage System (BESS), in Fairbanks, Alaska, powers the entire city and nearby rural communities during outages.

Example: Emergency Power System

An emergency power system is a standby generator which may include lighting, electric generators, fuel cells, uninterruptible power supplys and other apparatus, to provide backup power resources in a crisis or when regular systems fail.

They find uses in a wide variety of settings from residential homes to hospitals, scientific laboratories, data centers, telecommunication equipment and modern naval ships. Emergency power systems can rely on generators, deep cycle batteries, flywheel energy storage or hydrogen fuel cells. Finally, some homebrew emergency power systems use regular lead-acid car batteries

Mains power can be lost due to downed lines, malfunctions at a sub-station, inclement weather, planned blackouts or in extreme cases a grid-wide failure. In modern buildings, most emergency power systems have been and are still based on generators. Usually, these generators are Diesel engine driven, although smaller buildings may use a gasoline engine driven generator and larger ones a gas turbine. However, lately, more use is being made of deep cycle batteries and other technologies such as flywheel energy storage or fuel cells. These latter systems do not produce polluting gases, thereby allowing the placement to be done within the building. Also, as a second advantage, they do not require a separate shed to be built for fuel storage.

With regular generators, an automatic transfer switch is used to connect emergency power. One side is connected to both the normal power feed and the emergency power feed; and the other side is connected to the load designated as emergency. If no electricity comes in on the normal side, the transfer switch uses a solenoid to throw a triple pole, single throw switch. This switches the feed from normal to emergency power. The loss of normal power also triggers a battery operated starter system to start the generator, similar to using a car battery to start an engine. Once the transfer switch is switched and the generator starts, the building's emergency power comes back on (after going off when normal power was lost.)

Unlike emergency lights, emergency lighting is not a type of light fixture; it is a pattern of the building's normal lights that provides a path of lights to allow for safe exit, or lights up service areas such as mechanical rooms and electric rooms. Exit signs, Fire alarm systems (that are not on back up batteries) and the electric motor pumps for the fire sprinklers are almost always on emergency power.

Other equipment on emergency power may include smoke isolation dampers, smoke evacuation fans, elevators, handicap doors and outlets in service areas. Hospitals use emergency power outlets to power life support systems and monitoring equipment. Some buildings may even use emergency power as part of normal operations, such as a theater using it to power show equipment because "the show must go on.

INFLUENCE OF INTERFERENCE

One of the best examples is EMI. Electromagnetic compatibility (EMC) is the branch of electrical sciences which studies the unintentional generation, propagation and reception of electromagnetic energy with reference to the unwanted effects (Electromagnetic interference, or EMI) that such energy may induce. The goal of EMC is the correct operation, in the same electromagnetic environment, of different equipment which use electromagnetic phenomena, and the avoidance of any interference effects.

In order to achieve this, EMC pursues two different kinds of issues. Emission issues are related to the unwanted generation of electromagnetic energy by some source, and to the countermeasures which should be taken in order to reduce such generation and to avoid the escape of any remaining energies into the external environment. Susceptibility or Immunity issues, in contrast, refer to the correct operation of electrical equipment, referred to as the victim, in the presence of unplanned electromagnetic disturbances.

Interference mitigation and hence electromagnetic compatibility is achieved by addressing both emission and susceptibility issues, i.e., quieting the sources of interference and hardening the potential victims. The coupling path between source and victim may also be separately addressed to increase its attenuation

Conducted noise occurs when the signal is in electrical contact with a noise source. This is because the noise current source and our signal share a common path. This usually takes the form of a ground loop, an extra ground connection between 2 components with different internal potentials. The ground loop acts like an antenna, effectively picking up interference, leading to a system ground reference that

Figure 2. Layout a power supply with separator resistor between signal and power grounds

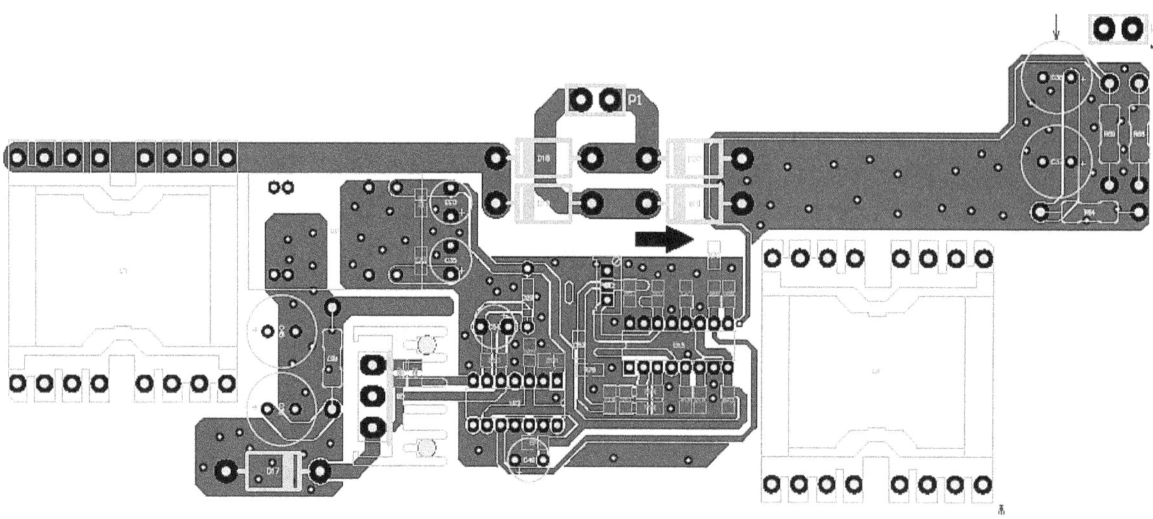

Figure 3. Two systems with common voltage reference

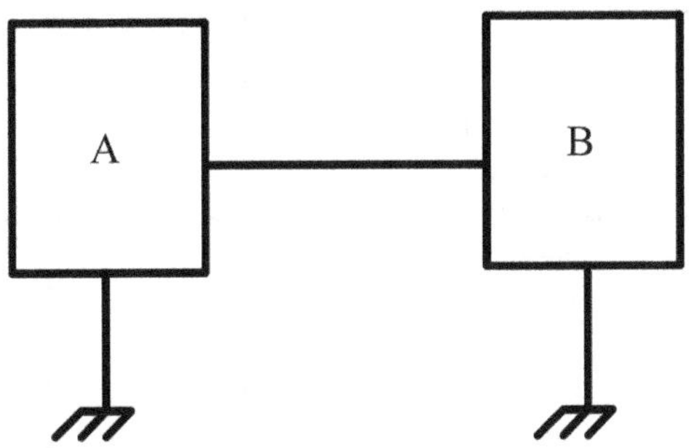

is no longer stable. However subtle, the differences in potential can induce lots of noise. Figure 2 shows a power supply board with separated grounds to prevent interference via ground.

Grounding

In electrical engineering, ground or earth can refer to the reference point in an electrical circuit from which voltages are measured, a common return path for electric current, or a direct physical connection to the Earth.

Electrical circuits may be connected to ground (earth) for several reasons. In mains powered equipment, exposed metal parts are connected to ground to prevent user contact with dangerous voltage if electrical insulation fails. Connections to ground limit the build-up of static electricity when handling flammable products or electrostatic-sensitive devices. In some telegraph and power transmission circuits, the earth itself can be used as one conductor of the circuit, saving the cost of installing a separate return conductor (see single-wire earth return).

For measurement purposes, the Earth serves as a (reasonably) constant potential reference against which other potentials can be measured. An electrical ground system should have an appropriate current-carrying capability to serve as an adequate zero-voltage reference level. In electronic circuit theory, a "ground" is usually idealized as an infinite source or sink for charge, which can absorb an unlimited amount of current without changing its potential. Where a real ground connection has a significant resistance, the approximation of zero potential is no longer valid. Stray voltages or earth potential rise effects will occur, which may create noise in signals or if large enough will produce an electric shock hazard

There are two primary reasons for grounding devices, cables, equipments, and systems. The first reason is to prevent shock and fire hazards in the event that an equipment frame or housing develops a high voltage due to lightning or an accidental breakdown of wiring or components. The second reason is to reduce EMI effects resulting from electromagnetic fields, common impedance, or other forms of interference coupling. Figure 3 shows a simplified diagram of two connected systems with a common potential reference.

Figure 4 shows a mechanism of voltage division between a hot line and chassis.

In an electrical system, a ground loop refers to a current, almost always unwanted, in a conductor connecting two points that are supposed to be at the same potential, often ground, but are actually at different potentials.

Figure 4. Voltage division between power line and chassis

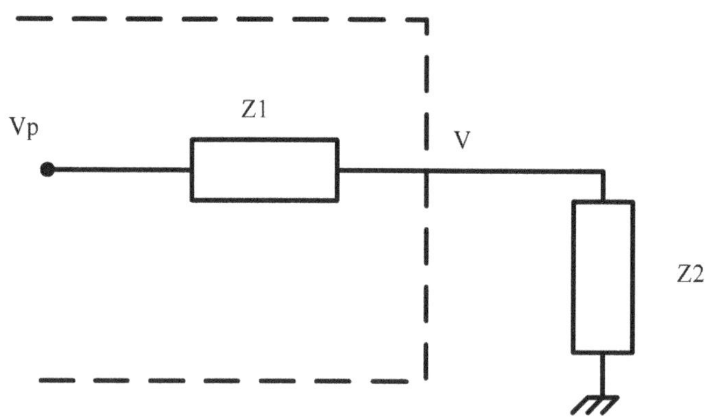

Ground loops are a major cause of noise, hum, and interference in audio, video, and computer systems. They can also create an electric shock hazard, since ostensibly "grounded" parts of the equipment, which are often accessible to users, are not at ground potential

Loops in the ground path can cause currents in signal cable grounds by two main mechanisms. The diagrams at right show a typical ground loop caused by a signal cable S connecting two grounded electronic components C1 and C2. The loop consists of the signal cable's ground conductor, which is connected through the components' metal chassis to the ground wires P in their "three wire" power cords, which are plugged into outlet grounds which are connected through the building's utility ground wire system G.

- Ground loop currents can be induced by stray AC magnetic fields (B, green) The ground loop constitutes a conductive wire loop which may have a large area of several square meters. It acts like a short circuited single-turn "transformer winding"; any AC magnetic flux passing through the loop, from nearby transformers, electric motors, or just adjacent power wiring, will induce currents in the loop by induction. Since its resistance is very low, often less than 1 ohm, the induced currents can be large.

Ground loop current caused by leakage currents in the building's ground wire system from an appliance A.

- Another source of ground loop currents is current leaking from the "hot" side of the power line into the ground system. In addition to resistive leakage, current can also be induced through low impedance capacitive or inductive coupling. The ground potential at different outlets may differ by as much as 10 to 20 volts due to voltage drops from these currents. Figure 2 shows leakage current from an appliance such as an electric motor A flowing through the building's ground system G to the neutral wire at the utility ground bonding point at the service panel. The ground loop between components C1 and C2 creates a second parallel path for the current. The current divides, with some passing through component C1, the signal cable S ground conductor, C2 and back through the outlet into the ground system G. The AC voltage drop across the cable's ground conductor from this current introduces hum or interference into component C2.

Figure 5 shows effect of difference between grounds of two connected systems.

The ground system for a collection of circuits within a system or facility can assume any one of several different configurations. Each of these configurations tends to be optimal under certain conditions and may contribute to EMI problems under other conditions. In general, the ground configurations will involve either a floating ground, a single point ground, a multipoint ground, or some hybrid combination of these. In Figure 6 system "B" is floated and can be connected to ground system "A" with a long return wire.

The single-point or star type of grounding scheme shown in the figure avoids problems of common-mode impedance coupling discussed in the previous section. The only common path is in the earth ground (for earth-based structures), but this usually consists of a substantial conductor of very-low impedance. Thus, as long as no or low ground currents flow in any low-impedance common paths, all subsystems or equipments are maintained at essentially the same reference potential. Figure 7 shows a system with single point ground.

An important advantage of the single-point configuration is that it helps control conductively coupled interference.

Rather than have an uncontrolled situation the other grounding alternative is multipoint grounding as illustrated in Figure 8. For the example shown in Figure 9 and 10, each equipment or subsystem is bonded as directly as possible to a common low-impedance ground plane to form a homogeneous, low-impedance path. Thus, common-mode currents and other EMI problems will be minimized. The ground plane then is earthed for safety purposes.

Earth Connection

Figures 11, 12, and 13 show various types of earth connection in a system. It is important to connect the earth internally to prevent voltage difference between grounds.

Figure 5. Noise voltage between different grounds

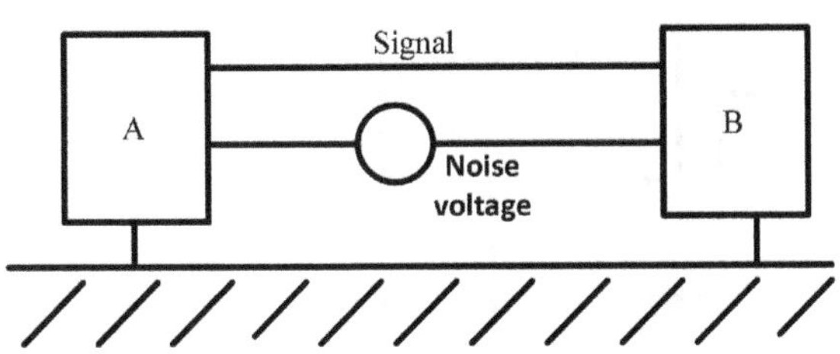

Figure 6. Long ground wire

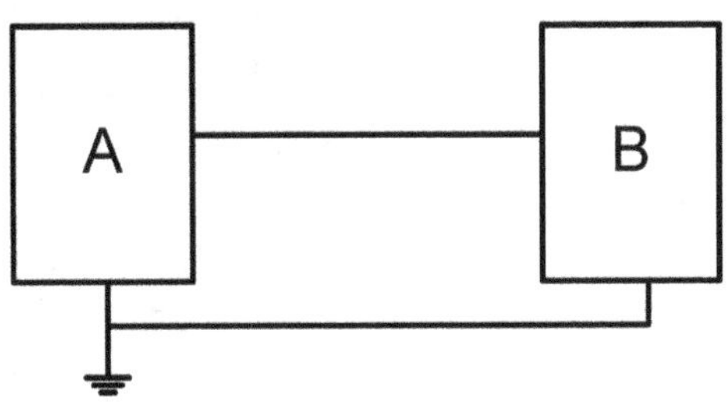

Figure 7. Single point ground connection

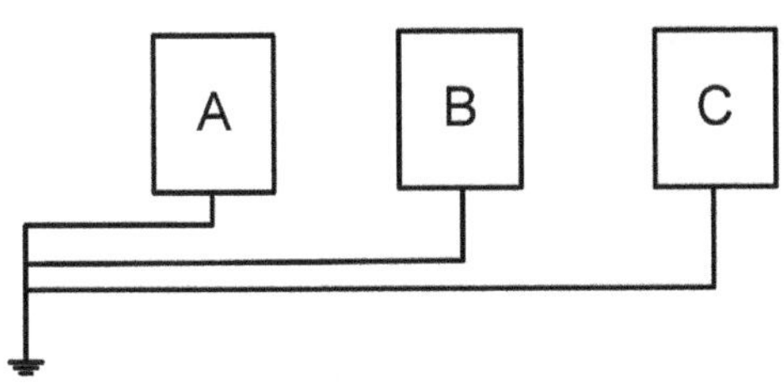

Figure 8. Multi point ground connection

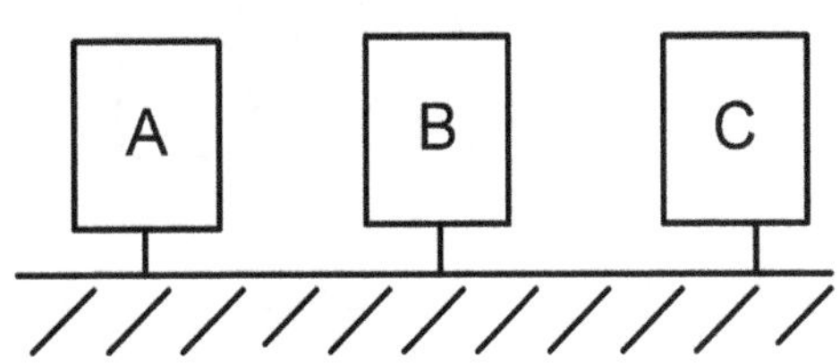

Figure 9. Grouding the system parts according to their noise behavior

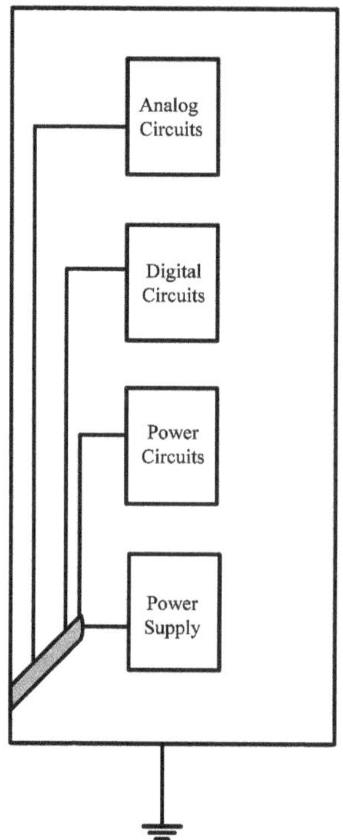

Figure 10. An example of separation in a PDU

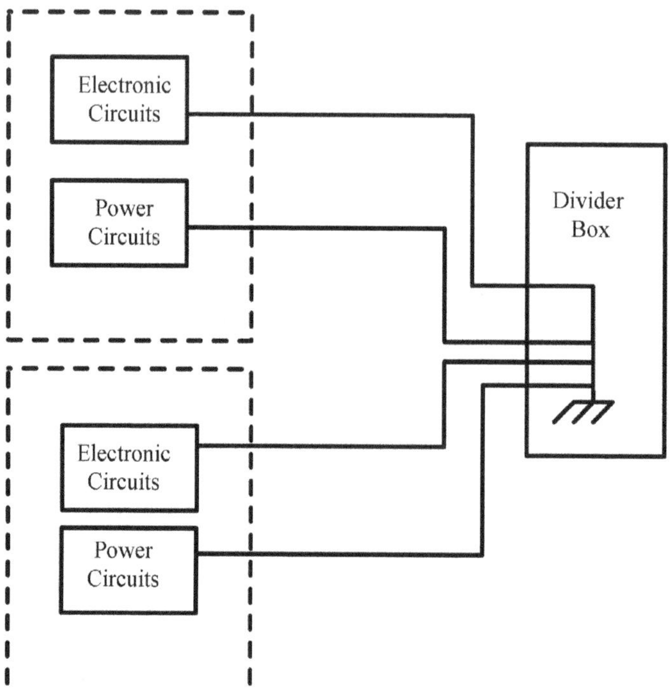

Figure 11. Three subsystems with a common earth connection

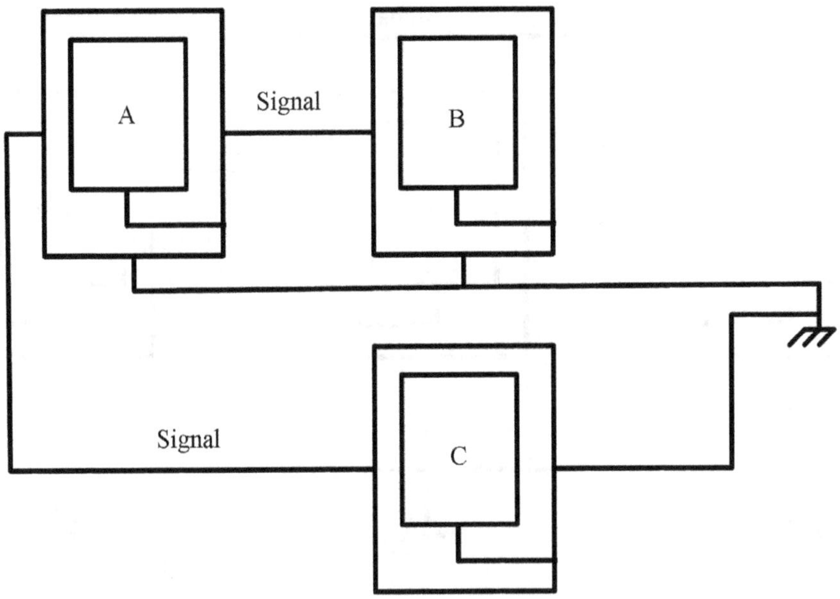

Figure 12. Three subsystems with separated earth connections

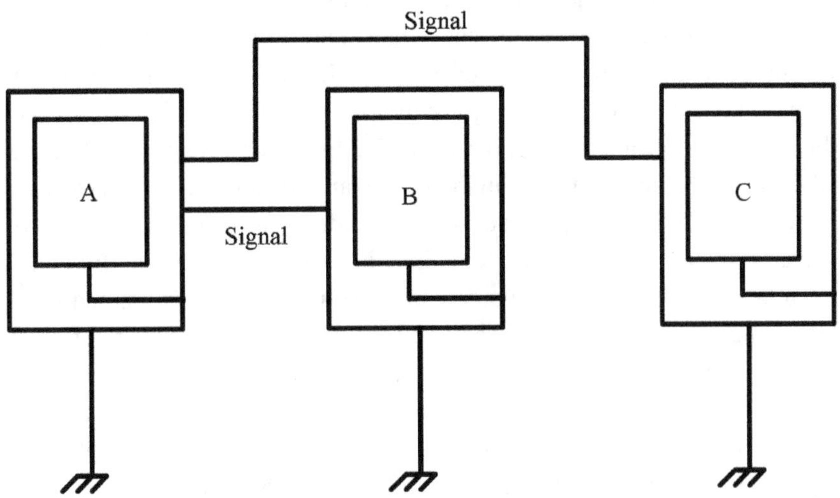

Shielding

Electromagnetic shielding is the practice of reducing the electromagnetic field in a space by blocking the field with barriers made of conductive or magnetic materials. Shielding is typically applied to enclosures to isolate electrical devices from the 'outside world', and to cables to isolate wires from the environment through which the cable runs. Electromagnetic shielding that blocks radio frequency electromagnetic radiation is also known as RF shielding.

Figure 13. Three subsystems with separated but externally connected earth connection

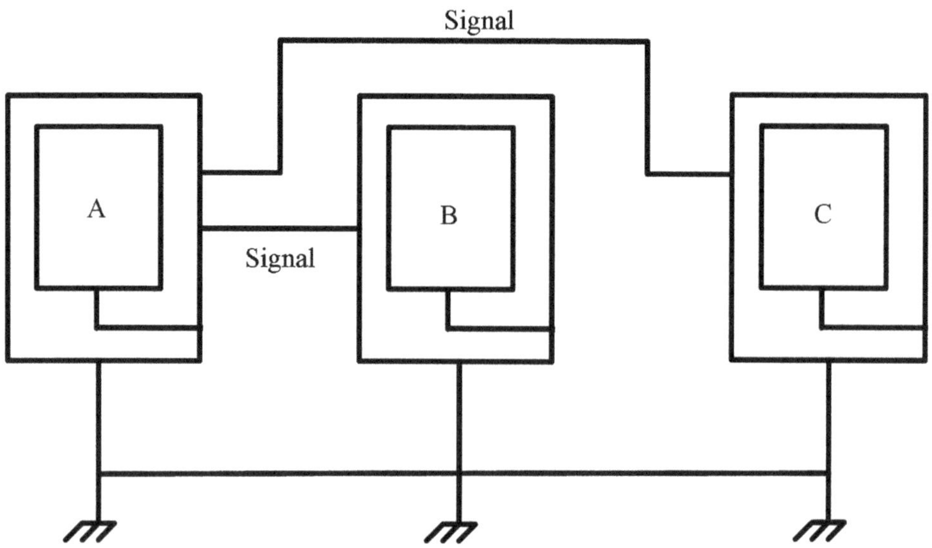

The shielding can reduce the coupling of radio waves, electromagnetic fields and electrostatic fields. A conductive enclosure used to block electrostatic fields is also known as a Faraday cage. The amount of reduction depends very much upon the material used, its thickness, the size of the shielded volume and the frequency of the fields of interest and the size, shape and orientation of apertures in a shield to an incident electromagnetic field. Figure 14 shows effect of shield on electric coupling via parasitic capacitor.

The primary reason for the use of isolated grounds (IG) is to provide a noise-free ground return, separate from the equipment grounding (EG) return. The EG circuit includes all of the metal conduit, outlet boxes, and metal enclosures that contain the wiring and must be grounded to provide a safe return path in case of fault currents. The IG' provides an insulated, separate ground path for the ground reference in electronic equipment, such as computers, hospital equipment, and audio equipment. IG helps eliminate the potential for a ground loop, which can cause noise, data errors, and disruptions to these systems. The IG is typically insulated and separate all the way back to the point of origin of the circuit, either a main panel or sub-panel. The IG, EG and neutral are all bonded together at that point. Due to the installation of a separate, insulated conductor and the associated special outlets required, IG circuits are more expensive to install than standard power circuits. Figure 15 shows application a transformer for isolating the ground. Figure 16 shows a power supply with EMC considerations.

An EMI filter, or electromagnetic interference filter, is an electronic passive device which is used in order to suppress conducted interference that is present on a signal or power line. EMI filters can be used to suppress interference that is generated by the device or by other equipment in order make a device more immune to electromagnetic interference signals present in the environment. Most EMI filters consist of components that suppress differential and common mode interference. Figure 17 shows a typical EMI filter. Its circuit diagram and specifications are shown in figures 18 and 19.

To carry out compliance measurements, standard regulations require a standard network, namely, "Line Impedance Stabilization Network (LISN)". LISN is an industrial network that must have the following characteristics in order to satisfy measurement conditions:

Figure 14. Operation principle of shielding

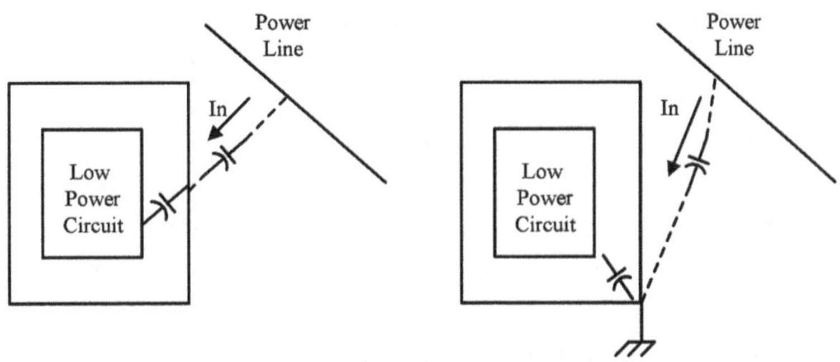

Figure 15. Application of transformer for ground separation, (right): MOSFET driver without isolation, (left): isolation in the input signal of MOSFET driver with transformer

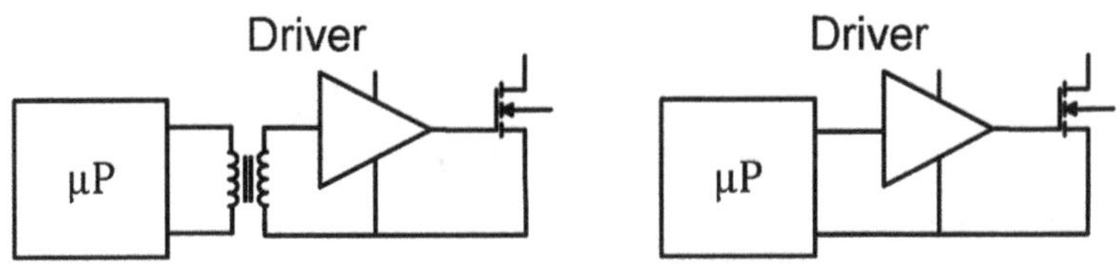

Figure 16. A power supply with EMC considerations

- Transferring power from source to converter,
- Providing a low impedance path for noise measurement,
- Reducing of the effect of source impedance on the noise measurement.

The analysis of an industrial LISN is too complicated. Therefore, a simple model with sufficient accuracy is used. A commonly used LISN, which has been employed in our study, is shown in Figure 4.

Since only the differential mode noise is of our concern, the common ground of the LISN is not considered in Figure 4 and only a series circuit between phases is assumed. With this structure and assuming that the power supply is a short circuit for a dynamic analysis 10, the equivalent noise circuit can be represented as in Figure 5, in which I_{ex} is the converter input current.

There are two ways for noise calculating, direct method and indirect method. In the direct method the LISN is added to the circuit in simulation and frequency spectrum of LISN output voltage (voltage across Z_n) is calculated. In the indirect method, the system function of noise equivalent circuit is utilized. According to the noise equivalent linear circuit the system function $H(j\omega)$ can be written as:

Figure 17. A practical EMI filter (IEC publication 61204 ed.1.0 Copyright © 1993 of publication IEC Geneva, Switzerland. www.iec.ch)

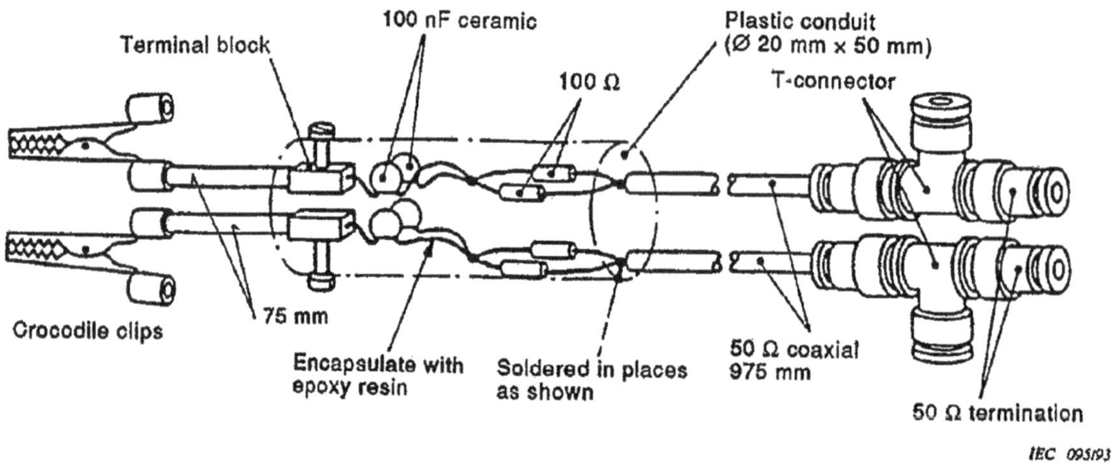

Figure 18. Circuit diagram of an EMI filter, Published with permission of International Rectifier

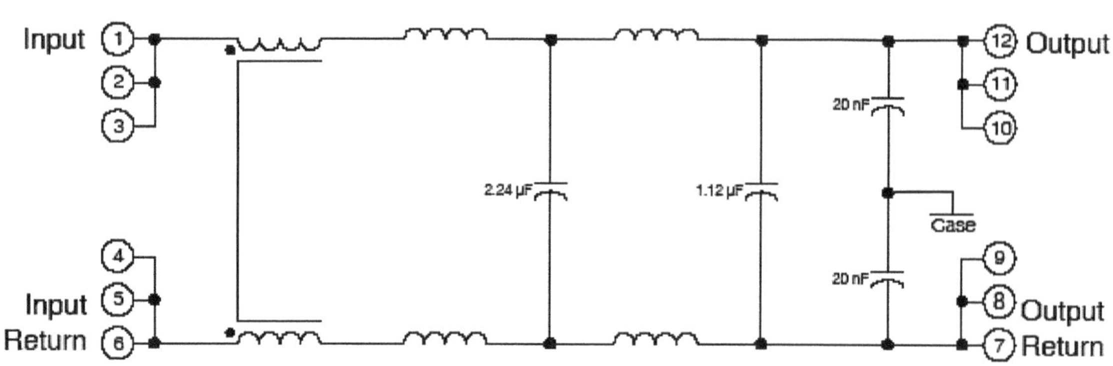

Figure 19. Specifications of an EMI filter, Published with permission of International Rectifier

Electrical Characteristics -55°C ≤ T_{CASE} ≤ +125°C, -400V ≤ V_{IN} ≤ +400V unless otherwise specified

Parameter	Group A Subgroups	Test Conditions	Min	Max	Unit
Leakage Current Note 3	1, 2, 3	± 500V DC Input Voltage	0	50	μA
DC Resistance Note 4	1, 2, 3	T_C = 25°C T_C = 125°C		400 700	mΩ
Noise Reduction	4, 5, 6	200 KHz - 500 KHz 500 KHz - 1 MHz 1 MHz – 10 MHz	35 60 65		dB
Isolation	1	Any Pin to Case, Tested @ 500VDC	100		MΩ
Capacitance	1 2, 3	Measured Between Any Pin and Case	34 30	56 62	nF

Notes to Specifications

1. Operation above maximum ratings may cause permanent damage to the device. Operation at maximum ratings may degrade performance and affect reliability.
2. Device can tolerate ± 720 Volt transient whose duration is ≤ 100 ms when R_S ≥ 0.5 Ω.
3. Derate Output Current linearly from 100% at 125°C to 0 at 135°C.
4. DC resistance is the total resistance of the device and includes the sum of the *input* to *output* resistance and the *return in* to *return out* resistance paths.

$$H(j\omega) = \frac{V_{Z_n}(j\omega)}{I_{ex}(j\omega)} \tag{1}$$

where $V_{z_n}(j\omega)$ and I_{ex} are LISN output voltage and converter input current respectively 10. The system function $H(s)$ is calculated as a function of the noise equivalent circuit parameters as:

$$H(s) = \frac{n_0 + n_1 s + n_2 s^2 + n_3 s^3 + n_4 s^4}{d_0 + d_1 s + d_2 s^2 + d_3 s^3 + d_4 s^4 + d_5 s^5} \tag{2}$$

where the coefficients n_i and d_j correspond to the parameters of the noise equivalent circuit.

In this method, $I_{ex}(t)$ can be computed by computer simulation of the converter. Then by applying FFT to $I_{ex}(t)$, $I_{ex}(j\omega)$ can be calculated. Therefore, $V_{z_n}(j\omega)$ will be computed using equation (1). Simulation indicates that there is no significant difference between the results obtained with these two methods 10. The method with system function is used which is more convenient for calculation of noise level

3-4-Software Methods

Example: Digital control of Power Electronic Converters

The conventional PWM is the most famous and popular method for regulating the output voltage. Most of integrated DC/DC converters usually are based on PWM control method. However, this method has a slow response time especially to load changes. Skip cycle modulation technique is a category of Bang-Bang controllers. Having fast response, this method has been used to control some DC/DC converters. This control technique is proposed for digital control of Flyback converters. This control technique regulates the output voltage based on presence and absence of power. This control scheme offers a faster dynamic response compared with PWM method and improves the efficiency by lowering the switch turn-off loss at the end of each power pulse. The basic control law is pulse by pulse control of the converter due to the output voltage.

In this paper, skip cycle modulation is used to control a resonant converter. Although this converter is a series loaded resonant converter, the results of this paper can be applied to other types of resonant converters. A pulse generator circuit is presented for zero current switching (ZCS). Simulation and experimental results are presented to show the features of this control method. Results show that the converter has a fast response and high efficiency over wide load range.

This method is a class of the Bang-Bang controllers that operates based on applying or canceling the trigger pulses of converter. Figure 20 shows a typical time diagram of this method. The trigger gate pulses have a constant duration in this method and control of the converter is performed by canceling a

Figure 20. Time diagram of the skip cycle modulation technique

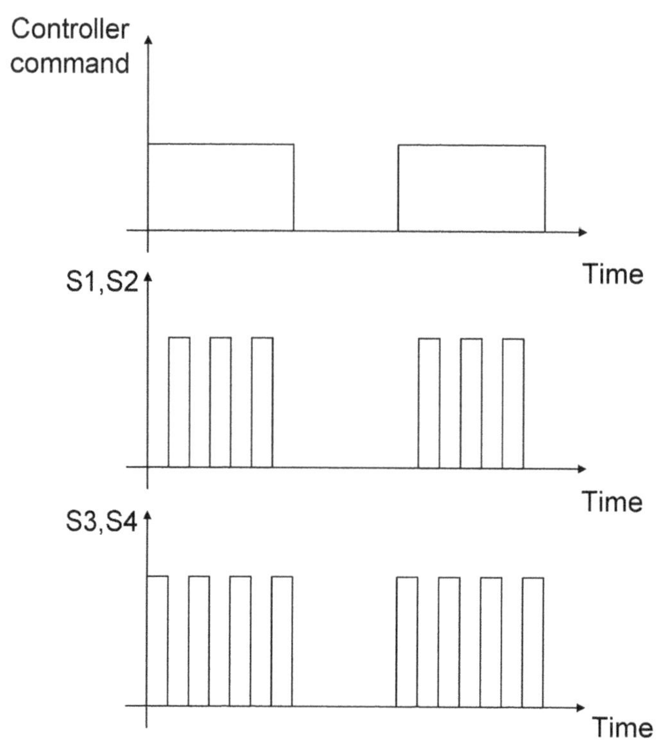

certain interval of the trigger pulses which is done by a control signal command. The control signal is generated by comparing desired variable of the converter (for example output voltage) with a specified reference value. The associated energy of each gate pulse is considered so that the incorrect applying a trigger pulse does not show considerable effect on the controlled variable.

Figure 21 shows the block diagram of proposed method applied to a series resonant converter. The output voltage is measured and compared with a reference value. The result of this stage is a digital signal (U(t)). Regarding to the state of this signal, the pulses of the converter would be applied or stopped. If the control signal, U(t), is stopped at the middle of a gate pulse this last pulse is applied completely to keep the ZCS condition in the converter. All of this logic is implemented in a FPGA and it controls the converter (Kaboli, Alavi, & Oraee, 2008).

Table 1 shows the characteristics of the converter used in simulations. Figure 22 shows the effect of control circuit on the output voltage. When the output voltage falls under the reference value, the control circuit will apply a full cycle resonant current and compensate the voltage reduction. Figure 23 shows the zero current switching condition in converter switches. This fact causes in omitting the switching loss and increasing the efficiency of the converter. Figure 24 shows the performance of the pulse generator. It can be seen that while the control signal U(t) reaches zero, the output gate pulses are applied completely in order to maintain the ZCS condition. Figure 25 shows the conduction of output diodes in ZCS condition. Figure 26 shows a good and short transient response of the converter to load changes. When the load is small, the gate pulses are applied in great time intervals. These intervals are small when the load is increased. Figure 27 shows the start up of the converter. Due to the high inrush current, a soft start circuit should be used for this converter.

Figure 21. Block diagram of proposed method

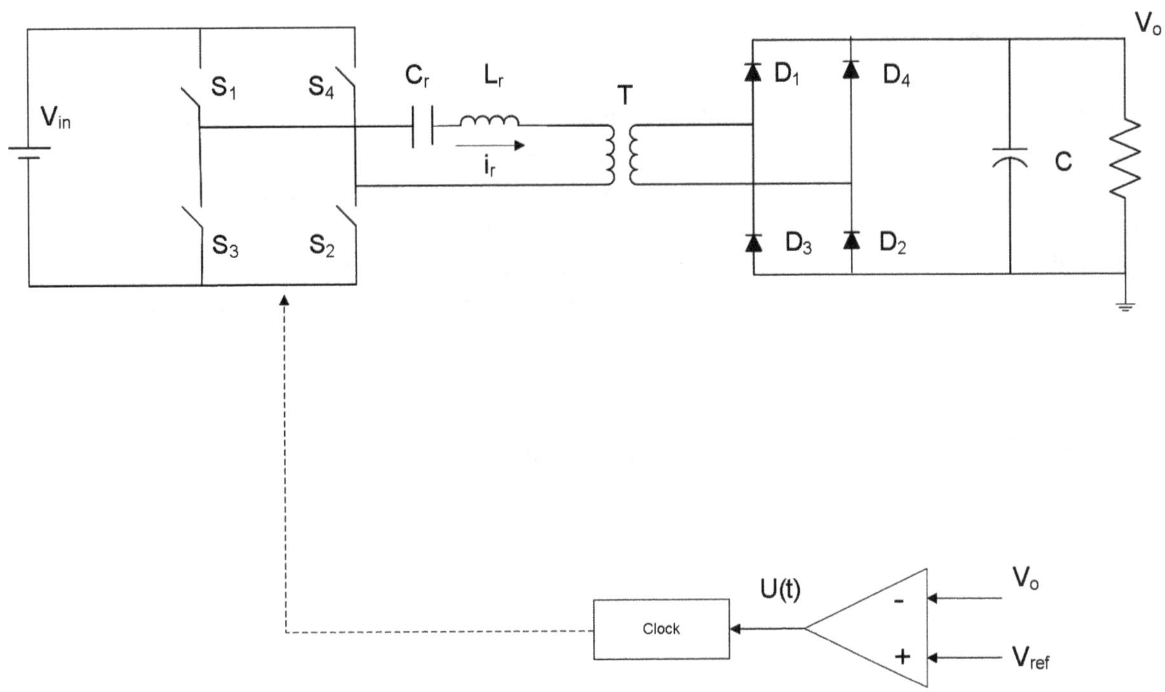

Table 1. Characteristics of converter

Parameter	value
Rated Power	4kW
Switching frequency	25 kHz
Output Voltage	20V
Vin	530V
Resonance capacitor	220 nf
Resonance Inductance	280 μH

To verify the computer simulation results, a resonant converter has been designed and built. The converter shown in Figure 28 has the same parameters as what is used in the simulations. Figures 29 and 30 show the current of resonant circuit for two cases: heavy or light converter load. It can be seen that in the heavy load the number of control pulses is increased to supply the load current. Figure 31 shows the current of resonant circuit in a narrow scale. This sinusoidal current guarantees the ZCS condition. Thus, the switching loss is omitted and the efficiency of the converter is increased. The measured efficiency of converter is approximately 94%. This efficiency is measured in other loads and it is shown in Figure 32. This figure shows that the converter has a high efficiency over wide load range because the constant loss of converter such as transformer core loss is omitted when the control signal, U(t), is stopped. Therefore, in light load the converter loss is also decreased and the converter efficiency is increased. Figure 33 shows the time response of the converter to a load change. It can be seen that the output voltage does not show considerable variation for load current change and this confirms the fast response of the proposed method. Figure 34 shows the sinusoidal waveform of converter that leads to ZCS condition in switching.

3-5-EMC Standards

The goal of electromagnetic compatibility, or EMC, is to design electronic systems that are electromagnetically compatible with their environment. EMC requirements exist so that electronic systems designers have a set of guidelines that explain the limits of what is considered electromagnetically compatible. There is not, however, one all-encompassing set of EMC guidelines. Instead, EMC guidelines are created by individual product manufacturers, and by the government. Requirements set forth by the government are legal requirements that products must meet, while the requirements set forth by the manufacturer are self-imposed and often more stringent than those set forth by the government.

The following list outlines a number of EMC standards which are known at the time of writing to be either available or have been made available for public comment. These standards attempt to standardize product EMC performance, in respect to radio interference for electrical equipment.

1. CISPR
2. IEC
3. ISO
4. ANSI

Figure 22. Effect of applied control pulse on the output voltage, Up: Output voltage, Down: Resonant current

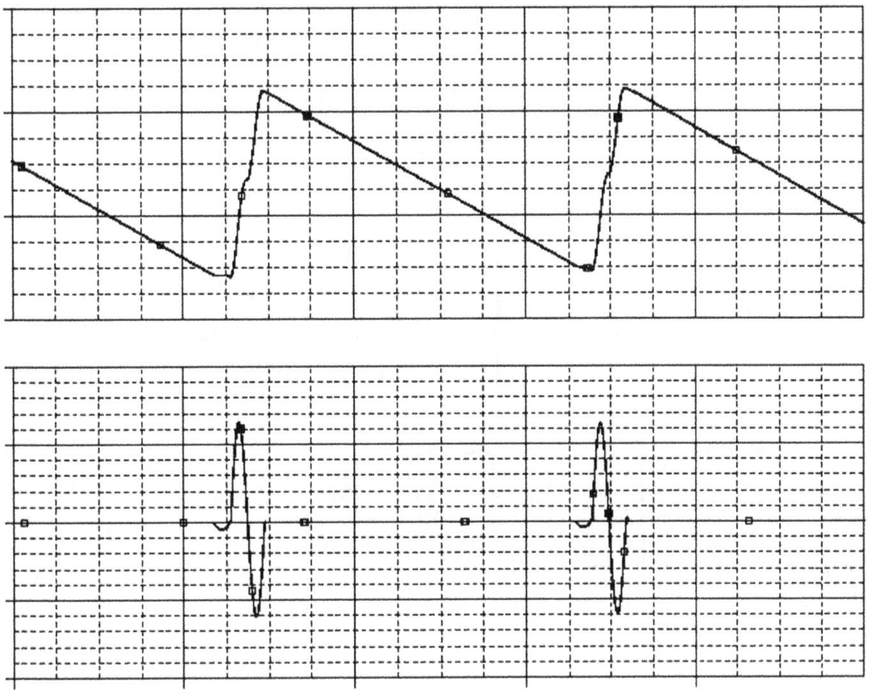

Figure 23. Zero current switching, Up: Switch current, Down: Switch voltage

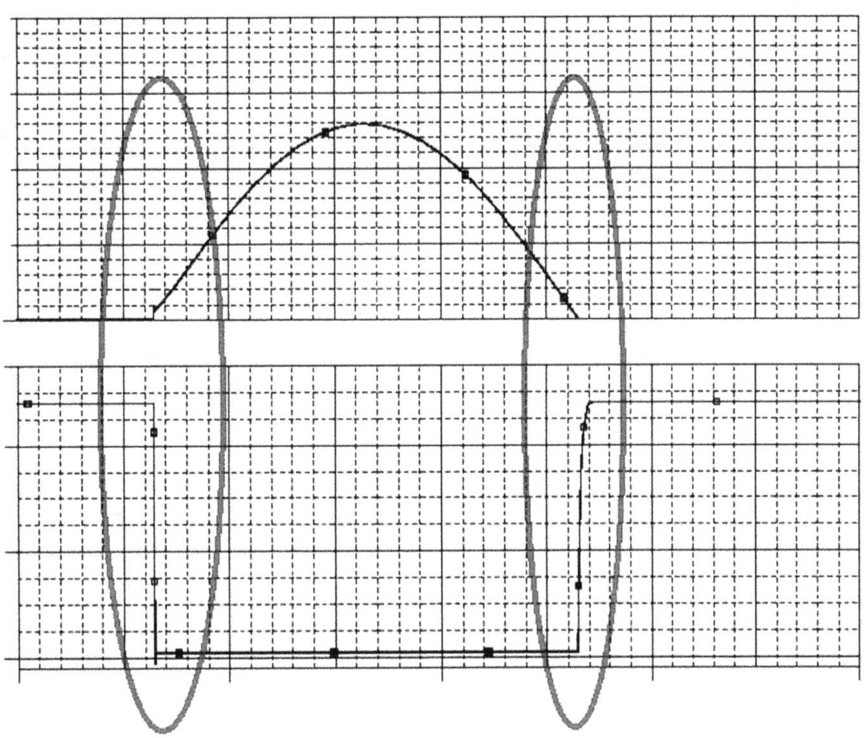

Figure 24. Control pulses, Up: U(t), Middle and Down: Gate pulses

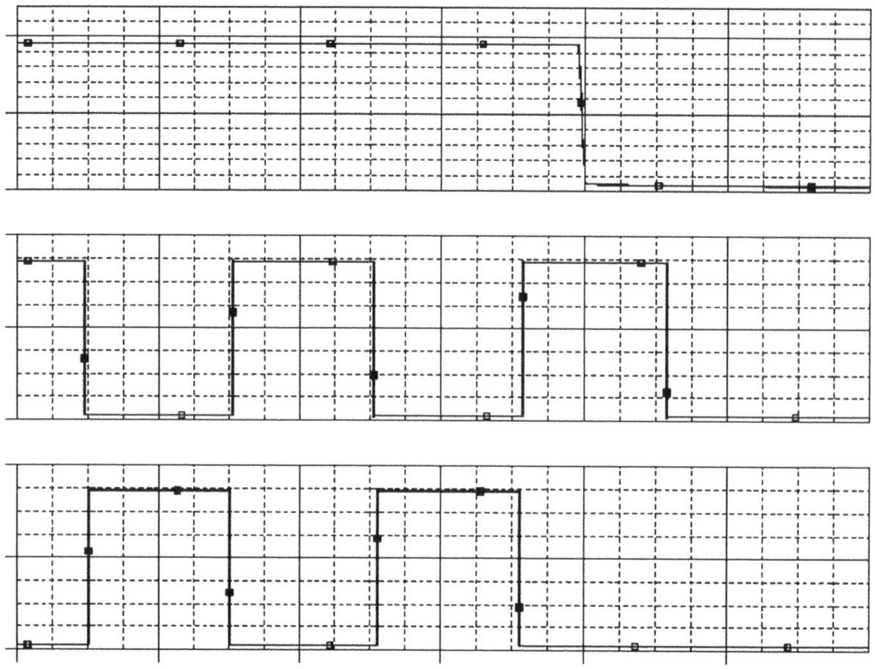

Figure 25. Current sharing of output diodes, Up: resonant current, Middle: Current of diodes D1 and D3, Down: Current of diodes D2 and D4

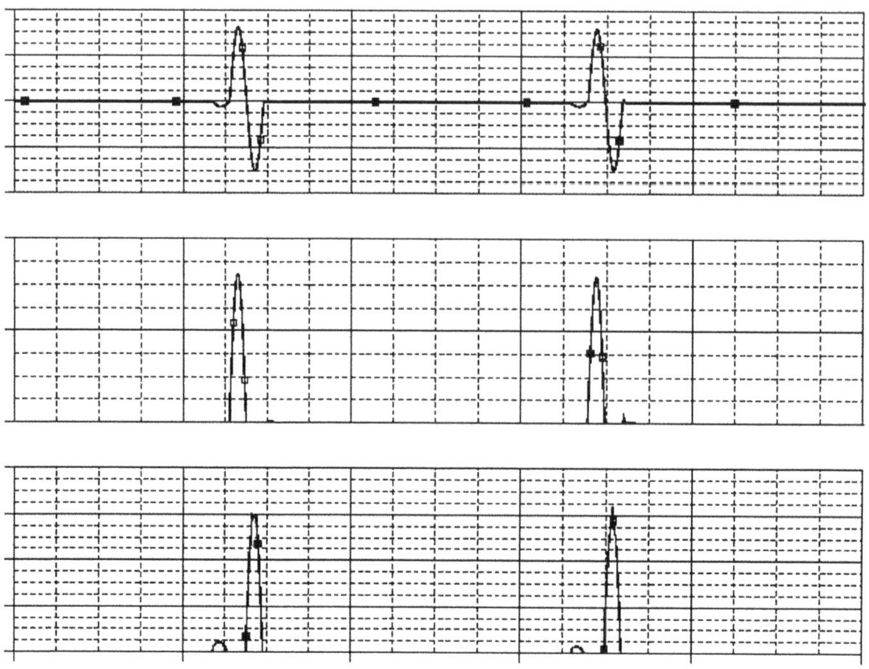

Figure 26. Converter response to load change, Up: Output current, Middle: Output voltage, Down: Resonant current

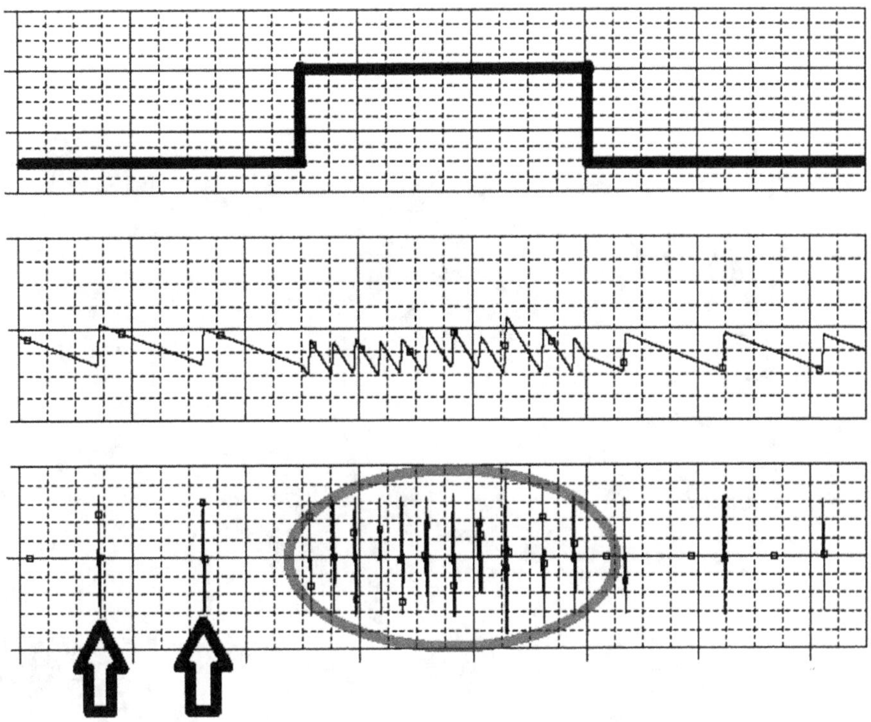

Figure 27. Converter Start up, Up: Output voltage, Down: Resonant current

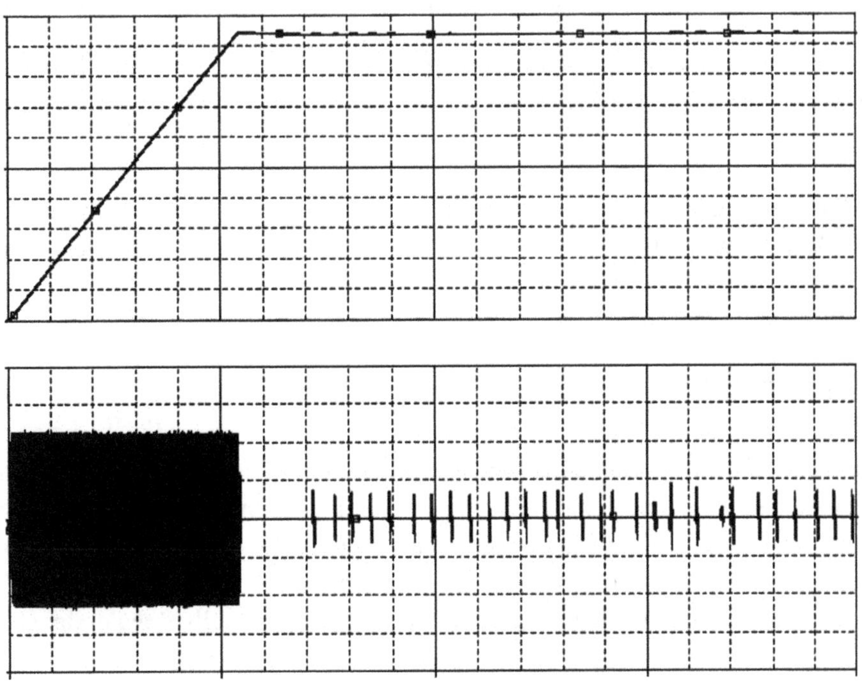

Figure 28. Experimental setup, A: Input rectifier, B: Inverter switches, C: Control circuit, D: Resonant capacitors, E: Transformer, F: Output diodes, H: Auxiliary power supply, G: Output capacitors

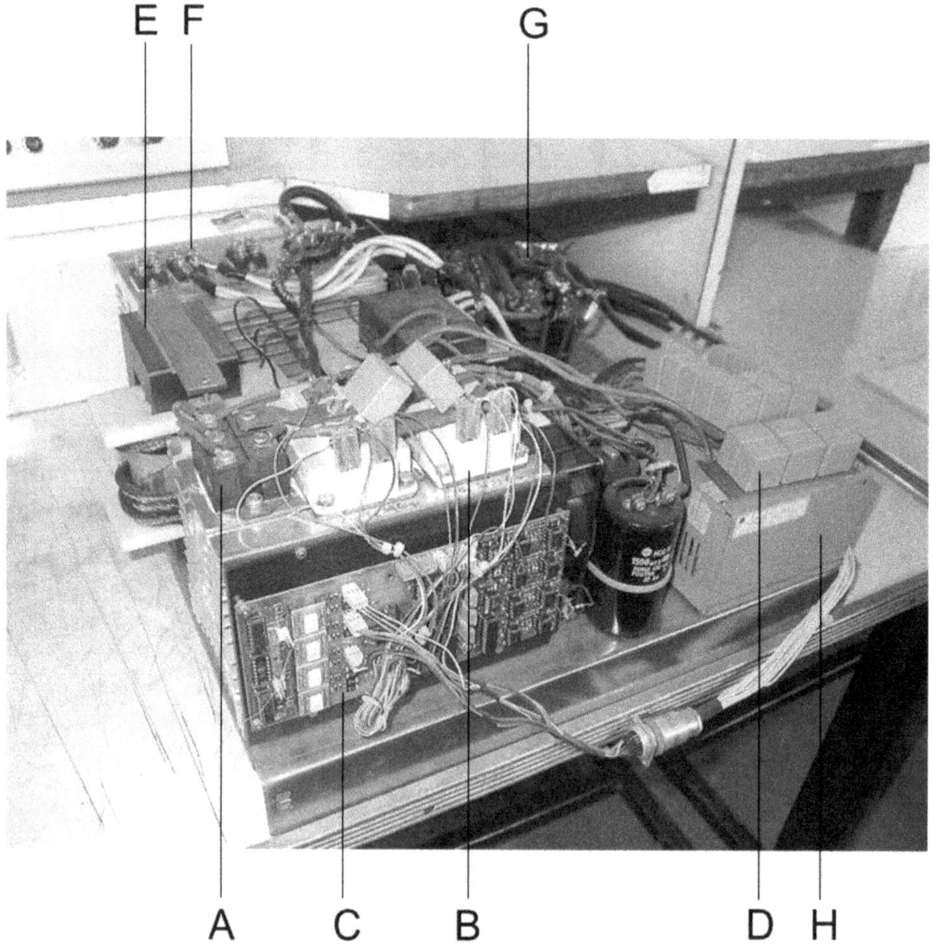

Figure 35 shows a sample of EMC levels in IEC standard. Figure 36 shows a switching ASD which is used for an induction motor. Figure 37 shows an equipment which is used for generating the certain harmonics. It is applied to EMC studies.

ALARM MANAGEMENT

Alarm management is the application of human factors (or ergonomics as the field is referred to outside the U.S.) along with instrumentation engineering and systems thinking to manage the design of an alarm system to increase its usability. Most often the major usability problem is that there are too many alarms annunciated in a plant upset, commonly referred to as alarm flood (similar to an interrupt storm), since it is so similar to a flood caused by excessive rainfall input with a basically fixed drainage output capac-

Figure 29. Current of resonant circuit for light load

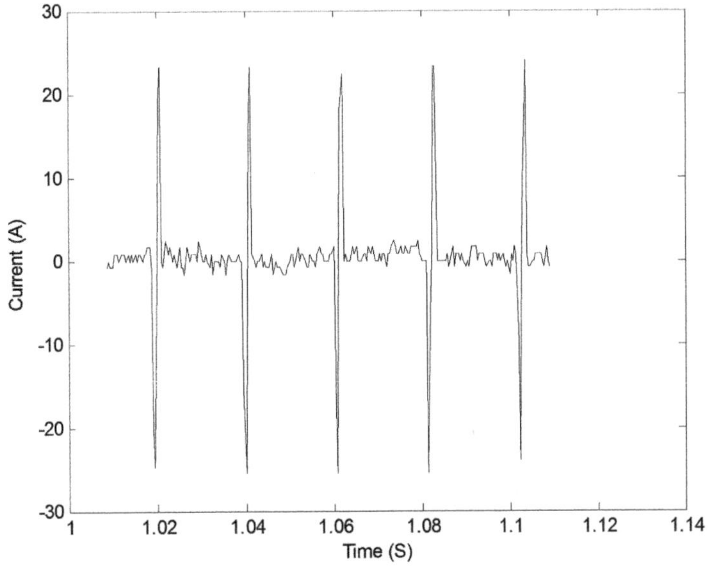

Figure 30. Current of resonant circuit for heavy load

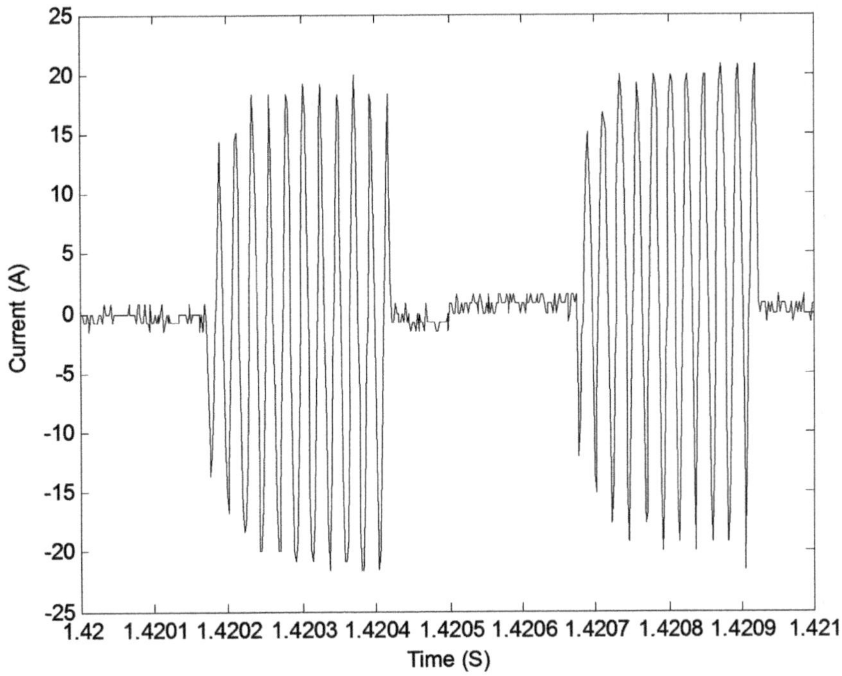

Figure 31. Zero current switching in converter switches

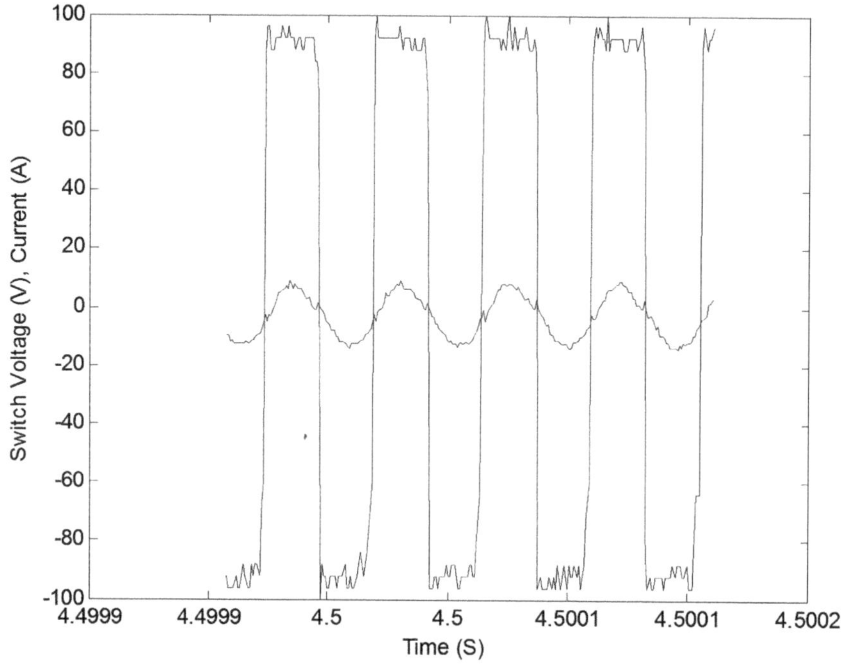

Figure 32. Variation of converter efficiency over load range

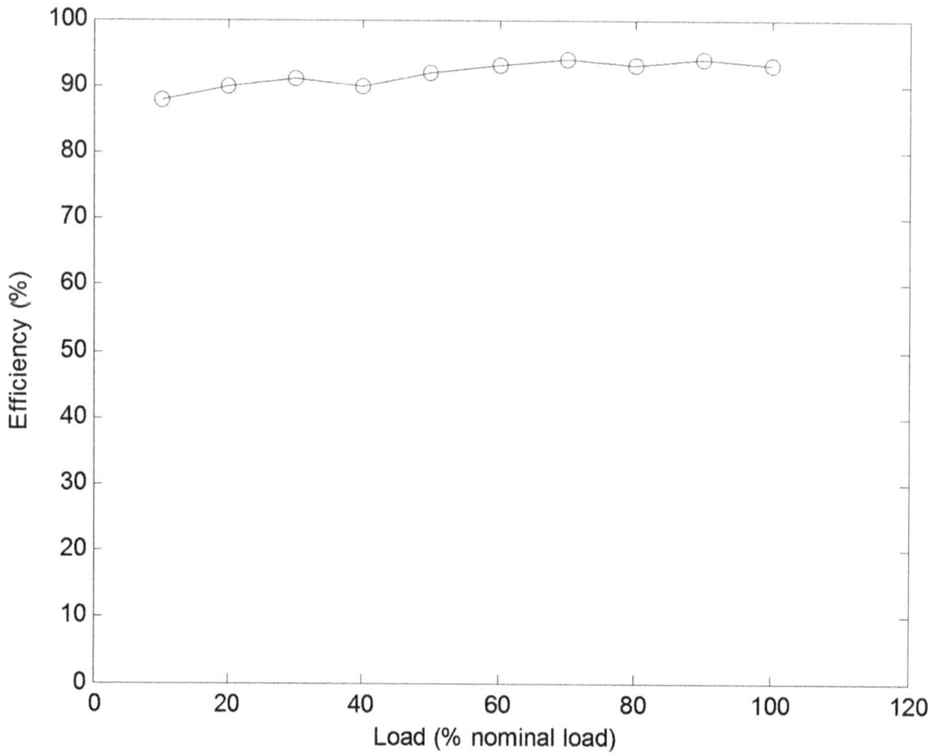

Figure 33. Time response of the converter to load change, Up: Load current, Down: Output voltage

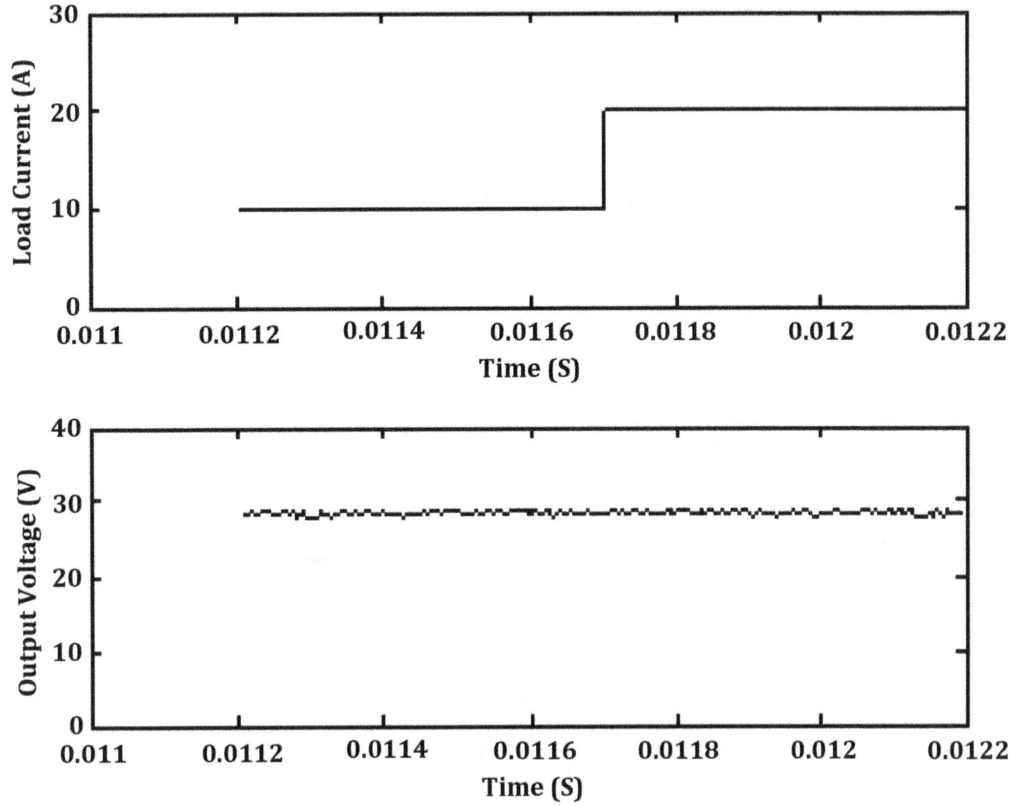

Figure 34. Current of resonant circuitFigure 35. Different harmonic levels for various classes in electric equipment (IEC publication 61000-3-2 ed.1.2 "Copyright © 2014 of publication IEC Geneva, Switzerland. www.iec.ch)

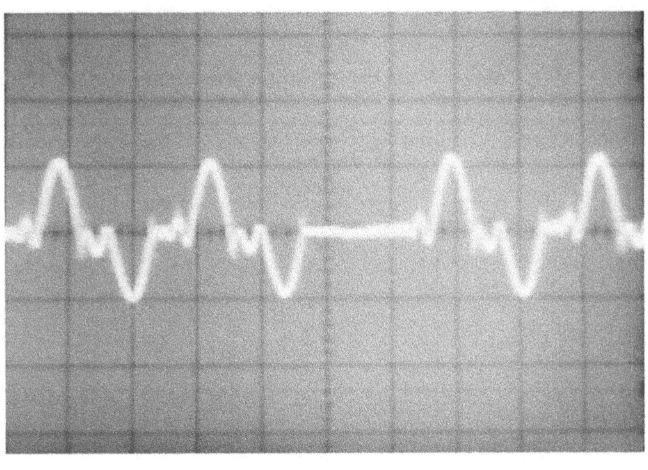

Figure 35. Different harmonic levels for various classes in electric equipment (IEC publication 61000-3-2 ed.1.2 Copyright © 2014 of publication IEC Geneva, Switzerland. www.iec.ch)

Table 1 – Limits for Class A equipment

Harmonic order n	Maximum permissible harmonic current A
Odd harmonics	
3	2,30
5	1,14
7	0,77
9	0,40
11	0,33
13	0,21
$15 \leq n \leq 39$	$0{,}15 \frac{15}{n}$
Even harmonics	
2	1,08
4	0,43
6	0,30
$8 \leq n \leq 40$	$0{,}23 \frac{8}{n}$

Table 2 – Limits for Class C equipment

Harmonic order n	Maximum permissible harmonic currrent expressed as a percentage of the input current at the fundamental frequency %
2	2
3	$30 \cdot \lambda$ *
5	10
7	7
9	5
$11 \leq n \leq 39$ (odd harmonics only)	3
* λ is the circuit power factor	

Table 3 – Limits for Class D equipment

Harmonic order n	Maximum permissible harmonic current per watt mA/W	Maximum permissible harmonic current A
3	3,4	2,30
5	1,9	1,14
7	1,0	0,77
9	0,5	0,40
11	0,35	0,33
$13 \leq n \leq 39$ (odd harmonics only)	$\frac{3{,}85}{n}$	See table 1

Figure 36. A practical ASD

Figure 37. A harmonic generator using for EMI tests

ity. However, there can also be other problems with an alarm system such as poorly designed alarms, improperly set alarm points, ineffective annunciation, unclear alarm messages, etc.

The fundamental purpose of alarm annunciation is to alert the operator to deviations from normal operating conditions, i.e. abnormal operating situations. The ultimate objective is to prevent, or at least minimize, physical and economic loss through operator intervention in response to the condition that was alarmed. For most digital control system users, losses can result from situations that threaten environmental safety, personnel safety, equipment integrity, economy of operation, and product quality control as well as plant throughput. A key factor in operator response effectiveness is the speed and accuracy with which the operator can identify the alarms that require immediate action.

By default, the assignment of alarm trip points and alarm priorities constitute basic alarm management. Each individual alarm is designed to provide an alert when that process indication deviates from normal. The main problem with basic alarm management is that these features are static. The resultant alarm annunciation does not respond to changes in the mode of operation or the operating conditions.

When a major piece of process equipment like a charge pump, compressor, or fired heater shuts down, many alarms become unnecessary. These alarms are no longer independent exceptions from normal operation. They indicate, in that situation, secondary, non-critical effects and no longer provide the operator with important information. Similarly, during startup or shutdown of a process unit, many alarms are not meaningful. This is often the case because the static alarm conditions conflict with the required operating criteria for startup and shutdown.

In all cases of major equipment failure, startups, and shutdowns, the operator must search alarm annunciation displays and analyze which alarms are significant. This wastes valuable time when the operator needs to make important operating decisions and take swift action. If the resultant flood of alarms becomes too great for the operator to comprehend, then the basic alarm management system has failed as a system that allows the operator to respond quickly and accurately to the alarms that require immediate action. In such cases, the operator has virtually no chance to minimize, let alone prevent, a significant loss.

In short, one needs to extend the objectives of alarm management beyond the basic level. It is not sufficient to utilize multiple priority levels because priority itself is often dynamic. Likewise, alarm disabling based on unit association or suppressing audible annunciation based on priority do not provide dynamic, selective alarm annunciation. The solution must be an alarm management system that can dynamically filter the process alarms based on the current plant operation and conditions so that only the currently significant alarms are annunciated.

The fundamental purpose of dynamic alarm annunciation is to alert the operator to relevant abnormal operating situations. They include situations that have a necessary or possible operator response to ensure:

- Personnel and Environmental Safety,
- Equipment Integrity,
- Product Quality Control.

The ultimate objectives are no different from the previous basic alarm annunciation management objectives. Dynamic alarm annunciation management focuses the operator's attention by eliminating extraneous alarms, providing better recognition of critical problems, and insuring swifter, more accurate operator response

Alarm management is usually necessary in a process manufacturing environment that is controlled by an operator using a control system, such as a DCS or a programmable logic controller (PLC). Such a system may have hundreds of individual alarms that up until very recently have probably been designed with only limited consideration of other alarms in the system. Since humans can only do one thing at a time and can pay attention to a limited number of things at a time, there needs to be a way to ensure that alarms are presented at a rate that can be assimilated by a human operator, particularly when the plant is upset or in an unusual condition. Alarms also need to be capable of directing the operator's attention to the most important problem that he or she needs to act upon, using a priority to indicate degree of importance or rank, for instance. Figure 38 shows the operator panel of a power supply with pre-alarm of over temperature. Figure 39 and 40 show two temperature sensors implemented on a transistor heat sink and a transformer, respectively. The first should be very fast but the second can be just a pre-alarm. Application of a temperature sensor in the first case with a lower activating temperature set point can be also used as pre-alarm as shown in figure 41.

MAINTAINABILITY

In engineering, maintainability is the ease with which a product can be maintained in order to:

Figure 38. An operator panel with two levels alarms

Figure 39. Pre-alarm temperature sensor on a heat sink

Figure 40. Pre-alarm temperature sensor on a transformer

Figure 41. Protecting temperature sensor on a heat sink

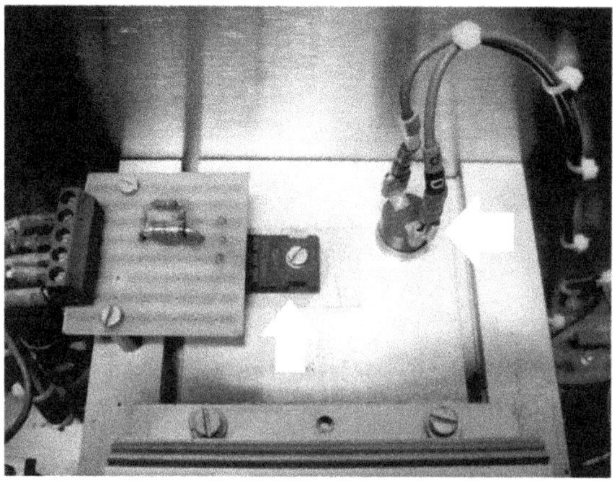

- isolate defects or their cause,
- correct defects or their cause,
- repair or replace faulty or worn-out components without having to replace still-working parts,
- prevent unexpected breakdowns,
- maximize a product's useful life,
- maximize efficiency, reliability, and safety,
- meet new requirements,
- make future maintenance easier, or
- cope with a changed environment.

In some cases, maintainability involves a system of continuous improvement - learning from the past in order to improve the ability to maintain systems, or improve reliability of systems based on maintenance experience.

In telecommunication and several other engineering fields, the term maintainability has the following meanings:

- A characteristic of design and installation, expressed as the probability that an item will be retained in or restored to a specified condition within a given period of time, when the maintenance is performed in accordance with prescribed procedures and resources.
- The ease with which maintenance of a functional unit can be performed in accordance with prescribed requirements

SUMMARY AND CONCLUSION

Protection systems save the converter against faults. However, a safe but out of service converter is not the goal of reliability. In this chapter, availability of an electric power converter is described. The main topics of this chapter are summarized as follows:

1. Some failure factors may not damage the system but interfere with its proper operation.
2. Noise is a common interfering factor in the systems. Interference mitigation and hence electromagnetic compatibility is achieved by addressing both emission and susceptibility issues, i.e., quieting the sources of interference and hardening the potential victims. The coupling path between source and victim may also be separately addressed to increase its attenuation. There are two methods of noise coupling: conducted and radiated.
3. Grounding has a key role for controlling the effect of noise and interference in circuits. The most important law in grounding is that all electric references must have the same potential in a power network.
4. An EMI filter, or electromagnetic interference filter, is an electronic passive device which is used in order to suppress conducted interference that is present on a signal or power line. EMI filters can be used to suppress interference that is generated by the device or by other equipment in order to make a device more immune to electromagnetic interference signals present in the environment.
5. Some fault alarms are not effective for immediate shutdown of systems.

Protection systems usually prevent damaging the converter due to faults. However, some faults can not be isolated from the converter and they cause damage the converter. In this case, the converter should remain in service from reliability point of view. In the next chapters, we present the methods for recovering a faulty converter with derated specifications.

REFERENCES

Kaboli, S., Alavi, S. M. H., & Oraee, H. (2008). Skip cycle controlled resonant converter with high efficiency over wide load range. In *Proceedings of 4th IET Conference on Power Electronics, Machines and Drives* (pp. 742-746). IEEE. doi:10.1049/cp:20080620

Kaboli, S., Mahdavi, J., & Agah, A. (2007). Application of Random PWM Technique for Reducing the Conducted Electromagnetic Emissions in Active Filters. *IEEE Transactions on Industrial Electronics*, *54*(4), 2333–2343. doi:10.1109/TIE.2007.899944

Kaboli, S., Rajaei, A. H., & Sheikhi, A. (2009). Application of random PWM techniques for reducing the electromagnetic interference of Vienna rectifiers in distribution power system. In *Proceedings of IEEE 6th International Power Electronics and Motion Control Conference* (pp.998-1003). IEEE. doi:10.1109/IPEMC.2009.5157530

Kwasinski, A. (2011). Quantitative Evaluation of DC Microgrids Availability: Effects of System Architecture and Converter Topology Design Choices. *IEEE Transactions on Power Electronics*, *26*(3), 835–851. doi:10.1109/TPEL.2010.2102774

Kwasinski, A., Krishnamurthy, V., Song, J., & Sharma, R. (2012). Availability Evaluation of Micro-Grids for Resistant Power Supply During Natural Disasters. *IEEE Transactions on Smart Grid*, *3*(4), 2007–2018. doi:10.1109/TSG.2012.2197832

Liu, T.-H., Fu, J.-R., & Lipo, T. A. (1993). A Strategy for Improving Reliability of Field-Oriented Controlled Induction Motor Drives. *IEEE Transactions on Industry Applications*, *29*(5), 910–918. doi:10.1109/28.245714

Mahdavi, J., Kaboli, S. H., & Toliyat, H. A. (1999). Conducted electromagnetic emissions in unity power factor AC/DC converters: comparison between PWM and RPWM techniques. In *Proceedings of 30th Annual IEEE Power Electronics Specialists Conference* (pp. 881-885). IEEE. doi:10.1109/PESC.1999.785615

Song, J., Krishnamurthy, V., Kwasinski, A., & Sharma, R. (2013). Development of a Markov-Chain-Based Energy Storage Model for Power Supply Availability Assessment of Photovoltaic Generation Plants. *IEEE Transactions on Sustainable Energy*, *4*(2), 491–500. doi:10.1109/TSTE.2012.2207135

ADDITIONAL READING

Ott, H. (2010). *Electromagnetic compatibility*. McGraw-Hill.

KEY TERMS AND DEFINITIONS

Alarm: A flag for announcement of fault occurrence.
EMC: Electromagnetic compatibility.
EMI: Electromagnetic interference between two systems.
Ground: Common potential reference of various parts in an electric network.
Noise: A variable (electrical) without information.

Section 5

Reliability in Operation Process

Chapter 10
Derating

ABSTRACT

In previous chapters, we discussed the converter with or without fault. The common similarity between them is that they continue to operate without reduction of their nominal specification. In this chapter, uninterrupted operation of a faulty power conversion system with catastrophic damages in some of its parts is investigated. It is shown that a faulty electric power converter can continue to work with degraded specifications. This algorithm is named derating for accessibility. This technique can be used for both a faulty system because of its uninterrupted operation and a normal system because of extensive life time. Algorithms for derating of a faulty electric machine and a power supply are described. Derating for increasing the useful life of a motor drive system is presented. A novel method for switching frequency selection in a switching power supply is proposed based on derating concept. Derating is introduced as a technique to compensate additional losses in an electric power converter operating in harsh environment (for example: a motor drive which is supplied with a non sinusoidal voltage waveform). Industrial examples are presented in details for better understanding of the derating concept. Some of the presented examples contain novel idea for derating and others are well known in industry.

INTRODUCTION: DERATING TO CONTINUE THE OPERATION

Protection techniques prevent catastrophic failures in electric power converters. However, as it was described before, protection systems can not prevent any fault. There are two main approaches to a faulty converter. Isolating the faulty converter is one of the approaches and it is used when the converter mission can be interrupted for repair and maintenance. If the mission of the converter is important at the time of failure, users should use another method. Derating is a method for allowing the faulty converter to continue its mission. Figure 1 shows the state of this chapter in the flowchart of the book.

Although, derating method is mainly a method for faulty systems, it can also be used for life extending in normal cases. Component failure rates generally decrease as applied stress levels decrease. Thus, derating or operating components at levels below their ratings (for current, voltage, power dissipation, temperature, etc.) will increase reliability. You can achieve this derating by circuit design (minimizing applied stress), component selection (using components with ratings well above the applied stress), and

DOI: 10.4018/978-1-4666-9429-3.ch010

Figure 1. State of chapter 10 in the flowchart of the book

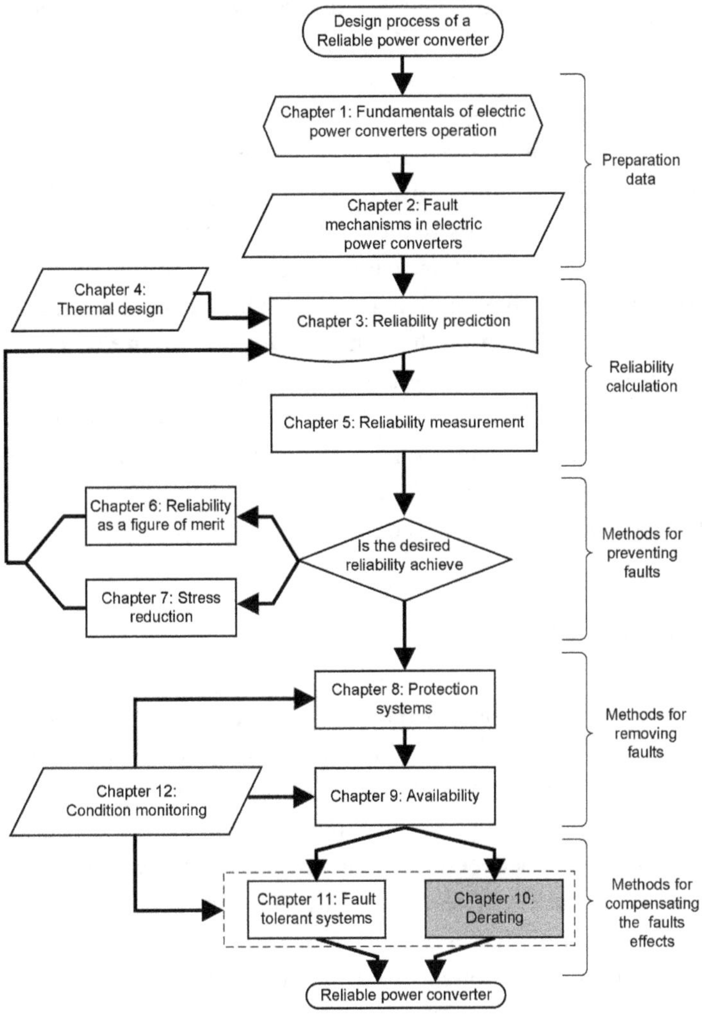

thermal design (reducing the component's operating temperature). Derating is a technique usually employed in electrical power and electronic devices, wherein the devices are operated at less than their rated maximum power dissipation, taking into account the case/body temperature, the ambient temperature and the type of cooling mechanism used. In this article, we will briefly explain the theoretical background of derating and how it is applied.

Derating increases the margin of safety between part design limits and applied stresses, thereby providing extra protection for the part. By applying derating in an electrical or electronic component, its degradation rate is reduced. The reliability and life expectancy are improved.

Intuitively, if a component or system is operated under its design limit, it will be more reliable than if it is operated at or above the design limit. Theoretically, the benefit of derating can be explained using load-strength interference theory.

Most estimates of failure rate are based on an operating temperature, input voltage, and output power. Using the converter at an output power or current different from that used for the reliability calculation can yield a reliability different from that predicted. We can use these data for operating the converter at an output power less than the maximum permissible value. The power dissipated in many internal components — including most of the high-power devices in the power conversion chain — is approximately proportional to the output power or current. This means that derating the output current by x% will reduce the component temperature rise by at least x% and by more for components such as MOSFETs and magnetics, where the power dissipation is proportional to the current squared. The lower power dissipation leads to a corresponding lower failure rate. Consequently, the power system designer can decrease the converter's failure rate by reducing its output power.

A necessary procedure for manufacturing that enhances reliability is burn-in. Burn-in keeps infant mortality in the factory, rather than allowing it in the field. This can be done at the part, board, or system level. All dc-dc converters should go through a burn-in process. Failure rates and times need to be recorded and analyzed to ensure the burn-in period is long enough to bring out all cases of infant mortality.

If you can't achieve the required level of predicted reliability solely by derating and thermal management of one dc-dc converter, then you can add more converters in parallel. Although more components are present to fail, the consequence of one failure may be completely overcome, requiring the system to have two failures before compromising functionality

LOAD-STRENGTH INTERFERENCE

Usually, failure happens when the applied load exceeds the strength. Load and strength should be considered in a general way. For electronic parts, "load" might refer to voltage, power or an internal stress such as junction temperature. "Strength" might refer to any resisting physical properties.

Electronic components of a given type are not identical. They have strength variability. This variability results from the differences between raw materials and between manufacturing processes. Even for components made from the same materials and by the same processes, differences still exist due to noise factors such as microscopic material defects or variations within a single manufacturing process. Therefore, the strength of a component is considered to be a random variable. The load applied to electronic parts, such as power, temperature or humidity, is also a random variable. Thus, statistical distributions are usually used to describe the load and strength. Figure 2 shows some typical derating curve for solid state power switches.

DERATING OF A FAULTY SYSTEM

Derating algorithms may be applied on the faulty power converters or even the power converters in normal condition for life time extension.

Example: Derating of a Faulty Turbogenerator

The stator winding in a large electric machine must be able to carry the rated current without exceeding specified thermal limits (Faiz, Ghazizadeh, & Oraee, 2015). Figure 3 shows direction of hydrogen flow in a plant turbogenerator. Figure 4 shows a cutaway of stator core with cooling path direction. A medium material like water is usually used to transfer the heat generated in stator bar. Therefore, it is

Figure 2. some derating curves for various switches, (a): permissible drain current of a MOSFET versus temperature, (b): permissible drain current of an IGBT versus temperature, (c): Recommended derating curve of a MOSFET, Published with permission of International Rectifier

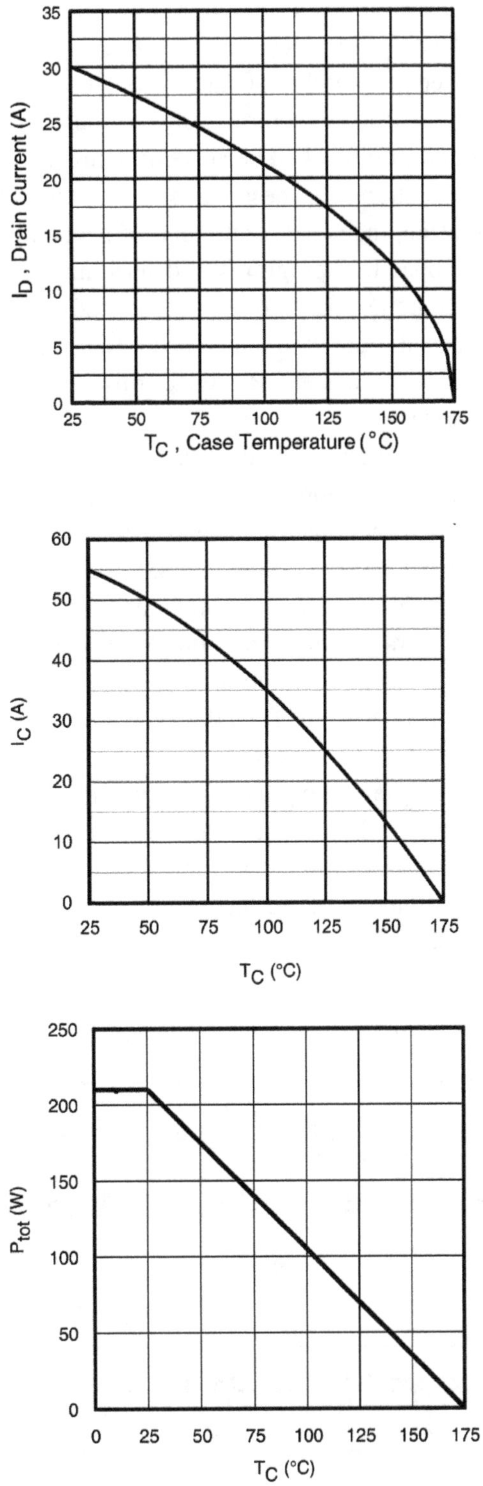

necessary to place some channels for cooling water inside the stator bar. On the other hand, An ac current carrying conductor embedded in a narrow slot in a magnetic material drives magnetic flux around itself and this alternating leakage flux induces alternating voltages along the length of the strips. If solid conductors are used, these voltages would cause circulating currents around the bar, resulting in unacceptable eddy current loss and heating. In order to minimize this effect, the conductor is divided into strips lightly insulated and arranged in a number of stacks along the bar width. The strips are transposed along the length of the bar using the Roebel method. Some of these strands are hollow type to pass the cooling water. These hollow strands are blocked because of many factors such as existence of external particles in water and oxidation of strand. In these cases, the water can not pass through channel and the temperature increases. Therefore, the steady state load of generator should be decreased to reduce the temperature. There are some operation programs to determine the load of generator in faulty case. But, it seems that these programs are not based on scientific study and comes from experience. Thus, the load of generator may be reduced more than that is need. In this section, an investigation is done to determine the effect of water channels blockage. Results of this study are used to prepare a program for operation of generator in faulty cases.

Transposition of strips causes heat being trapped in some parts of the stator bar and therefore leads to generation of local hot spots. In this section, the effects of conductor displacement on the generation of hot spots in stator bars in a large turbo-generator are investigated. The study is carried out through a thermal model considering the displacement of conductors. Simulation and on-site experimental results show that strip displacement leads to generation of some undesired local hot spots along the length of stator bar. These hot spots cause extra thermal stress on stator winding insulation. The thermal stress accelerates insulation aging, which leads to reduction of generator useful life 1.

Experiments were performed on a stator bar. The other tests were performed on an experimental setup because these experiments could not be done on a stator bar which is placed inside the turbogenerator. The experimental setup, whose block diagram and picture are shown in Figures 5 and 6, consists of a stator bar, a circulating water system, two coolers and a data acquisition system for temperature monitoring. To simulate the conditions of the tests set similar to the conditions of turbogenerator, the stator bar is placed inside a magnetic plate to simulate the magnetic flux of stator. The algorithm of water channels blockage is shown in Figures 7 and 8.

Figure 3. A turbogenerator cutaway and its cooling gas directions

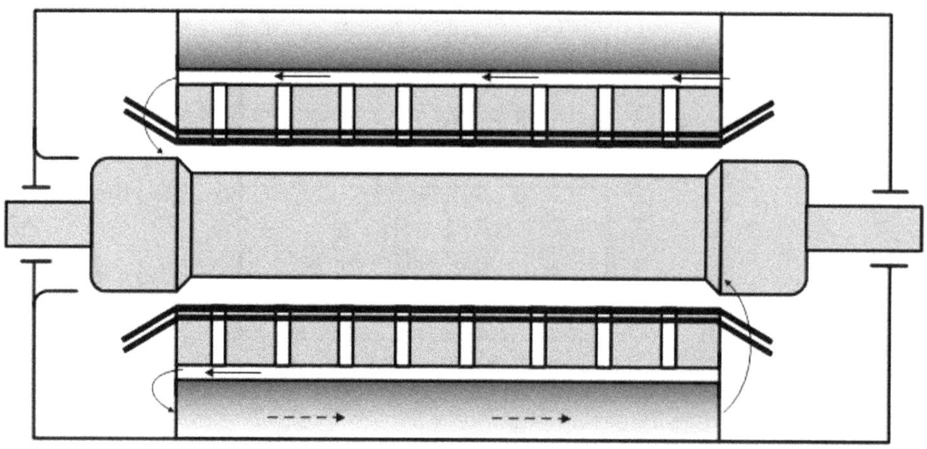

Figure 4. Stator cutaway of a turbogenerator and its cooling gas direction

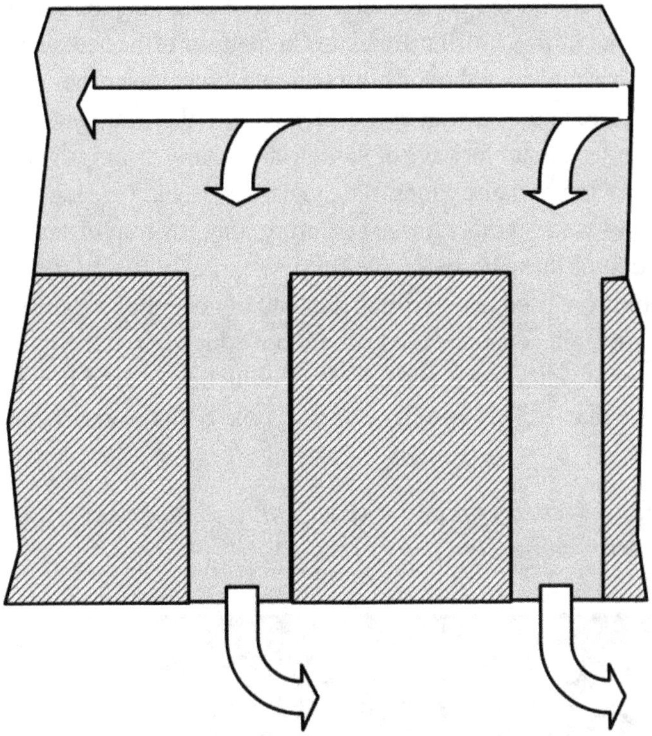

Figure 5. Block diagram of laboratory experimental setup

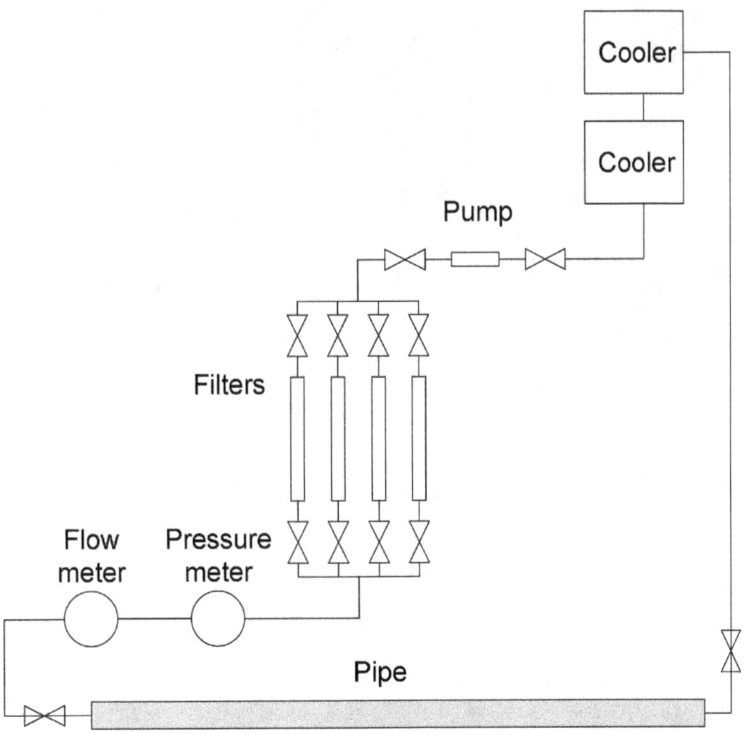

Tables 1 and 2 show the electrical and thermal specifications of the stator bar used in the simulation, respectively. Figure 9 shows the temperature profile of the hotspot along the length of the stator bar. There are two hotspots in this profile. Refer to the pattern of bar cross section these hotspots are generated when some consecutive solid conductors are placed at the corner of stator bar. Therefore, the heat generated by these conductors can not be transferred into the water and this leads to an increase in the temperature. It can be seen that blockage of some water channels helps generation of these hotspots.

Figure 10 shows water temperature along the stator bar. It can be seen that such hotspots can not be detected by monitoring water temperature. Therefore, the position of temperature sensors plays an important role in detection of these hotspots.

The results of this study are used to determine the load of generator in faulty case. When the hotspots of stator bar are considered, the generator load should be reduced even the number of blocked channels is small. If the outer layer of stator bar is consider, this reduction of load is started when the number of blocked channels is more than 3. This number is 7 when the outlet water is considered.

Figure 6. Laboratory experimental setup, (a): stator bar, (b): stator core, (c): temperature measuring equipment, (d): cooler, (f): water cycle

Figure 7. Different patterns for water channel blocking

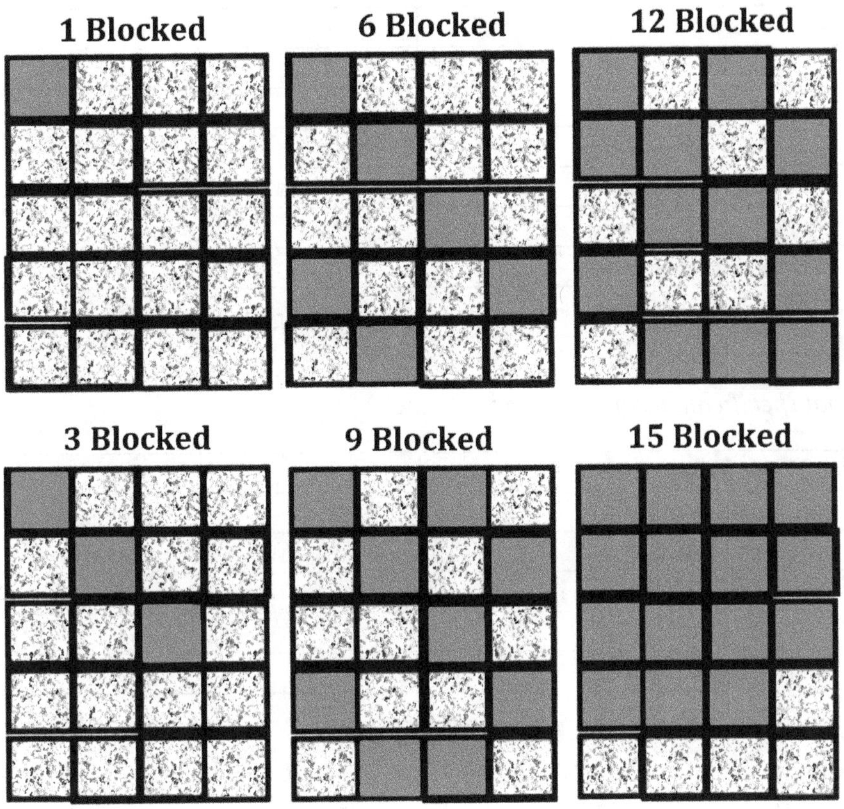

Figure 8. Method for water channel blocking

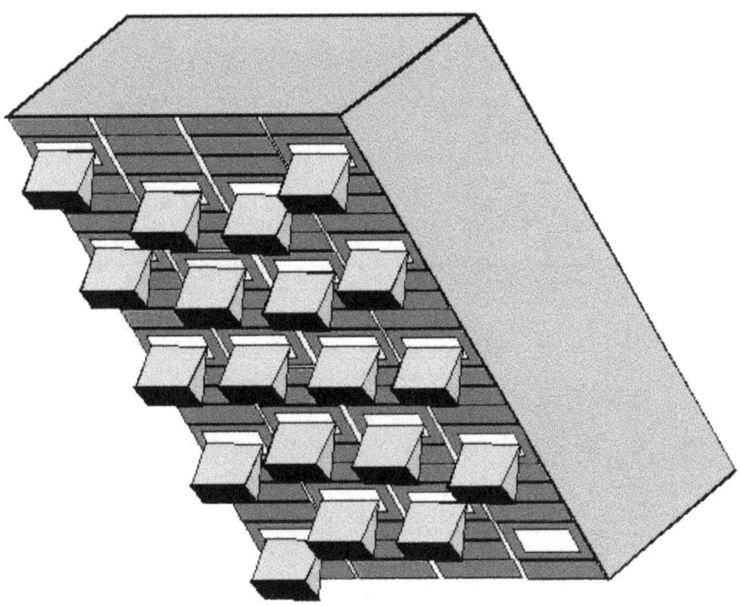

Table 1. Electrical specifications of stator bar

Parameter	Value
Rated current (A)	9150
Length (m)	7
Number of strands per column	19
Number of columns	4
Number of hollow conductors	20
Number of transpose nodes	38
Transposition angle (°)	360

Table 2. Thermal specifications of stator bar

Parameter	Value
Strand insulation	Paper
Column insulation	Mica
Cooling material	Water
Inlet water flow (m^3/min)	24
Inlet water temperature (°C)	40

Figure 9. Hotspots pattern along the bar length, down: normal, up: blockage of 10 water channels

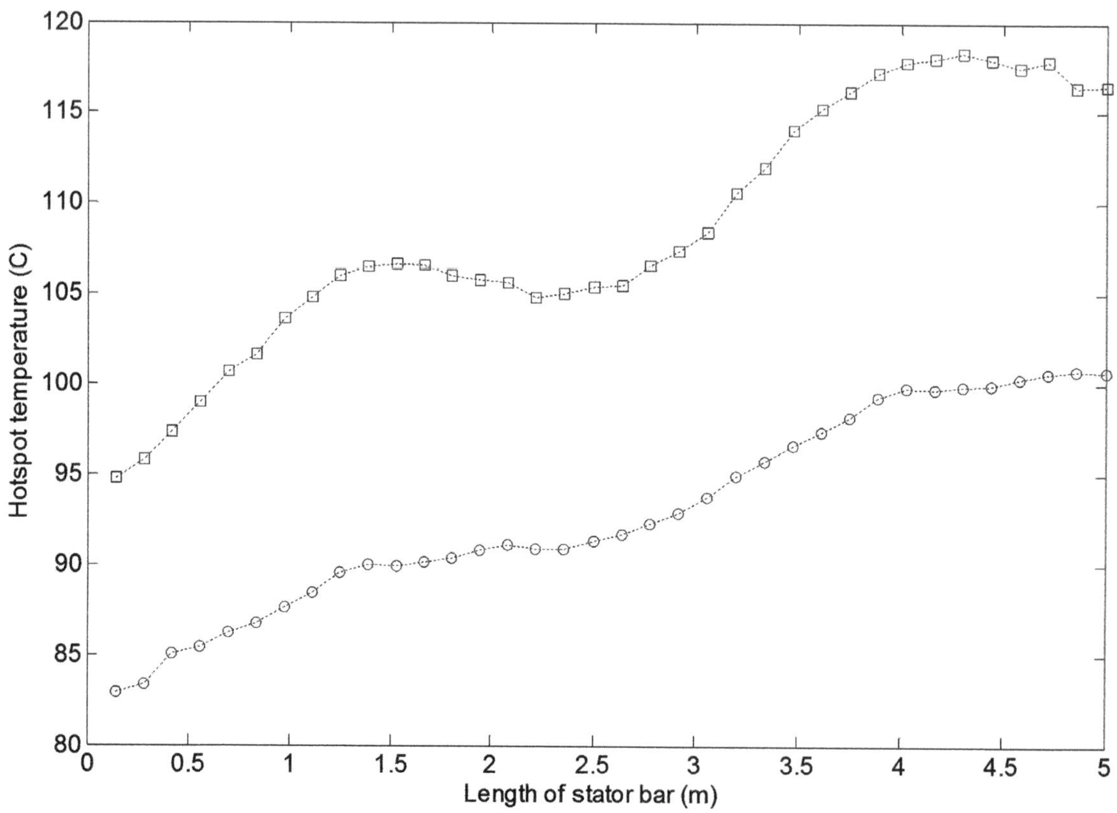

Figure 10. Water temperature pattern along the bar length, down: normal, up: blockage of 10 water channels

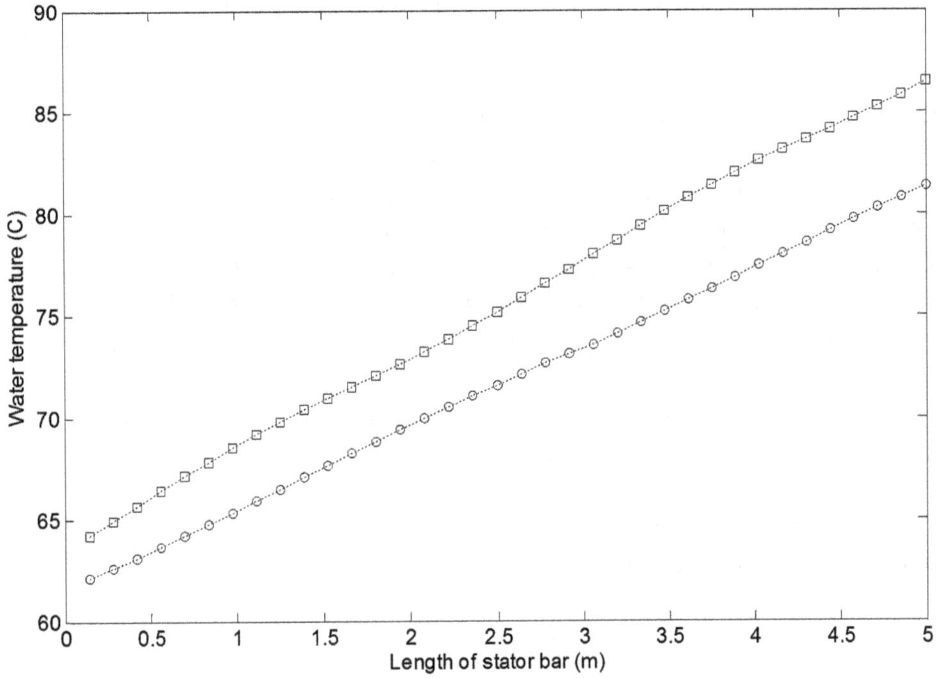

Figure 11. Load reduction when some of water channels are blocked (Air medium)

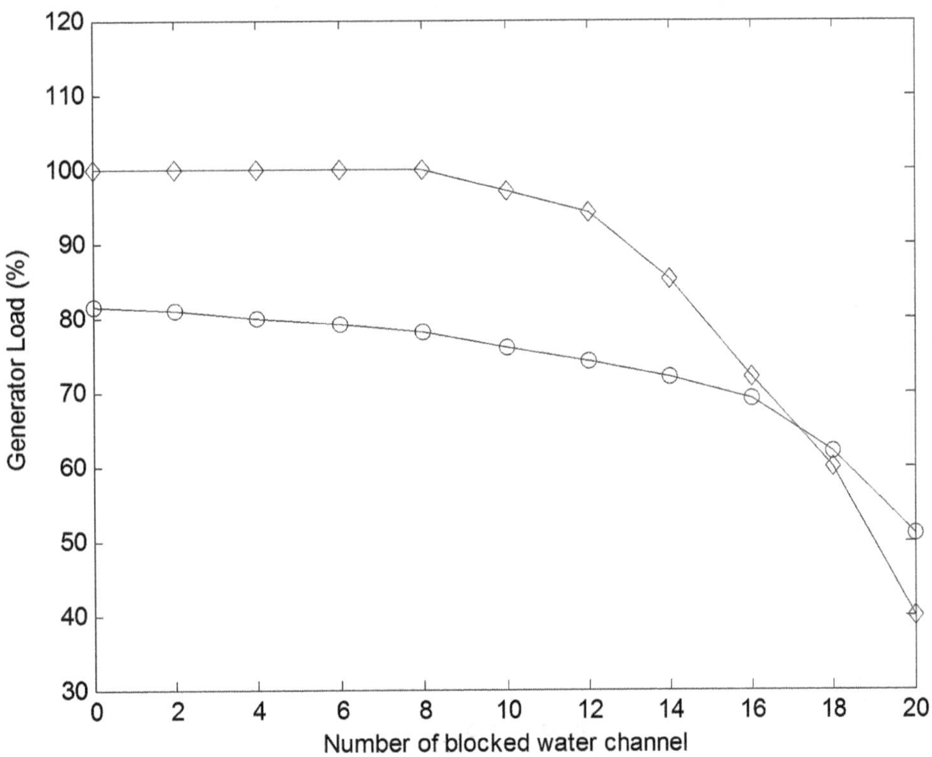

It can be seen that the hotspot of stator bar can not be detected in water temperature or bar average temperature. In addition, the difference between hotspot and bar temperatures is more than ten degrees and it leads to considerable reduction of generator useful life (Kaboli, Zolghadri, & Vahdati-Khajeh, 2007).

Figure 11 shows the effect of water cannels blockage on the allowable load of generator. It can be seen that the curve is very similar to the one obtained from simulation. Based on experimental results and with bar surface as the standard, the load of generator should be reduced when the water cannels are blocked. While based on outlet water temperature, till 9 channels can be blocked without need to reduction of generator load. Figure 12 shows the effect of cooling medium used inside the generator. It can be seen that when Hydrogen is used for this medium, the number of blocked channels which is necessary for load reduction is decreased to 4 while the stator bar can carry more current with the same surface temperature. Figures 13 and 14 show the effect of water cannel blockage on the surface temperature of bar and outlet water. These curves are used to determination of operation program of generator shown in Figure 15.

Therefore, derating algorithm of the generator can be performed as the following:

There are some choices for labeling the failure:

1- Hot spot temperature of stator bar
2- Maximum temperature of stator bar surface
3- Average temperature of stator bar surface
4- Maximum temperature of outlet cooling water

Figure 12. Load reduction when some of water channels are blocked (Hydrogen medium)

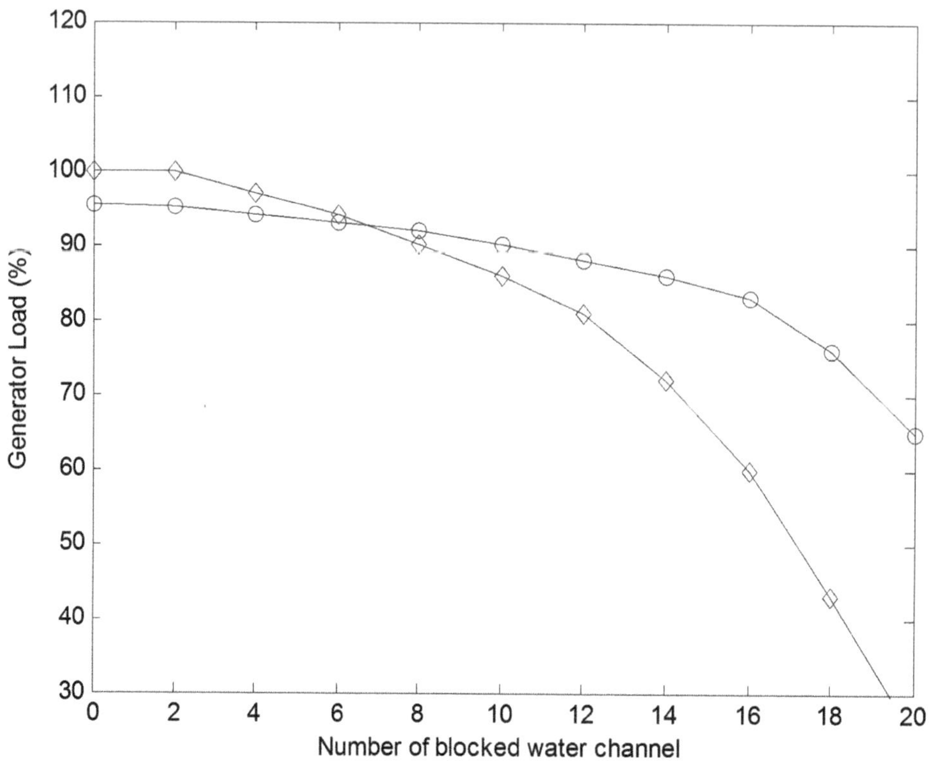

Figure 13. Stator bar surface temperature pattern when some of water channels are blocked

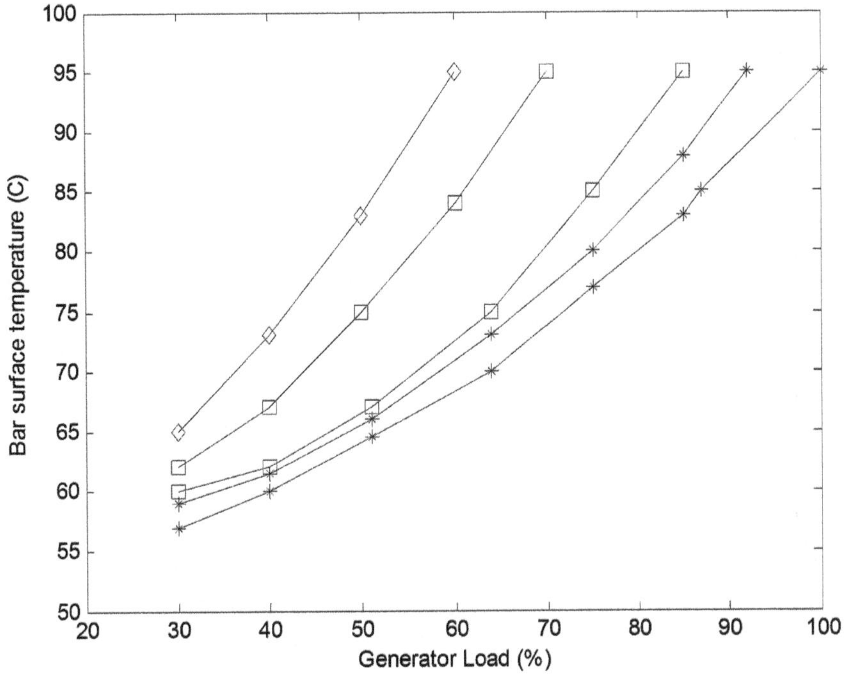

Figure 14. Outlet water temperature pattern when some of water channels are blocked

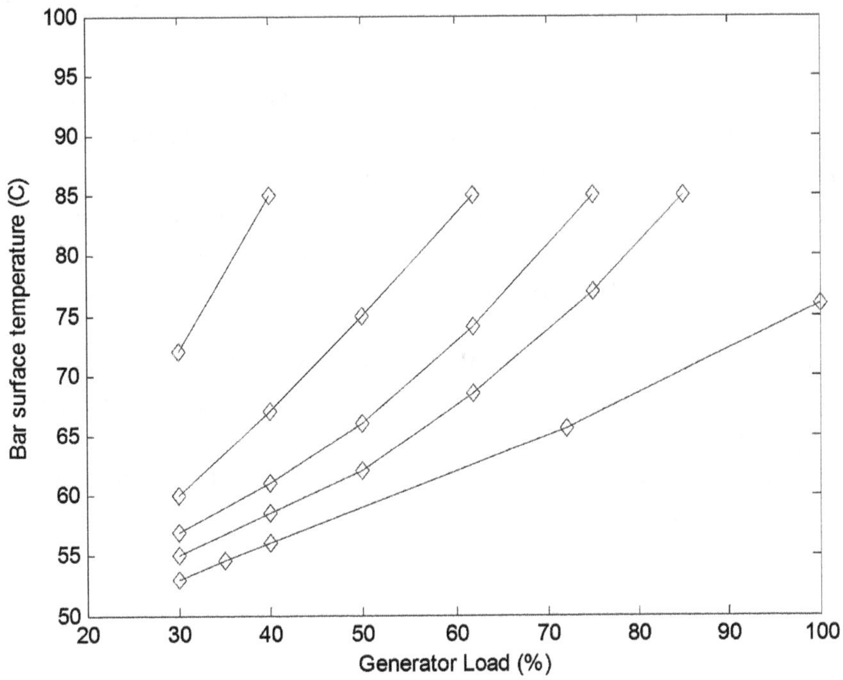

Figure 15. Proposed loading curve of generator

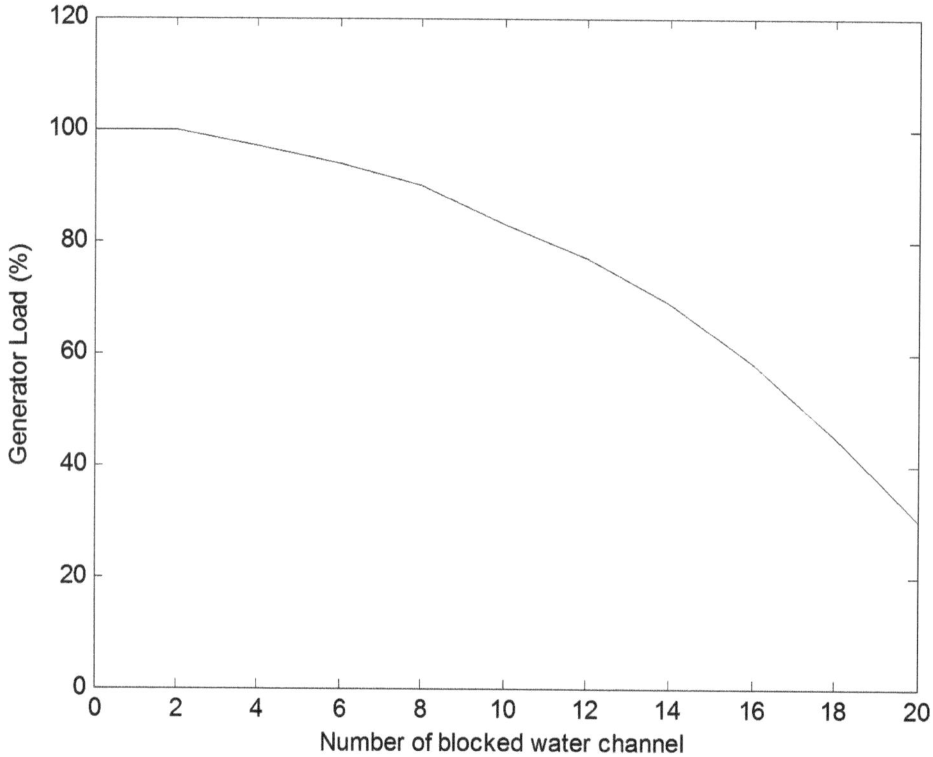

The results of computer simulation as well as laboratory and on-site experiments are summarized as follows:

- Transposition causes to generate some local hot spots at the length of stator bar.
- The temperature of these hot spots can be very high when some of hollow conductors are blocked.
- These hot spots can not be detected by the temperature rise of outlet cooling water.
- The position of protective thermal sensors of stator bar is important to detcct thc local hotspots.
- The load of generator should be decreased if some of the water cannels are blocked and the bar surface temperature is considered as standard.

Up to 3 water channels can be blocked without reduction of generator lead of if the outlet water temperature is considered as standard. After this number, the generator loading should be performed according the proposed curve.

Figure 16 shows finite element analysis results for blocking of stator bar cooling water. Generation of heat trap can be seen in this figure.

Figure 16. Various blockage pattern in stator bar and respective temperature profile

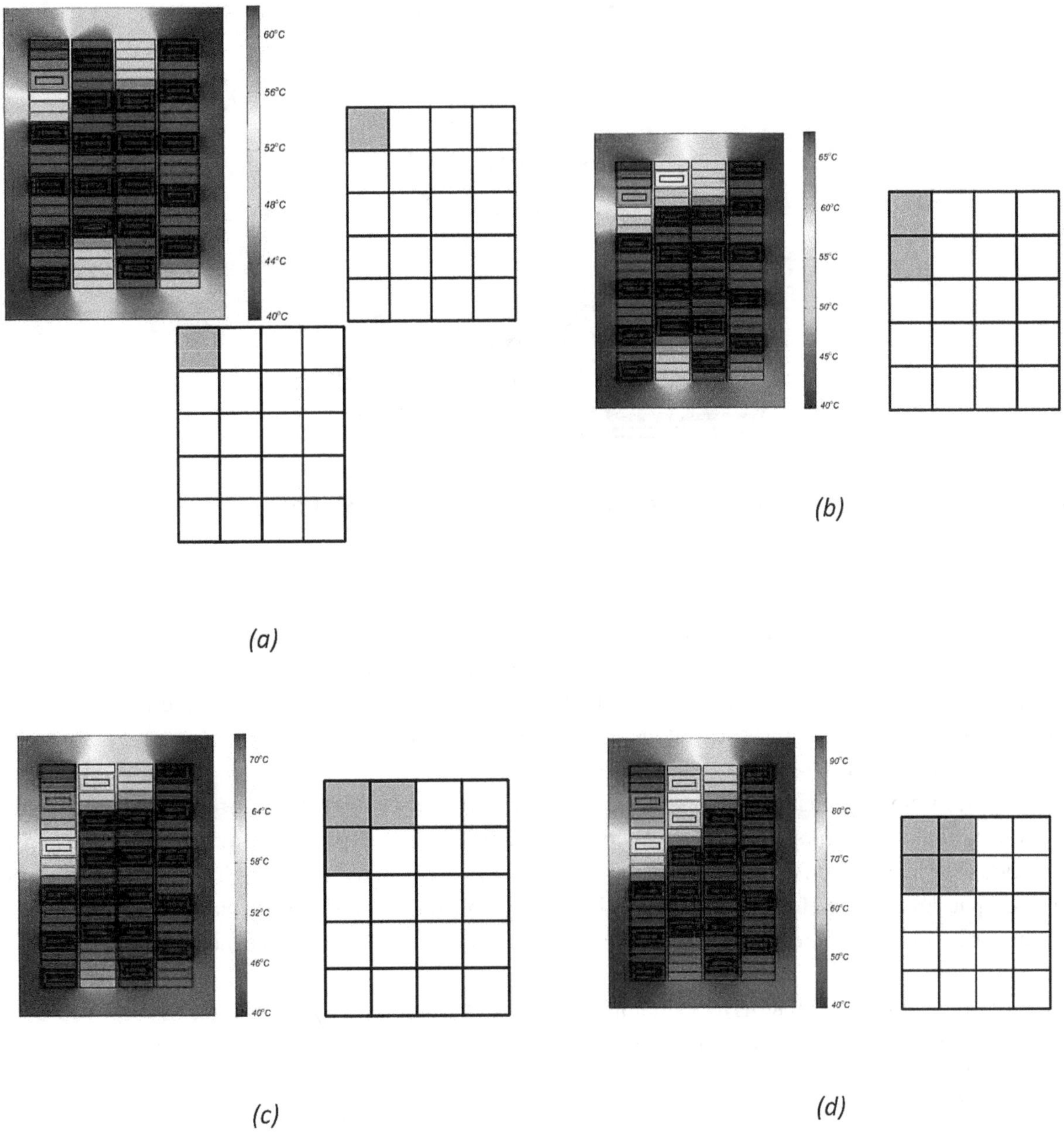

DERATING A NORMAL SYSTEM IN HARSH ENVIRONMENT

Harmonic information is necessary tools in estimation of motors life and performance. The conventional expression of harmonic amount is carried out by total Harmonic distortion. This distortion factor is only related to amplitude of harmonics while the motor loss is affected by frequency domain of them. In this paper a new definition of this factor has been presented. This definition includes the frequency of harmonics as well as their amplitudes. Experimental results verify validity of this new definition.

Figure 17. Three basic parts in a typical drive system

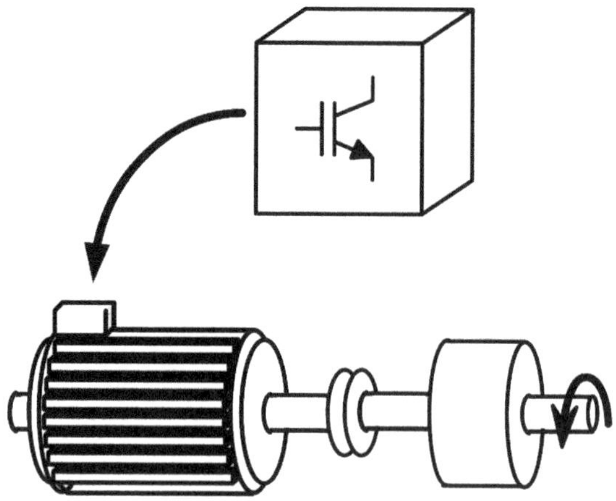

Several industrial and domestic loads comprising static power converter, such as arc melting furnaces, induction heating devices, switch mode power supplies inject current harmonics in the power system. Such phenomena as well as iron saturation in distribution transformers when operated at voltages exceeding the nominal value lead to voltage waveform distortion. Figure 17 shows a general diagram of a motor drive system.

Induction motors constitute the most popular energy converters and are sensitive to harmonic voltages. Both their efficiency and performance can be considerably affected by power quality of supply. On the other hand, a considerable proportion of induction motors are fed by voltage source inverters. The output voltage waveform of an inverter has a non sinosuidal shape and includes some harmonics. Estimation of motor life and its performance requires calculation of voltage harmonics and their effect on motor temperature rise.

One of the most popular tools for explaining amount of harmonics is total harmonic distortion (THD). The common definition for this parameter does not include frequency content of harmonics and only deals with amplitude. On the other hand, motor loss is a function of both frequency and amplitude of supply voltage. Therefore, definition of THD can not predict temperature rise of motor carefully 3.

In this paper a new definition of THD has been presented which is included not only the amplitude of harmonics but also the frequency of them. This new expression scales the amplitude of harmonics versus their frequency. This study has been performed on voltage waveform since access to voltage information of motor is simpler than its other parameters. Some experimental results will be presented that verify validity of this new expression of THD in estimation of motor temperature rise 5.

When a sinosuidal voltage V1(t) is applied to stator of an induction motor shown in Figure (1), relation between the motor and voltage parameters are presented by

$$E_1 = 4.44 N_s f B_1 A \tag{1}$$

where E1 and f are the amplitude and frequency of applied voltage, Ns is number of turns in stator winding, A is stator cross section and B1 is amplitude of magnetic flux density.

If another voltage, Vh(t) which has frequency equals to h*f and amplitude as Eh is applied to stator then relation (2) is changed to

$$E_h = 4.44N_s(h * f)B_h A \tag{2}$$

where Bh is the amplitude of flux density created by harmonic voltage Eh .

Figure (16). Induction motor and its power supply

Therefore, relation between the amplitude of flux density in this case versus amplitude of flux density for fundamental component (B1) is equal to

$$B_h = B_1 \frac{E_h}{h * E_1} \tag{3}$$

Core Loss for Voltage Harmonic

The hysteresis loss and eddy current loss are lumped together as core loss:

$$P_c = P_h + P_e \tag{4}$$

The power loss in core due to hysteresis effect is:

$$P_h = K_h B^n f \tag{5}$$

where B is maximum of flux density, Kh is a constant and f is the frequency.

The eddy current loss in a magnetic core subjected to a time varying flux is:

$$P_e = K_e B^2 f^2 \tag{6}$$

Thus, The hystersis loss of fundamental component, PH1, is given by

$$P_{H1} = K_H B_1^n f \tag{7}$$

The hystersis loss for h-th harmonic (Vh(t)), PHh, is given by

$$P_{Hh} = K_H B_h^n (h * f) \tag{8}$$

Therefore, hystersis loss for hth harmonic versus hystersis loss of fundamental component is calculated from (3), (7) and (8):

$$\begin{aligned} P_{Hh} &= K_H \left(B_1 \frac{E_h}{h * E_1} \right)^n (h * f) \\ &= \left(K_H B_1^n f \right) \left(\frac{E_h}{E_1} \right)^n \frac{1}{h^{n-1}} \\ &= P_{H1} \left(\frac{E_h}{E_1} \right)^n \frac{1}{h^{n-1}} \end{aligned} \tag{9}$$

Similarly, eddy current loss of fundamental component, PE1, is given by

$$P_{E1} = K_E B_1^2 f^2 \tag{10}$$

Eddy current loss for hth harmonic (Vh(t)), PEh, is given by

$$P_{Eh} = K_E B_h^2 (h * f)^2 \tag{11}$$

Therefore, eddy current loss for hth harmonic versus eddy current loss of fundamental component is calculated from (3), (10) and (11):

$$\begin{aligned} P_{Eh} &= K_E \left(B_1 \frac{E_h}{h * E_1} \right)^2 (h * f)^2 \\ &= \left(K_E B_1^2 f^2 \right) \left(\frac{E_h}{E_1} \right)^2 \\ &= P_{E1} \left(\frac{E_h}{E_1} \right)^n \end{aligned} \tag{12}$$

Therefore, harmonic core loss caused by h-th voltage component, Vh(t), is calculated from (9) and (12):

$$\begin{aligned} P_{Ch} &= P_{Hh} + P_{Eh} \\ &= P_{H1} \left(\frac{E_h}{E_1} \right)^n \frac{1}{h^{n-1}} + P_{E1} \left(\frac{E_h}{E_1} \right)^n \end{aligned} \tag{13}$$

Each of fundamental hysteresis loss, PH1, and fundamental eddy current loss, PE1, consist certain portion of fundamental core loss PCh as the following:

$$P_{H1} = a_1 * P_{C1}$$
$$P_{E1} = a_2 * P_{C1} \tag{14}$$

where a1 and a2 are two constants so that a1+a2=1.

Substituting of (14) into (13) leads to:

$$P_{Ch} = P_{C1} a_1 \left(\frac{E_h}{E_1}\right)^n \frac{1}{h^{n-1}} + a_2 \left(\frac{E_h}{E_1}\right)^n \tag{15}$$

Total harmonic distortion is defined for a waveform which has some harmonics on fundamental. For a voltage waveform THD is given by

$$THD = \frac{\sqrt{V_2^2 + V_3^2 + \ldots}}{V_1} \tag{16}$$

where V1 is the amplitude of fundamental component and V2, V3, … are the amplitudes of other harmonics.

Comparison of (15) and (16) shows that harmonic order h which is an important parameter in calculation of core loss does not enter in conventional definition of THD. Therefore, THD presented in relation (16) is not a suitable scale for evaluation of harmonic core loss.

Total harmonic core loss is the summation of harmonic core loss as follows:

$$P_{C-Harmonics} = \sum_h P_{Ch} = P_{C1} * \sum_h \left\{ a_1 \left(\frac{E_h}{E_1}\right)^n \frac{1}{h^{n-1}} + a_2 \left(\frac{E_h}{E_1}\right)^n \right\} \tag{17}$$

The fundamental core loss,Pc1, is constant for a motor and then modified definition of THD (MTHD) is given by:

$$\text{MTHD} = \sum_h \{ a_1 \left(\frac{E_h}{E_1}\right)^n \frac{1}{h^{n-1}} + a_2 \left(\frac{E_h}{E_1}\right)^n \} \tag{18}$$

In this definition, the most important parameters are coefficients of hysteresis and eddy current loss. It can be seen that these coefficients are supposed to be constant. However, these coefficients are functions of motor operating point.

To show this dependence some experiments have been performed on some ferromagnetic core. The experimental setup, shown in Figure 18 consists of a transformer, harmonic generator equipment, and measuring equipment for power loss. Two different transformer cores are used in tests. The specifications of transformers used in tests are shown in Table 3.

Figure 18. Experimental setup, (a) tests in the laboratory, (b): block diagram

(a)

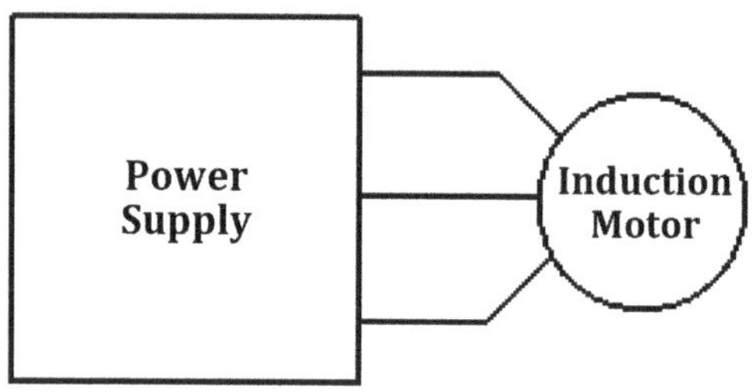

(b)

Figures 19 and 20 show variation of hysteresis loss coefficient versus frequency. Since distinct determination of K_h and n is difficult, the merged form of these coefficients, K_hB^n, is studied for a constant flux density (B=1T). It can be seen that these coefficients increase with frequency. Figure 21 and 22 show eddy current loss coefficient decreases with frequency. Therefore, these coefficients can not be assumed constant in modified THD definition.

To justify the theoretical results some tests are carried out an induction motor. The motor specifications are given in the following table.

Figure 23 shows two voltage waveforms applied to the motor. These waveforms have the same fundamental component and different harmonic content. Thus, fundamental core loss for both waveforms is equal but harmonic core loss is different. The conventional THD for waveform shown in Figure 23(a) is equal to 31% and for waveform shown in Figure 23(b) is equal to 42%. Therefore, it is predicted that motor temperature rise for the second waveform is more than the first waveform. But MTHD factor for the first waveform is greater than the second and it expresses that motor temperature rise for the first waveform is greater than the other voltage. Figure 24 shows motor temperature rise for these voltage waveforms. It is obvious that the first voltage waveform has more temperature rise. So its harmonic loss is more than the second voltage waveform.

Table 3. characteristics of transformer A and B

Parameter	Value
Rated Power	1 KW
Rated voltage	220/24
Shape	Toroid
Parameter	Value
Rated Power	100 W
Rated voltage	220/12
Shape	E-type

Figure 19. Variation of hysteresis loss coefficients with frequency in core A

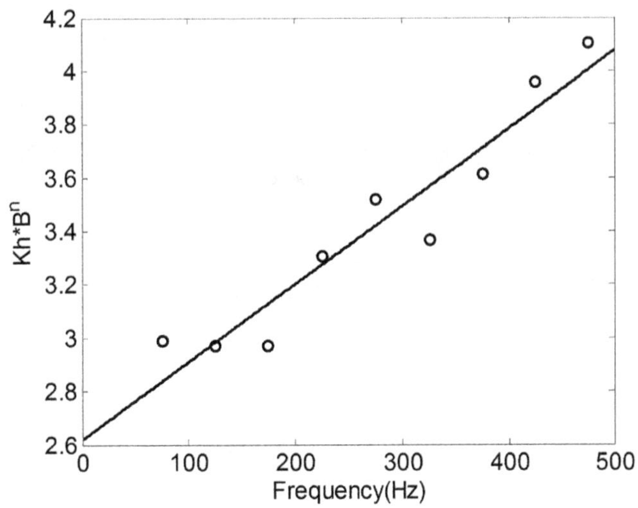

Figure 20. Variation of hysteresis loss coefficients with frequency in core B

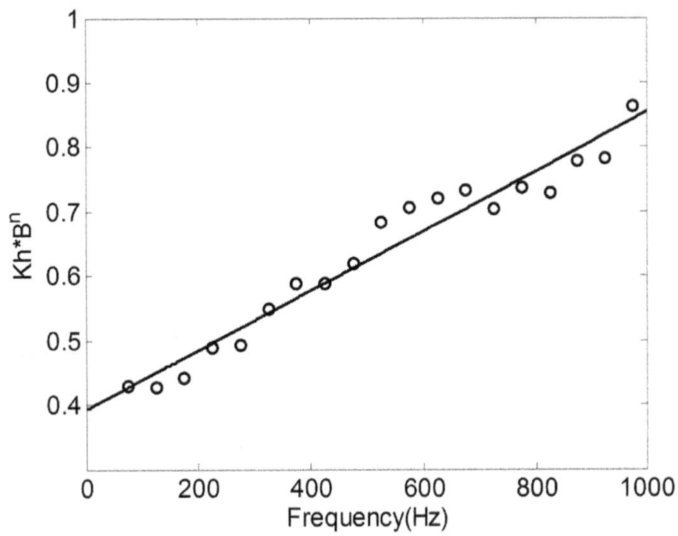

Figure 21. Variation of eddy loss coefficients with frequency in core A

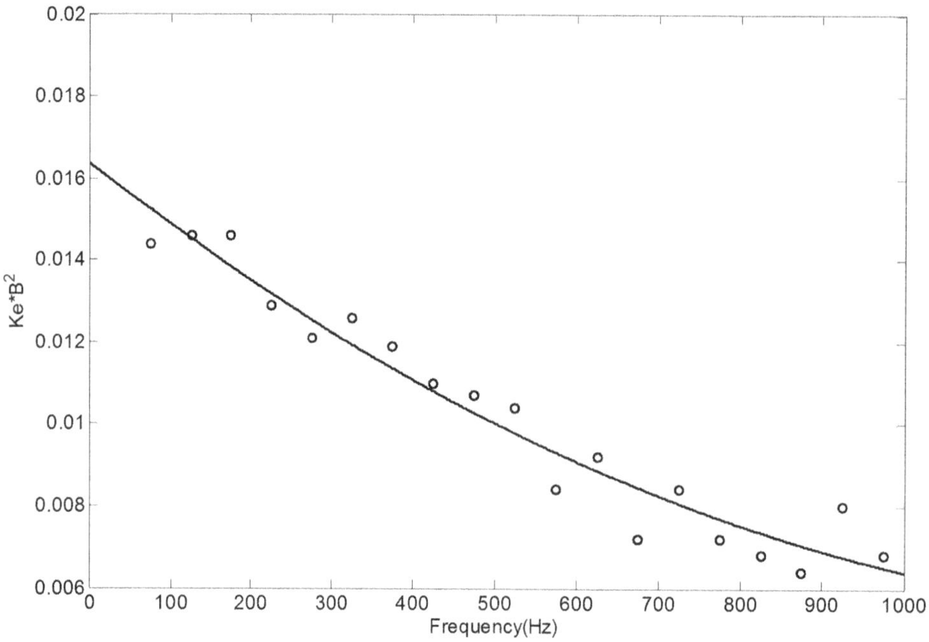

Figure 22. Variation of eddy loss coefficients with frequency in core B

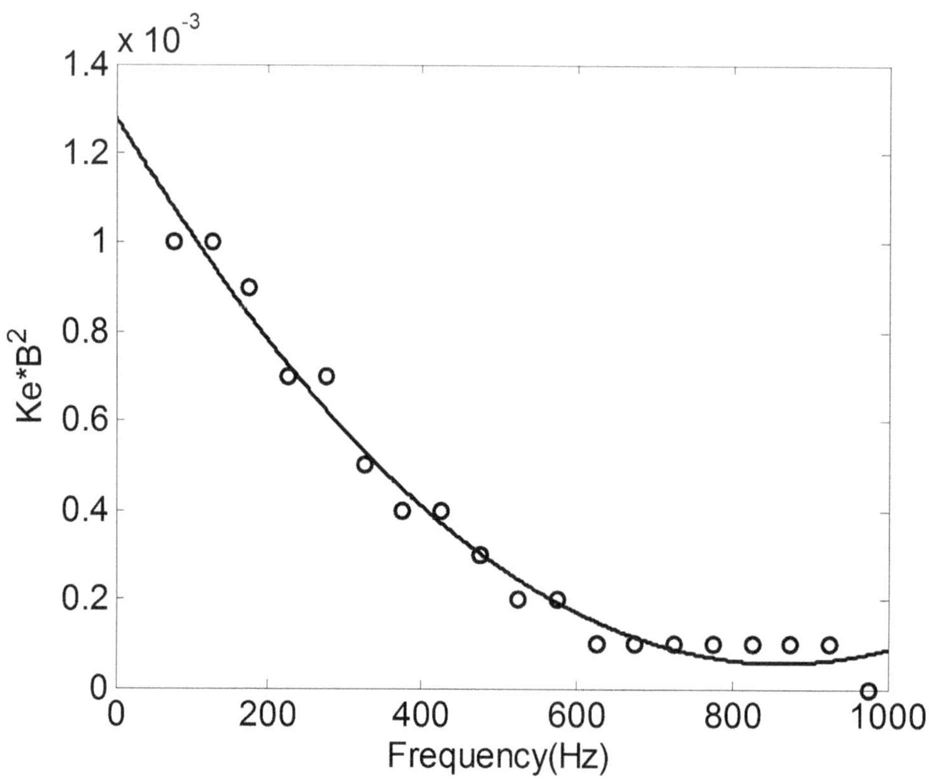

Table 4. Motor Specifications

Parameter	Value
Nomonal Power	2.2 KW
Nominal Voltage	220 V
Nominal Frequency	50 Hz
Number of Slots	30
Depth of slots	20 mm
Length of Slots	240 mm

Figure 23. Applied voltage waveform of motor, (a): high harmonic value, (b): low harmonic value

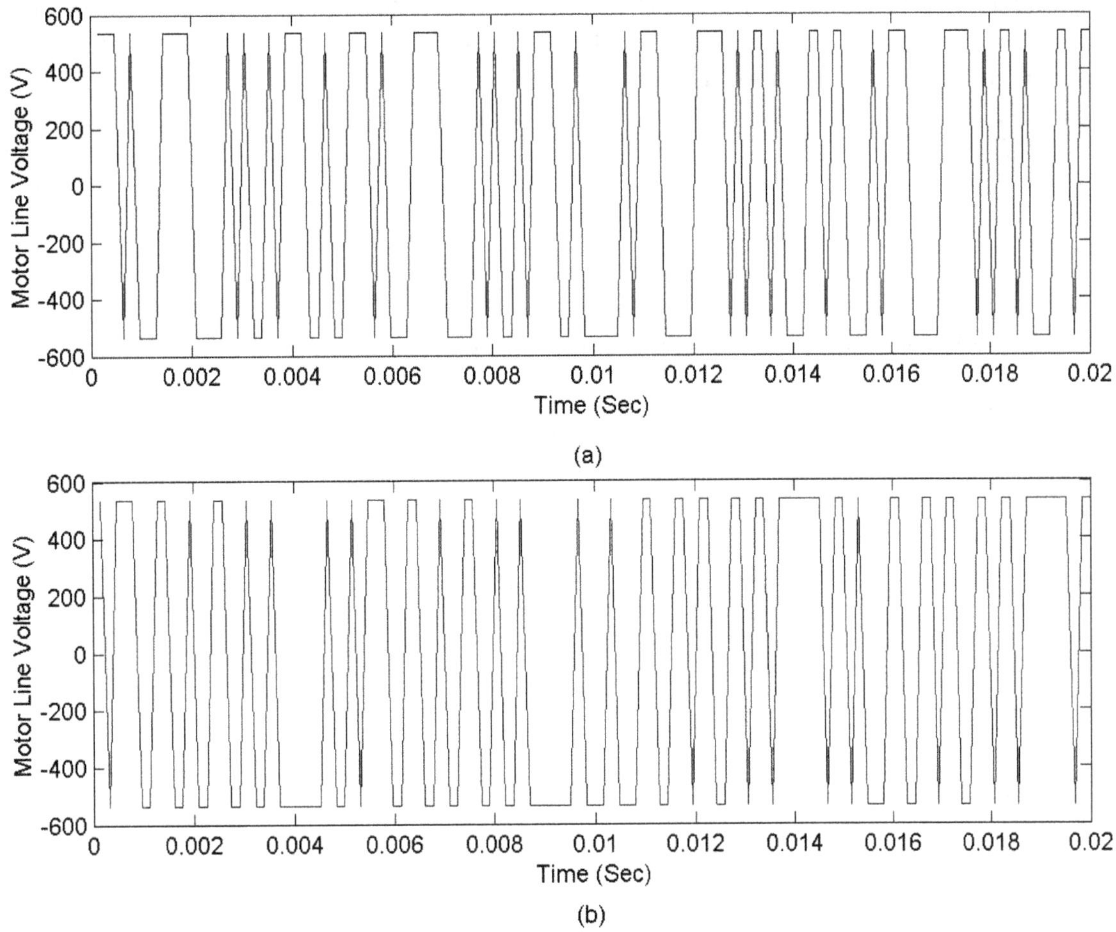

USEFUL LIFE EXTENSION FOR A NORMAL POWER ELECTRONIC CONVERTER

Example

Switched mode power supplies are usually work with a fixed switching frequency. However, it is not a good choice for all mode of operation. When the load of converter decreases, its fixed losses such as core loss of transformers and inductors remain about fixed but its dynamic losses such as switches conduction loss increase. This means that the converter can work with lower switching frequency and lower total loss. If the switching frequency of the converter is variable, it is possible to work with higher efficiency for all of load range. Figures 25 and 26 show the switching loss of the switch for a step down BUCK converter in two cases: 20KHz fixed switching frequency and variable one. It can be seen that the switching loss is higher in variable switching frequency algorithm. Now, note that the inductor core loss reduces obviously with variable switching frequency as shown in Figure 27. The conduction loss shown in Figure 28 is about constant. These results lead to higher efficiency and lower losses as shown in Figure 29,

Figure 24. Temperature rise of motor for two voltage waveform

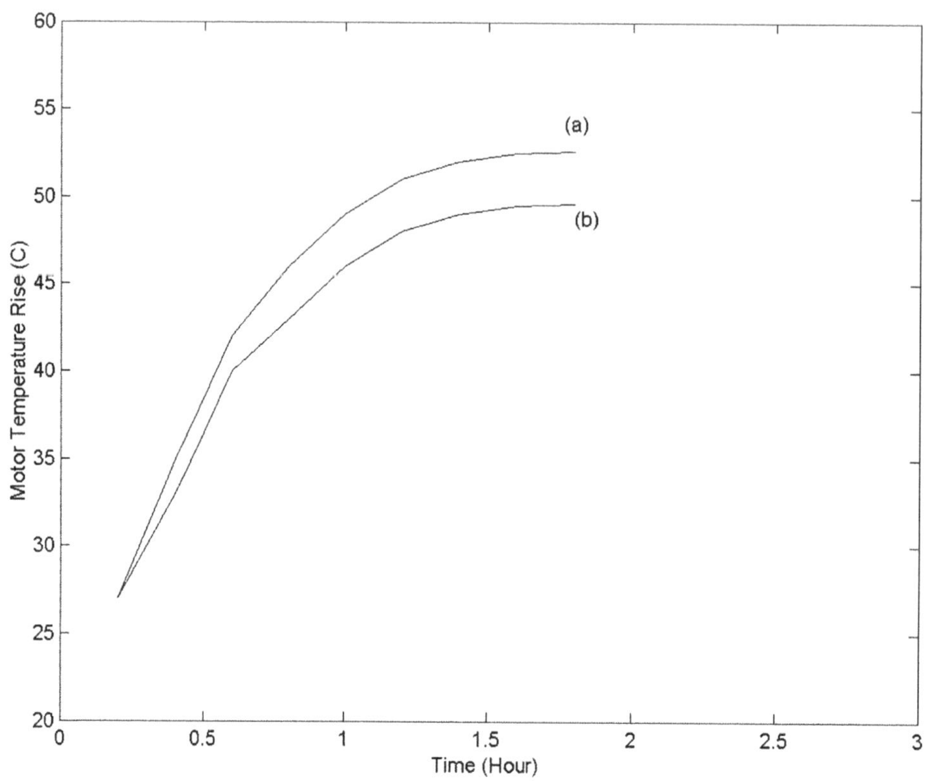

Example: Magnetic Derating of Electrical Motors

Usually motors are designed to have their maximum efficiency in their nominal operating point. Therefore, the value of motor flux should be readjusted when the load is less than nominal value. Adaptation of flux to load variations can be done in three ways: flux control as a function of torque, flux control based on loss model and flux control by a minimum loss search controller.

a. Flux Controller as a Function of Torque

This category was proposed by many authors. The most adapted approach to DTC sets the amplitude of flux as a function of torque as shown.

$$|\psi_S| = |\psi_{sn}|\sqrt{\frac{T}{T_n}}$$

This method is a well-known technique in industry because of its simplicity. However, it does not consider the power dissipation of drive and can not find the optimal value of flux. Therefore, it yields only a partial improvement in drive efficiency .

b. Loss Model Based Flux Controller

In the second category, the optimal flux is determined according to the loss model of induction motor. Of course, this method is very rapid and gives the optimum operating point of motor. But, the performance of this method depends highly on the accuracy in identification of motor parameters and their variations during operation .

c. Flux Search Controller

In the third category, one of the drive variables (usually input power) is minimized by adjusting the flux value. This method has a long time response (more than 10 second) but it is quite independent of drive parameters Therefore, it is the most well-adapted method for DTC because the other methods eliminate the independence of DTC performance from machine parameters.

As mentioned before in this family of flux controllers, one of the drive variables is selected as an objective function and then it is minimized by adjusting the flux value in consecutive steps. Figure 30 shows the principle of this method. The stator flux is moves from its initial value ($|\psi_s|_0$) toward its final value ($|\psi_s|_{opt}$), the value that minimizes the objective function. The control algorithm is presented in.

$$|\psi_s|_2 = |\psi_s|_1 + \Delta\psi$$

If $f(|\psi_s|_2) > f(|\psi_s|_1)$ Then $\Delta\psi = -\Delta\psi$

Table 5 shows the parameters of the induction motor used in this study. To verify the simulation results, the proposed control method has been applied to a DTC experimental test setup. The experimental

Figure 25. Variation of turn on switching loss with load variation in two different frequencies

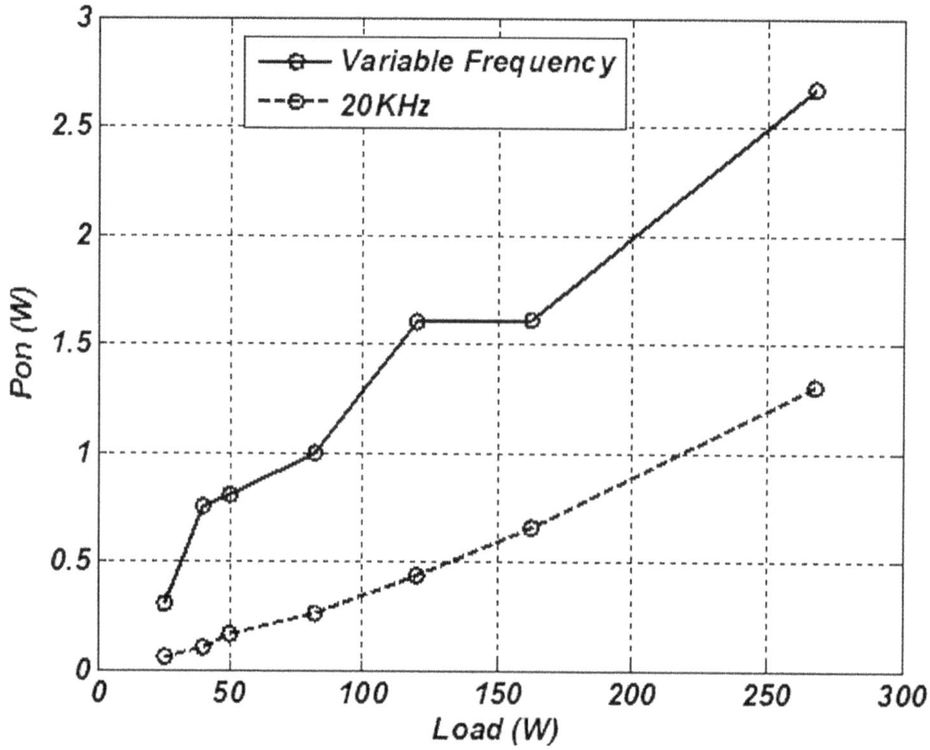

Figure 26. Variation of turn off switching loss with load variation in two different frequencies

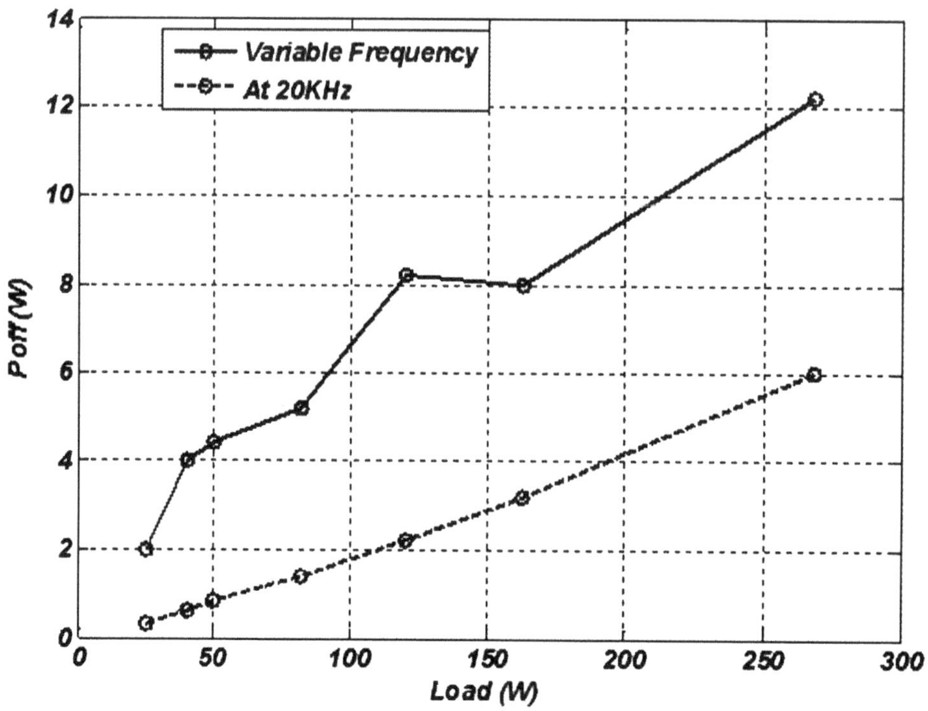

Figure 27. Variation of inductor core loss with load variation in two different frequencies

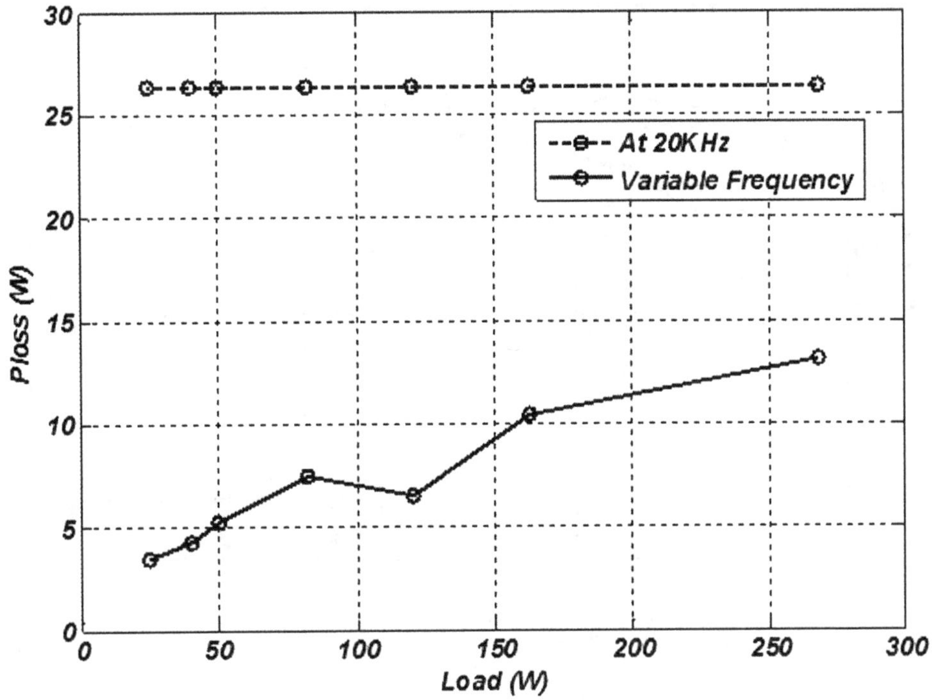

Figure 28. Variation of switch conduction loss with load variation in two different frequencies

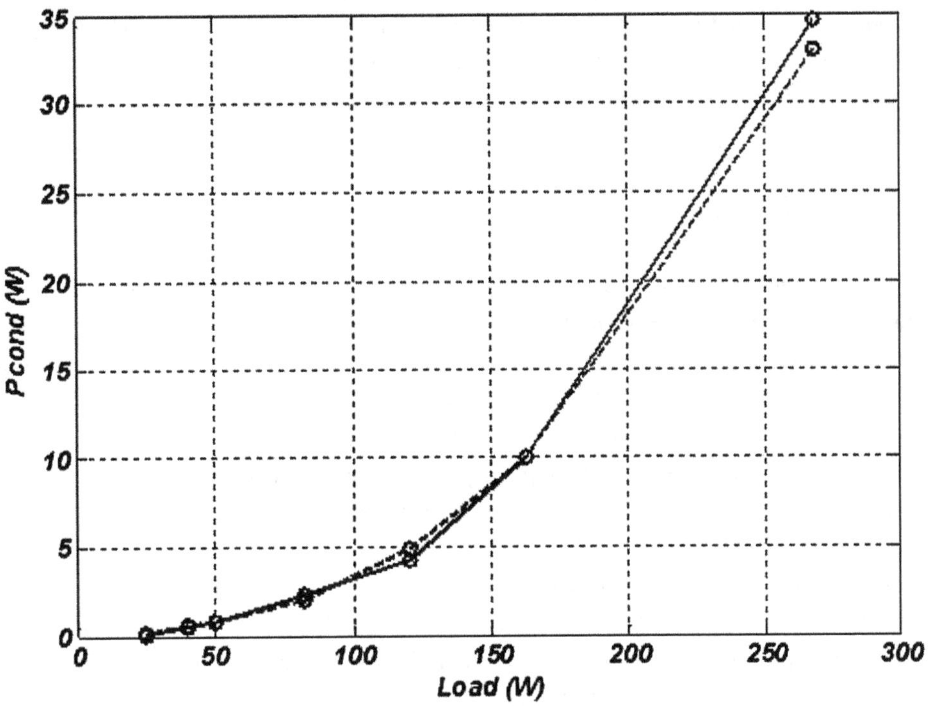

Figure 29. Variation of converter efficiency with load variation in two different frequencies

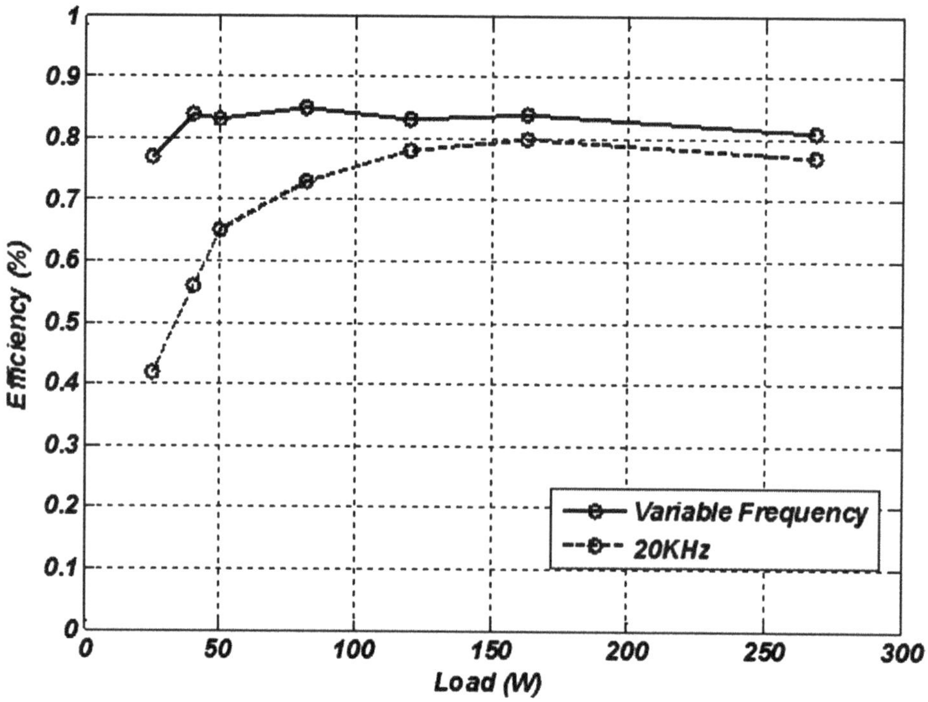

setup consists of an induction motor, insulated gate bipolar transistor (IGBT) based inverter, and digital signal processor (DSP) (TMS320C) based controller. Detailed characteristics of DSP controlled inverter are presented in Tables 6 The machine currents i_a and i_b and the dc bus voltage were interfaced into the controller through an analog to digital (A/D) converter built into the DSP board. The sampling time and motor speed are 133 μsec and 300 RPM, respectively. Figure 31 shows operation of search controller for reducing the operating flux for better efficiency than that found in nominal condition.

COMPONENT DERATING

Engineering best practices indicate that all designs should be derated for temperature. Which means a design should never force a device to operate at its maximum power dissipation on near its maximum power dissipation, without leaving some operating room. A circuit design should also never force a semiconductor to operate at or near its maximum junction temperature. Pushing the device up to its dissipation limits causes the semiconductor junction to heat up.

The problem is compounded as either the package temperature increases or the surrounding air temperature is increased. In general the life of a component transistor in this case reduces as the component temperature is increased junction temperature

Maintenance experts agree that excessive heat causes rapid deterioration of motor winding insulation. The common rule states that insulation life is cut in half for every 10 C of additional heat to the windings. As an example, if a motor that would normally last 20 years in regular service is running 40 C above rated temperature, the motor would have a life of about 1 year.

Figure 30. Algorithm of search control

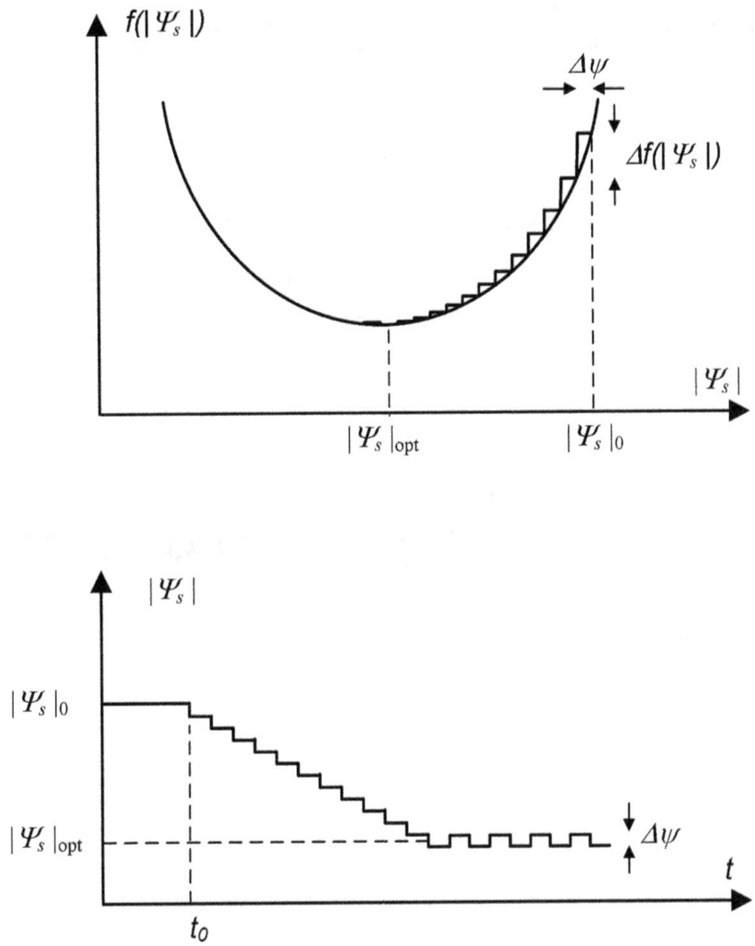

Table 5. Characteristics of induction motor

parameter	value
Rated Power (KW)	5.5
Number of Poles	4
Stator Resistance (Ω)	0.18
Stator Inductance (mH)	56
Magnetizing Inductance (mH)	53
Rotor Resistance (Ω)	0.50
Rotor Inductance (mH)	56
Nominal Torque (N.m)	35
Nominal stator flux (Wb)	0.65

Table 6. Characteristics of inverter switches and DSP board used in experiments

Parameters of inverter switches	Value	Parameters of DSP	Value
Nominal voltage (V)	600	Clock frequency (MHz)	40
Nominal current (A)	30	Maximum sampling frequency (KHz)	20

Figure 31. Stator flux behavior with small flux step with averaging algorithm

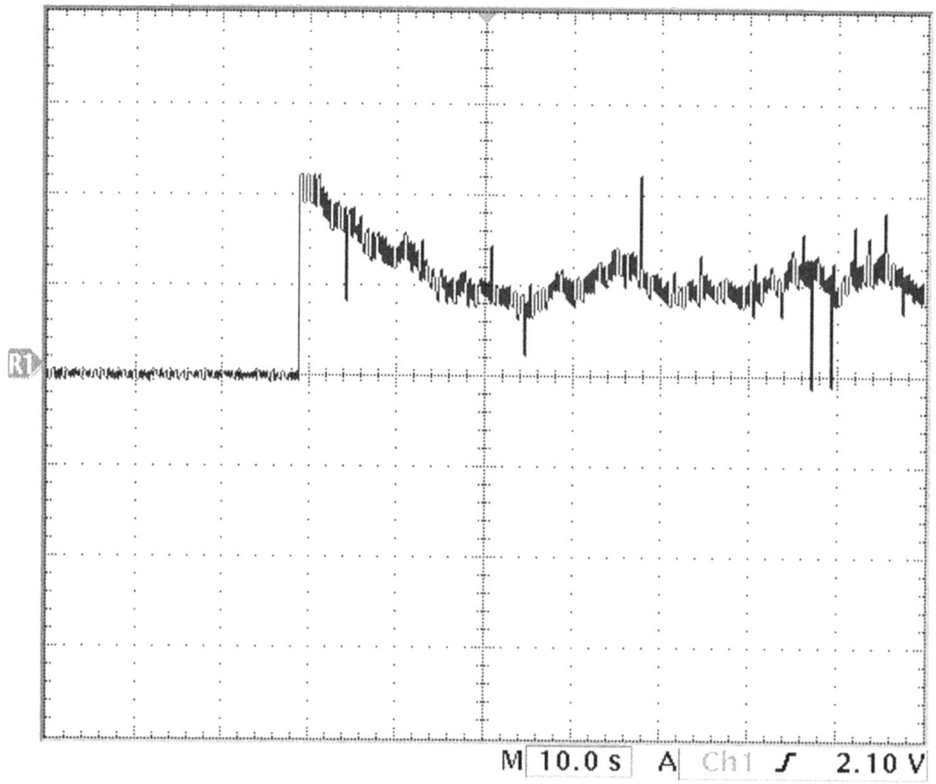

Leading standardization organizations have concluded that 30 percent of motor failures are attributed to insulation failure and 60 percent of these are caused by overheating. Articles have been published stating that a significant cause of bearing deterioration is overheating.

There are typically five main reasons for overheating—overload, poor power condition, high effective service factor, frequent stops and starts, and environmental reasons.

Derating of Electronic parts

In electronics, derating (or de-rating or de-tuning) is the operation of a device at less than its rated maximum power in order to prolong its life.

Power electronic devices have a maximum power dissipation rating usually quoted at a case temperature of 25 °C (77 °F). The datasheet for the device also includes a derating curve which indicates how much a device will dissipate without getting damaged at any given case temperature and this must be taken into account while designing a system.

As can be seen from the derating curve image for a hypothetical bipolar junction transistor, the device (rated for 100 W at 25 °C (77 °F)) cannot be expected to dissipate anything more than about 40 W if the ambient temperature is such that the temperature at which the device's case will stabilise (after heat-sinking) is 65 °C (149 °F). This final case temperature is a function of the thermal resistance between the device's case and the heat-sink; and the heat-sink and the ambient (this includes the heat-sink's temp/watt rating - with lower values implying better cooling characteristics).

In Electrical Installations

All dimmers rely on heat conduction and convection to keep the electronic components cool. Similarly, power wiring (e.g., house wiring) not surrounded by an air space (e.g., inside a conduit) needs to have its current-limiting device (e.g. circuit breaker or fuse) adjusted so as not to carry as much current through that circuit. Derating is the reduction of the maximum capacity (load) a unit can reliably handle when fins/side sections are removed.

Components can be treated in such a way as to enhance their life expectancy. 'Derating' is the name normally given to operating a component well inside its normal operating limits, in order to reduce the rate at which the component deteriorates. That derating is a practical means of reducing failures is supported by much published literature.

Conceptually, it is easy to see that, whilst the component may be specified to operate at high voltage and high temperature, applying those conditions simultaneously would probably be worse than applying either one or the other. Also that, if a component has a voltage rating such that it will start to fail at, say, 130% of maximum rating, reducing the voltage applied to substantially below the maximum permitted should reduce distress, and by doing so extend the life.

Also given that reactions are known to proceed at higher speeds at higher temperatures (an insight originally shared by Arrhenius), one would predict reduced degradation, and hence extended life, by running a component at lower than its maximum category temperature.

This is an activity where the information you collect will depend very much on the search terms you use – some sites will recommend derating as a matter of course; others concentrate on other meanings of the term.

Particularly with power devices, derating will be interpreted mostly in terms of power derating, that is reducing the device power as the ambient temperature rises, in order to keep the junction temperature at a safe level. As with all devices, the junction area runs substantially hotter than the case and ambient, depending on the thermal resistance between junction and ambient, produced by a series of poorly thermally conductive paths in the silicon, through the die bond, in the package, and finally through to any heat sink.

Manufacturers will sometimes point out that there is a strong relationship between junction temperature and failure rate, frequently modelling this as an Arrhenius curve, and predicting perhaps a 10:1 increase in failure rate for a rise in junction temperature from 130°C to 160°C, based on a 1eV activation energy.

There will be other evidence of derating, for example high-current devices may be recommended a 'soft start' circuit, in order to prevent damage from inrush current. When we did our own search, we found an interesting comment from Phillips Semiconductors that "Exposure to limiting values for extended periods may affect device reliability".

You may also have come across references to MIL-HDBK-217F, which predicts failure rates for different devices based on the severity of the application, generally using the Arrhenius model. At the same time you may have read material that leaves you far from convinced that the MIL-HDBK-217F model is fully applicable to temperature, and may well not apply at all to other sources of stress.

Lest you think that derating as a practice is not supported by theory, it is worth looking at an alternative view, which strongly supports at least a modicum of derating. This approach is explained in the Reliability Analysis Center's Mechanical Applications in Reliability Engineering. This refers to the 'strength' of a part, which is a random variable that can be represented by statistical distribution. Likewise, the stress applied to a part is random, changing the temperature, vibration, transients, shock and other environmental factors, and able also to be represented by a statistical distribution.

The classical approach has been to select every part to have enough 'strength' to handle the worse case stress conditions, thereby reducing to a minimum the intersection (shaded) areas of the graphs where there is a slight chance that the stress applied to a part will exceed its strength. More recent approaches take into account the probabilistic nature of this 'interference' between the two distributions.

Using this insight, the four basic strategies for stress derating can be seen to be:

- to increase the average strength
- to decrease the average stress
- to decrease the stress variation
- to decrease the strength variation.

All of these are possible, but variations are more difficult to control.

The purpose of derating is to protect against these variations, preventing small changes in operating characteristics (usually temperature) from creating large increases in failure rate. Given that the simplest approach to increase average strength, this will normally be done by procuring a more capable component. For example, choosing a 100V capacitor rather than the 63V type for operation on a 60V line.

The amount of derating that is needed will depend on how well the designer can predict the variation in operating parameters, both before the part is assembled and during the operating environment over the lifetime of the part. Because the sources of variation are extremely difficult to quantify, engineering estimates in past experience are often used to estimate the derating level needed.

Derating Standards

Several derating guidelines have been issued by military or other agencies, and the following are some examples:

- MIL-STD-975, published by NASA, focuses on selection of parts used in the design and construction of space flight hardware as well as mission-essential ground support equipment.
- MIL-STD-1547, published by the Department of Defense, is targeted to aid in the design, development and fabrication of electronic systems with long life and/or high reliability requirements while operating under the extreme conditions of space and launch vehicles.
- AS4613, published by the U.S. Navy, sets forth derating requirements for the reliable application of electronic and electromechanical parts.

- NAVSEA TE000-AB-GTP-010, published by the U.S. Navy, contains derating requirements and part selection and application information on the ten most commonly used electrical and electronic parts.
- ECSS-Q-30-11A, prepared and maintained under the authority of the Space Components Steering Board in partnership with the European Space Agency, contains derating requirements applicable to electronic, electrical and electromechanical components.
- MSFC-STD-3012, prepared by NASA's Marshall Space Flight Center, sets requirements for electrical, electronic and electromechanical parts selection, management and control for space flight and mission-essential ground support equipment for Marshall Space Flight Center programs.

In addition to the derating standards published by the military, many semiconductor manufacturers such as Freescale and Hitachi provide their own derating guidelines, which can be explored in the datasheets or application notes of electronic parts.

EFFECT OF ENVIRONMENT

Environments where the ambient temperature exceeds 104 Degrees F are potentially harmful to motors. Operations within such environments necessitate motors with higher horsepower ratings or special designs. Larger motors, if loaded below full capacity, stand less chance of overheating due to excessive ambient temperatures.

For operation in abnormally hot places, motors are designed with lower winding temperature rise to withstand the ill effects of intense heat. (If the ambient temperature exceeds 122 Degrees F, special lubricants are also implemented to prevent bearing damage.

To downgrade rated horsepower, the simplest and most common method of de-rating a motor for operation in high ambients, a temperature correction factor is applied.

To find the proper temperature correction factor, the Desired Ambient Temperature is projected on a graph. A vertical line is drawn from the point of Desired Ambient Temperature, on the x axis, to the point at which it intersects the Temperature Correction Factor Line. A horizontal line is drawn from this point to the appropriate point on the y axis, which indicates the de-rating factor for the motor

DERATING IS IN THE OPPOSITE OF AAT

All electrical appliances have set conditions under which they function at optimum levels. Any fluctuations in these conditions can cause the appliances to run at a lower efficiency. Power generators are no exception to this. Generators are typically designed to run most efficiently at or near sea level under standard conditions of temperature and pressure (STP). Any fluctuation from STP conditions can impair generators and cause decreased output. Under extreme circumstances, generators can cease to function entirely. For most applications, many of these factors are relatively minimal unless the generator set is run at elevations over 5000 feet above sea level, or has ambient temperatures that remain over 100 degrees Fahrenheit for a significant length of time. Special care should be taken to compensate for these types of extremes and is discussed a bit further below.

Application Example: Environmental Factors that Affect Generator Function

Ambient conditions of temperature are extremely crucial for proper ignition and functioning of a generator. All generators, irrespective of the fuel that powers them, require adequate air for combustion. Decreased air levels can lead to start-up failure. In diesel engines, air and fuel are infused together. The compressed air becomes hot and when peak temperature and pressure is achieved, diesel is injected, which then ignites under the given conditions. In generators using gasoline, an air and fuel mixture is introduced at once using a carburetor and a spark is induced to ignite the engine. However in both cases, adequate levels of air are required for proper start up and operation.

Altitude: In areas of high altitude, air pressure drops reducing the air density. This can create problems with generator start up if not accounted for since air is crucial for ignition in any type of generator. Another factor that gets affected is availability of ambient air to facilitate heat dissipation from the generator. A lot of heat is created during the combustion process and needs to be dissipated into the environment to reduce engine temperature. At high altitudes, due to the low air density, heat dissipation occurs at a much slower rate than it would at sea levels, resulting in high engine temperatures for a sustained period of time. The engine remains hot and overheating is a common problem in such cases.

Temperature: High temperatures are also associated with lower air density and can cause similar ignition problems due to inadequate air supply. This can burden the engine which pushes itself to deliver the power it is designed to. However, due to inadequate oxygen levels available for combustion, it fails to do so. In many such instances, the engine gets overheated and sometimes collapses altogether.

Humidity: Humidity is the measure of water content in a given volume of air. In conditions of extreme humidity, water vapor in the air displaces oxygen. Low oxygen levels impair ignition, since oxygen is the element in air that is ignited in an engine for the burning of fuel.

Rating Generators

Generators come in various sizes. Each of them is preset for particular output levels. Generators are selected and installed based on the power requirements of any facility. A typical generator is ideally set to run at 80% of its capacity for continuous usage. In an emergency, it may be utilized for 100% efficiency. Various companies manufacturing generators have now come up with standard ratings for these generators, which give the buyer an idea of actual generator capacity. As per the requirement of the consumer, he can then make a choice between available brands, since each brand capacity is standardized as per international standards.

Derating Generators

Derating a generator depends on the manufacturer of the unit. Different manufacturers design generators using materials from different sources. Also, design developments are non-similar and so is the technique in many cases. All these can contribute to overall generator efficiency. Therefore, derating a generator is dependant on the manufacturing process. Different brands have different derating factors for estimating generator output under non-standard ambient conditions.

However, a general formula can be used to calculate close estimates for the output levels. The standard derating formula states that for every 1000 ft above sea-level, a gasoline, diesel, or liquid propane generator usually should be derated by 2–3% of its standard output. In case of generators using natural gas, the derating factor is typically closer to 5%

SUMMARY AND CONCLUSION

In this chapter, derating of a faulty electric power converter was described. In these systems, continuous operation has no interruption even with occurrence of a fault but the system performance can be degraded. The main topics of this chapter are summarized as follows:

1. Derating is a tool for utilization of a faulty system with degraded performance.
2. Derating factor is affected by environmental and operating point of the system.
3. Derating algorithm can be used for increasing some system functional indices while decreasing some others (for example: magnetic derating).
4. Degree of derating is dependent on acceptable reduction in system performance.

Derating is a generally accepted method for operation of the converter without interruption. However, it is impossible to apply the derating method in some converters. In these converters, all nominal specifications should be maintained. In the next chapter, we present fault tolerant converters as an alternative solution for this problem.

REFERENCES

(2004). Online optimal flux controller for DTC based induction motor drives. In *Proceedings of 30th Annual Conference of IEEE Industrial Electronics Society* (pp.1391-1395). IEEE.

Sadati, N., Kaboli, S., Adeli, H., Hajipour, E., & Ferdowsi, M. (2009). Online Optimal Neuro-Fuzzy Flux Controller for DTC Based Induction Motor Drives. In *Proceedings of Twenty-Fourth Annual IEEE Applied Power Electronics Conference and Exposition* (pp.2215). IEEE doi:10.1109/APEC.2009.4802657

ADDITIONAL READING

Apsley, J. M. (2010). Derating of Multiphase Induction Machines Due to Supply Imbalance. *IEEE Transactions on Industry Applications*, *46*(2), 798–805. doi:10.1109/TIA.2010.2040056

Liu, W. (2014). The prevalent motor bearing premature failures due to the high frequency electric current passage. *Engineering Failure Analysis*, *45*, 118–127. doi:10.1016/j.engfailanal.2014.06.021

Oraee, A., Abdi, E., & McMahon, R. A. (2015). Converter rating optimization for a brushless doubly fed induction generator. *IET Renewable Power Generation*, *9*(4), 360–367. doi:10.1049/iet-rpg.2014.0249

Picazo-Ródenas, M. J., Royo, R., Antonino-Daviu, J., & Roger-Folch, J. (2013). Use of the infrared data for heating curve computation in induction motors: Application to fault diagnosis. *Engineering Failure Analysis*, *35*, 178–192. doi:10.1016/j.engfailanal.2013.01.018

KEY TERMS AND DEFINITIONS

Derating: Degradation of a system performance under faulty conditions.

Faulty System: A system with some damaged parts.

Heat Trap: Increasing the temperature at a local specific point.

Motor Drive: A motor which is controlled by a switching power electronic converter.

Voltage Harmonics: Sinusoidal voltages with higher frequency than fundamental term in a periodic but nonsinusoidal waveform.

Chapter 11
Fault Tolerant Systems

ABSTRACT

Another alternative derating, which was described in the previous chapter, is application of fault tolerant structures for the power converter. Fault tolerance is the property that enables a system to continue operating properly in the event of a failure of (or one or more faults within) some of its components. Fault tolerant systems are systems that can be operating after fault occurrence with no degraded performance in their basic functional requirements. This is the main difference between fault tolerant systems and derated systems. In this chapter, some methods for fault tolerance in electric power converters are presented. Fault tolerance is almost the only method for achieving a desired reliability in a converter that operates with non-zero fault possibility. There are two main approaches for this aim: re-configuration of the faulty system and using redundant systems. Redundancy is the provision of functional capabilities that would be unnecessary in a fault-free environment. Various types of redundant systems as passive and active redundancy are described and their application in power supply systems is presented. A new approach for a reliable and fault tolerant power supply is proposed and justified with experimental results. The concept of fault tolerance in electrical machines is presented.

INTRODUCTION: ROBUSTNESS AGAINST FAULTS

The presented methods of reliability improvement as well as reliability calculation techniques help to have a safe converter without catastrophic failures. Derating method is usually exclusive method of reliability improvement for a faulty converter. However, derating is derating! It usually means the derated converter continues to operate with new rated characteristics which are less than the converter original nominal specifications. In many cases, this is not acceptable and it is needed to keep the original nominal rating of converter. For example, consider a DC power distribution unit with several output voltage levels. In a power distribution unit, as it was presented for a satellite in chapter 1, there are several voltage regulators and they provide some output voltage levels from a common DC input voltage source such as a battery. In this system, failure in one of the output voltage levels causes to failure in the subsystem related to the failed output voltage level. It is true that just one of the output channels is failed and it is

DOI: 10.4018/978-1-4666-9429-3.ch011

not whole of the power distribution unit. But, the system does not operate properly even other output channels operate normally (Gorginpour, Jandaghi, & Oraee, 2013).

Fault tolerant methods are solution of this drawback and we will discuss about them in this chapter. Figure 1 shows the state of this chapter in the flowchart of the book. Fault tolerance is the property that enables a system to continue operating properly in the event of the failure of (or one or more faults within) some of its components. If its operating quality decreases at all, the decrease is proportional to the severity of the failure, as compared to a naïvely designed system in which even a small failure can cause total breakdown. Fault tolerance is particularly sought after in high-availability or life-critical systems. A fault-tolerant design enables a system to continue its intended operation, possibly at a reduced level, rather than failing completely, when some part of the system fails. Recovery from errors in fault-tolerant systems can be characterised as either roll-forward or roll-back. When the system detects that it has made an error, roll-forward recovery takes the system state at that time and corrects it, to be able to move forward. Roll-back recovery reverts the system state back to some earlier, correct version, for

Figure 1. State of chapter 11 in the flowchart of the book

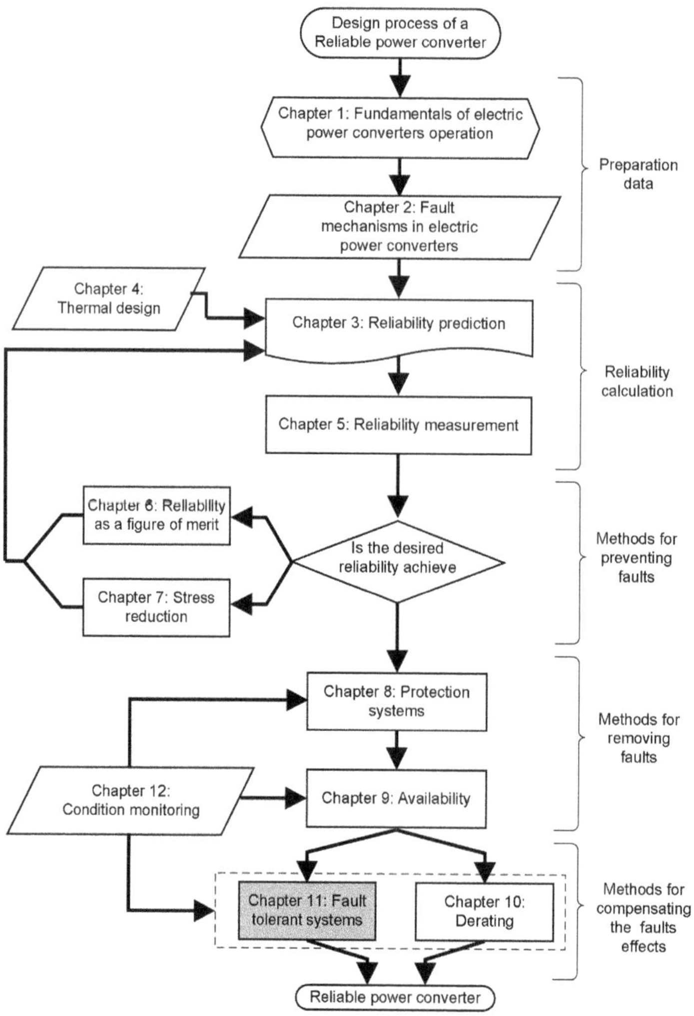

example using checkpointing, and moves forward from there. Roll-back recovery requires that the operations between the checkpoint and the detected erroneous state can be made idempotent. Some systems make use of both roll-forward and roll-back recovery for different errors or different parts of one error.

Fault tolerance can be achieved by anticipating exceptional conditions and building the system to cope with them, and, in general, aiming for self-stabilization so that the system converges towards an error-free state. However, if the consequences of a system failure are catastrophic, or the cost of making it sufficiently reliable is very high, a better solution may be to use some form of duplication. In any case, if the consequence of a system failure is so catastrophic, the system must be able to use reversion to fall back to a safe mode. If each component, in turn, can continue to function when one of its subcomponents fails, this will allow the total system to continue to operate as well.

REDUNDANCY

Redundancy is the provision of functional capabilities that would be unnecessary in a fault-free environment (Hao, Covic, & Boys, 2014). This can consist of backup components which automatically "kick in" should one component fail. The idea of incorporating redundancy in order to improve the reliability of a system was pioneered by John von Neumann in the 1950s.

Providing fault-tolerant design for every component is normally not an option. Associated redundancy brings a number of penalties: increase in weight, size, power consumption, cost, as well as time to design, verify, and test. Therefore, a number of choices have to be examined to determine which components should be fault tolerant:

- How critical is the component? In a car, the radio is not critical, so this component has less need for fault tolerance.
- How likely is the component to fail? Some components, like the drive shaft in a car, are not likely to fail, so no fault tolerance is needed.
- How expensive is it to make the component fault tolerant? Requiring a redundant car engine, for example, would likely be too expensive both economically and in terms of weight and space, to be considered.

The basic characteristics of fault tolerance require:

1. **No single point of failure:** If a system experiences a failure, it must continue to operate without interruption during the repair process.
2. **Fault isolation to the failing component:** When a failure occurs, the system must be able to isolate the failure to the offending component. This requires the addition of dedicated failure detection mechanisms that exist only for the purpose of fault isolation. Recovery from a fault condition requires classifying the fault or failing component. The National Institute of Standards and Technology (NIST) categorizes faults based on locality, cause, duration, and effect.
3. **Fault containment to prevent propagation of the failure:** Some failure mechanisms can cause a system to fail by propagating the failure to the rest of the system. An example of this kind of failure is the "rogue transmitter" which can swamp legitimate communication in a system and cause overall

system failure. Firewalls or other mechanisms that isolate a rogue transmitter or failing component to protect the system are required.

Spare components address the first fundamental characteristic of fault tolerance in three ways:

- **Replication:** Providing multiple identical instances of the same system or subsystem, directing tasks or requests to all of them in parallel, and choosing the correct result on the basis of a quorum;
- **Redundancy:** Providing multiple identical instances of the same system and switching to one of the remaining instances in case of a failure (failover);
- **Diversity:** Providing multiple *different* implementations of the same specification, and using them like replicated systems to cope with errors in a specific implementation.

Fault-tolerant design's advantages are obvious, while many of its disadvantages are not:

- **Interference with fault detection in the same component:** To continue the above passenger vehicle example, with either of the fault-tolerant systems it may not be obvious to the driver when a tire has been punctured. This is usually handled with a separate "automated fault-detection system". In the case of the tire, an air pressure monitor detects the loss of pressure and notifies the driver. The alternative is a "manual fault-detection system", such as manually inspecting all tires at each stop.
- **Interference with fault detection in another component:** Another variation of this problem is when fault tolerance in one component prevents fault detection in a different component. For example, if component B performs some operation based on the output from component A, then fault tolerance in B can hide a problem with A. If component B is later changed (to a less fault-tolerant design) the system may fail suddenly, making it appear that the new component B is the problem. Only after the system has been carefully scrutinized will it become clear that the root problem is actually with component A.
- **Reduction of priority of fault correction:** Even if the operator is aware of the fault, having a fault-tolerant system is likely to reduce the importance of repairing the fault. If the faults are not corrected, this will eventually lead to system failure, when the fault-tolerant component fails completely or when all redundant components have also failed.
- **Test difficulty:** For certain critical fault-tolerant systems, such as a nuclear reactor, there is no easy way to verify that the backup components are functional. The most infamous example of this is Chernobyl, where operators tested the emergency backup cooling by disabling primary and secondary cooling. The backup failed, resulting in a core meltdown and massive release of radiation.
- **Cost:** Both fault-tolerant components and redundant components tend to increase cost. This can be a purely economic cost or can include other measures, such as weight. Manned spaceships, for example, have so many redundant and fault-tolerant components that their weight is increased dramatically over unmanned systems, which don't require the same level of safety.
- **Inferior components:** A fault-tolerant design may allow for the use of inferior components, which would have otherwise made the system inoperable. While this practice has the potential to mitigate the cost increase, use of multiple inferior components may lower the reliability of the system to a level equal to, or even worse than, a comparable non-fault-tolerant system

Figure 2. A basic power system with redundancy

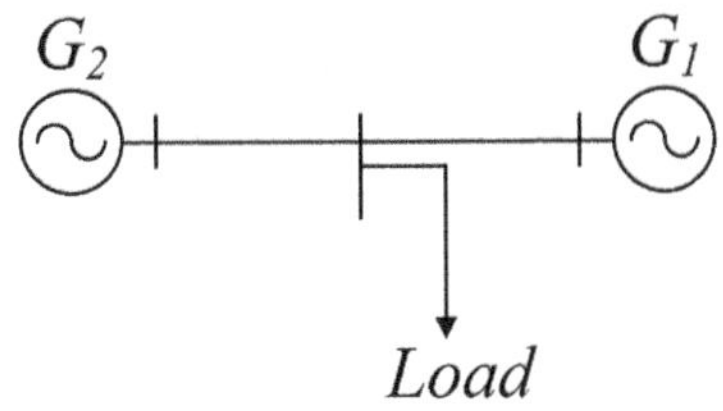

Figure 3. A solar plant with redundancy

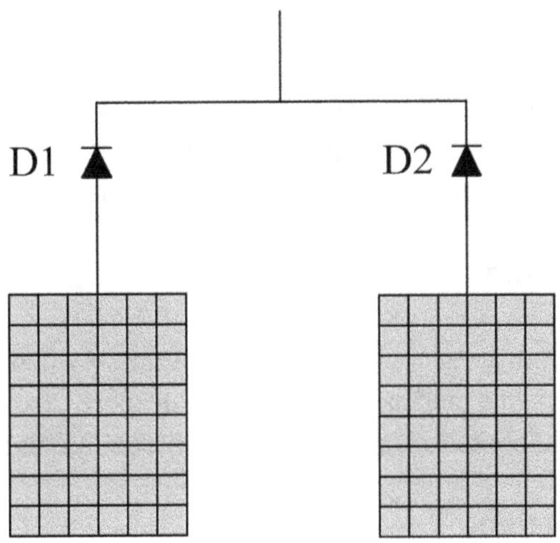

Figure 4. A converter with redundant power IC

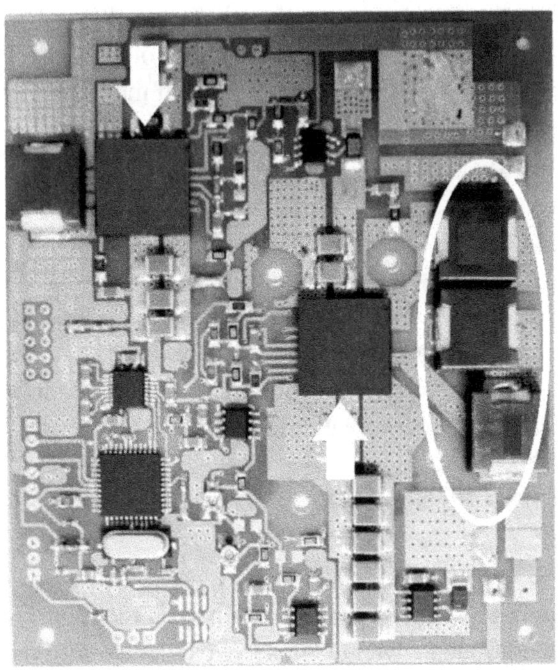

Figure 5. A converter and its redundant module with isolating switches

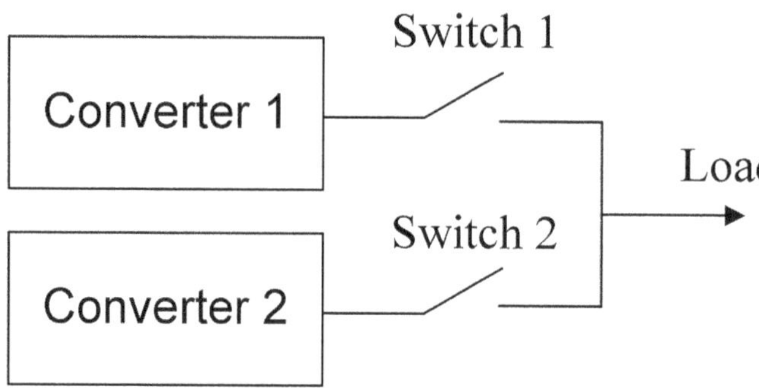

Figure 2 shows a power network with two generators that can operate as redundant. Figure 3 shows a PV system with redundancy. Figure 4 shows a power supply with a redundant DC/DC converter. Figure 5 shows a power supply with two converters and isolating switches.

Active Redundant Systems

Active redundancy in active components requires reconfiguration when failure occurs. Computer programming must recognize the failure and automatically reconfigure to restore operation. Figure 6 shows two states of a power supply in hot redundancy. Isolator switches are used to bypass the damaged unit.

Passive Redundant Systems

In this case we have a device that is doing something and a device that is idle and ready to take over the task of the other device in case it fails (Zhang, Xu, Enjeti, Li, Hawke, & Krishnamoorthy, 2014). This means that there must be some switching system that takes care of starting the idle device and stopping the active device and switch both on the input side and on the output side. Figure 7 shows a power supply with standby redundant. One of the units operates at each moment. Figure 8 shows a three various states of a power supply with redundancy and isolator switches. Figure 9 shows a power supply with passive isolator elements (fuses).

Fault Tolerant Power Electronic Converter

Fault tolerance is the property that enables a system to continue operating properly in the event of the failure of (or one or more faults within) some of its components. If its operating quality decreases at all, the decrease is proportional to the severity of the failure, as compared to a naively designed system in which even a small failure can cause total breakdown.

Example: Reconfigurable PDU

Power distribution unit (PDU) is one of the most critical systems in electrical equipments since all the subsystems need power supply to maintain their function. The main purpose of the PDU is to provide continuous, regulated and conditioned power to all electrical device subsystems within the mission time

Figure 6. Two possible states for a converter with active redundancy, (left): both converters are normal, (right): converter 1 was failed

Figure 7. Two possible states for a converter with standby redundancy, (left): converter 2 operates and converter 1 is OFF, (right): converter 2 was failed and converter 1 operates

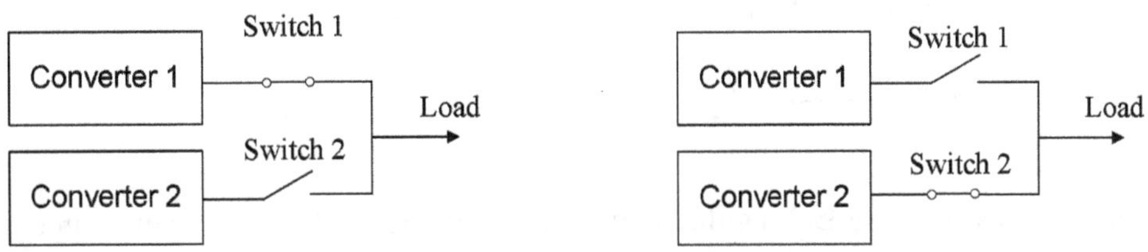

Figure 8. Three possible state of a power converter with redundancy during operation

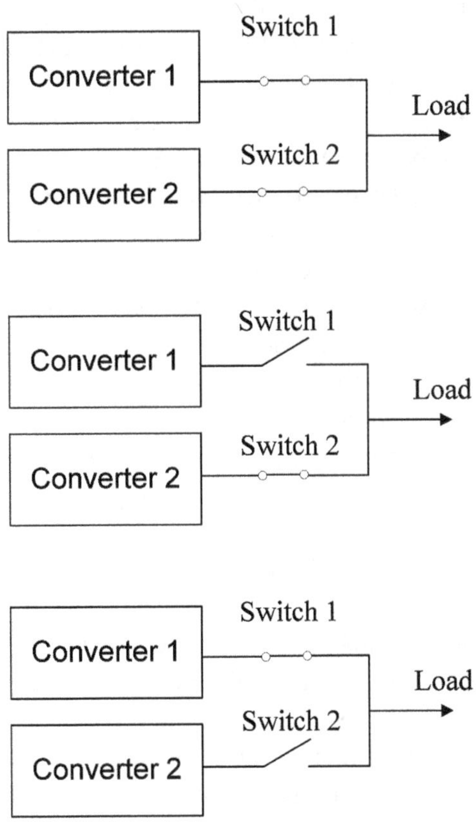

Figure 9. Application of fuses as isolator switches

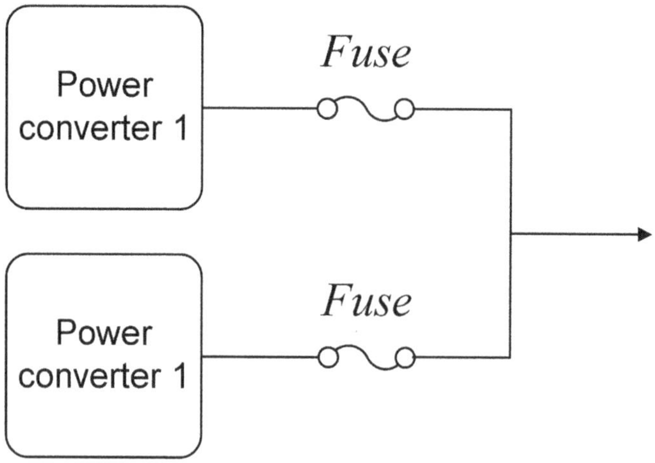

[6]. The PDU consists of many DC regulators which receive power from a power source and convert into many regulated voltage levels. A satellite PDU and block diagram of a PDU with two output voltage level are shown in Figure 10. Reliability of the PDU for electrical devices is very important in some applications such as telecommunications systems in space and remote areas in which a change of faulty components are difficult or impossible. Many solutions have been proposed to improve the reliability of convertors and thus, improve the reliability of PDU. Some papers have attempted to increase the reliability of dc/dc regulators by reducing component's stress or use of simpler circuit structures. Although this technique improves reliability somewhat, but if we have a breakdown of components for any reason, then the regulator will fail. Therefore, we need to design a system where failure of a component does not result in the failure of electrical devices. To achieve this goal, fault tolerant operation methods are proposed. The fault tolerant operation methods are recognized in two ways, use of control strategies, and use of redundancy for components with more failure probability. Control strategies are used to reconfiguring a failed structure into a new structure in order for the regulator continues to work. Redundancy is defined as the use of additional components or sub-systems beyond the number actually required for the system to operate reliably. Redundancies can be categorized as active (parallel) or passive (standby) redundancy as shown in Figure 11. In systems with active redundancy, all redundant components are in operation and upon failure of one component, the surviving components carry the load. In passive or standby system, using a switching device, the redundant component starts to operate only when one or more of components fails.

Another approach in order to cover all fails in PDU is the use of a complete regulator as redundancy for each main regulator. In this structure, if one of the regulators and its standby units fail, then the PDU fails, and other standbys from other regulators are not used hereunto.

In this section, in order to have a PDU with high reliability, a use of adjustable standby regulators is suggested. In this approach, each standby unit can be adjusted for any voltage levels, and replaced or paralleled to a failed or overloaded regulator. In this structure, PDU fails only when all of the standby units are used, or all of the switches that adjusts regulators or connect the standby regulators to outputs fail. If a regulator is overloaded, and the overload is not removed, the regulator quickly burns. Also, if a standby regulator is replaced or paralleled to a failed or overloaded regulator, then that standby regula-

Figure 10. Typical application of a PDU, (left): Application of a PDU in satellite power system, (right): a PDU with two output voltage levels

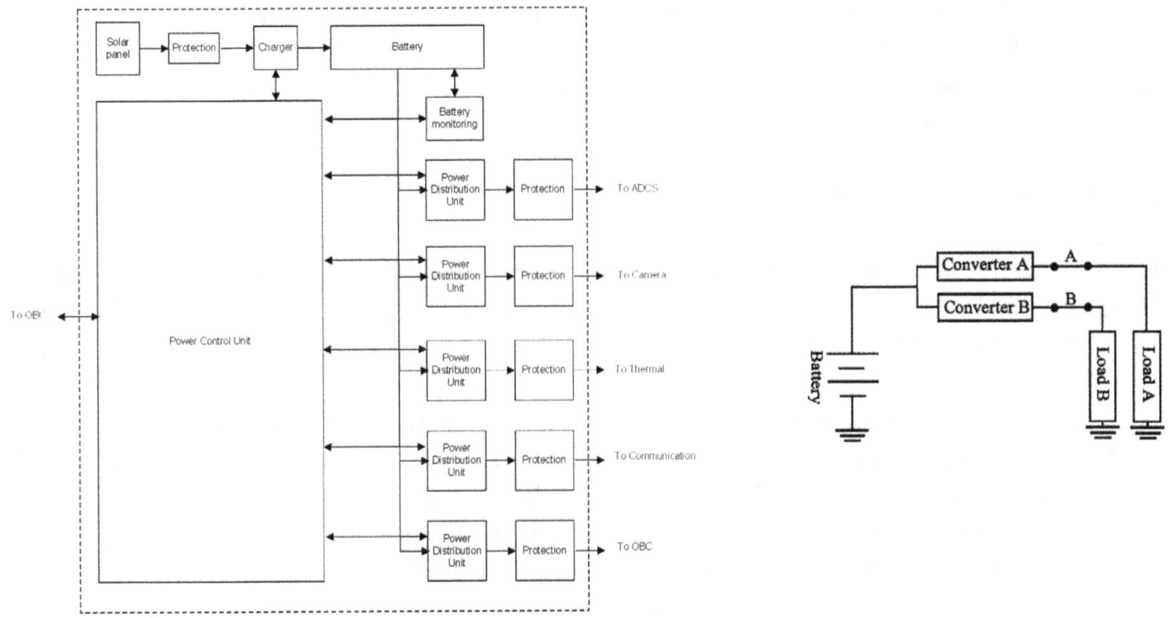

Figure 11. (a) A system with parallel redundancy (b) A system with standby redundancy

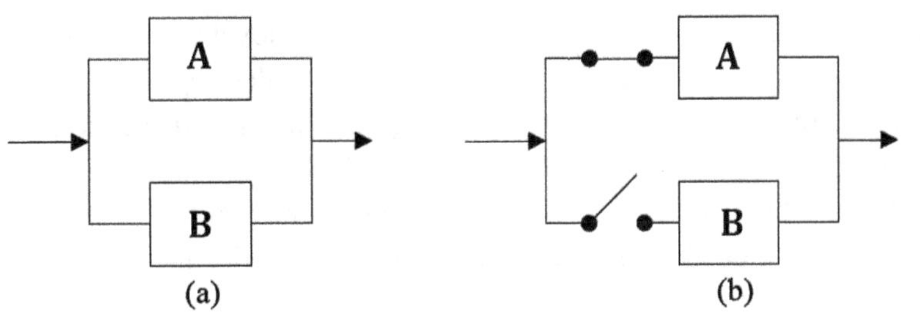

tor is used. Therefore, for the simplicity of this paper, we assume that a failed regulator is equivalent to an overloaded regulator.

There are many approaches for reliability analyses of systems with standby redundancies. An analytical method based on sequential binary decision diagrams for combinatorial reliability evaluation of non-repairable cold-standby systems is proposed. A methodology for the reliability evaluation of a system with dissimilar unit and non-repairable cold-standby redundant by use of Markov proses is introduced. The reliability of a system is introduced with a repairable unit and warm standby unit, assuming a constant failure probability of switching device and exponential lifetime for subsystems using the Markov model. In this project, it is assumed that the switching failure probability is not a constant value. As a consequence, if switching failure probability time variation is considered, the traditional Markov models are not valid. This time dependence of the switching failure probability leads us to use a non-homogeneous Markov model. The non-homogeneous Markov model is extended for the reliability evaluation of a Markov system with time-dependent state transition.

In order to test the reliability of the proposed structure, we use the accelerated aging test (AAT) at overloading condition of regulator to increase their temperature. Accelerated aging test uses exasperated conditions of heat, humidity, radiation, vibration, etc. for failure in much shorter time than the normal operating conditions. Real date reliability can be achieved by use of date reliability in this condition and empirical mathematical equations or models. AAT at high temperature condition is used for reliability testing of many electronic and power devices.

Conventional Structure

In the conventional structures proposed for PDUs, in order to cover all fails in the regulators of the PDU, the use of complete regulators as standby redundancies for each regulator at each voltage level was proposed. In Figure 12, a PDU with two voltage levels and one standby regulator for each voltage level is shown. In this structure, if regulator and its standby units at one voltage level fail, then the PDU fails. Also, if sensing and switching devices that replace the standby units to failed regulator fail then the PDU could not produce that voltage level, so the PDU fails. In this structure, if all the standby units of a regulator at one voltage level fail, then the PDU will fail, and other standbys from other regulators are not used hereunto. This state is very probable to occur, because the power consumption in each voltage level. The power losses of regulators at each voltage level in the PDU are different, and since their component are almost similar, so their temperatures will be different. Also, since failure rate of electrical devices is a function of temperature, then the failure rate of them will be different.

In this paper, the use of adjustable regulators as standby units is proposed. In this structure the PDU fails, only when all of standby unit are used, or all of switches that replace the failed regulator to standby regulator fail. A PDU with this structure and two output voltage level and two standby units is shown in Figure 13. In this structure, each regulator before replacing to failed regulator must be adjusted for that voltage level. Adjusting the regulator for a voltage level using a resistance change within the adjustable regulators by use of many switches is done. Adjustable regulators have more complexity and so they have more failure rate. Also in this structure we need more power switches to replace each standby regulator to each voltage level, and we need many low power switches as adjusting switches. In this structure, in

Figure 12. A PDU with one standby redundancy for each voltage regulator

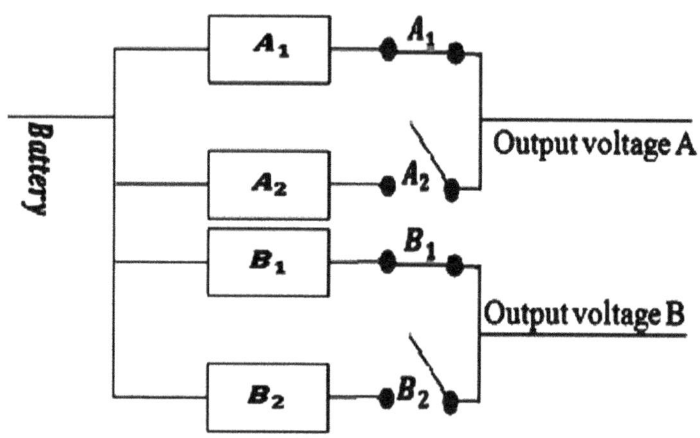

Figure 13. A PDU with adjustable redundancy

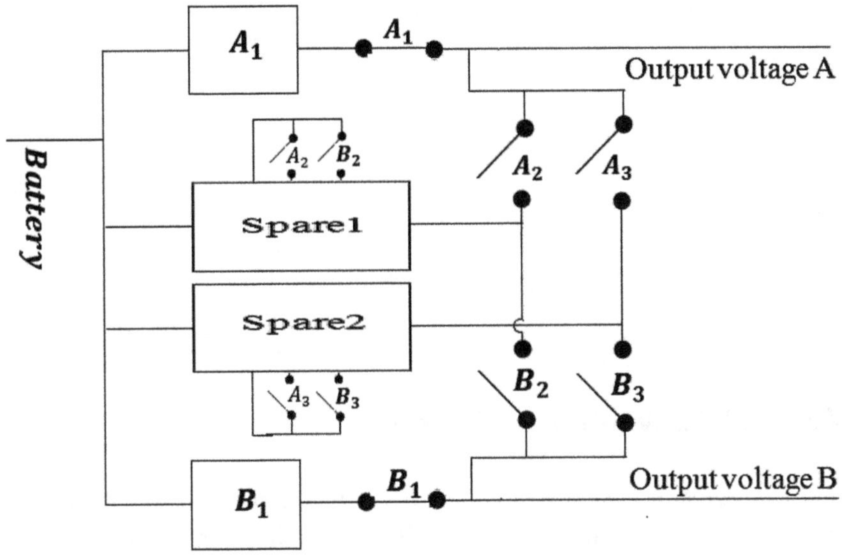

order to replace an adjustable regulator to a failed regulator, the adjusting switch and the switch that connect the standby regulator to that voltage level must work properly.

If X is a stochastic process then, X is a Markov chain, if for all $j \in E$,

$$\Pr\{X_{n+1} = j|X_0, X_1, ..., X_n\} = \Pr\{X_{n+1} = j|X_n\} \tag{1}$$

Here $E = \{1, 2, ..., x\}$ and $n \in \mathbb{N}$. The $p_n(i, j), (n \in \mathbb{N}, i, j \in E)$ is the transition probability function of chain X in n-th subinterval. Multiple-step transition probabilities are defined by:

$$P_{n,m}(i,j) = \Pr\{X_m = j \mid X_n = i\} = \begin{cases} \left[\prod_{k=n}^{m-1} P_k\right](i,j), & for \quad m > n \geq 0, \\ 1, & for\ i = j, \quad m = n \geq 0, \\ 0, & for\ i \neq j, \quad m = n \geq 0. \end{cases} \tag{2}$$

$$p_{n,n+1}(i,j) = p_n(i,j) \tag{3}$$

If $p_n(i, j)$ does not depend on n, the Markov chain is time homogeneous.

The state probability vector in the Markov chain of X is defined by its α and $p_n(n \in \mathbb{N})$, where α is initial distribution of X and p_n is the transition probability matrix at n-th subinterval.

$$P_j(n) = \Pr\{X_n = j\} = \left[\prod_{k=n}^{m-1} P_k\right](j) \tag{4}$$

$$P(n) = \alpha . p_{0,n} \tag{5}$$

Also *MTTF* is equal to:

$$MTTF = \sum_{n=1}^{\infty} R(n).\Delta t \tag{6}$$

Reliability Model of Conventional Structure

In this structure, each regulator has only one standby regulator, and that standby only can be replaced or paralleled to it by use of one switch. Therefore, if the switch doesn't work correctly or there is no standby regulator for that voltage level, then that voltage level doesn't produce. In Markov model of this structure, in order to investigate the failure due to switches and failure due to regulators, we use the two separated states for failures in the Markov model as shown in Figure 14. In this figure, dashed line shows the failure due to switches. Ma shows number of main regulators that work correctly.

Reliability Model of the Proposed Structure

In this structure each standby regulator can be adjusted by using one switch for one voltage level and replaced to failed regulator with another switch. Hence, if adjusting switch and the switch that connect the standby regulator to failed regulator don't work correctly, then that regulator can't replace the failed regulator. So, if none of the standby regulators replace the failed regulators, then the PDU can't produce that voltage level. In this structure, if we have N main regulators and M standby regulators, then the Markov model will be as shown in Figure 15. In this model, in order to investigate the PDU failure

Figure 14. The Markov model of a regulator with its standby

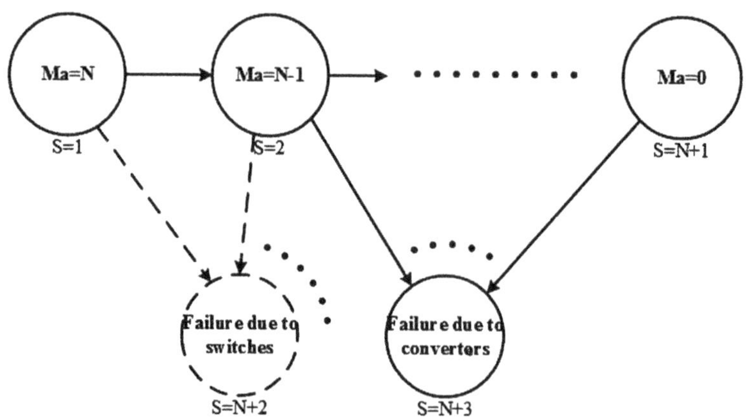

due to switches and failure due to regulators, we use two separate states for fail state. In this figure, the dashed line shows failure due to switching devices, S shows the number of states, st shows the number of existing standbys at each state, and Ma shows the number of main regulators that work correctly. The transition probabilities of each state to other states are as follows.

If standby regulators exist (st ≠ 0) then:

Transition probability matrix is shown in table. I. In this matrix, for $s=1$ to $s=M+1$:

$$p_n(s, s+1) = (M - s + 1).\lambda.\Delta t.P_{sw} \tag{7}$$

Figure 15. The Markov model of the proposed PDU

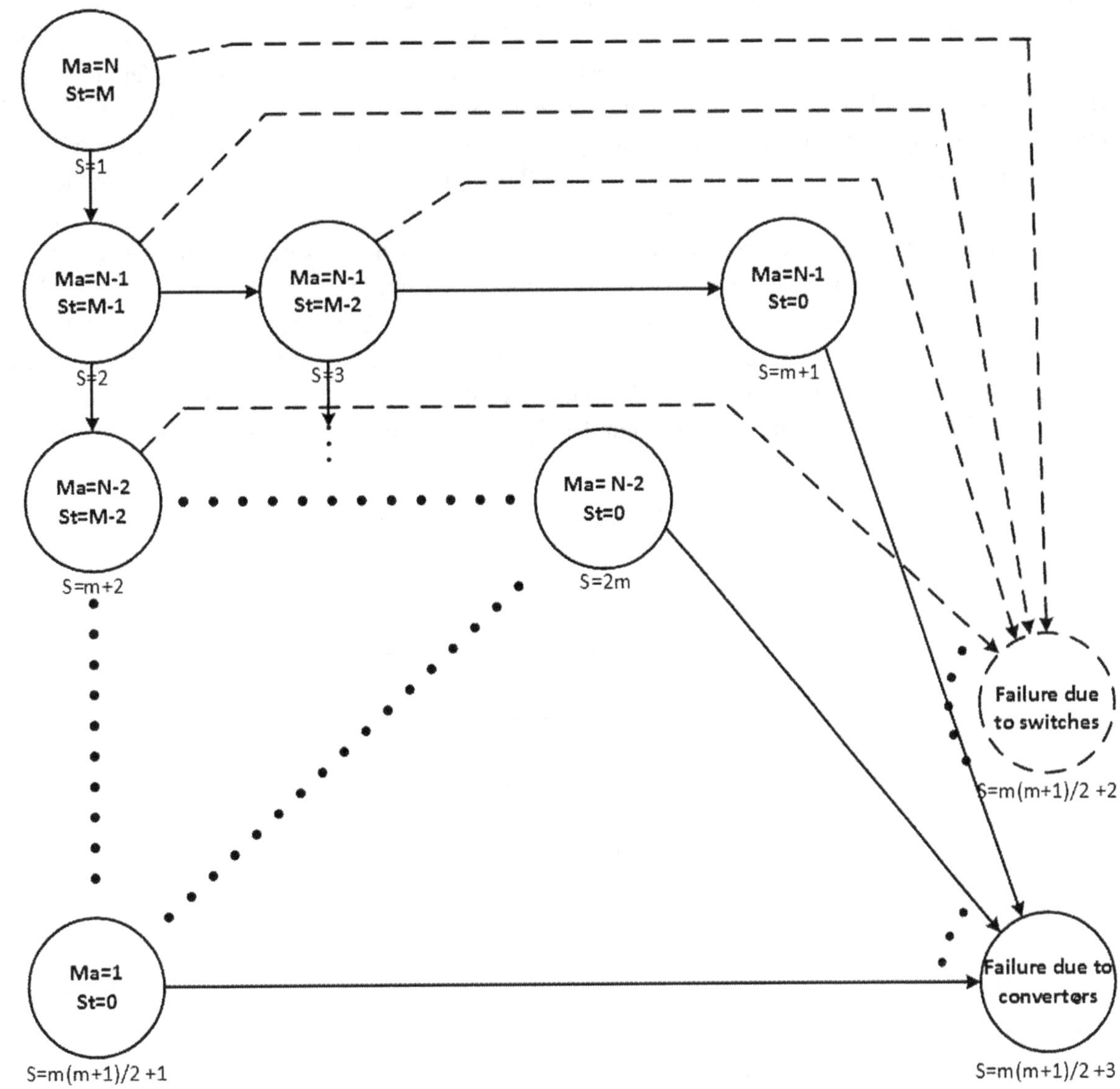

$$p_n(s, M+3) = (M - s + 1).\lambda.\Delta t.(1 - P_{sw}) \quad (8)$$

For $s=2$ to $s=M$:

$$p_n(s, M+2) = (s+1).\lambda.\Delta t.P_{sw} \quad (9)$$

The transition probabilities to all of the other states are zeros, except $p_n(s, s)$ that is equal to:

$$p_n(s,s) = 1 - \sum_{i \neq s} p_n(s,i) \quad (10)$$

In this paper, in order to compare the proposed structure and the conventional structure, we use two systems. Each of the standby regulators only can replace one of the main regulators, and other system has adjustable standbys regulators that each of them can replace each main regulator. Here, in order to numerically compare these two structures, we use of LM317 reliability data for main and adjustable regulators, ATMEGA32 reliability data for sensing and control device and reliability data of SPDT MILLON relays for reliability data of switching devices. Because the other components that are used in the circuit are highly reliable and for simplicity, we assume that they are fully reliable. Each of the adjustable regulators are as shown in Figure 16. When these regulators fail, the contact between the pins of them opened or short to each other, also R2 and R3 are high resistors, so if the switch that separates the regulator from the load works correctly, then that failed regulator does not dissipate noticeable power. It is not necessary to separate the regulator from the input bus voltage. The reliability information of the relays, ATMEGA32, and LM317T are documented in Tables 1, 2, and 3.

Here, we assume Δt is equal to 1000 hours. The reliability curve of these two structures is shown in Figure 17. For conventional structure MTTF is equal to 192000 hours, and for proposed structure is equal to 236430 hours, that are not that different. But if assumed reliability for safe operation time must

Figure 16. An adjustable regulator

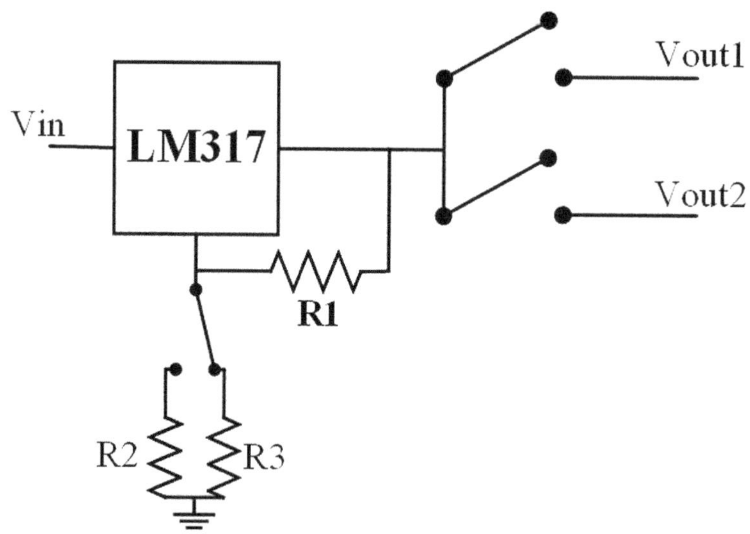

Table 1. Linear Regulator LM317T MTTF Tabulation

Parameter Name	Description	Value	Comments
C1	Die complexity	0.010	Linear device, <100 transistors
π_T	Temperature Factor.	58	Worst case junction operating temperature (125 °C) and Linear Bipolar Device
C2	Package Failure Rate	0.0012	Non-hermetic DIP, 3 pins
π_E	Environment Factor	2.0	Ground Fixed
π_Q	Quality Factor	10	Commercial
π_L	Learning Factor	1.0	3 years, 1/06 – 2/10
Entire design:	$\lambda p = (C_1 \pi_T .+ C_2 \pi_E) \pi_Q \pi_L$	5.824	Failures/10 hours
		19.6 years	MTTF

Table 2. Relay MTTF Tabulation

Parameter Name	Description	Value	Comments
λ_b	Base Failure Rate	0.0084	Ambient Temperature +25°C
π_C	Contact Form Factor	1.75	SPDT
π_A	Load Stress Factor	1.06	Resistance with 0.2 Rated Current
π_Q	Quality Factor	2.9	Commercial
π_E	Environment Factor	2	Ground Fixed
π_{CYC}	Cycling Factor	1	Cycle Rate 10
π_F	Application and Construction Factor	5	General Purpose with Balanced Armature Construction Type
Entire design:	$\lambda p = (\pi_C \pi_A \pi_E \pi_Q \pi_F \pi_{CYC}) \lambda_b$	0.451	Failures/10 hours
		252.6 years	MTTF

Table 3. ATmega32 8-bit microcontroller MTTF Tabulation

Parameter Name	Description	Value	Comments
C1	*Die complexity*	*0.14*	*8-bit microcontroller*
π_T	*Temperature Factor.*	*0.19*	*Worst case operating temperature (40 °C) and Digital MOS*
C2	*Package Failure Rate*	*0.029*	*44-pin TQFP, used eq. 3*
π_E	*Environment Factor*	*2.0*	*Ground Fixed*
π_Q	*Quality Factor*	*10*	*Commercial*
π_L	*Learning Factor*	*1.0*	*Years in production* ≥ 2.0
Entire design:	$\lambda p = (C_1 \pi_T .+ C_2 \pi_E) \pi_Q \pi_L$	*1.01*	*Failures/10 hours*
		113 years	*MTTF*

Figure 17. Reliability curves of proposed structure and conventional structure of PDU with two output voltage levels

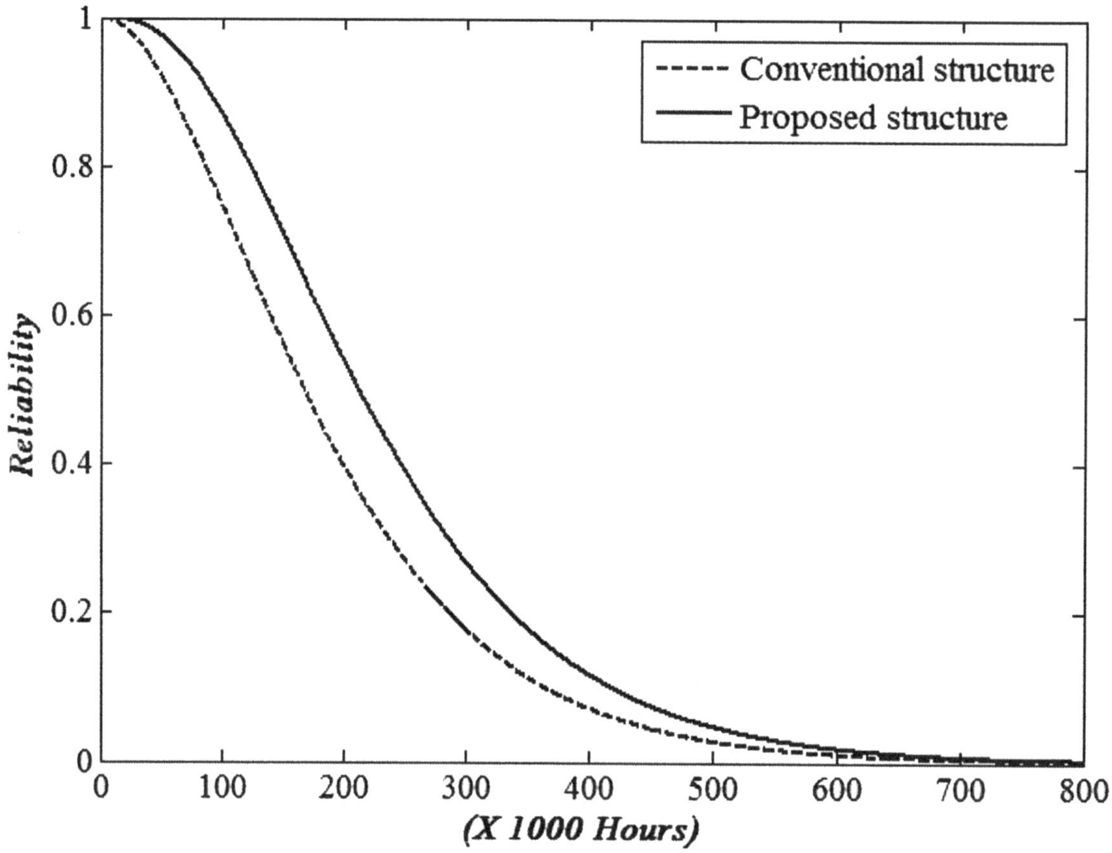

be above 0.99 then for conventional structure safe operation time is about 17000 hours, and for proposed structure is about 36000 hours. This is the most important effectiveness of the proposed structure, and this means that we can warranty up that the PDU works correctly for longer time.

Example: Fault Tolerant High Voltage Switch

Figure 18 shows a high voltage pulser used for driving a microwave tube. This pulser consists of series pulsers that can continue its operation with failure in some of them. In normal condition, the output voltage of the pulser is divided among *N* pulsers. If a fault occurs in one of the pulsers, the output voltage of the faulty pulser become zero. To hold the nominal voltage of the chathode pulse voltage, all of *N-1* reminded pulsers take a higher output voltage and compensate the missed voltage of the faulty pulser.

Fault Tolerant Electrical Machine

Example: Fault Tolerant Transformer

Figure 19 shows a Delta to Wye transformer and its resistive load. This transformer can be supply the three phase voltages even one of its legs is damaged. In this case, the three phase voltage is generated but with reduced nominal power.

Figure 18. High voltage pulser with active redundant modules

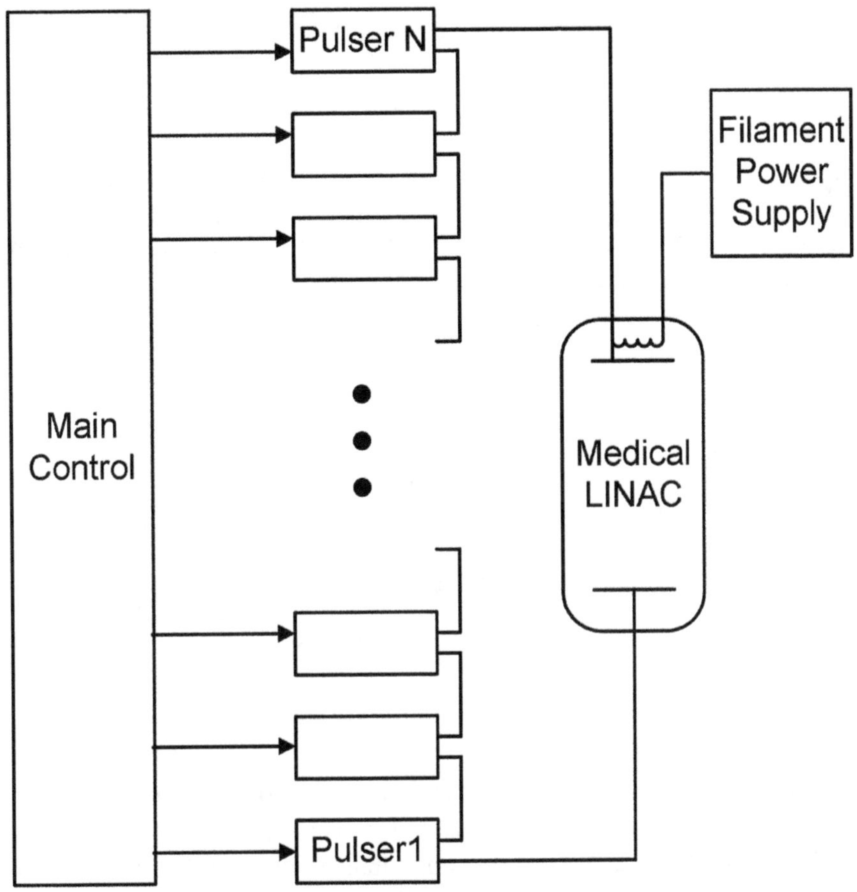

Example: Fault Tolerant Machines

The result of adding three 120-degrees phased sine waves on the axis of the motor is a single rotating vector. The rotor has a constant magnetic field. The N pole of the rotor will move toward the S pole of the magnetic field of the stator, and vice versa. This magneto-mechanical attraction creates a force which will drive the rotor to follow the rotating magnetic field in a synchronous manner [8].

A permanent magnet in such a field will rotate so as to maintain its alignment with the external field. This effect was utilized in early alternating current electric motors. A rotating magnetic field can be constructed using two orthogonal coils with a 90 degree phase difference in their AC currents. However, in practice such a system would be supplied through a three-wire arrangement with unequal currents. This inequality would cause serious problems in the standardization of the conductor size. In order to overcome this, three-phase systems are used where the three currents are equal in magnitude and have a 120 degree phase difference. Three similar coils having mutual geometrical angles of 120 degrees will create the rotating magnetic field in this case. The ability of the three phase system to create the rotating field utilized in electric motors is one of the main reasons why three phase systems dominate in the world electric power supply systems [9].

Figure 19. Delta connected transformers as a fault tolerant power supply

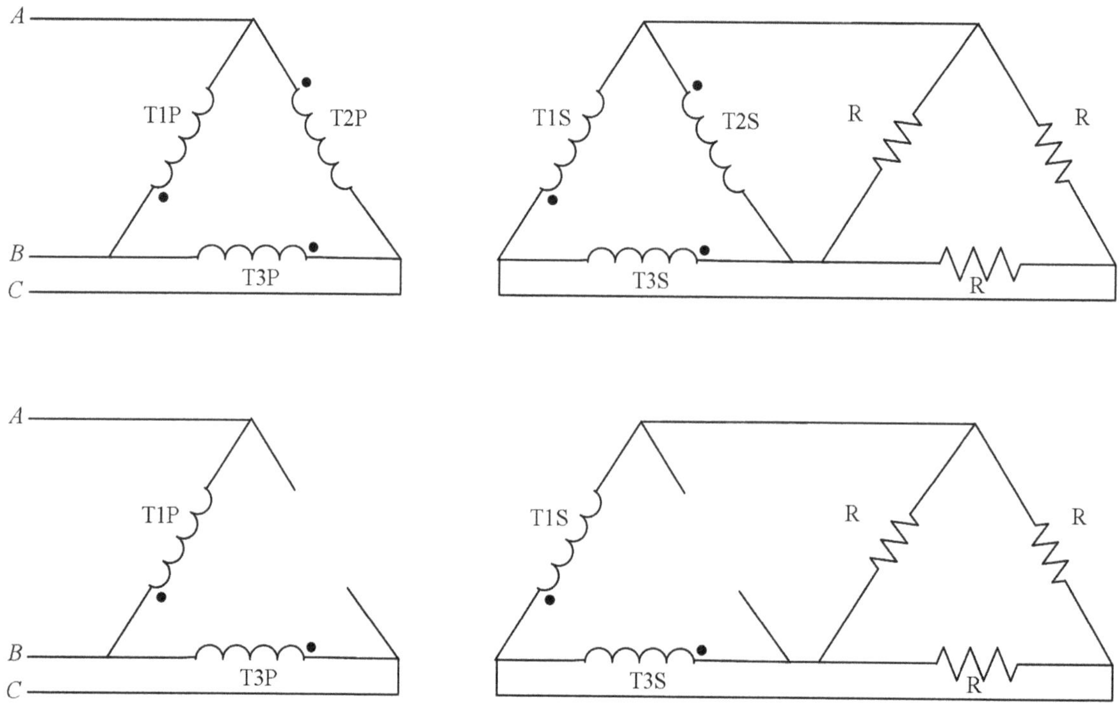

Rotating magnetic fields are also used in induction motors. Because magnets degrade with time, induction motors use short-circuited rotors (instead of a magnet) which follow the rotating magnetic field of a multicoiled stator. In these motors, the short circuited turns of the rotor develop eddy currents in the rotating field of the stator which in turn move the rotor by Lorentz force. These types of motors are not usually synchronous, but instead necessarily involve a degree of 'slip' in order that the current may be produced due to the relative movement of the field and the rotor.

A symmetric rotating magnetic field can be produced with as few as two polar wound coils driven at 90 degrees phasing. However, 3 sets of coils are nearly always used because it is compatible with a symmetric 3-phase AC sine current system. The three coils are driven with each set driven 120 degrees in phase from the others. For the purpose of this example, the magnetic field is taken to be the linear function of the coil's current. Figures 20 and 21 show two typical two phases stator of induction motor.

RECONFIGURATION

Fault tolerant operation methods only cover some common failures, and there is not a control strategy for each failure or a redundancy for each component. Hence, failing of many components could not be modified and the regulator fails.

Figure 20. A Three phases AC machine

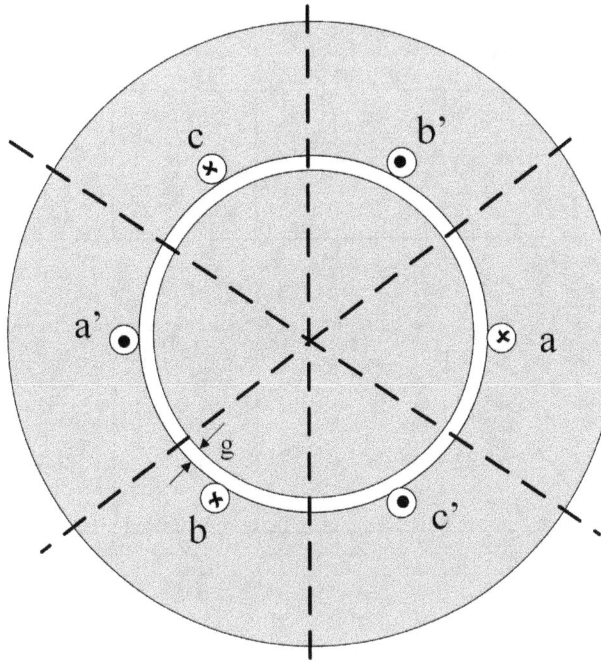

Figure 21. A two phases AC machine

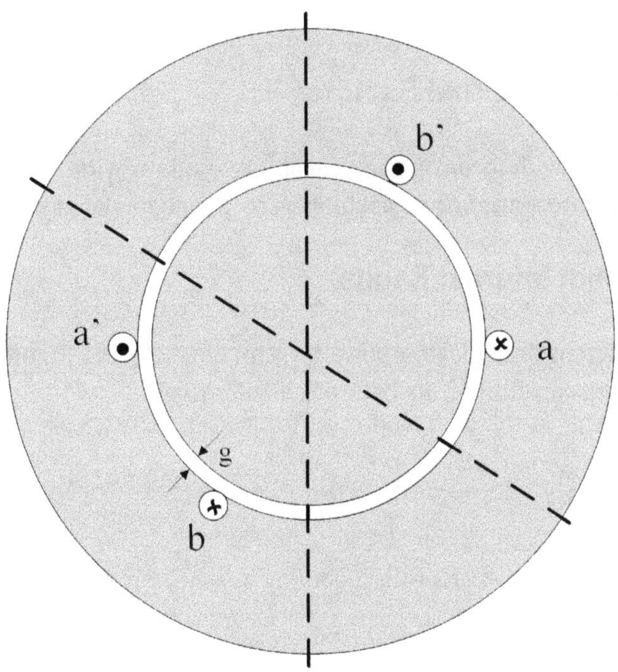

Figure 22. A reconfigurable converter, up: original scheme, middle: step down operation, down: step up operation

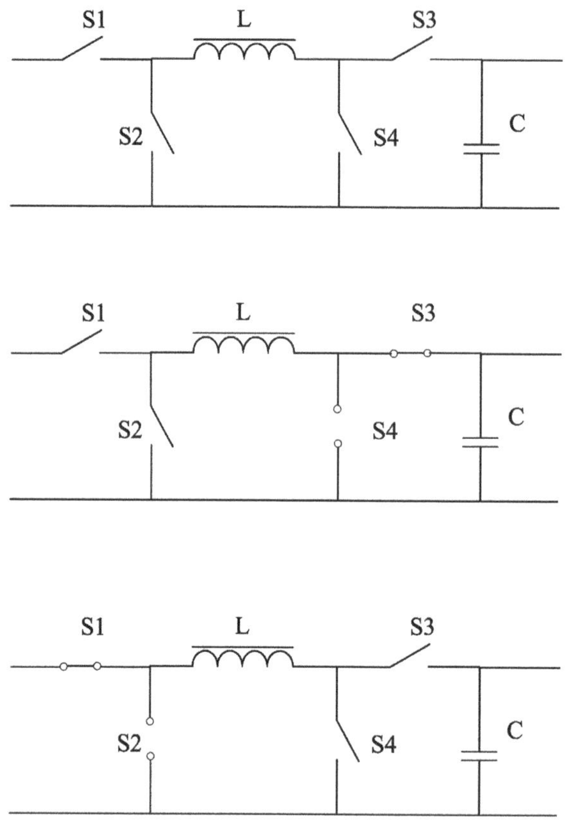

Reconfiguring Against External Factors

Figure 22 shows a converter with reconfiguring capability used for wide range of input voltage. Figure 23 shows a PV system with reconfiguring capability – by passing a damaged panel-.

Reconfiguring Against Internal Faults

Figure 24 shows two different strategies for power system of a satellite. Figure 25 shows power system of a satellite with changing capability if its batteries are damaged.

MULTI STAGE ALARMS

Figure 26 shows a protection alarm chain in a cooling system. Figure 27 shows an over temperature pre-alarm sensor on a heat sink.

Figure 23. A fault tolerant solar plant

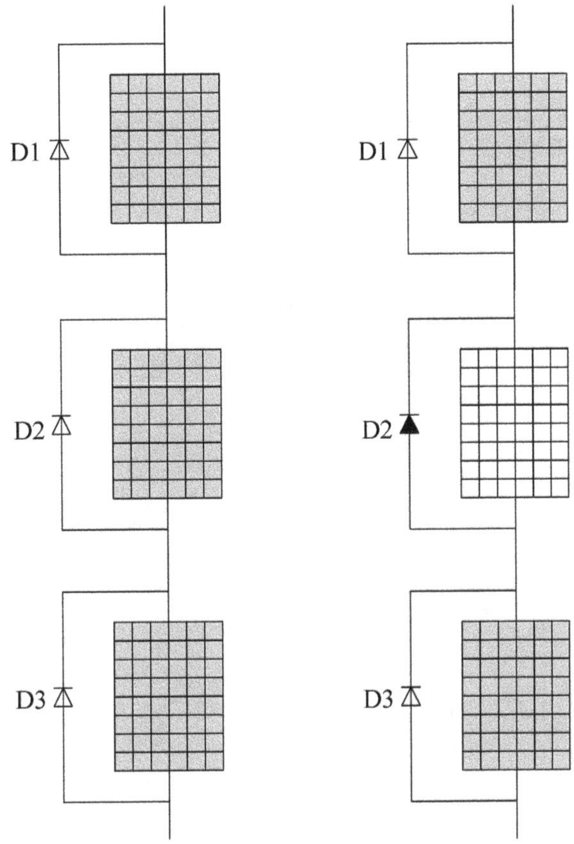

OVER DESIGN

Figure 28 shows a high voltage switch used as high voltage pulser with series connection of low voltage switches that are operating in a lower voltage than their nominal voltage. Figure 29 shows a DC power supply with parallel operation of low power DC power supplies that are operate under their nominal power.

SUMMARY AND CONCLUSION

In this chapter, fault tolerant electric power converters were described. In these systems, continuous operation with usually nominal specification has no interruption even with occurrence of a fault. The main topics of this chapter are summarized as follows:

1. There are two main approaches in a fault tolerant system: Using redundant systems and re-configuration.
2. Reconfiguring can be used by normally operating converters to compensate the I/O variations of the converter (for example: shadowing in PV systems)

Figure 24. Two different architecture for satellite power supply, (up): parallel scheme, (down): series scheme

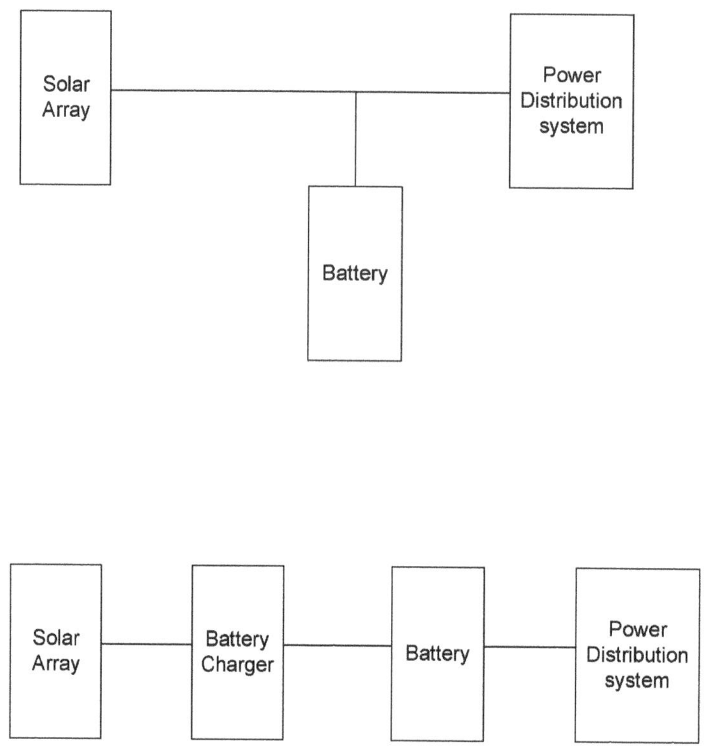

Figure 25. A fault tolerant satellite power supply

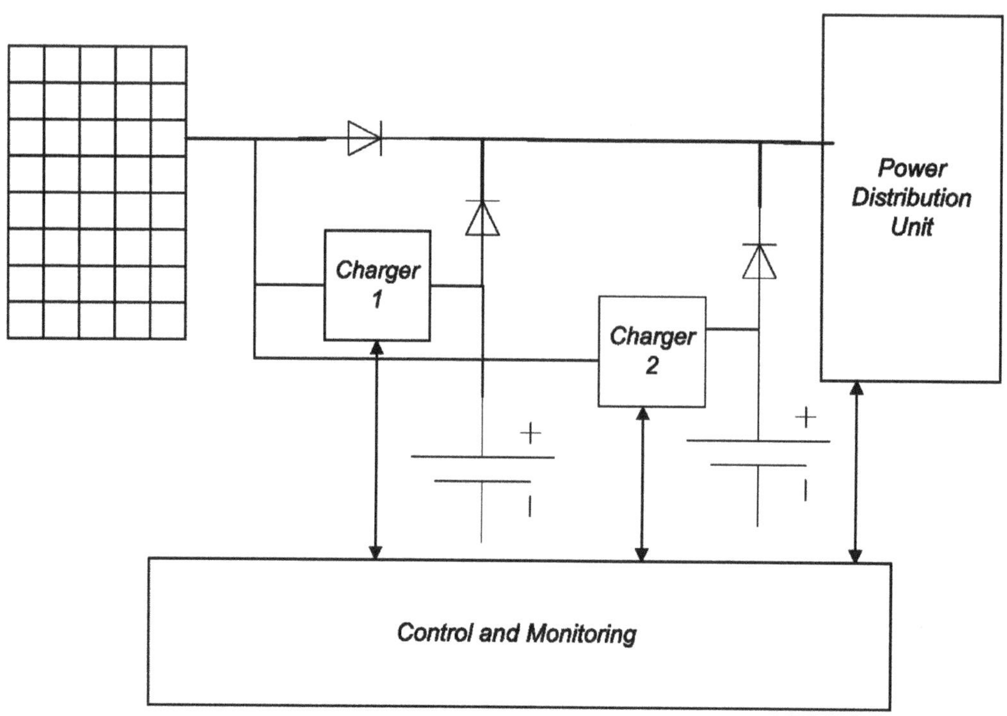

Figure 26. A multi-stage alarm connection in a cooling system

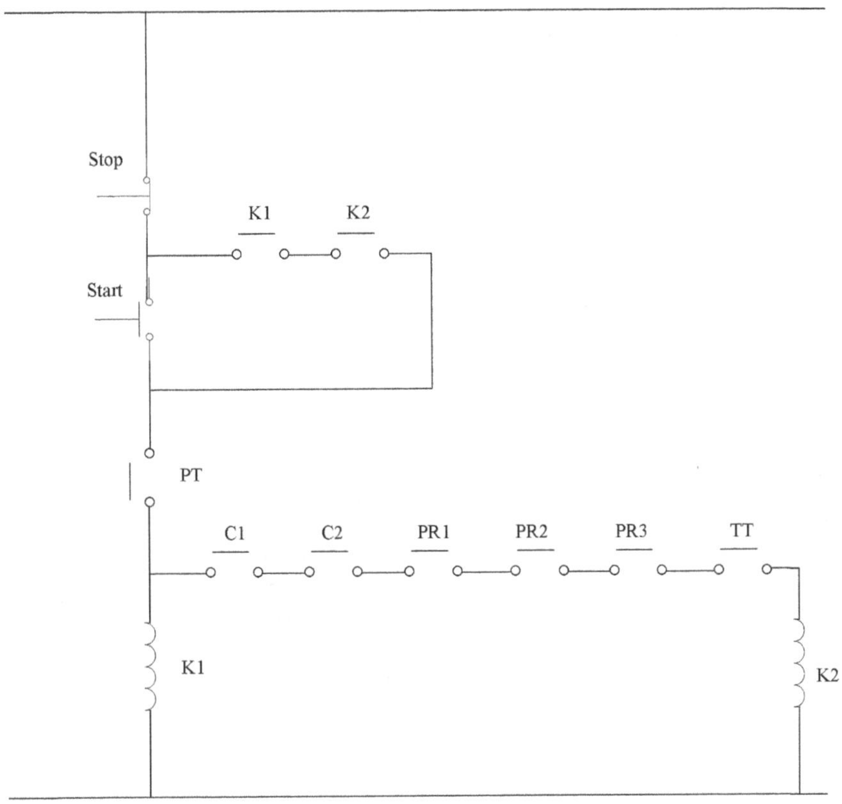

Figure 27. A pre-alarm thermal sensor (right) for an IGBT (left) which is mounted on a heat sink

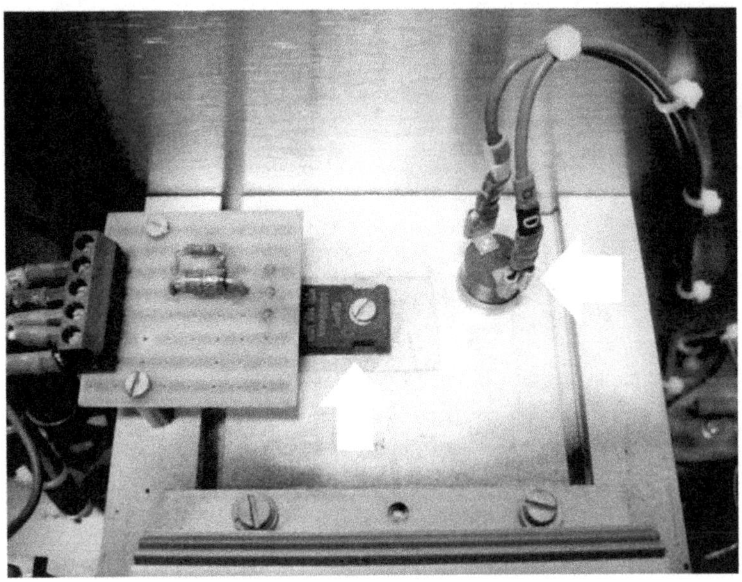

Figure 28. A high voltage switch constructed by series connection of low voltage switches

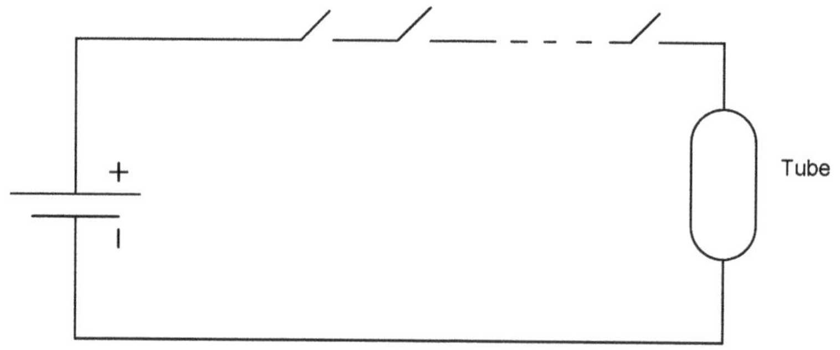

Figure 29. A power supply with power sharing between modules

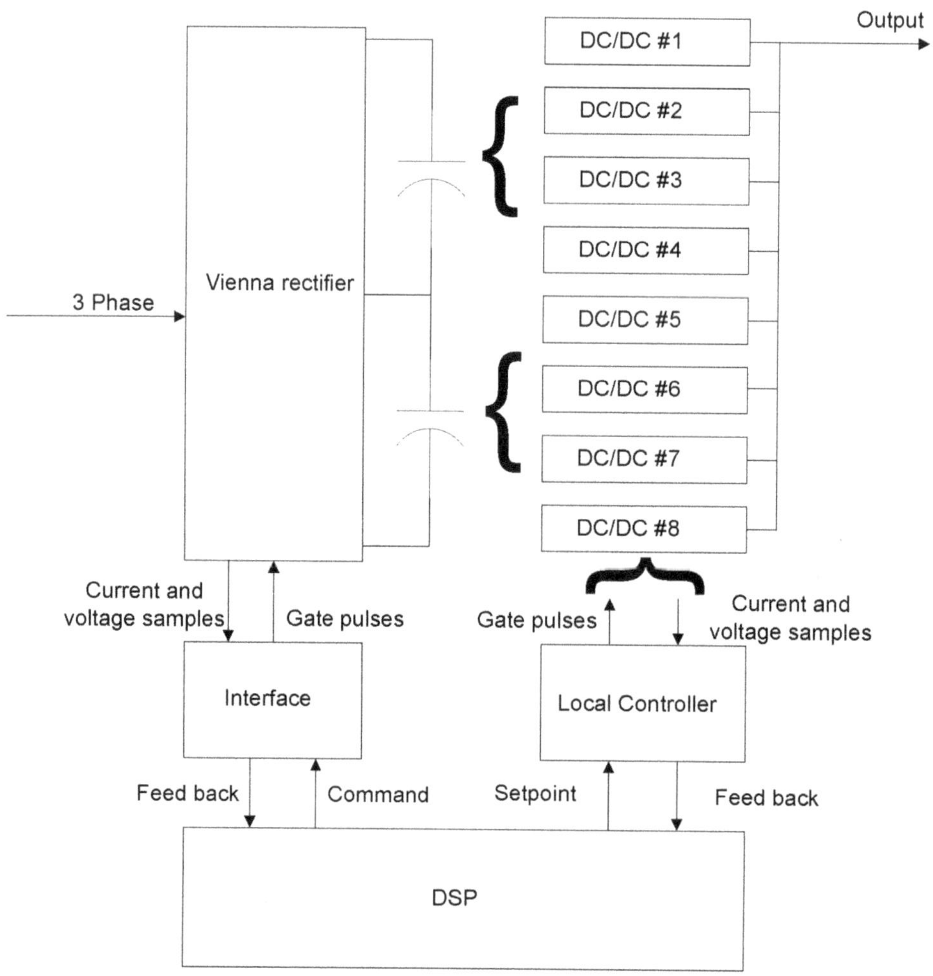

3. Method of redundancy can consist of backup components which automatically "kick in" should one component fail. Redundancy can take as hot and standby forms. There is no delay in hot redundant systems. However, all of redundant parts continue operation under stress in this type. Standby systems need a delay time for stable operation instead of faulty parts but they are free from stress before failure in the main parts.
4. Multi phasing is a commonly used method for increasing reliability in electric machines.

We presented various methods for reliability improvement in chapters 7 to 11. Important part for implementing these methods is a data acquisition system about present condition of the converter. The last chapter of the book is dedicated to condition monitoring in electric power converters.

REFERENCES

Freire, N. M. A., & Cardoso, A. J. M. (2014). A Fault-Tolerant Direct Controlled PMSG Drive for Wind Energy Conversion Systems. *IEEE Transactions on Industrial Electronics*, *61*(2), 821–834. doi:10.1109/TIE.2013.2251734

Gorginpour, H., Jandaghi, B., & Oraee, H. (2013). A novel rotor configuration for brushless doubly-fed induction generators. *IET Electric Power Applications*, *7*(2), 106–115. doi:10.1049/iet-epa.2012.0194

Guo, H., Wang, W., Xing, W., & Li, Y. (2010). Design of Electrical/Mechanical Hybrid 4-Redundancy Brushless DC Torque Motor. *Chinese Journal of Aeronautics*, *23*, 2215.

Hao, H., Covic, G. A., & Boys, J. T. (2014). A Parallel Topology for Inductive Power Transfer Power Supplies. *IEEE Transactions on Power Electronics*, *29*(3), 1140–1151. doi:10.1109/TPEL.2013.2262714

Ribeiro, E., Cardoso, A. J. M., & Boccaletti, C. (2013). Fault-Tolerant Strategy for a Photovoltaic DC-DC Converter. *IEEE Transactions on Power Electronics*, *28*(6), 3008–3018. doi:10.1109/TPEL.2012.2226059

Song, Y., & Wang, B. (2013). Analysis and Experimental Verification of a Fault-Tolerant HEV Powertrain. *IEEE Transactions on Power Electronics*, *28*(12), 5854–5864. doi:10.1109/TPEL.2013.2245513

Tabbache, B., Benbouzid, M., Kheloui, A., Bourgeot, J.-M., & Mamoune, A. (2013). An improved fault-tolerant control scheme for PWM inverter-fed induction motor-based EVs. *ISA Transactions*, *52*(6), 862–869. doi:10.1016/j.isatra.2013.07.004 PMID:23916869

Zhang, W., Xu, D., Enjeti, P. N., Li, H., Hawke, J. T., & Krishnamoorthy, H. S. (2014). Survey on Fault-Tolerant Techniques for Power Electronic Converters. *IEEE Transactions on Power Electronics*, *29*(12), 6319–6331. doi:10.1109/TPEL.2014.2304561

ADDITIONAL READING

Cabanas, M. F., Pedrayes, F., Rojas, C. H., Melero, M. G., Norniella, J. G., Orcajo, G. A., & Fuentes, D. R. et al. (2010). A New Portable, Self-Powered, and Wireless Instrument for the Early Detection of Broken Rotor Bars in Induction Motors. *IEEE Transactions on Industrial Electronics*, *58*(10), 4917–4930. doi:10.1109/TIE.2011.2109340

Choi, S., Akin, B., Rahimian, M. M., & Toliyat, H. A. (2012). Performance-Oriented Electric Motors Diagnostics in Modern Energy Conversion Systems. *IEEE Transactions on Industrial Electronics*, *59*(2), 1266–1277. doi:10.1109/TIE.2011.2158037

Ding, S. X. (2013). *Model-Based Fault Diagnosis Techniques*. Berlin, Germany: Springer. doi:10.1007/978-1-4471-4799-2

Drif, M., & Cardoso, A. J. M. (2008). Airgap-Eccentricity Fault Diagnosis, in Three-Phase Induction Motors, by the Complex Apparent Power Signature Analysis. *IEEE Transactions on Industrial Electronics*, *55*(3), 1404–1410. doi:10.1109/TIE.2007.909076

Dubrova, E. (2013). *Fault-Tolerant Design*. Berlin, Germany: Springer. doi:10.1007/978-1-4614-2113-9

Estima, J. O., & Cardoso, A. J. M. (2013). A New Algorithm for Real-Time Multiple Open-Circuit Fault Diagnosis in Voltage-Fed PWM Motor Drives by the Reference Current Errors. *IEEE Transactions on Industrial Electronics*, *60*(8), 3496–3505. doi:10.1109/TIE.2012.2188877

Ginart, A. E., Kalgren, P. W., Roemer, M. J., Brown, D. W., & Abbas, M. (2010). Transistor Diagnostic Strategies and Extended Operation Under One-Transistor Trigger Suppression in Inverter Power Drives. *IEEE Transactions on Power Electronics*, *25*(2), 499–506. doi:10.1109/TPEL.2009.2026750

Jalilian, G. V. I., Perera, B. S. P., & Cooper, P. (1999). Double Chamber Calorimeter (DCC): A New Approach to Measure Induction Motor Harmonic Losses. *IEEE Transactions on Energy Conversion*, *14*(3), 680–685. doi:10.1109/60.790935

Ma, D., & Bondade, R. (2013). *Reconfigurable Switched-Capacitor Power Converters*. Berlin, Germany: Springer. doi:10.1007/978-1-4614-4187-8

Myers, A. (2010). *Complex System Reliability*. Berlin, Germany: Springer. doi:10.1007/978-1-84996-414-2

Nelson, A. L., & Chow, M.-Y. (2002). Characterization of Coil Faults in an Axial Flux Variable Reluctance PM Motor. *IEEE Transactions on Energy Conversion*, *17*(3), 340–348. doi:10.1109/TEC.2002.801730

Rajagopal, M. S., Seetharamu, K. N., & Ashwathnarayana, P. A.RAJAGOPA. (1998). TRANSIENT THERMAE ANALYSIS OF INDUC'IION MOTORS. *IEEE Transactions on Energy Conversion*, *13*(1), 62–69. doi:10.1109/60.658205

Song, Y., & Wang, B. (2013). Analysis and Experimental Verification of a Fault-Tolerant HEV Powertrain. *IEEE Transactions on Power Electronics*, *28*(12), 5854–5864. doi:10.1109/TPEL.2013.2245513

Xuejun, P., Songsong, N., Yu, C., & Yong, K. (2006). Open-Circuit Fault Diagnosis and Fault-Tolerant Strategies for Full-Bridge DC–DC Converters. *IEEE Transactions on Power Electronics*, *27*(5), 2550–2556.

KEY TERMS AND DEFINITIONS

Electric Parallel Connection: Current division among some parts.

Electric Series Connection: Voltage division among some parts.

Fault Tolerant System: A system with reconfiguring capability.

Reconfiguration: Change in inter connections of system with the aim of disabling effect of fault on functional requirements of the system.

Redundant Systems: Systems with provision of functional capabilities that would be unnecessary in a fault-free environment.

Rotating Magnetic Field: Magnetic field of a multi phase electric machine with a rotating axis.

Chapter 12
Condition Monitoring

ABSTRACT

Implementation of all previous methods for reliability improvement needs to have enough information about condition of the converter. This is the topic of the last chapter of this book. Condition monitoring is the process of monitoring a parameter of condition in machinery (vibration, temperature etc.), in order to identify a significant change which is indicative of a developing fault. The use of conditional monitoring allows maintenance to be scheduled, or other actions to be taken to prevent failure and avoid its consequences. In this chapter, commonly used methods for condition monitoring the converters and electric machines are presented. The aim of this task is producing an alarm in converter before failure factor damage the system. Sensor based and sensor less methods for converter and motor parameter monitoring are described. The data obtained from sensor based methods is real but sensor is a weakness point in a converter. On the other hand, sensorless methods give estimated information but they are reliable. Temperature as the most important parameter from reliability point of view is a common parameter for monitoring in all systems. Other parameters like vibration, harmonics can be used for monitoring of various faults inside the system. Many typical cases are presented to demonstrate the techniques.

INTRODUCTION: PREDICTION OF FAILURE

Implementation of the methods of reliability improvement which were describes in the previous chapters needs an important tool: condition monitoring of power converter. Condition monitoring allows to inform about growth of failure in power converters. Our decision about this failure is related to the data which are obtained from condition monitoring system. For example monitoring of a long term high output current beyond nominal specification of a power electronic converter usually cause to operate the protection system and make it out of service (Kaboli, Zolghadri, Roye, Guiraud, Schanen, 2004). As another alternative, monitoring of a high temperature hotspot in this converter may leads to application of derating scenario and it holds the converter in service. In this chapter, we present the basic approaches for condition monitoring of electric power converters. There are two general goals for monitoring the state of an electric power converter:

DOI: 10.4018/978-1-4666-9429-3.ch012

1. Monitoring for controlling a variable (usually output voltage of power converter)
2. Condition monitoring for preventing a catastrophic failure

Monitoring is a key function during implementation of any control process. All of closed loop control schemes work based on monitoring of output variable of the process. However, converter condition monitoring methods are also used for informing about state of the converter form failure point of view. Monitoring is the common term of all previous chapters in this book. Many reliability improvement techniques need to have a view about the state of electric power converter. In this chapter, we describe the commonly used methods for condition monitoring of power converters. Condition monitoring is the technique of monitoring a parameter in power converter in order to identify a considerable change which is index of a developing fault. Conditional monitoring of power converters have many benefits for the converter. Condition monitoring is important in certain conditions that would shorten normal lifespan can be informed before they lead to a major failure. Condition monitoring allows scheduling the maintenance to prevent failure and avoid its consequences. Figure 1 shows the state of this chapter in the book.

Figure 1. State of chapter 12 in the flowchart of the book

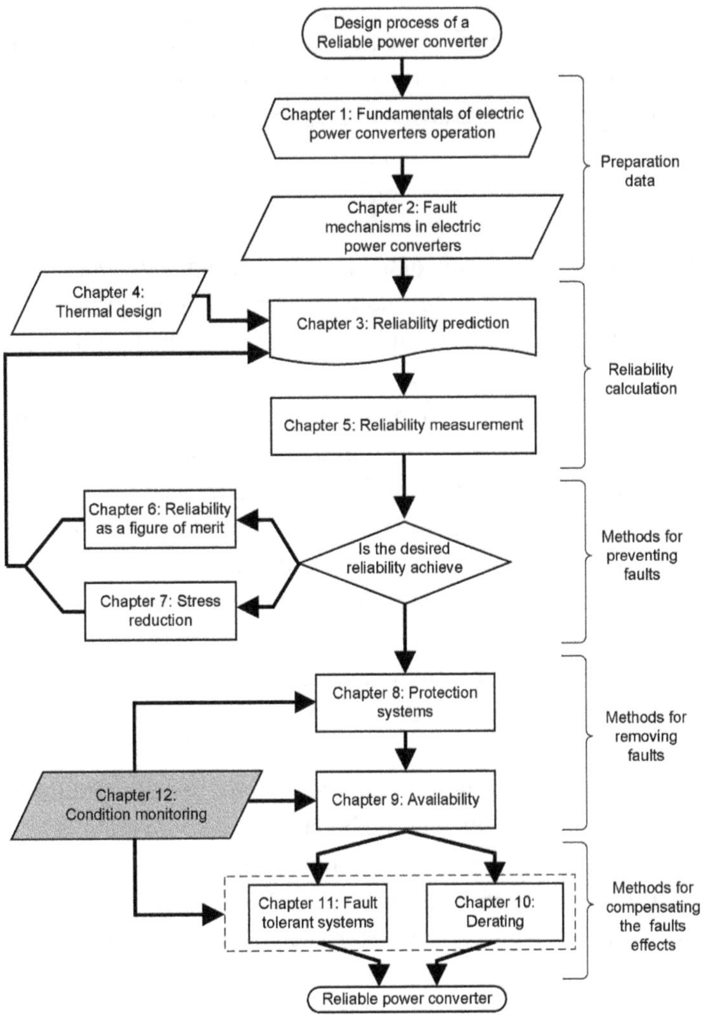

There are two main approaches for condition monitoring 1-5: Sensorless and sensor based methods. In sensor based methods, a physical sensor is used to measure and monitor of the desired parameter or variable. In sensorless method, the job is done based on calculations.

SENSOR BASED METHODS

Condition monitoring techniques are very important for reducing in-service failures and unscheduled down-time and allow the implementation of intelligent maintenance. Therefore, the main goal of all approaches for condition monitoring is monitoring of parameters and variables of a power converter. The mechanism of this monitoring is different in various techniques of monitoring. One technique is using an individual sensor for each parameters which should be monitored. As an example, temperature sensors gives valuable data about temperature profile of a converter. These techniques are reliant on a range of sensors based principally on piezoelectric, electrodynamic, eddy current, inductive, magnetic and thermal technologies.

Advantages

Sensor based method are historically the first method for condition monitoring. They have some features and advantages that causes to still use this method in power converter. Faster access to data derived from online measurements provides for a more efficient response should a remedial action prove necessary.

Disadvantages

Although the benefits and even the necessity of using online sensors are recognized, several factors need to be considered prior to their installation in industrial applications. Typical industrial environments often involve fluids, temperatures, pressures and flows that could be harmful to the delicate electronic sensor components.

It is important to note that as the number of monitored parameters increases, the number of necessary sensors increases. Thus, there is a limitation for the number of sensors. Because the sensor response signals need to be collected and transmitted with minimal loss or interference from the surrounding equipment. The challenge then lies in constructing sensors rugged enough to withstand the rigors of the application and cost-effective enough to allow for their broad deployment throughout industrial enterprises. Therefore, minimum number of sensors for sufficient data for converter monitoring is very important. Some information can help to reduction the sensor number. The main approach of these information is recognizing the points of converter which are under considerable stress or are more sensitive to stress. Some examples of these information are listed as follows:

- Recognizing the most important point of converter from over temperature point of view. It is important to note that it is usual to consider the hotspot of the converter as this point. Although, this is a correct assumption in many cases, but it is not always the most important point. For example, stator winding of a motor in an inverter driven ASD may be the hotspot of ASD. But, the most important point of ASD is junction of inverter switches because they are more sensitive to over temperature than motor winding. To check the sensitivity of a part to temperature, reliability calculation presented in chapter 3 is used.

- Recognizing the most important point of converter from over voltage point of view. Similar to temperature, the most important point of a converter from over voltage point of view is a sensitive point to over voltage not a point with maximum over voltage in the converter. Again, presented charts of reliability calculation in chapter 3 can be used for the analysis. For example, semiconductor switches are more sensitive to over voltage than transformer in a switch mode DC power supply.
- Analyzing the mechanical forces and finding the weakness of the converter.

Application Example: Vibration in Electrical Motor

One of the important factors in operation of electrical machines is vibration and many methods have been developed to measure it. Figure 2 shows the block diagram of a motor setup for measuring the vibration of its structure. An acceleration sensor is mounted on the bearing of the load. The output of this sensor is the vibration response of the shaft. A signal processor is used for analyzing the output data of this sensor. Figure 3 shows the radial mechanical vibration of motor which is measured by an acceleratrometer. This figure shows the normalized harmonic spectrum of vibrations for a typical induction motor. This plot is obtained by applying the Fourier transform to the output data of the acceleration sensor. It can be seen that the vibration harmonics have a high amplitude in a specific frequency band. This is the mechanical resonance phenomena and it can lead to mechanical damage of the motor shaft or shaft accelerated aging.

It can be seen that great mechanical vibration is generated by high frequency harmonics of stator current. Using the predicting method reduces the amplitude of these harmonics and shifts them to higher frequencies as shown in Figure 4. Figure 5 shows the harmonic spectrum of vibration in a DC motor with lower resonance frequency that that found in Figure 3.

The measurement of vibration is a complex subject. Companies have optimized its selection of vibration sensors to obtain the best performance, in a variety of industrial applications, from its range of vibration instrumentation systems

Application Example: Temperature Monitoring

Various machine operating conditions concerning temperature extremes, magnetic fields, vibration range, frequency range, electromagnetic compatibility (EMC) and electrostatic discharge (ESD) conditions and the required signal quality necessitate the need for a variety of sensors. As it was mentioned in the previous chapters, over temperature is a common reason of failure in power systems. Hence, temperature monitoring is an useful tool to prevent thermal damage of converters. Figure 6 shows a temperature sensor which is implemented on the winding of a current transformer to detect over temperature fault. Conductive grease is used to reduce the thermal resistance between sensor and the winding. This figure shows the difficulties associated with implementation of a sensor for condition monitoring. It is important to note that the designer is forced to implement on the outside of the transformer winding. It means there is not easy access to inside the winding and all of decisions should be applied based on a typical temperature from outside of the winding.

Figure 2. A typical diagram for vibration measurement in a drive system

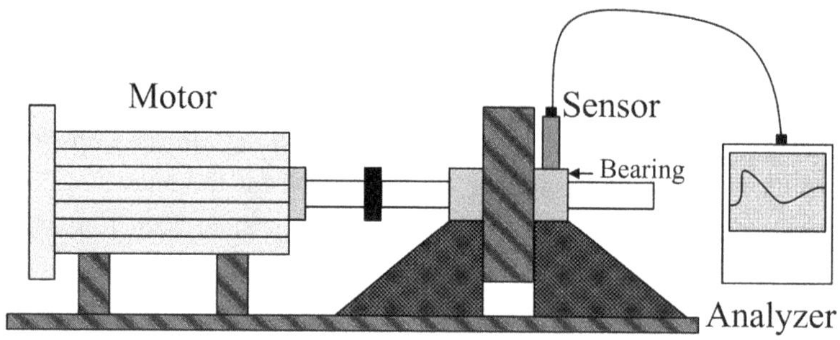

Figure 3. Harmonic spectrum of vibration in an electrical motor

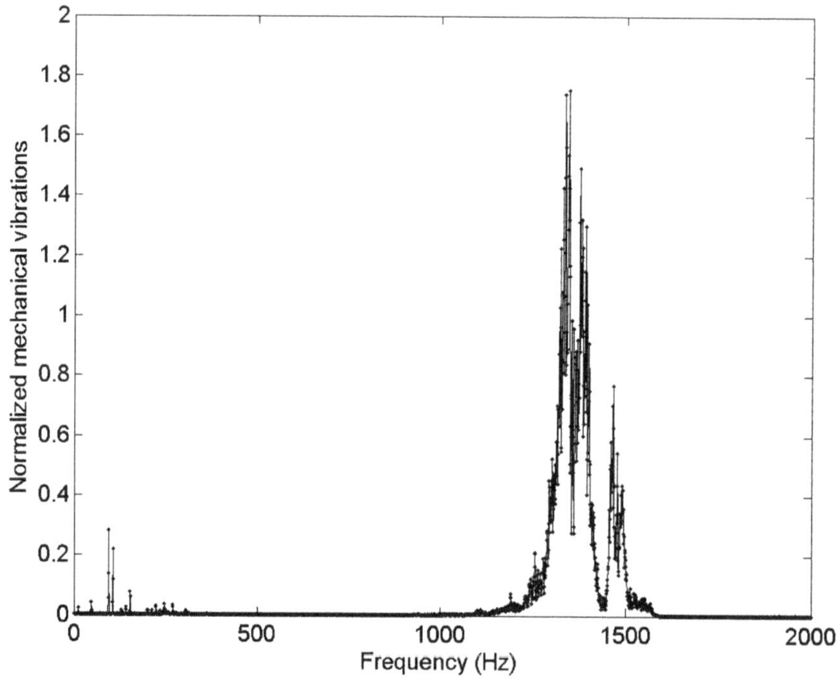

The PT100 sensor is used for measuring the temperature primarily in windings of electric motors, generators and transformers, and has been specifically designed for thermal protection of machines. Figure 7 shows a temperature switch used for temperature protection of a dry-type transformer in traction applications. This sensor is mounted on the transformer core. The critical temperature for converter shutdown is resulted from experimental tests. Figure 8 shows a typical temperature sensors. This temperature sensors acts as a switch. When the contact temperature reaches to the specified temperature of the sensor its built in switch changes from ON state to OFF state or vice versa. The switch types of temperature sensors are reliable but they can not use for temperature monitoring. Another alternative for temperature sensor is analog temperature sensor. AD590 is a typical analog temperature sensor with quasi-linear characteristic. Figure 9 shows position of temperature sensor for monitoring of winding surface temperature which is used in a plant turbogenerator. Figure 10 shows application of temperature sensor for measuring an air flow temperature of a fan.

Figure 4. Improved vibration spectrum by using predictive control algorithm

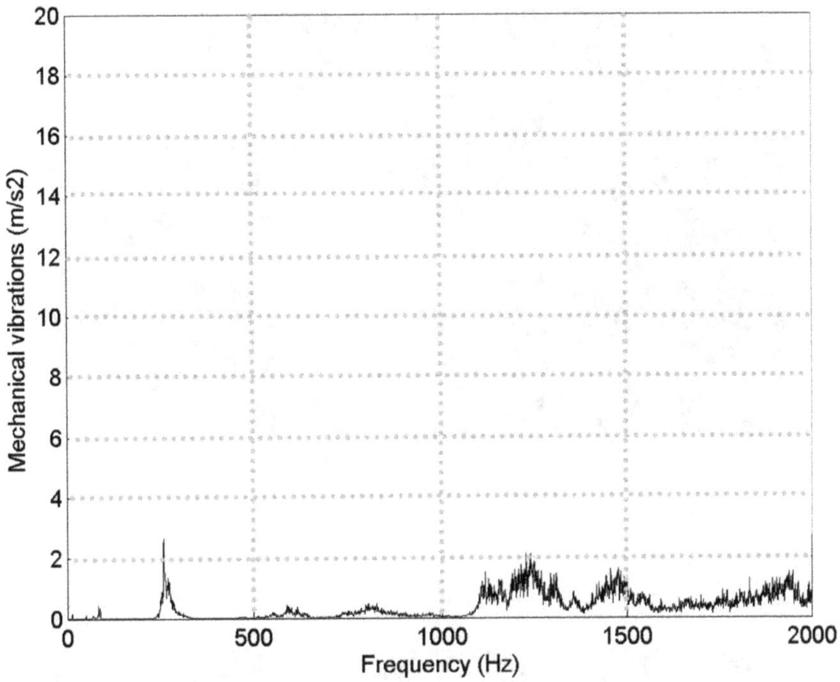

Figure 5. Harmonic spectrum of a DC electric motor

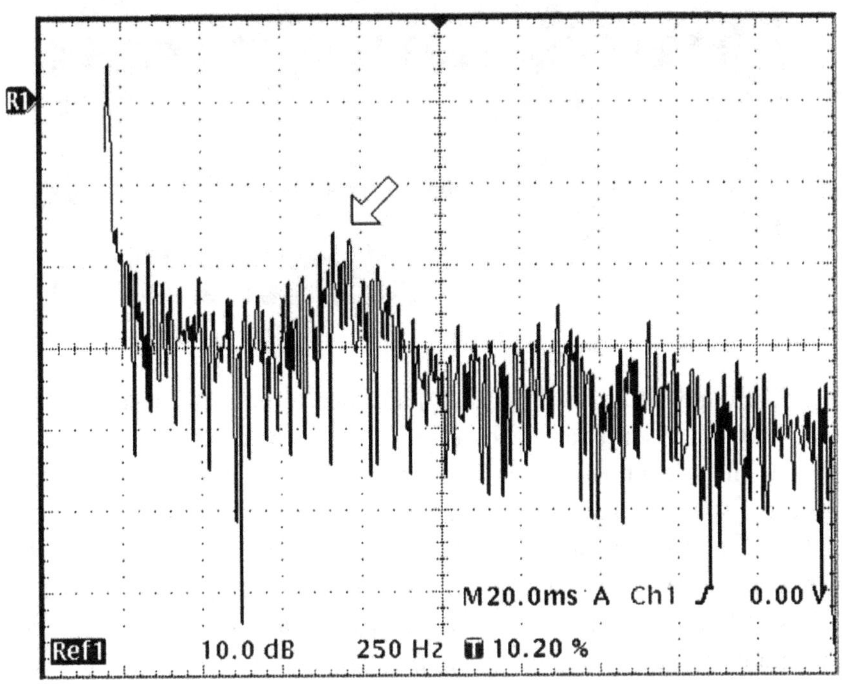

Figure 6. Application of temperature sensor for condition monitoring

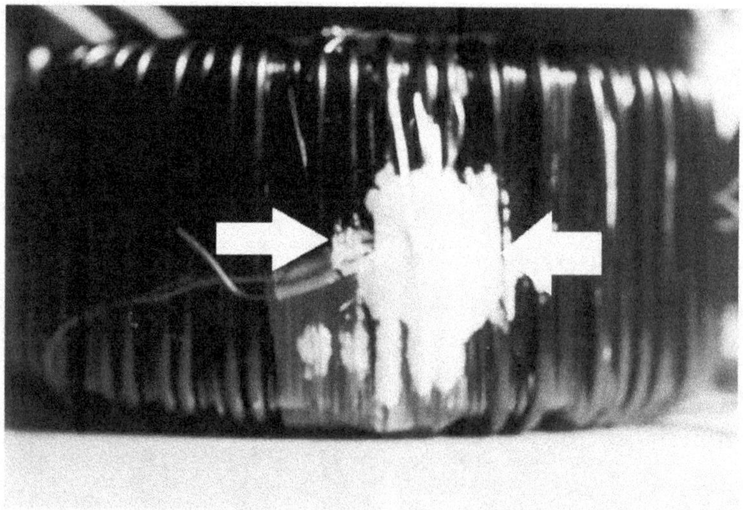

Figure 7. Temperature monitoring of a 3-phase transformer which is mounted on the core of transformer

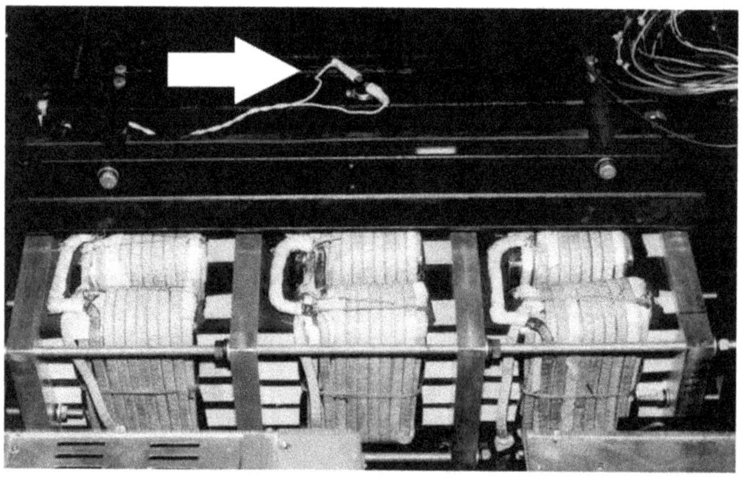

The Stator winding temperature detector (RTD) are used to measure winding temperature of large Motors, Generators etc. These sensor are sandwiched between the windings of Motors/Generators. Unlike on/off devices, It allows continuous measurement of winding temperature. The NEMA recognizes this kind of sensor as a standard protection for motor and generator Insulation. Normally, these RTDs are of wirewound and sensing element extends throughout the length of the detector. These feature will provide average temperature reading. Normally six sensors are recommended for a three phase motor, Two per Phase. This RTD are generally flat type in construction and are available in various sizes. Figure 11 shows a temperature sensor for monitoring winding temperature of an electric motor.

Figure 8. A typical thermal sensor with switching operation

Figure 9. Position of temperature sensor on a stator bar

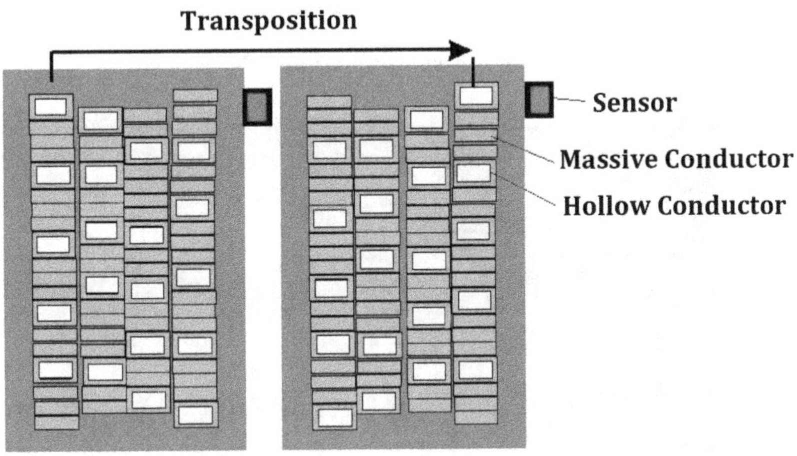

User can insert these thin, laminated stator RTDs in winding slots to detect high temperatures before insulation damage occurs. Stator winding temperature sensors continuously monitor conditions and provide the long term trend data that is necessary for making adjustments before unexpected alarms occur 7.

With the proper sensor to supply the critical operating information, the machine operates in a safer condition for both the machine as well as the personnel operating the machine.

Application Example: Thermography

If the temperature sensor is not located on the proper position, the monitoring system can not recognize the local hotspots in a converter. It is one of the problems in temperature monitoring. Thermography is a solution for this problem. Temperature variations across a surface can be seen with visual inspection and testing with thermography. Heat is indicative of failing components, especially degrading electrical

Figure 10. Air flow temperature measurement

Figure 11. Temperature monitoring of a motor stator winding

contacts and terminations. Thermography can also be successfully applied to high-speed bearings, fluid couplings, conveyor rollers, and storage tank internal build-up.

Application Example: Bearing Condition

It is important to keep in mind that more than 80% of all complex mechanical equipment fail accidentally and without any relation to their life-cycle period. As another example, a degrading rolling-element bearing will usually exhibit increasing vibration signals at specific frequencies as it wears. Special analysis instruments can detect this wear before failure, giving ample warning to schedule replacement before a failure which could cause a much longer down-time.

SENSORLESS SYSTEM IDENTIFICATION

One of the results of chapter 3 is that the reliability of a converter generally decreases with increasing the number of converter components. Sensorless methodology is a solution for eliminating the speed sensor and other physical sensors need for control process. It is historically applied to servomotor systems for controlling goals. However, sensorless methods are used for normal condition monitoring too. Monitoring the state of mechanical machine tool components is of growing importance for increasing machine tool availability and reducing inspection efforts and costs. Open numerical controls offer a new opportunity for getting access to signals of drives as well as of the control itself and to integrate the end-user's individual applications. Because the actual drive signals contain information on the mechanical behaviour of the drive chain components, their use for condition monitoring mechanical drive components is a reliable and cost effective approach.

Application Example: Motor Parameters Identification

Stator resistance PI compensator is one of the simplest methods proposed to estimate stator resistance. This method is based on the real time stator resistance correction regarding the variations of the stator current magnitude, which must be a constant value when the stator flux and motor torque are constant. This reference value can be calculated using the flux and torque commands and motor parameters. In this section, it is shown that for direct torque control systems using digital controller with limited sampling frequency, stator current vector amplitude varies as a function of speed, sampling period, and switching strategy. Effects of sampling frequency, motor speed, and switching strategy on this current error are investigated and their effects on performance of the stator resistance PI compensator are considered 8-13.

Direct torque control (DTC) of induction motor due to its simple structure and ability to achieve fast response of flux and torque has attracted more and more interest in recent years. The basic idea of the method is to calculate the instantaneous values of the flux and torque from the stator variables only. Control is made by an appropriate selection of the inverter switching configurations. Stator resistance is the only motor parameter, which is used for the estimation of the stator flux. The stator resistance changes due to the temperature variations and stator frequency variation that deteriorate the drive performance by introducing errors in the estimated magnitude and position of the stator flux vector. This in turn affects the estimation of the electromagnetic torque and degrades the performance of the DTC system. At low speed, this effect is important and if the value of the stator resistance which is used in controller be less than its actual value, the developed flux and torque will be decreased. Using greater value of the stator resistance in controller than its real value leads to instability.

To overcome this problem, several control schemes have been proposed. One of these methods is using a real time PI estimator for tuning the stator resistance. This method has simple structure and is proper for practical implementation. Steady state current reference is calculated using the motor parameters and torque and flux references. It is shown that even in the steady state there is an error between the current reference and its measured value. This error is a function of the sampling time, motor speed and switching method which is used. In this paper, effects of sampling frequency, speed of motor, and switching strategy on the performance of the PI estimator for tuning the stator resistance are considered.

Variations of the stator resistance due to temperature and frequency cause incorrect estimation of the stator flux and electromagnetic torque. At high speeds, $R_s i_s$, voltage drop in stator resistance, is negligible compared to the stator voltage; but, at low speeds, a stator resistance error causes the estimated stator flux to deviate from its theoretical trajectory that degrades DTC performance. In addition, if stator resistance that is used in controller be more than its real value, the DTC system will be unstable.

In DTC of induction motor, it is shown that if stator flux and motor torque be equal to their reference values, the magnitude of the stator current vector is a constant value that can be calculated with these reference values and motor parameters as

$$\left|\bar{i}_s^*\right| = \sqrt{((i_{sd}^*)^2 + (i_{sq}^*)^2)}$$

$$i_{sq}^* = \frac{2}{3}\frac{2}{P}\frac{T_e^*}{\left|\bar{\psi}_s^*\right|}$$

and i^*_{sd} is calculated from the following equation.

$$L_s(i_{sd}^*)^2 - \left|\bar{\psi}_s^*\right|(1 - \frac{L_s L_r}{L_m^2 - L_s L_r})i_{sd}^* + L_s(i_{sq}^*)^2 - \frac{(\left|\bar{\psi}_s^*\right|)^2 L_r}{L_m^2 - L_s L_r} = 0$$

In above equations, variables that are marked with * show the reference values. L_s, L_r, and L_m are stator self inductance, rotor self inductance, and stator and rotor mutual inductance, respectively. In addition, it can be shown that the magnitude of the stator current vector changes due to the variations of the stator resistance. On the other hand, the variations of the stator resistance change the flux and torque of the motor which leads to variations of the stator current vector magnitude. Therefore, the real time stator resistance estimation for correction of the flux and torque estimation is one of the proposed methods based on the fact that variations of the stator current vector magnitude from its reference value are referred to the stator resistance changes.

To eliminate high frequency component of the stator current error and estimated resistance, low pass filters are used. Output of the PI controller is variations of the stator resistance. This stator resistance variation, ΔR_s, continuously is added to the previous estimated resistance value, R_{so}. Then, it enters to the low pass filter and a limiter to produce estimated stator resistance ($\hat{R}_s$). It is clear that low pass filters time constants must be lower than the stator resistance variations rate. For tuning the estimator, gradient decent method or trial and error method can be used.

The PI resistance estimator performance is investigated using simulation results. To achieve more similarity to the practical systems, a digital controller is used that has adjustable sampling frequency.

As it was mentioned before, when the stator resistance that is using in DTC controller is equal to its real value and motor develops its reference flux and torque, the stator current must be constant and equal to a calculable reference value. However, simulation results show that in systems with digital controllers and finite sampling frequency, the magnitude of the stator current vector is not precisely equal to its reference value and there is an error that changes with motor and controller operation conditions.

To investigate the performance of the PI controller on the DTC system, a simulation was run in which the stator resistance was changed as shown in Figure 12. During 7 seconds, the stator resistance changes from 100% to 200% of its nominal value. Then, after 4 seconds and during 7 seconds, it changes to its nominal value. The estimated value of the resistance is shown in Figure 12 as well. The figure shows that the estimated stator resistance follows closely the actual stator resistance value. The stator flux and electromagnetic torque of the motor are shown in Figures 13 and 14, respectively. Attention to the waveforms of the flux and torque shows that the system has not good performance and there are fluctuations on the flux and torque. The flux is shifted from its main trajectory and has some oscillations that their frequencies are proportional to the speed of the motor.

Figure 12. Real value and estimated value of the stator resistance

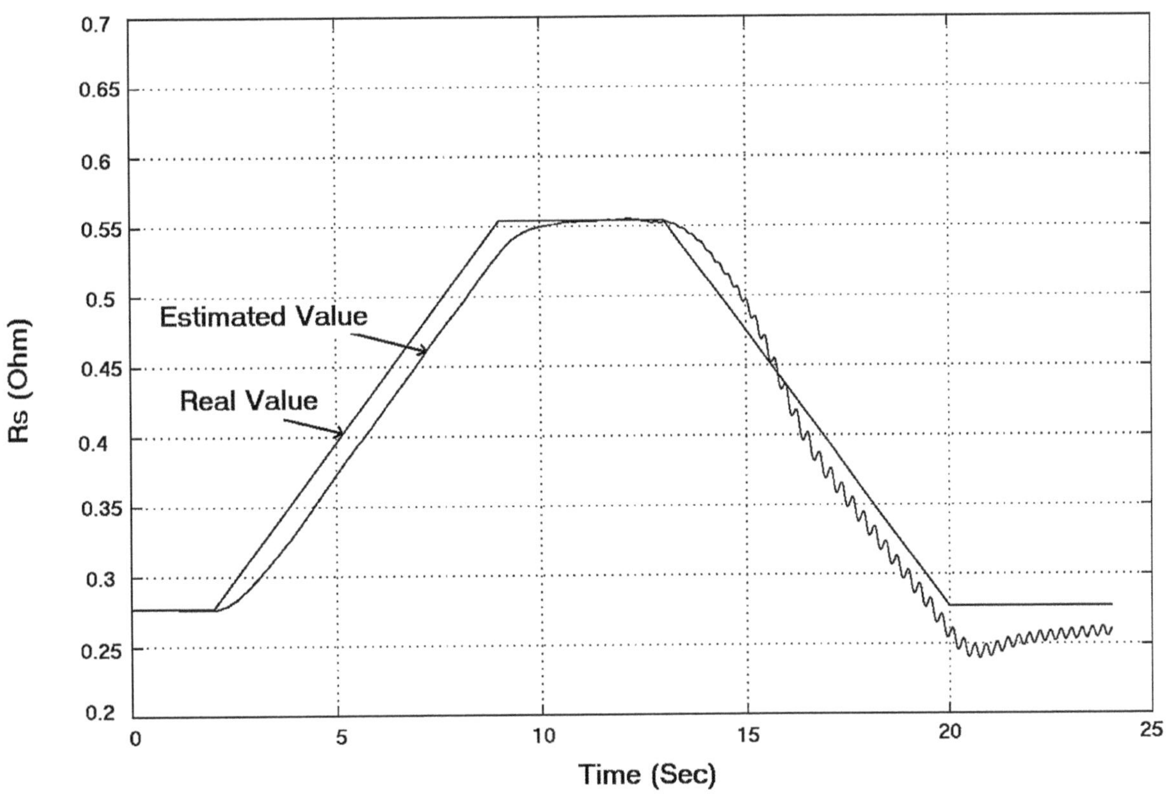

Figure 13. Magnitude of the stator flux vector and its reference value

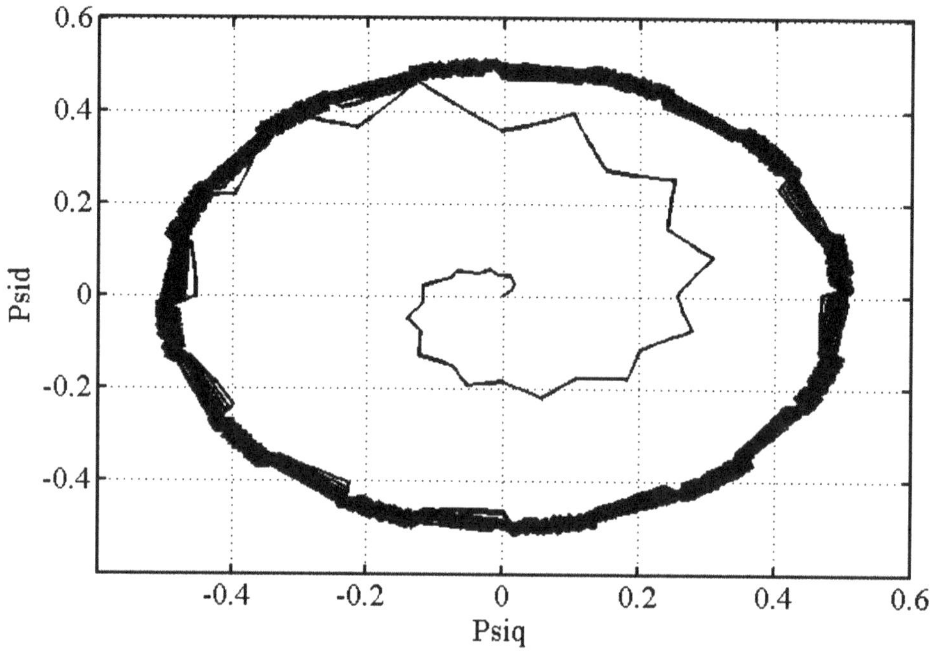

Figure 14. Electromagnetic torque and its reference

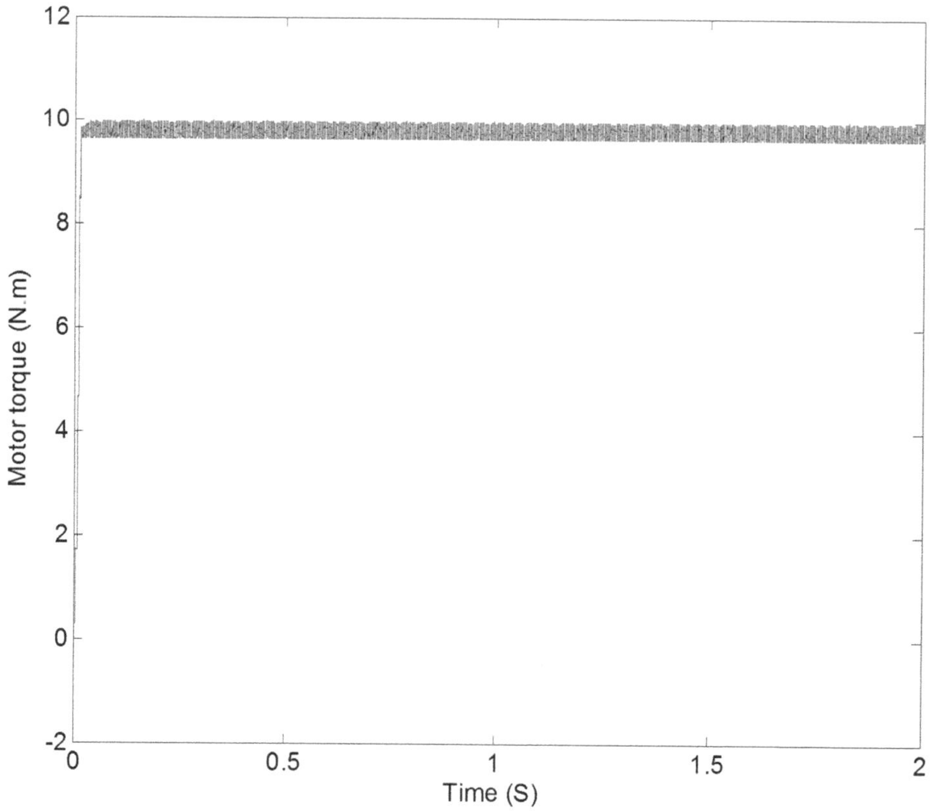

In this simulation, sampling frequency was set to 20 kHz and switching strategy D (with eliminated zero stator voltage vector) was used in the controller. The speed of the motor was 100 rpm and motor developed its rated flux and torque. In this case, K_p was selected 0.1 and K_I was selected 0.005. Cut frequency of the current error low pass filter was set to 2 Hz and for stator resistance low pass filter was selected as 0.24 Hz.

Application Example: Predictive Algorithms

Some predictive controllers have been proposed for prediction of motor torque. The time diagram of conventional DTC algorithm performed by DSP is shown in Figure 15. The stator current is measured at the beginning of sampling period (t_m), the stator flux and torque are calculated using these measured values (t_{cf} and T_c) and the voltage vector respective to these calculations is applied at the beginning of the next sampling time (t_A). It may exist a free time (t_{f1}) especially in high power drives where the switching frequency is low. Therefore, the application of voltage vector at the beginning of each sampling time is done based on stator current measured at the beginning of the previous sampling time. Although the voltage vector is constant during a sampling interval the stator current varies and it causes variations in the stator flux and torque. Thus, estimation of torque at the instance $t+T_c$ is not done based on online variables. Its lag of actual torque is shown in Figure 16. On the other hand, applying a forward (zero) vector causes a great jumping of motor torque at low (high) speed because of fast torque response of the DTC. When the motor torque jumps out of the hysteresis band during a sampling interval, the DSP may not detect it in the same period. Therefore, the DSP processor continues to apply the incorrect vector in the next sampling time and this causes an extra torque ripple 17-21.

Figure 15. Low speed torque response of motor in conventional DTC

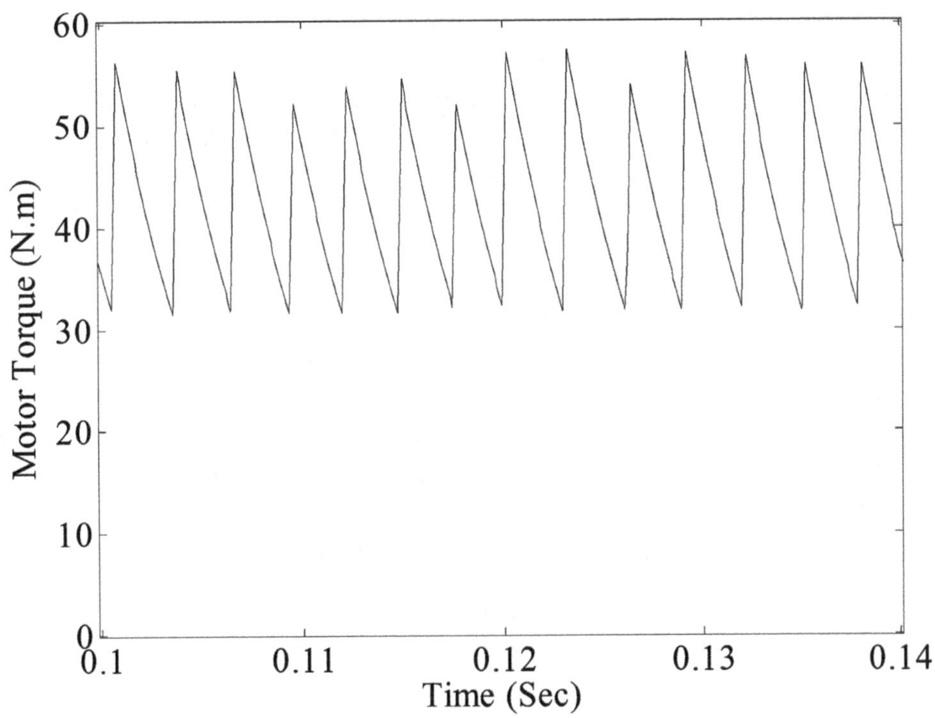

Figure 16. Low speed torque response of motor in modified DTC

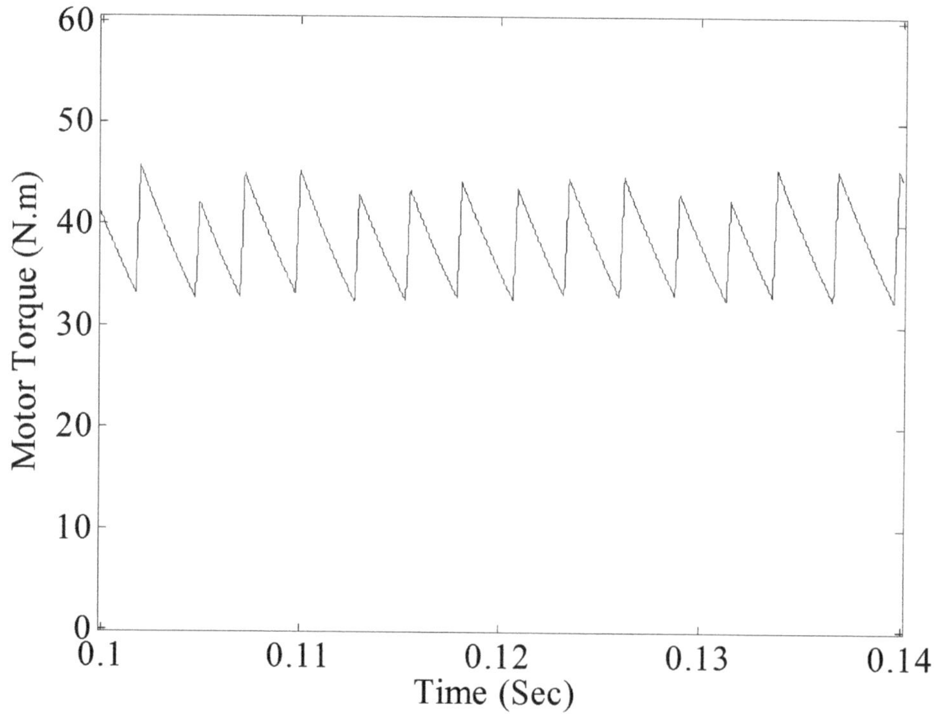

The time diagram of predictive controller is shown in Figure 17. In the beginning of each sampling interval, the stator current is measured two times with a specific predefined value between these two measurements. Assuming a linear variation for current during a sampling interval, the expected value of stator current in the end of period can be extrapolated.

The predicted value of electromagnetic torque is calculated according to the predicted value of stator current and the selection of the voltage vector is performed using the predicted value of torque. In this algorithm, the DC voltage of the inverter is assumed constant during the sampling period.

The sampling time and motor speed are 133 μsec and 300 RPM, respectively. The set point for the motor torque is 10 N.m. The torque ripple factor defined here is used for evaluating the improvement of torque ripple.

T is the fundamental period of torque ripple and τ_{av} is the mean value of motor torque.

Figures 15 and 16 compare the low speed operation of new controller and conventional DTC. Conventional DTC shows inherent time delay of torque response and it makes a great torque ripple with ripple factor equals to 35%. Predictive controller can remove this delay and reduce the ripple factor to 22%. This Improvement is also seen in high speed behavior of motor which are shown in Figures 15 and 16. The new controller reduces the torque ripple factor from 34% (in conventional DTC) to 16% (in modified DTC).

Figures 15 to 16 show the steady state torque response of the machine under conventional DTC at high speed. A great torque ripple is seen in the motor torque. The value of torque ripple factor is 44%. It can be seen that the torque ripple is decreased in the beginning of the sectors. But the improving effect

Figure 17. Schematic diagram of bearing position in an electric motor

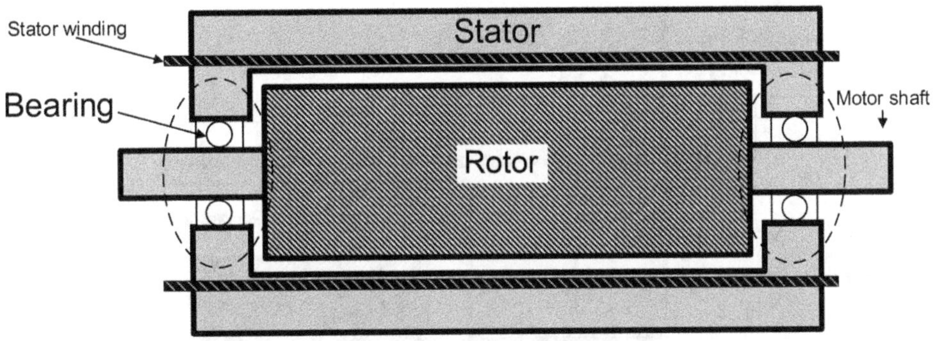

of new switching method is less than its effect at low speed so that the value of torque ripple factor only reduced 5% (Kaboli, Zolghadri, Roye, Guiraud, & Crebier, 2004).

Application Example: Bearing Condition

In modern drives, stator winding is usually controlled by a high frequency PWM controller. This leads to a capacitive parasitic current in stray capacitors between winding and motor core. In addition, bearings are the only mechanical connection between rotor and stator in an electric motor. Therefore, bearings are affected by high frequency current and the resulted losses causes to failure in bearings. Bearing temperature and current help to detect the problem before failure. Figure 17 shows position of bearing in an electric motor. Figure 18 shows the high frequency current path. This current passes through parasitic capacitor between stator winding and rotor.

DATA ACQUISITION SYSTEMS

Data acquisition is the process of sampling signals that measure real world physical conditions and converting the resulting samples into digital numeric values that can be manipulated by a computer. Data acquisition applications are controlled by software programs developed using various general purpose programming languages such as LabVIEW.

Data Loggers

A data logger is an electronic device that records data over time or in relation to location either with a built in instrument or sensor or via external instruments and sensors. Increasingly, but not entirely, they are based on a digital processor (or computer). They generally are small, battery powered, portable, and equipped with a microprocessor, internal memory for data storage, and sensors. Some data loggers interface with a personal computer and utilize software to activate the data logger and view and analyze the collected data, while others have a local interface device (keypad, LCD) and can be used as a stand-alone device.

Figure 18. High frequency harmonic current path through bearing

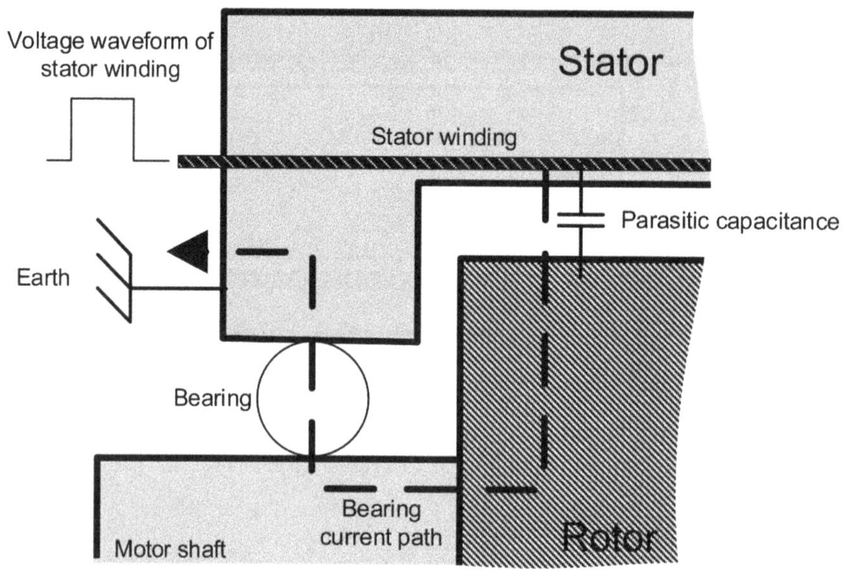

Data loggers vary between general purpose types for a range of measurement applications to very specific devices for measuring in one environment or application type only. It is common for general purpose types to be programmable; however, many remain as static machines with only a limited number or no changeable parameters. Electronic data loggers have replaced chart recorders in many applications.

One of the primary benefits of using data loggers is the ability to automatically collect data on a 24-hour basis. Upon activation, data loggers are typically deployed and left unattended to measure and record information for the duration of the monitoring period. This allows for a comprehensive, accurate picture of the environmental conditions being monitored, such as air temperature and relative humidity

Figure 19. A data logger board, input analog inputs (up), high impedance input buffer (middle), I/O LAN connector (left), isolated digital switch (bottom)

Figure 20. User manual of a data logger software, history record (1), start to next record (2)

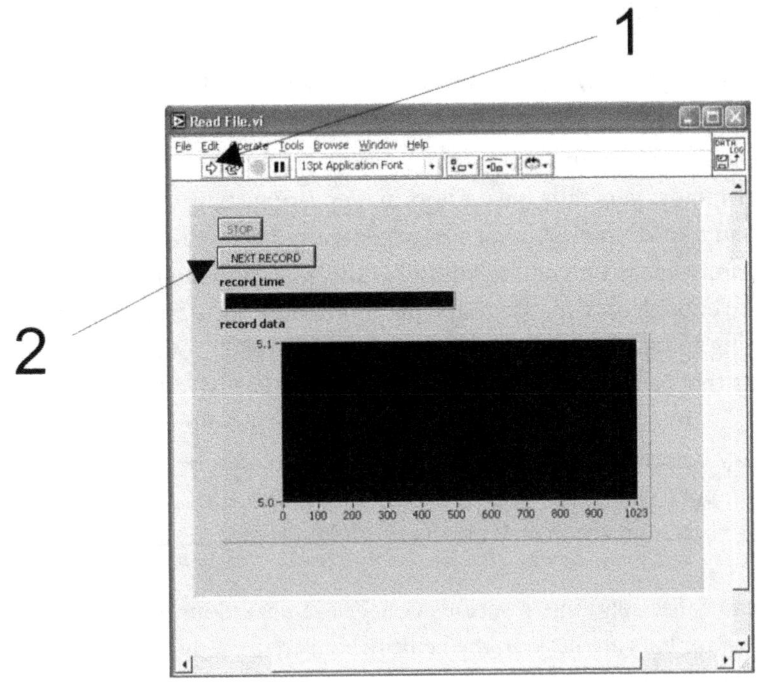

Figure 19 shows a typical data logger board. It contains some analog input channels. This data is obtained from various sensors which are distributed inside the converter. The output resistance of these sensors is usually high and they need to be buffered at the first step. Therefore, there are some input buffer on this board which are implemented by operational amplifiers. These buffered analog inputs are analyzed by a microprocessor. The result of this analysis activates some isolated digital switch or send commend via a LAN connection.

Event Recording

A sequence of events recorder is an intelligent standalone microprocessor based system, which monitors external inputs and records the time and sequence of the changes. Sequence of events recorders usually have an external time source. When wired inputs change state, the time and state of each change is recorded.

Figure 20 shows user page of a data logger software used for event recording in an industrial site. This software is associated with LABview software and a data logger beard as its hardware.

SIGNAL PROCESSING TOOLS

With advancements in digital electronics and reduced component costs in recent years, monitoring instruments for use in condition-based maintenance programs have become more cost-effective and dependable. Machinery does not need to be taken out of service as many tests are done online, and in many cases very little expertise is required for testing and data interpretation. This enables the user to make well-informed decisions for planning maintenance and repairs, which ultimately leads to increased productivity. There are two categories in signal processing: digital and analog signal processing.

Application Example: Consequences of Broken Rotor Bars

Rotor windings in squirrel cage induction motors are manufactured from aluminum alloy, copper, or copper alloy. Larger motors generally have rotors and end-rings fabricated out of these whereas motors with ratings less than a few hundred horsepower generally have die-cast aluminum alloy rotor cages. Broken rotor bars rarely cause immediate failures, especially in large multi-pole (slow-speed) motors. However, if there are enough broken rotor bars, the motor may not start as it may not be able to develop sufficient accelerating torque. Regardless, the presence of broken rotor bars precipitates deterioration in other components that can result in time-consuming and expensive fixes 22-24.

Replacement of the rotor core in larger motors is costly; therefore, by detecting broken rotor bars early, such secondary deterioration can be avoided. The rotor can be repaired at a fraction of the cost of rotor replacement, not to mention averting production revenue losses due to unplanned downtime.

Some of the more common secondary effects of broken rotor bars are:

- Broken bars can cause sparking, a serious concern in hazardous areas.
- If one or more rotor bars are broken, the healthy bars are forced to carry additional current leading to rotor core damage from persistent elevated temperatures in the vicinity of the broken bars and current passing through the core from broken to healthy bars.
- Broken bars cause torque and speed oscillations in the rotor, provoking premature wear of bearings and other driven components.
- Large air pockets in die-cast aluminum alloy rotor windings can cause nonuniform bar expansion leading to rotor bending and imbalance that causes high vibration levels from premature bearing wear.
- As the rotor rotates at high radial speed, broken rotor bars can lift out of the slot due to centrifugal force and strike against the stator winding causing a catastrophic motor failure.
- Rotor asymmetry (the rotor rotating off-center), both static and dynamic, could cause the rotor to rub against the stator winding leading to rotor core damage and even a catastrophic fault.

Motor current signature analysis technology has existed for many years to help diagnose problems in induction motors related to broken rotor bars, air gap eccentricity, drive-train wear analysis, and shaft misalignment. The technology relies on the fact that each of these problems produces recognizable frequency patterns in the motor load current that can be predicted by using empirical formulae and measured. These problems give rise to magnetic asymmetry in the rotor air gap that produces current components at specific frequencies in the load current.

A trace of the motor supply current is obtained by using a clamp-on current probe either from one of the main phase leads to the motor or from the secondary side of a motor CT. A Fast Fourier Transform is performed on the time-domain data to obtain a frequency spectrum. Depending on the device used, this can be done either by the datalogger itself or by computer software.

Once the frequency spectrum is obtained and stored, empirical formulae are used to look for frequency signatures in the spectrum within various frequency ranges depending on the problem to be diagnosed. For example, broken rotor bar frequencies (also called sidebands or pole-passing frequencies) usually can be found within ±5 Hz of the motor supply frequency; for air gap eccentricity a wider range is required for the search, from a few hundred Hz up to a few kHz. If the predicted frequency patterns are present in the spectrum, a positive diagnosis is returned.

In all cases, accurate estimate of the operating slip of the motor is a prerequisite to reliable diagnosis as the predictor equations require operating slip as one of the input parameters. In an induction motor, slip is dependent on the load and increases with increased load. In most cases, the only knowledge a tester would have regarding slip is that at full load; the motor nameplate data contains the rated speed at rated horsepower and the slip can therefore be easily derived when the motor is running at full rated load. However, as motors rarely operate at exactly full load, determining the operating slip becomes a challenge.

Figure 30 shows harmonic spectrum of an induction motor line current in normal operation. Figure 31 shows effect of break in one of the rotor bars on this harmonic spectrum. It can be seen that breaking the motor bars causes to generate high frequency harmonics in the motor current. Harmonic spectrum analyzing the motor current can be an index for detection of fault type. Figures 23 shows variation of line current harmonics with different numbers of motor broken bar.

MEASUREMENT TOOLS

The above mentioned methods need to measurement devices for operating. Regarding to fast and comprehensive application of sensorless methods, the most important measuring devices are voltage and current sensors. In addition, thanks to critical role of over temperature in failure process, temperature measurement are preferred to perform sensorbased. A sensor is a device that converts a physical property into a corresponding electrical signal. An acquisition system to measure different properties depends on the sensors that are suited to detect those properties. Signal conditioning may be necessary if the signal from the transducer is not suitable for hardware. The signal may need to be filtered or amplified in most cases.

Thermal Measurement

There are many techniques for temperature measurement. In case of condition monitoring, small size as well as acceptable accuracy are important factors. Based on this view some of the most famous temperature sensors are listed as follows:

- A thermocouple is a temperature-measuring device consisting of two dissimilar conductors that contact each other at one or more spots, where a temperature differential is experienced by the different conductors (or semiconductors). It produces a voltage when the temperature of one of the spots differs from the reference temperature at other parts of the circuit.
- Thermistors are thermally sensitive resistors whose prime function is to exhibit a large, predictable and precise change in electrical resistance when subjected to a corresponding change in body temperature. Negative Temperature Coefficient (NTC) thermistors exhibit a decrease in electrical resistance when subjected to an increase in body temperature and Positive Temperature Coefficient (PTC) thermistors exhibit an increase in electrical resistance when subjected to an increase in body temperature.

Voltage and Current Measurement

Current

A current transformer (CT) is used for measurement of alternating electric currents. Current transformers, together with voltage transformers (VT) (potential transformers (PT)), are known as instrument transformers. When current in a circuit is too high to apply directly to measuring instruments, a current transformer produces a reduced current accurately proportional to the current in the circuit, which can be conveniently connected to measuring and recording instruments. A current transformer isolates the measuring instruments from what may be very high voltage in the monitored circuit. Current transformers are commonly used in metering and protective relays in the electrical power industry. Figure 22 shows a CT mounted on a high voltage circuit breaker. This CT has a high voltage isolation between high voltage line and the measured current signal.

Figure 23 shows application of high frequency CT for monitoring of high frequency current of an inverter. These sensors are hall effect type of CTs and they can used for a high frequency current. This is an important characteristic of this type of CTs. However, hall effect CTs needs auxiliary power supply to work. Passive CTs usually have a lower frequency bandwidth than hall effects types of CTs. Figure 24 shows a circuit diagram of a AC/DC power supply which is controlled with a Triac in AC side. Figure 25 shows its measured line current obtained from a passive CT in nonsinusoidal waveform. In this case, application of passive CT is possible because of low frequency behavior of the converter line current. Figure 28 shows a Hall-effect current sensor used for measuring DC-link current of an inverter. The DC-link current is a direct current and conventional passive CTs based on transformers can not be used.

Instrument transformers are high accuracy class electrical devices used to isolate or transform voltage or current levels. The most common usage of instrument transformers is to operate instruments or metering from high voltage or high current circuits, safely isolating secondary control circuitry from the high voltages or currents. The primary winding of the transformer is connected to the high voltage or high current circuit, and the meter or relay is connected to the secondary circuit.

Figure 21. Current spectrum for a broken bar induction motor (one bar)

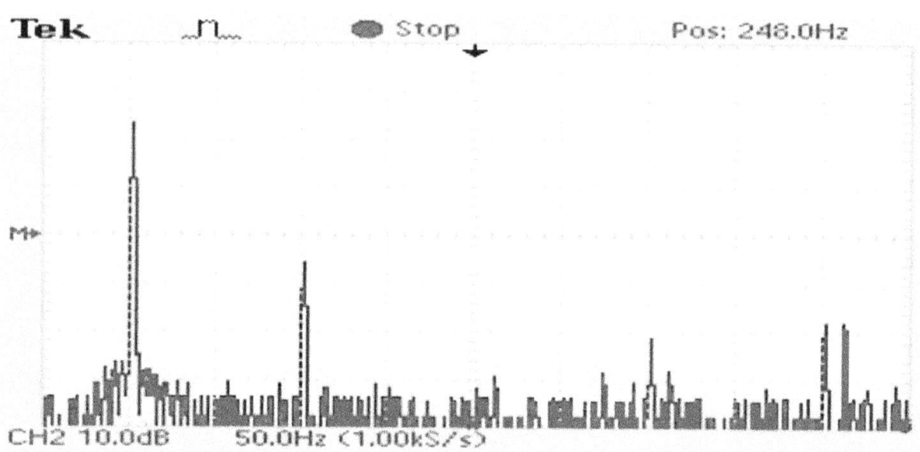

Figure 22. Application of CT for current sensing

Figure 23. Application of high frequency CT for current sensing

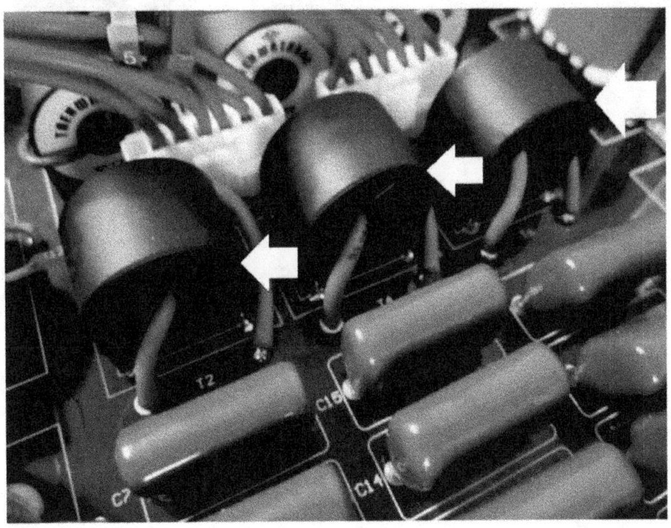

Figure 24. A sample low frequency distortion in current waveforms of a power electronic converters

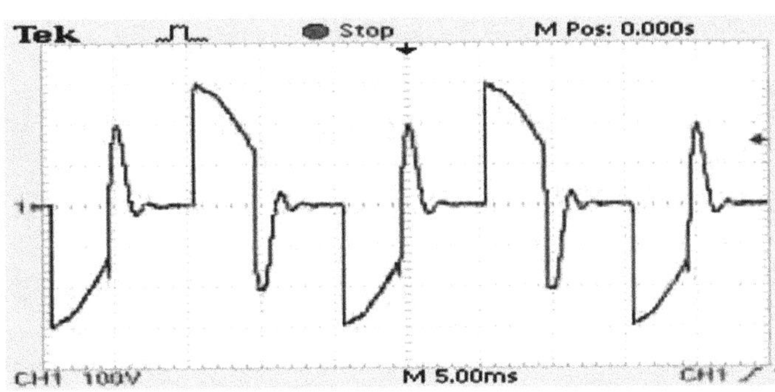

Figure 25. A Hall-Effect current sensor which is used for current measuring in an inverter DC bus

Figure 26. Application of PT for voltage measurement in a high frequency inverter

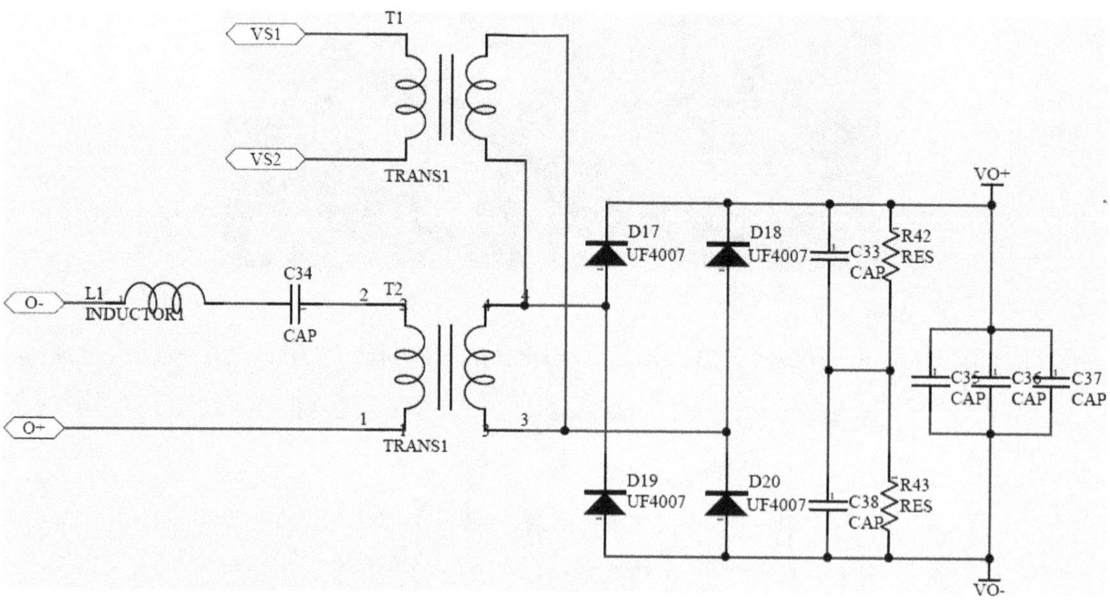

Figure 27. Application of PT for voltage measurement

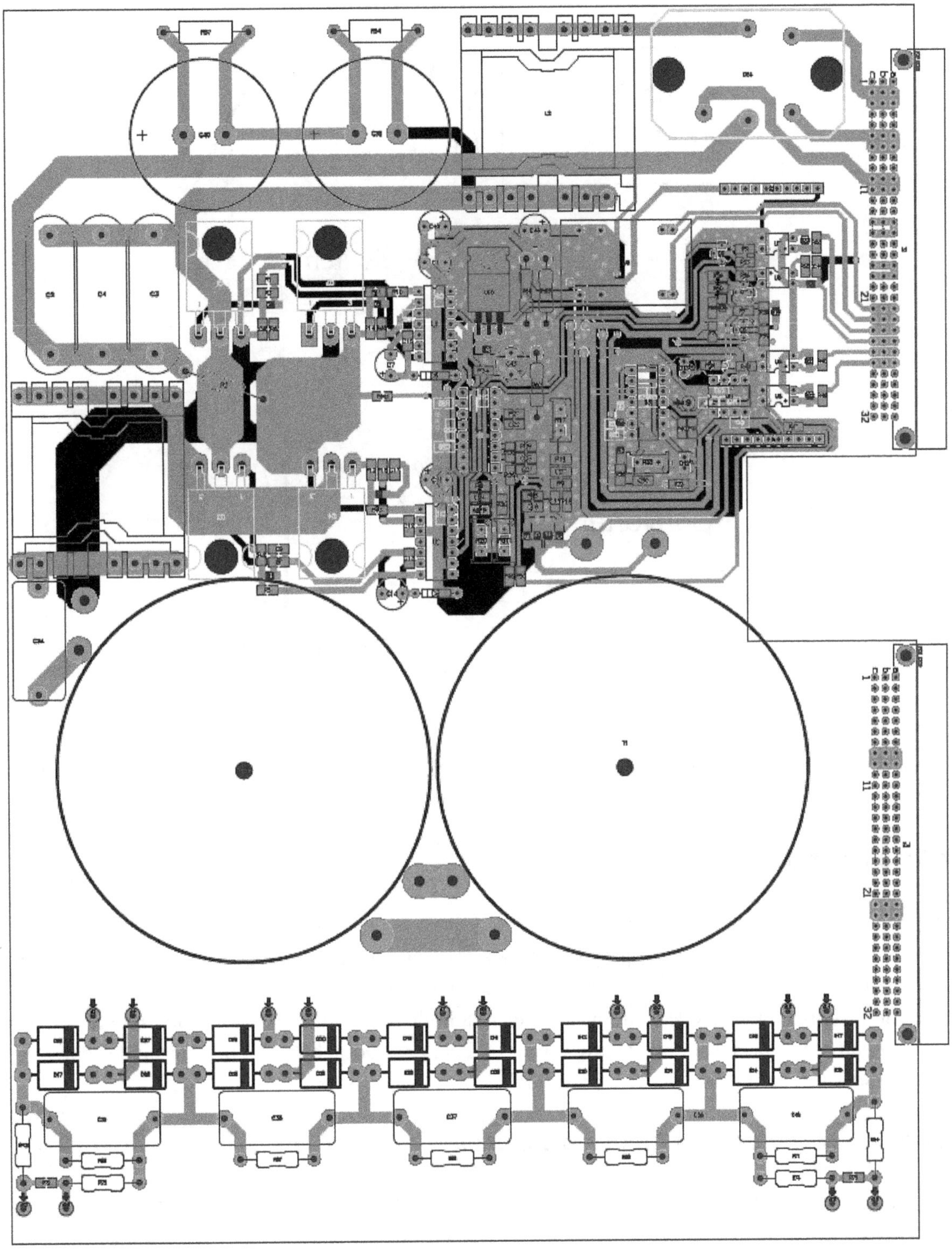

Figure 28. A wind sensor, normal operation (up), operation during a fault in fan (down)

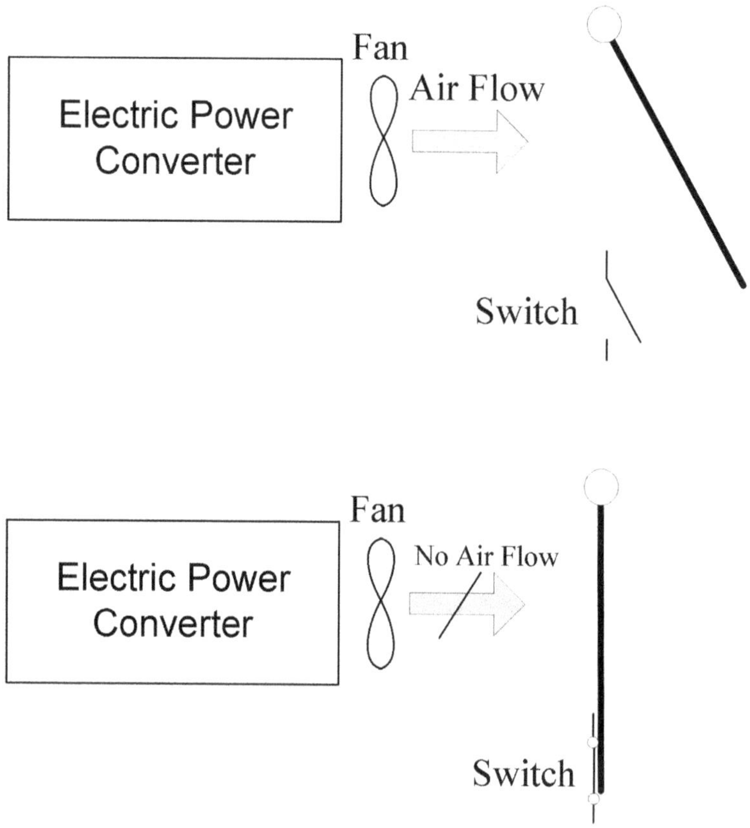

Instrument transformers may also be used as an isolation transformer so that secondary quantities may be used in phase shifting without affecting other primary connected devices

Voltage Measurement

Potential transformers (PT) are a parallel connected type of instrument transformer. They are designed to present negligible load to the supply being measured and have an accurate voltage ratio and phase relationship to enable accurate secondary connected metering. Figure 26 shows circuit diagram of a PT connection to high voltage side of a transformer in a high voltage DC power supply. Figure 27 shows its PCB board with this PT. In this board, there is a 20KV isolation between primary and secondary of the transformer in both main transformer and PT.

Condition Monitoring in Auxiliary Devices

An anemometer is a device used for measuring wind speed, and is a common weather station instrument. Figure 28 shows a simplified diagram of a wind sensor. In normal state, the fan air flow causes to remove the plate from the switch and the switch is in OFF state. If a fault occurs in the fan, the plate causes to change the switch state to ON state.

Figure 29. Application of fiber optic for isolated condition monitoring

Figure 30. Application of high voltage opto isolator for isolated condition monitoring

Figure 31. Water flow switches in a cooling system

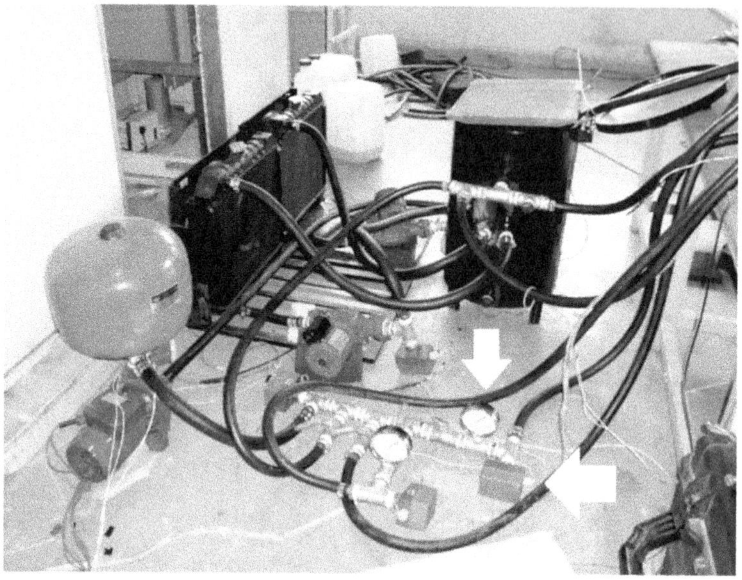

Isolation

Measuring devices must be reliably isolated because of safety and EMC problems. One of the most safe methods is application of optical isolators. An optical fiber is a flexible, transparent fiber made of high quality extruded glass or plastic. It can function as a waveguide to transmit light between the two ends of the fiber. Power over Fiber optic cables can also work to deliver an electric current for low-power electric devices. The field of applied science and engineering concerned with the design and application of optical fibers is known as fiber optics.

Optical fibers are widely used in fiber-optic communications, where they permit transmission over longer distances and at higher bandwidths (data rates) than wire cables. Fibers are used instead of metal wires because signals travel along them with less loss and are also immune to electromagnetic interference. Fibers are also used for illumination, and are wrapped in bundles so that they may be used to carry images, thus allowing viewing in confined spaces. Specially designed fibers are used for a variety of other applications, including sensors and fiber lasers

Optical fiber cables carry the information over light waves which travel in the fibers due to the properties of the fiber materials, similar to the light traveling in free space. The light waves (one form of electromagnetic radiation) are unaffected by other electromagnetic radiation nearby. The optical fiber is electrically non-conductive, so it does not act as an antenna to pick up electromagnetic signals which may be present nearby. So the information traveling inside the optical fiber cables is immune to electromagnetic interference e.g. radio transmitters, power cables adjacent to the fiber cables, or even electromagnetic pulses generated by nuclear devices

Optical fibers are made and drawn from silica glass which is nonconductor of electricity and so there are no ground loops and leakage of any type of current. Optical fibers are thus laid down along with high voltage cables on the electricity poles due to its electrical insulator behavior. Figure 29 shows application of fiber optic link for measuring a high voltage DC voltage. In this converter, there is a float-

Figure 32. A temperature-humidity control used in a cabinet with communication application

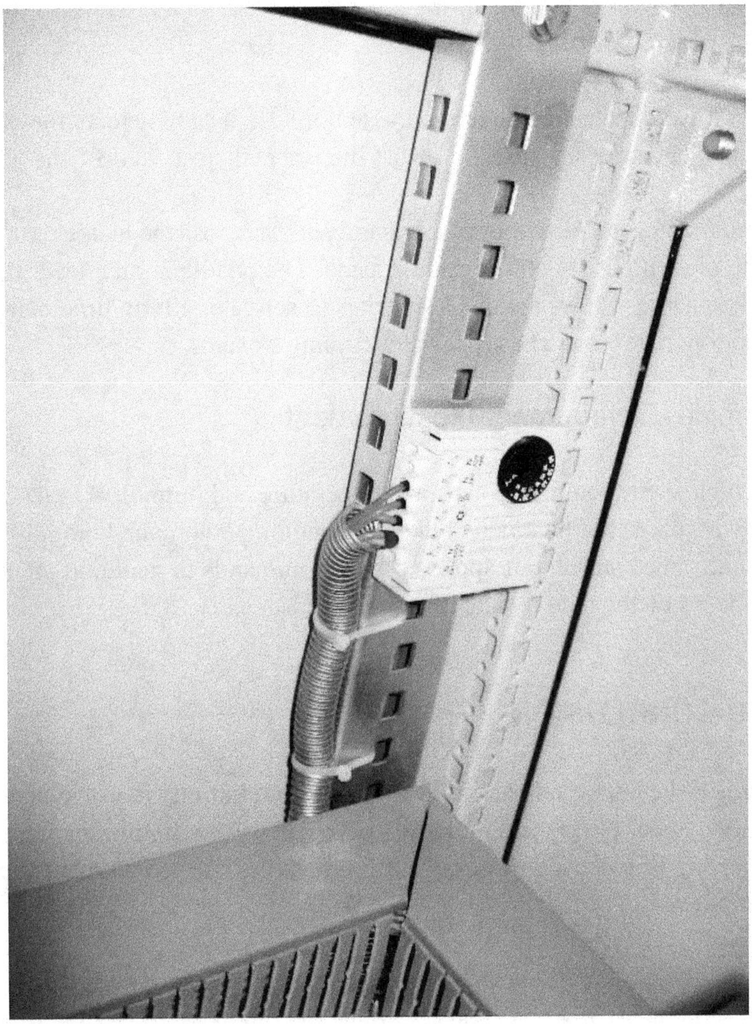

ing DC voltage which is biased to high voltage power supply. To measure this floating DC voltage, it is converted to a digital code by a microprocessor in high voltage side and sent to low voltage side via fiber optic link. It is converted again to an analog signal in the low voltage side by a digital to analog converter. Figure 30 shows an infra-red link to produce an alarm in faulty cases in a high voltage switch. In this figure, any fault or abnormal condition is converted to a digital flag in high voltage side and sent to low voltage side via infrared link.

Measurement of Environmental Parameters

Application Example: Cooling System

The waterflow switch is a mechanical/electrical assembly. As it only detects the water flowing in the pipes, the waterflow switch doesn't turn on or off the water. It just sets off the alarms by activating electrical switches.

To detect the water's movement, the assembly has a paddle across the inside of the pipe. When water starts moving in the pipe, it pushes on the paddle aside. The paddle is on a lever that, when the water has pushed aside the paddle, allows a pair of switches to activate, after a time delay. Figure 31 shows application of some water flow switches in a water cooling system.

Application Example: Environment Measurement

Figure 31 shows a typical environmental sensor (temperature and humidity) in a communication rack. In this cabinet, the environmental parameters like ambient temperature and humidity are measured and sent to the control unit. The control unit applies proper commands to maintain the ambient parameters in a specific range based on the data resulted from this sensor.

SUMMARY AND CONCLUSION

Condition monitoring is the important part of the methods for reliability improvement. The output of this part is real information about the present condition of the converter. In this chapter, various techniques for condition monitoring of electric power converters was described. The main topics of this chapter are summarized as follows:

1. There are two main approaches for condition monitoring: sensor based and sensor less methods.
2. Sensor based methods usually give a precise value of certain parameter but they suffer from some drawbacks such as a need for some space for mounting, sensitivity to noise and physically damaging.
3. Sensorless methods give the desired parameters based on some calculations which start from a minimum number of measured parameters. They have no need to a physical space and can be quite immune to noise. However, they are dependent on the initial measured parameters. Accuracy is always a concern in these methods.
4. Nowadays, sensorless methods are popular in controlling electric machines. They are used not only for condition monitoring but also for controlling a desired parameter like electromagnetic torque.
5. Time response of these methods can be slow for a parameter like temperature and fast for prediction of electromagnetic torque in a motor.

REFERENCES

Haghbin, S., Zolghadri, M. R., Kaboli, S., & Emadi, A. (2003). Performance of PI stator resistance compensator on DTC of induction motor. In *Proceedings of The 29th Annual Conference of the IEEE Industrial Electronics Society* (pp.425-430). IEEE. doi:10.1109/IECON.2003.1280018

Kaboli, S., Zolghadri, M. R., Roye, D., Guiraud, J., & Crebier, J. C. (2004). Design and implementation of a predictive controller for reducing the torque ripple in direct torque control based induction motor drives. In *Proceedings of IEEE 35th Annual Power Electronics Specialists Conference* (pp476-481). IEEE.

Kaboli, S., Zolghadri, M. R., Roye, D., Guiraud, J., & Schanen, J. L. (2004). Design and implementation of a predictive controller for reducing the torque ripple in direct torque control based high frequency induction motor drives. In *Proceedings of IEEE 35th Annual Power Electronics Specialists Conference* (pp.1169-1174). IEEE. doi:10.1109/PESC.2004.1355588

Sheng, H., Wang, F., & Tipton, C. W. (2012). A Fault Detection and Protection Scheme for Three-Level DC–DC Converters Based on Monitoring Flying Capacitor Voltage. *IEEE Transactions on Power Electronics, 27*(2), 685–697. doi:10.1109/TPEL.2011.2161333

Yazdani, A., Sepahvand, H., Crow, M. L., & Ferdowsi, M. (2011). Fault Detection and Mitigation in Multilevel Converter STATCOMs. *IEEE Transactions on Industrial Electronics, 58*(4), 1307–1315. doi:10.1109/TIE.2010.2050415

ADDITIONAL READING

Akar, M. (2013). Detection of a static eccentricity fault in a closed loop driven induction motor by using the angular domain order tracking analysis method. *Mechanical Systems and Signal Processing, 34*(1-2), 173–182. doi:10.1016/j.ymssp.2012.04.003

Alavi, M., Wang, D., & Luo, M. (2014). Short-Circuit Fault Diagnosis for Three-Phase Inverters Based on Voltage-Space Patterns. *IEEE Transactions on Industrial Electronics, 61*(10), 5558–5569. doi:10.1109/TIE.2013.2297298

AlKazzaz, S. A. S., & Singh, G. K. (2003). Experimental investigations on induction machine condition monitoring and fault diagnosis using digital signal processing techniques. *Electric Power Systems Research, 65*(3), 197–221. doi:10.1016/S0378-7796(02)00227-4

Alsofyani, I. M., & Idris, N. R. N. (2013). A review on sensorless techniques for sustainable reliability and efficient variable frequency drives of induction motors. *Renewable & Sustainable Energy Reviews, 24*, 111–121. doi:10.1016/j.rser.2013.03.051

Antonino-Daviu, J. A., Climente-Alarco´, V., Pons-Llinares, J., Puche, R., & Pineda-Sa'nchez, M. (2013). Transient-based analysis for the detection of broken damper bars in synchronous motors. *Mechanical Systems and Signal Processing, 34*(1-2), 367–377. doi:10.1016/j.ymssp.2012.06.007

Ben, S. A. Y., Gu, F., Li, Y., & Ball, A. D. (2006). A validated model for the prediction of rotor bar failure in squirrel-cage motors using instantaneous angular speed. *Mechanical Systems and Signal Processing*, *20*(7), 1572–1589. doi:10.1016/j.ymssp.2005.09.010

Blanke, M., Kinnaert, M., Lunze, J., & Staroswiecki, M. (2006). *Diagnosis and Fault-Tolerant Control*. Berlin, Germany: Springer.

Brombin, M., Boldrin, M., Dalla Palma, M., Ghiraldelli, R., Molon, F., & Pomaro, N. et al. (2012). Optical Transmission of Thermal Measurements from High Voltage Devices in High Vacuum Conditions. *IEEE Transactions on Plasma Science*, *40*(3), 724–729. doi:10.1109/TPS.2011.2181957

Cabanas, M. F., Pedrayes, F., Rojas, C. H., Melero, M. G., Norniella, J. G., & Orcajo, G. A. et al. (2010). A New Portable, Self-Powered, and Wireless Instrument for the Early Detection of Broken Rotor Bars in Induction Motors. *IEEE Transactions on Industrial Electronics*, *58*(10), 4917–4930. doi:10.1109/TIE.2011.2109340

Catelani, M., & Ciani, L. (2012). Experimental tests and reliability assessment of electronic ballast system. *Microelectronics and Reliability*, *52*(9-10), 1833–1836. doi:10.1016/j.microrel.2012.06.077

Choi, S., Akin, B., Rahimian, M. M., & Toliyat, H. A. (2012). Performance-Oriented Electric Motors Diagnostics in Modern Energy Conversion Systems. *IEEE Transactions on Industrial Electronics*, *59*(2), 1266–1277. doi:10.1109/TIE.2011.2158037

Concari, C., Franceschini, G., Tassoni, C., & Toscani, A. (2013, September). Validation of a Faulted Rotor Induction Machine Model with an Insightful Geometrical Interpretation of Physical Quantities. *IEEE Transactions on Industrial Electronics*, *60*(9), 4074–4083. doi:10.1109/TIE.2012.2228142

Da, Y., Shi, X., & Krishnamurthy, M. (2013). A New Approach to Fault Diagnostics for Permanent Magnet Synchronous Machines Using Electromagnetic Signature Analysis. *IEEE Transactions on Power Electronics*, *28*(8), 4104–4112. doi:10.1109/TPEL.2012.2227808

Dn Arun, D., & Chelliah, T. R. (2014). Analysis of field-oriented controlled induction motor drives under sensor faults and an overview of sensorless schemes. *ISA Transactions*, *53*(5), 1680–1694. doi:10.1016/j.isatra.2014.04.008 PMID:24981890

Drif, M., & Cardoso, A. J. M. (2008). Airgap-Eccentricity Fault Diagnosis, in Three-Phase Induction Motors, by the Complex Apparent Power Signature Analysis. *IEEE Transactions on Industrial Electronics*, *55*(3), 1404–1410. doi:10.1109/TIE.2007.909076

Drif, M., & Cardoso, A. J. M. (2009). The Use of the Instantaneous-Reactive-Power Signature Analysis for Rotor-Cage-Fault Diagnostics in Three-Phase Induction Motors. *IEEE Transactions on Industrial Electronics*, *56*(11), 4606–4614. doi:10.1109/TIE.2009.2027922

Duque-Perez, O., Garcia-Escudero, L.-A., Morinigo-Sotelo, D., Gardel, P.-E., & Perez-Alonso, M. (2015). Analysis of fault signatures for the diagnosis of induction motors fed by voltage source inverters using ANOVA and additive models. *Electric Power Systems Research*, *121*, 1–13. doi:10.1016/j.epsr.2014.11.021

Duygu, B., & Serhat, S. (2013). Wavelet based Neuro-Detector for low frequencies of vibration signals in electric motors. *Applied Soft Computing*, *13*(5), 2683–2691. doi:10.1016/j.asoc.2012.11.019

Duygu, B., & Serhat, S. (2013). Wavelet based Neuro-Detector for low frequencies of vibration signals in electric motors. *Applied Soft Computing*, *13*(5), 2683–2691. doi:10.1016/j.asoc.2012.11.019

Estima, J. O., & Cardoso, A. J. M. (2013). A New Algorithm for Real-Time Multiple Open-Circuit Fault Diagnosis in Voltage-Fed PWM Motor Drives by the Reference Current Errors. *IEEE Transactions on Industrial Electronics*, *60*(8), 3496–3505. doi:10.1109/TIE.2012.2188877

Foucher, B., Boulli, J., Meslet, B., & Das, D. (2002). A review of reliability prediction methods for electronic devices. *Microelectronics and Reliability*, *42*(8), 1155–1162. doi:10.1016/S0026-2714(02)00087-2

Freire, N. M. A., Estima, J. O., & Cardoso, A. J. M. (2014). A Voltage-Based Approach Without Extra Hardware for Open-Circuit Fault Diagnosis in Closed-Loop PWM AC Regenerative Drives. *IEEE Transactions on Industrial Electronics*, *61*(9), 4960–4970. doi:10.1109/TIE.2013.2279383

Fulzele, A. G., Arajpure, V. G., Holay, P. P., & Patil, N. M. (2012). Condition monitoring of shaft of single-phase induction motor using optical sensor. *Mechanical Systems and Signal Processing*, *29*, 428–435. doi:10.1016/j.ymssp.2011.12.002

García-Escudero, L. A., Duque-Perez, O., Morinigo-Sotelo, D., & Perez-Alonso, M. (2011). Robust condition monitoring for early detection of broken rotor bars in induction motors. *Expert Systems with Applications*, *38*(3), 2653–2660. doi:10.1016/j.eswa.2010.08.055

Georgakopoulos, I. P., Mitronikas, E. D., & Safacas, A. N. (2011). Detection of Induction Motor Faults in Inverter Drives Using Inverter Input Current Analysis. *IEEE Transactions on Industrial Electronics*, *58*(9), 4356–4373. doi:10.1109/TIE.2010.2093476

Ghorbanian, V., & Faiz, J. (2015). A survey on time and frequency characteristics of induction motors with broken rotor bars in line-start and inverter-fed modes. *Mechanical Systems and Signal Processing*, *54-55*, 427–456. doi:10.1016/j.ymssp.2014.08.022

Ginart, A. E., Kalgren, P. W., Roemer, M. J., Brown, D. W., & Abbas, M. (2010). Transistor Diagnostic Strategies and Extended Operation Under One-Transistor Trigger Suppression in Inverter Power Drives. *IEEE Transactions on Power Electronics*, *25*(2), 499–506.

Guerrero, V., Pontt, J., Dixon, J., & Rebolledo, J. (2013). A Novel Noninvasive Failure-Detection System for High-Power Converters Based on SCRs. *IEEE Transactions on Industrial Electronics*, *60*(2), 450–458. doi:10.1109/TIE.2012.2188251

Guzinski, J., Diguet, M., Krzeminski, Z., Lewicki, A., & Abu-Rub, H. (2009). Application of Speed and Load Torque Observers in High-Speed Train Drive for Diagnostic Purposes. *IEEE Transactions on Industrial Electronics*, *56*(1), 248–256. doi:10.1109/TIE.2008.928103

Hutzler, A., Tokarski, A., & Schletz, A. (2013). Extending the lifetime of power electronic assemblies by increased cooling temperatures. *Microelectronics and Reliability*, *53*(9-11), 1774–1777. doi:10.1016/j.microrel.2013.07.029

Isermann, R. (2006). *Fault-Diagnosis Systems*. Berlin, Germany: Springer.

James, R. E., & Su, Q. (2008). Condition Assessment of High Voltage Insulation in Power System Equipment. London, UK: The Institution of Engineering and Technology. doi:10.1049/PBPO053E

Kanekawa, N., Ibe, E. H., Suga, T., & Uematsu, Y. (2011). *Dependability in Electronic Systems*. Berlin, Germany: Springer. doi:10.1007/978-1-4419-6715-2

Manana, M., Arroyo, A., Ortiz, A., Renedo, C. J., Perez, S., & Delgado, F. (2011). Field winding fault diagnosis in DC motors during manufacturing using thermal monitoring. *Applied Thermal Engineering*, *31*(5), 978–983. doi:10.1016/j.applthermaleng.2010.11.023

Mattila, T. T., & Paulasto-Kröckel, M. (2011). Toward comprehensive reliability assessment of electronics by a combined loading approach. *Microelectronics and Reliability*, *51*(6), 1077–1091. doi:10.1016/j.microrel.2011.03.022

Mohagheghi, S., Harley, R. G., Habetler, T. G., & Divan, D. (2009). Condition Monitoring of Power Electronic Circuits Using Artificial Neural Networks. *IEEE Transactions on Power Electronics*, *24*(10), 2363–2367. doi:10.1109/TPEL.2009.2017806

Nelson, A. L., & Chow, M.-Y. (2002). Characterization of Coil Faults in an Axial Flux Variable Reluctance PM Motor. *IEEE Transactions on Energy Conversion*, *17*(3), 340–348. doi:10.1109/TEC.2002.801730

Picazo-Ródenas, M. J., Royo, R., Antonino-Daviu, J., & Roger-Folch, J. (2013). Use of the infrared data for heating curve computation in induction motors: Application to fault diagnosis. *Engineering Failure Analysis*, *35*, 178–192. doi:10.1016/j.engfailanal.2013.01.018

Qi, H., Ganesan, S., & Pecht, M. (2008). No-fault-found and intermittent failures in electronic products. *Microelectronics and Reliability*, *48*(5), 663–674. doi:10.1016/j.microrel.2008.02.003

Qun-Tao, A., Li-Zhi, S., Ke, Z., & Li, S. (2011). Switching Function Model-Based Fast-Diagnostic Method of Open-Switch Faults in Inverters without Sensors. *IEEE Transactions on Power Electronics*, *26*(1), 119–126. doi:10.1109/TPEL.2010.2052472

Rezazadeh Mehrjou, M., Mariun, N., Hamiruce Marhaban, M., & Misron, N. (2011). Rotor fault condition monitoring techniques for squirrel-cage induction machine- A review. *Mechanical Systems and Signal Processing*, *25*(8), 2827–2848. doi:10.1016/j.ymssp.2011.05.007

Ribeiro, E., Cardoso, A. J. M., & Boccaletti, C. (2013). Fault-Tolerant Strategy for a Photovoltaic DC-DC Converter. *IEEE Transactions on Power Electronics*, *28*(6), 3008–3018. doi:10.1109/TPEL.2012.2226059

Roesch, W. J. (2006). Historical review of compound semiconductor reliability. *Microelectronics and Reliability*, *46*(8), 1218–1227. doi:10.1016/j.microrel.2006.02.008

Salah, M., Bacha, K., & Chaari, A. (2014). An improved spectral analysis of the stray flux component for the detection of air-gap irregularities in squirrel cage motors. *ISA Transactions*, *53*(3), 816–826. doi:10.1016/j.isatra.2014.02.001 PMID:24582300

Seera, M., Lim, C. P., Nahavandi, S., & Loo, C. K. (2014). Condition monitoring of induction motors: A review and an application of an ensemble of hybrid intelligent models. *Expert Systems with Applications*, *41*(10), 4891–4903. doi:10.1016/j.eswa.2014.02.028

Serhat, S., & Emine, A. (2003). Feature extraction related to bearing damage in electric motors bywavelet analysis. *Journal of the Franklin Institute*, *340*(2), 125–134. doi:10.1016/S0016-0032(03)00015-2

Seshadrinath, J., Singh, B., & Panigrahi, B. K. (2014). Investigation of Vibration Signatures for Multiple Fault Diagnosis in Variable Frequency Drives Using Complex Wavelets. *IEEE Transactions on Power Electronics, 29*(2), 936–945. doi:10.1109/TPEL.2013.2257869

Song, J., Krishnamurthy, V., Kwasinski, A., & Sharma, R. (2013). Development of a Markov-Chain-Based Energy Storage Model for Power Supply Availability Assessment of Photovoltaic Generation Plants. *IEEE Transactions on Sustainable Energy, 4*(2), 491–500. doi:10.1109/TSTE.2012.2207135

Song, Y., & Wang, B. (2013). Survey on Reliability of Power Electronic Systems. *IEEE Transactions on Power Electronics, 28*(1), 591–604. doi:10.1109/TPEL.2012.2192503

Songsong, N., Xuejun, P., Yu, C., & Yong, K. (2014). Fault Diagnosis of PWM DC–DC Converters Based on Magnetic Component Voltages Equation. *IEEE Transactions on Power Electronics, 29*(9), 4978–4988. doi:10.1109/TPEL.2013.2283881

Stroud, C. E. (2002). *A Designer's Guide to Built-In Self-Test*. NJ, USA: KLUWER ACADEMIC PUBLISHERS.

Tandon, N., Ramakrishna, K. M., & Yadava, G. S. (2007). Condition monitoring of electric motor ball bearings for the detection of grease contaminants. *Tribology International, 40*(1), 29–36. doi:10.1016/j.triboint.2006.01.024

Tandon, N., Yadava, G. S., & Ramakrishna, K. M. (2007). A comparison of some condition monitoring techniques for the detection of defect in induction motor ball bearings. *Mechanical Systems and Signal Processing, 21*(1), 244–256. doi:10.1016/j.ymssp.2005.08.005

Tang, W. H., & Wu, Q. H. (2011). *Condition Monitoring and Assessment of Power Transformers Using Computational Intelligence*. Berlin, Germany: Springer. doi:10.1007/978-0-85729-052-6

Tang, W. H., & Wu, Q. H. (2011). *Condition Monitoring and Assessment of Power Transformers Using Computational Intelligence*. Berlin, Germany: Springer. doi:10.1007/978-0-85729-052-6

Tavner, P., & Ran, L. (2008). Condition Monitoring of Rotating Electrical Machines. London, UK: The Institution of Engineering and Technology. doi:10.1049/PBPO056E

Tavner, P., Ran, L., Penman, J., & Sedding, H. (2008). Condition Monitoring of Rotating Electrical Machines. London, UK: The Institution of Engineering and Technology. doi:10.1049/PBPO056E

Torkaman, H., & Afjei, E. (2013). Comprehensive Detection of Eccentricity Fault in Switched Reluctance Machines Using High-Frequency Pulse Injection. *IEEE Transactions on Power Electronics, 28*(3), 1382–1390. doi:10.1109/TPEL.2012.2205947

Torkaman, H., & Afjei, E. (2013). Comprehensive Detection of Eccentricity Fault in Switched Reluctance Machines Using High-Frequency Pulse Injection. *IEEE Transactions on Power Electronics, 28*(3), 1382–1390. doi:10.1109/TPEL.2012.2205947

Tshilidzi, M. (2012). *Condition Monitoring Using Computational Intelligence Methods*. Berlin, Germany: Springer.

Tshilidzi, M. (2012). *Condition Monitoring Using Computational Intelligence Methods*. Berlin, Germany: Springer.

Wu, R., Blaabjerg, F., Wang, H., & Liserre, M. (2013). Overview of catastrophic failures of freewheeling diodes in power electronic circuits. *Microelectronics and Reliability*, *53*(9-11), 1788–1792. doi:10.1016/j.microrel.2013.07.126

Xiang, D., Ran, L., Tavner, P., Bryant, A., Yang, S., & Mawby, P. (2011). Monitoring Solder Fatigue in a Power Module Using Case-Above-Ambient Temperature Rise. *IEEE Transactions on Industry Applications*, *47*(6), 2578–2591. doi:10.1109/TIA.2011.2168556

Xuejun, P., Songsong, N., & Yong, K. (2015). Switch Short-Circuit Fault Diagnosis and Remedial Strategy for Full-Bridge DC-DC Converters. *IEEE Transactions on Power Electronics*, *30*(2), 996–1004. doi:10.1109/TPEL.2014.2310201

Xuejun, P., Songsong, N., Yu, C., & Yong, K. (2012). Open-Circuit Fault Diagnosis and Fault-Tolerant Strategies for Full-Bridge DC–DC Converters. *IEEE Transactions on Power Electronics*, *27*(5), 2550–2556.

Yang, S., Bryant, A., Mawby, P., Xiang, D., Ran, L., & Tavner, P. (2011). An Industry-Based Survey of Reliability in Power Electronic Converters. *IEEE Transactions on Industry Applications*, *47*(3), 1441–1451. doi:10.1109/TIA.2011.2124436

Yang, S., Dawei, X., Bryant, A., Mawby, P., Ran, L., & Tavner, P. (2010). Condition Monitoring for Device Reliability in Power Electronic Converters: A Review. *IEEE Transactions on Power Electronics*, *25*(11), 2734–2752. doi:10.1109/TPEL.2010.2049377

Yang, W., Tavner, P. J., & Court, R. (2013). An online technique for condition monitoring the induction generators used in wind and marine turbines. *Mechanical Systems and Signal Processing*, *38*(1), 103–112. doi:10.1016/j.ymssp.2012.03.002

Young, K. S., Kwanghee, N., Hong-Seok, S., & Ho-Gi, K. (2008). Fault Diagnosis of a ZVS DC–DC Converter Based on DC-Link Current Pulse Shapes. *IEEE Transactions on Industrial Electronics*, *55*(3), 1491–1494. doi:10.1109/TIE.2007.910627

Younsi, K., Neti, P., Shah, M., Zhou, J. Y., Krahn, J., Weeber, K., & Whitefield, C. (2005). On-line Capacitance and Dissipation Factor Monitoring of AC Stator Insulation. *IEEE Transactions on Dielectrics and Electrical Insulation*, *17*(5), 1441–1452. doi:10.1109/TDEI.2010.5595545

Zhang, P., Li, W., Li, S., Wang, Y., & Xiao, W. (2013). Reliability assessment of photovoltaic power systems: Review of current status and future perspectives. *Applied Energy*, *104*, 822–833. doi:10.1016/j.apenergy.2012.12.010

KEY TERMS AND DEFINITIONS

Condition Monitoring: Producing a pre-alarm in the system before failure.
Data Logger: A device for continuous recording of a variable or parameter.
Fourier Transform: A mathematical operation for frequency analysis of a time domain variable.
Prediction Algorithm: A method for calculating the coming value of a parameter or variable.
Sensor: A device for converting a parameter to an electric signal.

Compilation of References

Abdi, E., McMahon, R., Malliband, P., Shao, Mathekga, M.E., Tavner, P., ... Tatlow, M. (2013). Performance analysis and testing of a 250 kW medium-speed brushless doubly-fed induction generator. *IET Renewable Power Generation, 7*(6), 63638.

Abdi, S., Abdi, E., Oraee, A., & McMahon, R. (2014). Equivalent Circuit Parameters for Large Brushless Doubly Fed Machines (BDFMs). *IEEE Transactions on Energy Conversion, 29*(3), 706–715. doi:10.1109/TEC.2014.2311736

Abdi, S., Abdi, E., Oraee, A., & McMahon, R. (2015). Optimization of Magnetic Circuit for Brushless Doubly Fed Machines. *IEEE Transactions on Energy Conversion, 30*(4), 1611–1620. doi:10.1109/TEC.2015.2468063

Abusara, M. A., & Sharkh, S. M. (2013). Design and Control of a Grid-Connected Interleaved Inverter. *IEEE Transactions on Power Electronics, 28*(2), 748–764. doi:10.1109/TPEL.2012.2201505

Aghaei, M., Karimi, Y., & Kaboli, S. (2014). Analysis of Phase-Shifted Full-Bridge Based dc-dc Converter considering transformer parasitic elements in discontinuous current mode. In *Proceedings of 5th Power Electronics, Drive Systems and Technologies Conference* (pp. 366-372). IEEE. doi:10.1109/PEDSTC.2014.6799402

Bak-Jensen, B., Bech, J., Bjerregaard, C. G., & Jensen, P. R. (1999). Models for probabilistic power transmission system reliability calculation. *IEEE Transactions on Power Systems, 14*(3), 1166–1171. doi:10.1109/59.780952

Barati, F., & Oraee, H. (2010). Vector model utilization for nested-loop rotor Brushless Doubly-Fed Machine analysis, control and simulation. In *Proceedings of 1st Power Electronic & Drive Systems & Technologies Conference* (pp. 295–301). IEEE. doi:10.1109/PEDSTC.2010.5471806

Barati, F., Oraee, H., Abdi, E., Shiyi, Sh., & McMahon, R. (2008). The Brushless Doubly-Fed Machine Vector Model in the rotor flux oriented reference frame. In *Proceedings of 34th Annual Conference of IEEE Industrial Electronics* (pp.1415-1420). IEEE. doi:10.1109/IECON.2008.4758161

Barati, F., Shao, S., Abdi, E., Oraee, H., & McMahon, R. (2011). Generalized Vector Model for the Brushless Doubly-Fed Machine With a Nested-Loop Rotor. *IEEE Transactions on Industrial Electronics, 58*(6), 2313–2321. doi:10.1109/TIE.2010.2064279

Bhuiyan, F. A., & Yazdani, A. (2010). Reliability assessment of a wind-power system with integrated energy storage. *IET Renewable Power Generation, 4*(3), 211–220. doi:10.1049/iet-rpg.2009.0070

Bilgin, B., Emadi, A., & Krishnamurthy, M. (2013). Comprehensive Evaluation of the Dynamic Performance of a 6/10 SRM for Traction Application in PHEVs. *IEEE Transactions on Industrial Electronics, 60*(7), 2564–2575. doi:10.1109/TIE.2012.2196015

Boglietti, A., Cavagnino, A., Lazzari, M., & Pastorelli, M. (2003). A simplified thermal model for variable-speed self-cooled industrial induction motor. *IEEE Transactions on Industry Applications, 39*(4), 945–952. doi:10.1109/TIA.2003.814555

Boglietti, A., Cavagnino, A., Staton, D., Shanel, M., Mueller, M., & Mejuto, C. (2009). Evolution and Modern Approaches for Thermal Analysis of Electrical Machines. *IEEE Transactions on Industrial Electronics*, *56*(3), 871–882. doi:10.1109/TIE.2008.2011622

Boglietti, A., El-Refaie, A. M., Drubel, O., Omekanda, A. M., Bianchi, N., Agamloh, E. B., & Bartolo, J. B. et al. (2014). Electrical Machine Topologies: Hottest Topics in the Electrical Machine Research Community. *IEEE Industrial Electronics Magazine*, *8*(2), 18–30. doi:10.1109/MIE.2013.2294077

Chiesa, N., & Høidalen, H. K. (2010). Novel Approach for Reducing Transformer Inrush Currents: Laboratory Measurements, Analytical Interpretation and Simulation Studies. *IEEE Transactions on Power Delivery*, *25*(4), 2609–2616. doi:10.1109/TPWRD.2010.2045772

Cipriano, E., Jacobina, C. B., da Silva, E. R. C., & Rocha, N. (2012). Single-Phase to Three-Phase Power Converters: State of the Art. *IEEE Transactions on Power Electronics*, *27*(5), 2437–2452. doi:10.1109/TPEL.2011.2175751

Collins, E. R. (1992). Torque and slip behavior of single-phase induction motors driven from variable-frequency supplies. *IEEE Transactions on Industry Applications*, *28*(3), 710–715. doi:10.1109/28.137461

Eric, W., & Lehn, P. W. (2006). Digital Current Control of a Voltage Source Converter with Active Damping of LCL Resonance. *IEEE Transactions on Power Electronics*, *21*(5), 1364–1373. doi:10.1109/TPEL.2006.880271

Freire, N. M. A., & Cardoso, A. J. M. (2014). A Fault-Tolerant Direct Controlled PMSG Drive for Wind Energy Conversion Systems. *IEEE Transactions on Industrial Electronics*, *61*(2), 821–834. doi:10.1109/TIE.2013.2251734

Gerada, D., Mebarki, A., Brown, N. L., Gerada, C., Cavagnino, A., & Boglietti, A. (2014). High-Speed Electrical Machines: Technologies, Trends, and Developments. *IEEE Transactions on Industrial Electronics*, *61*(6), 2946–2959. doi:10.1109/TIE.2013.2286777

Gorgin, H., Sotoodeh, P., & Oraee, H. (2010). Dynamic model for Brushless Doubly-Fed Machine with stator winding faults. In *Proceedings of 1st Power Electronic & Drive Systems & Technologies Conference* (pp. 316–322). IEEE. doi:10.1109/PEDSTC.2010.5471801

Gorginpour, H., Jandaghi, B., & Oraee, H. (2011). Finite element analysis of Brushless Doubly-Fed Machine under stator winding faults. In *Proceedings of 2nd Power Electronics, Drive Systems and Technologies Conference* (pp.169-174). IEEE doi:10.1109/PEDSTC.2011.5742411

Gorginpour, H., Jandaghi, B., & Oraee, H. (2011). Time and space harmonics in brushless doubly-fed machine. In *Proceedings of 19th Iranian Conference on Electrical Engineering* (pp.1-6). IEEE.

Gorginpour, H., Jandaghi, B., & Oraee, H. (2013). A novel rotor configuration for brushless doubly-fed induction generators. *IET Electric Power Applications*, *7*(2), 106–115. doi:10.1049/iet-epa.2012.0194

Gorginpour, H., Jandaghi, B., Oraee, H., & Abdi, E. (2014). Magnetic equivalent circuit modelling of brushless doubly-fed induction generator. *IET Renewable Power Generation*, *8*(3), 334–346. doi:10.1049/iet-rpg.2012.0383

Gorginpour, H., Oraee, H., & Abdi, E. (2014). Calculation of Core and Stray Load Losses in Brushless Doubly Fed Induction Generators. *IEEE Transactions on Industrial Electronics*, *61*(7), 3167–3177. doi:10.1109/TIE.2013.2279357

Gorginpour, H., Oraee, H., & McMahon, R. A. (2013). A Novel Modeling Approach for Design Studies of Brushless Doubly Fed Induction Generator Based on Magnetic Equivalent Circuit. *IEEE Transactions on Energy Conversion*, *28*(4), 902–912. doi:10.1109/TEC.2013.2278486

Gorginpour, H., Oraee, H., & McMahon, R. A. (2014). Electromagnetic-Thermal Design Optimization of the Brushless Doubly Fed Induction Generator. *IEEE Transactions on Industrial Electronics*, *61*(4), 1710–1721. doi:10.1109/TIE.2013.2267705

Guo, H., Wang, W., Xing, W., & Li, Y. (2010). Design of Electrical/Mechanical Hybrid 4-Redundancy Brushless DC Torque Motor. *Chinese Journal of Aeronautics*, *23*, 2215.

Haghbin, S., Zolghadri, M. R., Kaboli, S., & Emadi, A. (2003). Performance of PI stator resistance compensator on DTC of induction motor. In *Proceedings of The 29th Annual Conference of the IEEE Industrial Electronics Society* (pp.425-430). IEEE. doi:10.1109/IECON.2003.1280018

Hamzeh, M., Karimi, Y., Asadi, E., & Oraee, H. (2014). Design and implementation of a single phase grid-connected PV inverter with a new active damping strategy. In *Proceedings of 5th Power Electronics, Drive Systems and Technologies Conference* (pp.72-77). IEEE doi:10.1109/PEDSTC.2014.6799347

Hanna, R. A., Hiscock, W., & Klinowski, P. (2007). Failure Analysis of Three Slow-Speed Induction Motors for Reciprocating Load Application. *IEEE Transactions on Industry Applications*, *43*(2), 429–435. doi:10.1109/TIA.2006.889969

Hanna, R., & Schmitt, D. W. (2012). Failure Analysis of Induction Motors: Magnetic Wedges in Compression Stations. *IEEE Industry Applications Magazine*, *18*(4), 40–46. doi:10.1109/MIAS.2012.2191340

Hao, H., Covic, G. A., & Boys, J. T. (2014). A Parallel Topology for Inductive Power Transfer Power Supplies. *IEEE Transactions on Power Electronics*, *29*(3), 1140–1151. doi:10.1109/TPEL.2013.2262714

Hasanzadeh, A., Zolghadri, M. R., Kaboli, S. H., & Homaifar, A. (2003). A genetic algorithm based programmed PWM optimum switching pattern calculation. In *Proceedings of The Fifth International Conference on Power Electronics and Drive Systems* (pp.1081-1085). IEEE. doi:10.1109/PEDS.2003.1283122

Jang, D. H. (2013). Problems Incurred in a Vector-Controlled Single-Phase Induction Motor, and a Proposal for a Vector-Controlled Two-Phase Induction Motor as a Replacement. *IEEE Transactions on Power Electronics*, *28*(1), 526–536. doi:10.1109/TPEL.2012.2199772

Jassal, A., Polinder, H., & Ferreira, J. A. (2012). Literature survey of eddy-current loss analysis in rotating electrical machines. *IET Electric Power Applications*, *6*(9), 743–752. doi:10.1049/iet-epa.2011.0335

Javadian, V., & Kaboli, S. (2013). Reliability assessment of some high side MOSFET drivers for buck converter. In *Proceedings of 3rd International Conference on Electric Power and Energy Conversion Systems* (pp.1-6). IEEE. doi:10.1109/EPECS.2013.6713092

Kaboli, S. H., & Zolghadri, M. R. (2003). Flux harmonic spectrum processing of direct torque controlled induction motor. In *Proceedings of the 2003 International Symposium on Circuits and Systems* (pp. III-383 - III-386). IEEE. doi:10.1109/ISCAS.2003.1205036

Kaboli, S. H., Zolghadri, M. R., & Homaifar, A. (2003). Effects of sampling time on the performance of direct torque controlled induction motor drive. In *Proceedings of IEEE International Symposium on Industrial Electronics* (pp.1049-1052). IEEE. doi:10.1109/ISIE.2003.1267968

Kaboli, S., Alavi, S. M. H., & Oraee, H. (2008). Skip cycle controlled resonant converter with high efficiency over wide load range. In *Proceedings of 4th IET Conference on Power Electronics, Machines and Drives* (pp. 742-746). IEEE. doi:10.1049/cp:20080620

Kaboli, S., Haddadi, A., & Khaligh, A. (2008). Axiomatic design of step down DC/DC converter. In *Proceedings of IEEE International Conference on Industrial Technology* (pp.1-5). IEEE. doi:10.1109/ICIT.2008.4608313

Kaboli, S., Mahdavi, J., & Agah, A. (2007). Application of Random PWM Technique for Reducing the Conducted Electromagnetic Emissions in Active Filters. *IEEE Transactions on Industrial Electronics*, *54*(4), 2333–2343. doi:10.1109/TIE.2007.899944

Kaboli, S., Moayedi, A., & Oraee, H. (2008). Application of Random PWM technique for reducing the high frequency harmonics in class-D amplifier. In *Proceedings of 4th IET Conference on Power Electronics, Machines and Drives* (pp.406-410). IEEE doi:10.1049/cp:20080553

Kaboli, S., Rajaei, A. H., & Sheikhi, A. (2009). Application of random PWM techniques for reducing the electromagnetic interference of Vienna rectifiers in distribution power system. In *Proceedings of IEEE 6th International Power Electronics and Motion Control Conference* (pp.998-1003). IEEE. doi:10.1109/IPEMC.2009.5157530

Kaboli, S., Vahdati-Khajeh, E., & Zolghadri, M. R. (2006). Probabilistic voltage harmonic analysis of direct torque controlled induction motor drives. *IEEE Transactions on Power Electronics*, *21*(4), 1041–1052. doi:10.1109/TPEL.2006.876838

Kaboli, S., Vahdati-Khajeh, E., Zolghadri, M. R., & Homaifar, A. (2005). A fast optimal flux search controller with improved steady state behavior for DTC based induction motor drives. In *Proceedings of IEEE International Conference on Electric Machines and Drives* (pp.1732-1736). IEEE. doi:10.1109/IEMDC.2005.195954

Kaboli, S., & Zolghadri, M. R. (2003). Effect of motor speed on the optimum operating point of direct torque controlled induction motor. In *Proceedings of Sixth International Conference on Electrical Machines and Systems* (pp.619-622). IEEE.

Kaboli, S., Zolghadri, M. R., & Emadi, A. (2003). Hysteresis band determination of direct torque controlled induction motor drives with torque ripple and motor-inverter loss considerations. In *Proceedings of IEEE 34th Annual Power Electronics Specialist Conference* (pp. 1107-1111). IEEE. doi:10.1109/PESC.2003.1216604

Kaboli, S., Zolghadri, M. R., Haghbin, S., & Emadi, A. (2003). Torque ripple minimization in DTC of induction motor based on optimized flux value determination. In *Proceedings of The 29th Annual Conference of the IEEE Industrial Electronics Society* (pp.431-435). IEEE. doi:10.1109/IECON.2003.1280019

Kaboli, S., Zolghadri, M. R., Haghbin, S., & Homaifar, A. (2004). A control strategy for reducing the torque ripple in low speed operation of direct torque controlled induction motor. In *Proceedings of IEEE International Symposium on Industrial Electronics* (pp.1203-1208). IEEE. doi:10.1109/ISIE.2004.1571984

Kaboli, S., Zolghadri, M. R., & Khaligh, A. (2008). Reliability Improvement of Industrial Drives Using Multi-Objective Optimization. In *Proceedings of IEEE Industry Applications Society Annual Meeting* (pp.1-6). IEEE. doi:10.1109/08IAS.2008.198

Kaboli, S., Zolghadri, M. R., Roye, D., Guiraud, J., & Crebier, J. C. (2004). Design and implementation of a predictive controller for reducing the torque ripple in direct torque control based induction motor drives. In *Proceedings of IEEE 35th Annual Power Electronics Specialists Conference* (pp476-481). IEEE.

Kaboli, S., Zolghadri, M. R., Roye, D., Guiraud, J., & Schanen, J. L. (2004). Design and implementation of a predictive controller for reducing the torque ripple in direct torque control based high frequency induction motor drives. In *Proceedings of IEEE 35th Annual Power Electronics Specialists Conference* (pp.1169-1174). IEEE. doi:10.1109/PESC.2004.1355588

Kaboli, S., Zolghadri, M. R., Vahdati-Khajeh, E., & Homaifar, A. (2005). On the Performance of Optimal Flux Search Controller for DTC Based Induction Motor Drives. In *Proceedings of IEEE International Conference on Electric Machines and Drives* (pp.1752-1756). IEEE. doi:10.1109/IEMDC.2005.195957

Kastha, D. K., & Bose, B. K. (1994). Investigation of fault modes of voltage-fed inverter system for induction motor drive. *IEEE Transactions on Industry Applications*, *30*(4), 1021038. doi:10.1109/28.297920

Kolar, J. W., & Friedli, T. (2013). The Essence of Three-Phase PFC Rectifier Systems - Part I. *IEEE Transactions on Power Electronics*, *28*(1), 176–198. doi:10.1109/TPEL.2012.2197867

Kwasinski, A. (2011). Quantitative Evaluation of DC Microgrids Availability: Effects of System Architecture and Converter Topology Design Choices. *IEEE Transactions on Power Electronics*, *26*(3), 835–851. doi:10.1109/TPEL.2010.2102774

Kwasinski, A., Krishnamurthy, V., Song, J., & Sharma, R. (2012). Availability Evaluation of Micro-Grids for Resistant Power Supply During Natural Disasters. *IEEE Transactions on Smart Grid*, *3*(4), 2007–2018. doi:10.1109/TSG.2012.2197832

Lahyani, A., Venet, P., Grellet, G., & Viverge, P.-J. (1998). Failure prediction of electrolytic capacitors during operation of a switchmode power supply. *IEEE Transactions on Power Electronics*, *13*(6), 1199–1207. doi:10.1109/63.728347

Levi, E., Bojoi, R., Profumo, F., Toliyat, H. A., & Williamson, S. (2007). Multiphase induction motor drives - a technology status review. *IET Electric Power Applications*, *1*(4), 489–516. doi:10.1049/iet-epa:20060342

Li, J.-P., & Thompson, G. (2005). A method to take account of inhomogeneity in mechanical component reliability calculations. *IEEE Transactions on Reliability*, *54*(1), 159–168. doi:10.1109/TR.2004.837702

Liu, T.-H., Fu, J.-R., & Lipo, T. A. (1993). A Strategy for Improving Reliability of Field-Oriented Controlled Induction Motor Drives. *IEEE Transactions on Industry Applications*, *29*(5), 910–918. doi:10.1109/28.245714

Lu, M. W., & Rudy, R. J. (2001). Laboratory reliability demonstration test considerations. *IEEE Transactions on Reliability*, *50*(1), 12–16. doi:10.1109/24.935011

Mahdavi, J., Kaboli, S. H., & Toliyat, H. A. (1999). Conducted electromagnetic emissions in unity power factor AC/DC converters: comparison between PWM and RPWM techniques. In *Proceedings of 30th Annual IEEE Power Electronics Specialists Conference* (pp. 881-885). IEEE. doi:10.1109/PESC.1999.785615

Melo, A. C. G., Pereira, M. V. F., & Leite da Silva, A. M. (1993). A conditional probability approach to the calculation of frequency and duration indices in composite reliability evaluation. *IEEE Transactions on Power Systems*, *8*(3), 1118–1125. doi:10.1109/59.260886

Morren, J., & de Haan, S. W. H. (2007). Short-Circuit Current of Wind Turbines With Doubly Fed Induction Generator. *IEEE Transactions on Energy Conversion*, *22*(1), 174–180. doi:10.1109/TEC.2006.889615

Motor Reliability Working Group. (1985). Report of Large Motor Reliability Survey of Industrial and Commercial Installations, Part I. *IEEE Transactions on Industry Applications*, *21*(4), 853–864.

Movahed, S. R., Oraee, H., Mirzamani, S. H., Rajabi, A., & Daneshvar, H. (2010). Estimation of insulation life of inverter- fed induction motors. In *Proceedings of 1st Power Electronic & Drive Systems & Technologies Conference* (pp. 335–339). IEEE. doi:10.1109/PEDSTC.2010.5471797

Nag, S. S., & Mishra, S. (2014). Current-Fed Switched Inverter. *IEEE Transactions on Industrial Electronics*, *61*(9), 4680–4690. doi:10.1109/TIE.2013.2289907

Najmi, V., Ebrahimi, S., & Oraee, H. (2012). Output voltage quality intensification of diode clamped multilevel inverters using FM PWM technique. In *Proceedings of 3rd Power Electronics and Drive Systems Technology* (p. 9102). IEEE. doi:10.1109/PEDSTC.2012.6183305

Nasirian, V., Davoudi, A., Kaboli, S., & Edrington, C. S. (2013). Excitation Shifting: A General Low-Cost Solution for Eliminating Ultra-Low-Frequency Torque Ripple in Switched Reluctance Machines. *IEEE Transactions on Magnetics*, *49*(9), 5135–5149. doi:10.1109/TMAG.2013.2251651

Nasirian, V., Kaboli, S., & Davoudi, A. (2012). Output power maximization and optimal symmetric freewheeling excitation for Switched Reluctance Generators. In *Proceedings of IEEE/IAS 48th Industrial & Commercial Power Systems Technical Conference* (pp.1-10). IEEE. doi:10.1109/ICPS.2012.6229595

Nasirian, V., Kaboli, S., & Davoudi, A. (2013). Output Power Maximization and Optimal Symmetric Freewheeling Excitation for Switched Reluctance Generators. *IEEE Transactions on Industry Applications*, *49*(3), 1031042. doi:10.1109/TIA.2013.2253438

Nasirian, V., Kaboli, S., Davoudi, A., & Moayedi, S. (2013). High-Fidelity Magnetic Characterization and Analytical Model Development for Switched Reluctance Machines. *IEEE Transactions on Magnetics*, *49*(4), 1505–1515. doi:10.1109/TMAG.2012.2222427

Oraee, A., Abdi, E., Abdi, S., McMahon, R., & Tavner, P. J. (2015). Effects of Rotor Winding Structure on the BDFM Equivalent Circuit Parameters. *IEEE Transactions on Energy Conversion*, *30*(4), 1660–1669. doi:10.1109/TEC.2015.2432272

Oraee, H. (2000). A quantative approach to estimate the life expectancy of motor insulation systems. *IEEE Transactions on Dielectrics and Electrical Insulation*, *7*(6), 790–796. doi:10.1109/94.891990

Park, K. B., Choi, S. W., Kim, Ch. E., Moon, G. W., & Youn, M. J. (2009). An AC-PDP Single Sustaining Driver Employing the Voltage Stress Reduction Technique. *IEEE Transactions on Power Electronics*, *24*(4), 1124–1128. doi:10.1109/TPEL.2008.2008993

Pellegrino, G., Vagati, A., Boazzo, B., & Guglielmi, P. (2012). Comparison of Induction and PM Synchronous Motor Drives for EV Application Including Design Examples. *IEEE Transactions on Industry Applications*, *48*(6), 2322–2332. doi:10.1109/TIA.2012.2227092

Poshtan, M., Kaboli, S., & Mahdavi, J. (2006). On the Suitability of Modeling Approaches for Power Electronic Converters. In *Proceedings of IEEE International Symposium on Industrial Electronics* (pp.1486-1491). IEEE. doi:10.1109/ISIE.2006.295691

Rahman, K. M., & Hiti, S. (2005). Identification of machine parameters of a synchronous motor. *IEEE Transactions on Industry Applications*, *41*(2), 557–565. doi:10.1109/TIA.2005.844379

Rajaei, A. H., Kaboli, S., & Emadi, A. (2008). Sliding-mode control of z-source inverter. In *Proceedings of 34th Annual Conference of IEEE Industrial Electronics* (pp. 947-952). IEEE.

Ribeiro, E., Cardoso, A. J. M., & Boccaletti, C. (2013). Fault-Tolerant Strategy for a Photovoltaic DC-DC Converter. *IEEE Transactions on Power Electronics*, *28*(6), 3008–3018. doi:10.1109/TPEL.2012.2226059

Richardeau, F., & Pham, T. T. L. (2013). Reliability Calculation of Multilevel Converters: Theory and Applications. *IEEE Transactions on Industrial Electronics*, *60*(10), 4225–4233. doi:10.1109/TIE.2012.2211315

Sadati, N., Kaboli, S., Adeli, H., Hajipour, E., & Ferdowsi, M. (2009). Online Optimal Neuro-Fuzzy Flux Controller for DTC Based Induction Motor Drives. In *Proceedings of Twenty-Fourth Annual IEEE Applied Power Electronics Conference and Exposition* (pp.2215). IEEE doi:10.1109/APEC.2009.4802657

Saha, T. K. (2003). Review of modern diagnostic techniques for assessing insulation condition in aged transformers. *IEEE Transactions on Dielectrics and Electrical Insulation*, *10*(5), 903–917. doi:10.1109/TDEI.2003.1237337

Sheikhi, A., Oraee, H., Kaboli, S., & Dorkhah, M. (2009). A new configuration of switched reluctance motor for reducing the torque ripple. In *Proceedings of International Conference on Electric Power and Energy Conversion Systems* (pp.1-5). IEEE.

Sheng, H., Wang, F., & Tipton, C. W. (2012). A Fault Detection and Protection Scheme for Three-Level DC–DC Converters Based on Monitoring Flying Capacitor Voltage. *IEEE Transactions on Power Electronics*, *27*(2), 685–697. doi:10.1109/TPEL.2011.2161333

Song, J., Krishnamurthy, V., Kwasinski, A., & Sharma, R. (2013). Development of a Markov-Chain-Based Energy Storage Model for Power Supply Availability Assessment of Photovoltaic Generation Plants. *IEEE Transactions on Sustainable Energy*, *4*(2), 491–500. doi:10.1109/TSTE.2012.2207135

Song, Y., & Wang, B. (2013). Analysis and Experimental Verification of a Fault-Tolerant HEV Powertrain. *IEEE Transactions on Power Electronics*, *28*(12), 5854–5864. doi:10.1109/TPEL.2013.2245513

Song, Y., & Wang, B. (2013). Survey on Reliability of Power Electronic Systems. *IEEE Transactions on Power Electronics*, *28*(1), 59604. doi:10.1109/TPEL.2012.2192503

Sorrentino, E., & Fernández, S. (2011). Comparison of six steady-state models for single-phase induction motors. *IET Electric Power Applications*, *5*(8), 61617. doi:10.1049/iet-epa.2011.0001

Staton, D., Boglietti, A., & Cavagnino, A. (2005). Solving the More Difficult Aspects of Electric Motor Thermal Analysis in Small and Medium Size Industrial Induction Motors. *IEEE Transactions on Energy Conversion*, *20*(3), 620–628. doi:10.1109/TEC.2005.847979

Sutanto, D., Outhred, H. R., & Lee, Y. B. (1989). Probabilistic power system production cost and reliability calculation by the Z-transform method. *IEEE Transactions on Energy Conversion*, *4*(4), 559–566. doi:10.1109/60.41714

Tabbache, B., Benbouzid, M., Kheloui, A., Bourgeot, J.-M., & Mamoune, A. (2013). An improved fault-tolerant control scheme for PWM inverter-fed induction motor-based EVs. *ISA Transactions*, *52*(6), 862–869. doi:10.1016/j.isatra.2013.07.004 PMID:23916869

Thomson, W. T., & Fenger, M. (2001). Current signature analysis to detect induction motor faults. *IEEE Industry Applications Magazine*, *7*(4), 26–34. doi:10.1109/2943.930988

Tohidi, S., Oraee, H., Zolghadri, M. R., Shiyi, Sh., & Tavner, P. (2013). Analysis and Enhancement of Low-Voltage Ride-Through Capability of Brushless Doubly Fed Induction Generator. *IEEE Transactions on Industrial Electronics*, *60*(3), 1146–1155. doi:10.1109/TIE.2012.2190955

Tohidi, S., Zolghadri, M. R., Oraee, H., & Oraee, A. (2012). Dynamic modeling of a wind turbine with brushless doubly fed induction generator. In *Proceedings of 3rd Power Electronics and Drive Systems Technology* (pp. 490–494). IEEE. doi:10.1109/PEDSTC.2012.6183379

Tohidi, S., Zolghadri, M. R., Oraee, H., Tavner, P., Abdi, E., & Logan, T. (2012). Performance of the brushless doubly-fed machine under normal and fault conditions. *IET Electric Power Applications*, *6*(9), 621–627. doi:10.1049/iet-epa.2012.0101

Wang, X., Zhong, H., Yang, Y., & Mu, X. (2010). Study of a Novel Energy Efficient Single-Phase Induction Motor with Three Series-Connected Windings and Two Capacitors. *IEEE Transactions on Energy Conversion*, *25*(2), 433–440. doi:10.1109/TEC.2009.2039218

Wei, H., Ghosh, S., Velusamy, S., Sankaranarayanan, K., Skadron, K., & Stan, M. R. (2006). HotSpot: A compact thermal modeling methodology for early-stage VLSI design. *IEEE Transactions on Very Large Scale Integration (VLSI). Systems*, *14*(5), 501–513.

Wen, C., Baosheng, H., Wei, S., Xuetao, S., Kai, W., Okamoto, T., & Suzuoki, Y. (2014). An approach for economic assessment on oil-paper insulation diagnosis through accelerated aging experiments. *IEEE Transactions on Dielectrics and Electrical Insulation*, *21*(4), 1842–1850. doi:10.1109/TDEI.2014.004370

Wheeler, P. W., Rodriguez, J., Clare, J. C., Empringham, L., & Weinstein, A. (2002). Matrix converters: A technology review. *IEEE Transactions on Industrial Electronics*, *49*(2), 276–288. doi:10.1109/41.993260

Wong, P., Albrecht, P., Allan, R., Billinton, R., Chen, Q., & Fong, C. et al. (1999). The IEEE Reliability Test System-1996. A report prepared by the Reliability Test System Task Force of the Application of Probability Methods Subcommittee. *IEEE Transactions on Power Systems*, *14*(3), 1010–1020. doi:10.1109/59.780914

Yang, Sh., Bryant, A., Mawby, P., Xiang, D., Ran, L., & Tavner, P. (2011). An Industry-Based Survey of Reliability in Power Electronic Converters. *IEEE Transactions on Industry Applications*, *47*(3), 1441451.

Yazdani, A., Sepahvand, H., Crow, M. L., & Ferdowsi, M. (2011). Fault Detection and Mitigation in Multilevel Converter STATCOMs. *IEEE Transactions on Industrial Electronics*, *58*(4), 1307–1315. doi:10.1109/TIE.2010.2050415

Yuancheng, R., Ming, X., Zhou, J., & Lee, F. C. (2006). Analytical loss model of power MOSFET. *IEEE Transactions on Power Electronics*, *21*(2), 310–319. doi:10.1109/TPEL.2005.869743

Zhang, W., Xu, D., Enjeti, P. N., Li, H., Hawke, J. T., & Krishnamoorthy, H. S. (2014). Survey on Fault-Tolerant Techniques for Power Electronic Converters. *IEEE Transactions on Power Electronics*, *29*(12), 6319–6331. doi:10.1109/TPEL.2014.2304561

Index

A

B

C

D

E

F

G

H

I

K

L

M

N

O

P

R

S

T

U

V

W

www.ingramcontent.com/pod-product-compliance
Lightning Source LLC
LaVergne TN
LVHW060353300926
849162LV00006B/1255